The Fate of Persistent Organic Pollutants in the Environment

NATO Science for Peace and Security Series

This Series presents the results of scientific meetings supported under the NATO Programme: Science for Peace and Security (SPS).

The NATO SPS Programme supports meetings in the following Key Priority areas: (1) Defence Against Terrorism; (2) Countering other Threats to Security and (3) NATO, Partner and Mediterranean Dialogue Country Priorities. The types of meeting supported are generally "Advanced Study Institutes" and "Advanced Research Workshops". The NATO SPS Series collects together the results of these meetings. The meetings are co-organized by scientists from NATO countries and scientists from NATO's "Partner" or "Mediterranean Dialogue" countries. The observations and recommendations made at the meetings, as well as the contents of the volumes in the Series, reflect those of participants and contributors only; they should not necessarily be regarded as reflecting NATO views or policy.

Advanced Study Institutes (ASI) are high-level tutorial courses intended to convey the latest developments in a subject to an advanced-level audience

Advanced Research Workshops (ARW) are expert meetings where an intense but informal exchange of views at the frontiers of a subject aims at identifying directions for future action

Following a transformation of the programme in 2006 the Series has been re-named and re-organised. Recent volumes on topics not related to security, which result from meetings supported under the programme earlier, may be found in the NATO Science Series.

The Series is published by IOS Press, Amsterdam, and Springer, Dordrecht, in conjunction with the NATO Public Diplomacy Division.

Sub-Series

A.	Chemistry and Biology	Springer
B.	Physics and Biophysics	Springer
C.	Environmental Security	Springer
D.	Information and Communication Security	IOS Press
E.	Human and Societal Dynamics	IOS Press

http://www.nato.int/science
http://www.springer.com
http://www.iospress.nl

Series C: Environmental Security

The Fate of Persistent Organic Pollutants in the Environment

edited by

Ebru Mehmetli
Chemistry and Environment Institute
Gebze, Kocaeli
Turkey

and

Bogdana Koumanova
University of Chemical Technology and Metallurgy
Sofia, Bulgaria

Springer

Published in cooperation with NATO Public Diplomacy Division

Proceedings of the NATO Advanced Research Workshop on
The Fate of Persistent Organic Pollutants in the Environment
Istanbul, Turkey
25–27 April 2007

A C.I.P. Catalogue record for this book is available from the Library of Congress.

ISBN 978-1-4020-6641-2 (PB)
ISBN 978-1-4020-6640-5 (HB)
ISBN 978-1-4020-6642-9 (e-book)

Published by Springer,
P.O. Box 17, 3300 AA Dordrecht, The Netherlands.

www.springer.com

Printed on acid-free paper

CONTENTS

Part V HEALTH EFFECTS

Part VI ADVANCES IN THE MEASUREMENT OF POPs

PREFACE

Persistent organic pollutants (POPs) are organic compounds of natural or anthropogenic origin that resist photolytic, chemical, and biological degradation. They are characterized by low water solubility and high lipid solubility, resulting in bioaccumulation in fatty tissues of living organisms. These properties of unusual high persistence and semivolatility, coupled with other characteristics, have resulted in the presence of POPs all over the world, even in regions where they have never been used. They have been measured on every continent, at sites representing every major climatic zone and geographic region throughout the world. These include remote regions, where no significant local sources exist and the only reasonable explanation for their presence is long-range transport from other parts of the globe. POPs have been found, on a global scale, in soils, sediments, in the fat of fish, and terrestrial animals, as well as in human breast milk. Some of the highest levels have been recorded in the polar areas of both the hemispheres.

POPs persist in the environment, bioaccumulate through the food web, and pose a risk of causing adverse effects to human health and the environment. Laboratory and field-based studies suggest the following human health effects from exposure to POPs: allergies, hypersensitivity, nervous system damage, reproductive and immune dysfunction, neurobehavioral and developmental disorders, endocrine disruption, and cancer. Many POPs are considered possible human carcinogens by the International Agency for Research on Cancer (IARC) of the World Health Organization. Hormone system dysfunction associated with POPs includes damage to the reproductive system, sex-linked disorders, and shortened lactation periods for nursing mothers.

With the evidence of long-range transport of these substances to regions, the international community has now, at several occasions, called for urgent global actions to reduce and eliminate releases of these chemicals. Many countries have already banned or severely restricted the production and use of the 12 POPs in recent decades. Yet they remain a serious problem around the world. Because these chemicals have the ability to travel long distances from their original sources, relying on national-level action alone is ultimately a futile effort. For this reason, the United Nations Environment Program (UNEP) has initiated international negotiations toward a legally binding global agreement to reduce or eliminate these 12 POPs and set criteria for including new chemicals in the agreement in the future.

The Stockholm Convention on Persistent Organic Pollutants obliges Parties to take measures to reduce, and where possible to eliminate, emissions of POPs as well as their by-products. It also provides a framework for doing so. However, there is limited practical experience with successful implementation of measures under conditions that prevail in many developing countries and countries with economies in transition. In addition, the framework established for doing this under the Stockholm Convention is new and unique in many of its aspects.

In order to discuss detailed knowledge and understanding of the effectiveness of existing legislation, any continuing risk to human health and ecosystems, and an appreciation of the additional control measures already being implemented by countries and recent developments in POPs science in a platform, the workshop organizers proposed to the NATO Science for Peace Programme to hold an Advanced Research Workshop (ARW) on this topic. After receving the grant, the NATO ARW on The Fate of Persistent Organic Pollutants in the Environment was held successfully in Istanbul, Turkey, from 25 to 27 April 2007.

For this success, the editors and the organizers are indebted to many who helped to stage the workshop, as listed in the Acknowledgment.

The achievements and deliverables of the workshop will amount to a project of substantial value and will be fundamental for important, subsequent applications. The expected step following the completion of the workshop will be the initiation of another international collaboration on POPs at a project level.

The editors:

Ebru Mehmetli
Bogdana Koumanova

ACKNOWLEDGMENT

The NATO Advanced Research Workshop on The Fate of Persistent Organic Pollutants in the Environment, held in Istanbul, Turkey, during 25–27 April 2007, was funded by a NATO award provided by the Assistant Secretary General for Public Diplomacy upon consideration by the Advisory Panel on Environmental Security and the Program Director Dr. Deniz Beten.

The codirectors are grateful to all participants who made this workshop a memorable interactive learning experience for all.

Special thanks are due to Dr. Deniz Beten, Program Director, Environmental Security, NATO, who provided liasion between the codirectors and NATO, and personally assisted with many tasks.

The editors are grateful to the Publishing editor of the book, Mrs. Annelies Kersbergen, and to the personnel of Springer/NATO Publishing Unit for their collaboration and help for the publication of the results.

LIST OF PARTICIPANTS

Adam GROCHOWALSKI
Cracow University of Technology,
Faculty of Chemical Engineering and Technology,
Institute of Inorganic Chemistry and Technology
Warszawska 24, PL 31-155,
Krakow
Poland

Anna CUMANOVA
State Hydrometeorological Service
193 Grenoble St., MD - 2043
Chisinau
Republic of Moldova

Bogdana KOUMANOVA
Department of Chemical Engineering,
University of Chemical Technology and Metallurgy
8 Kliment Ohridski blvd., 1756
Sofia
Bulgaria

Boris REVICH
Center for Demography and Human Ecology of Institute for Forecasting,
Russian Academy of Sciences
ul.Krasikova 32, 117418
Moscow
Russia

Ethel ELJARRAT
C.S.I.C., Dept. Environmental Chemistry
IIQAB-CSIC Jordi Girona 18-26, 8034
Barcelona
Spain

Gheorghe DUCA
Academy of Sciences of Moldova and Moldovan Research and Development Association
60 Mateevici str., 2009 MD
Chisinau
Republic of Moldova

Giampiero SCORTICHINI
Istituto Zooprofilattico dell'Abruzzo e del Molise "G. Caporale"
Via Campo Boario, 64100
Teramo
Italy

Heidelore FIEDLER
UNEP Chemicals Branch, DTIE,
International Environment House,
11-13, chemin des Anémones, CH-1219
Châtelaine (GE)
Switzerland

Irakli LEGASHVILI
Department of Waste and Chemical Substances Management,
Ministry of Environment Protection and Natural Resources of Georgia
68a Kostava Str.
Tbilisi
Georgia

Irena TWARDOWSKA
Polish Academy of Sciences,
Institute of Environmental Engineering
34 M. Sklodowska-Curie St., 41-819
Zabrze
Poland

Ivan HOLOUBEK
RECETOX
National POPs Centre
Central and Eastern Regional POPs Centre
Masaryk University
Kamenice 126/3, 62500
Brno
Czech Republic

Juan CORNEJO
Instituto de Recursos Naturales y Agrobiología de Sevilla.CSIC
Av. Reina Mercedes 10, E-41080
Sevilla
Spain

Leondios LEONDIADIS
National Centre for Scientific Research "Demokritos"
Ag. Paraskevi, 15310
Athens
Greece

Leyla TOLUN
TUBITAK MRC, Chemistry and Environment Institute
Gebze, 41470
Kocaeli
Turkey

Marat ISHANKULOV
Ministry of Environmental Protection
38, Bukei Khan Street St., 473000
Astana
Republic of Kazakhstan

Martin SCHERINGER
Institute for Chemical and Bioengineering Room HCI G 127,
Swiss Federal Institute of Technology
ETH Hönggerberg, CH-8093
Zuerich
Switzerland

Mu'taz AL-ALAWI
Mu'tah University, P.O. Box 3, 61710
Karak
Jordan

Müfit BAHADIR
Institute of Ecological Chemistry and Waste Analysis,
Braunschweig University of Technology
Hagenring 30, D-38106
Braunschweig
Germany

Nicola SENESI
Agro-Forestale ed Ambientale, Università Degli Studi Di Bari
Via G. Amendola 165/A, 70126
Bari
Italy

Nicolae OPOPOL
State Medical and Pharmaceutical University "Nicolae Testemitsanu"
165 Stefan-cel-Mare Bul, MD-2004
Chisinau
Republic of Moldova

Oleksiy KHARLAMOV
Institute for Problems of Materials Science of National Academia of Science
3, Krjijanovskogo str., 3680
Kiev
Ukraine

Ott ROOTS
Estonian Environmental Research Institute
Marja 4D, 10 617
Tallinn
Estonia

Perihan Kurt KARAKUS
Center for Atmospheric Research Experiments (CARE) (Environment Canada)
6248 Eighth Line Egbert, L0L1N0
Ontario
Canada

Sergey KAKAREKA
Institute for Problems of Natural Resources Use &Ecology
Skoriny 10, 220114
Minsk
Belarus

Sönmez DAĞLI
TUBITAK MRC, Chemistry and Environment Institute
Gebze, 41470
Kocaeli
Turkey

Tamara KUKHARCHYK
Institute for Problems of Natural Resources Use &Ecology
Skoriny 10, 220114
Minsk
Belarus

Tsvetanka G. DIMCHEVA
Ministry of Environment and Water
67, William Gladstone, 1000
Sofia
Bulgaria

Volodymyr KUZNYETSOV
Ukranian Scientific Research Institute of Ecological Problems
6, Bakulina Str, 61166
Kharkiv
Ukraine

Volodia BOGILLO
Department of Antarctic Geology and Geoecology,
Institute of Geological Sciences,
National Academy of Sciences
55B O. Gonchara Str., 1054
Kiev
Ukraine

Part I

REGULATORY ASPECTS

STOCKHOLM CONVENTION ON POPs: OBLIGATIONS AND IMPLEMENTATION

HEIDELORE FIEDLER*
UNEP Chemicals Branch, DTIE, 11-13 chemin des Anémones, CH-1219 Châtelaine (GE), Switzerland

Abstract: The Stockholm Convention on Persistent Organic Pollutants requires Parties to implement certain measures to protect human health and the environment from persistent organic pollutants by eliminating or reducing production, use, and releases of these chemicals. The global treaty involves politics and economics, but also science and technology to resolve global environmental problems. It is a living treaty and will evolve with time. First results to tackle a global problem jointly by developed and developing countries can already be seen. The Stockholm Convention also calls for cooperation between intergovernmental organizations.

Keywords: Persistent Organic Pollutants (POPs), Stockholm Convention, POPs pesticides, unintentional POPs, National Implementation Plan, Action Plan, Inventory

1. Introduction

The Stockholm Convention on Persistent Organic Pollutants (POPs) is a global treaty with the aim to protect humans and the environment from adverse effects from POPs. The Stockholm Convention entered into force on 17 May 2004 and presently – March 2007 – has 144 Parties and 152 signatories; an updated list can be viewed at www.pops.int (SSC, no date). The Secretariat of the Convention is located in Geneva, Switzerland.

The Convention adopts the "precautionary approach," so that where there are threats of serious or irreversible damage, the lack of full scientific certainty shall not be used as a reason for postponing cost-effective measures to prevent environmental degradation.

* To whom correspondences should be addressed. E-mail: hfiedler@unep.ch

E. Mehmetli and B. Koumanova (eds.),
The Fate of Persistent Organic Pollutants in the Environment, 3–12.

With entry-into-force, the Convention requires Parties to develop and submit to the Secretariat within 2 years, national implementation plans (NIPs) and to designate national focal points for exchanging information on POPs and their alternatives.

Parties are required to implement measures to reduce and eliminate the release of an initial 12 POPs. Of these, eight are pesticides, two are industrial chemicals (whereby hexachlorobenzene (HCB) has been used as a pesticide as well and, together with polychlorinated biphenyls (PCB), is a by-product), and two unintentionally formed by-products (whereby additionally, PCB and HCB are also included in the list of unintentionally generated chemicals). Table 1 lists the 10 POPs and their classification.

TABLE 1. The initial 12 POPs of the Stockholm Convention

Chemical	Intentional production		Unintentionall y formed (by-product)
	Pesticide	Industrial chemical	
Aldrin	X		
Chlordane	X		
DDT	X		
Dieldrin	X		
Endrin	X		
Heptachlor	X		
Mirex	X		
Toxaphene	X		
Hexachlorobenzene (HCB)	X	X	X
Polychlorinated biphenyls (PCBs)		X	X
Polychlorinated dibenzo-*para*-dioxins (PCDD)			X
Polychlorinated dibenzofurans (PCDFs)			X

The governing body of the Stockholm Convention is the Conference of the Parties (COP) that during the initial phase after entry-into-force meets annually and from the 4th meeting in 2009 will meet biannually. The COP, among others, approves the budget and the workplan.

2. Obligations and Implementation

2.1. GENERAL ISSUES

The Stockholm Convention, like other multilateral environmental agreements is a legally binding instrument for its Parties. It starts with 12 initial chemicals and recognizes that special efforts may sometimes be needed to phase out certain chemicals for certain uses and seeks to ensure that such effort is made. It also channels resources to assist countries that do not have the financial, human, or technical means to implement in full the obligations of the Convention.

The Convention calls upon governments to report regularly on efforts to implement the treaty and the Conference of the Parties will evaluate the effectiveness of the measures undertaken. Governments are encouraged to undertake research on POPs, monitor health effects, and exchange information.

The following lists some of the obligations and Section 3 gives some preliminary results and examples achieved to date.

2.2. NEW POPs

The Convention starts with 12 initial POPs but has a provision to add more chemicals to any of the intentionally produced chemicals or the unintentionally generated chemicals (Article 8). Each Party can submit a proposal to include a chemical – that meets the criteria of POPs – to the Convention. Criteria to be met by the chemical are (Annex D):

- Chemical identity, e.g., including trade names, synonyms, numbers of isomers, IUPAC and CAS numbers, and structure.
- Persistence, e.g., half-life in water greater than 2 months or 6 months in soil or sediment or other evidence to justify consideration
- Bioaccumulation, e.g., bioconcentration factor or bioaccumulation factor in aquatic species for the chemical is greater than 5,000 or, in the absence of such data, that the log K_{OW} is greater than 5 or monitoring data in biota indicate potential for bioaccumulation.
- Potential for long-range transport, e.g., measured levels of the chemical in locations distant from the sources of its release that are of potential concern; for a chemical that migrates significantly through the air, its half-life in air should be greater than 2 days.
- Evidence for toxicity, e.g., evidence of adverse effects to human health or to the environment that justifies consideration of the chemical within the scope of this Convention; or toxicity or ecotoxicity data that indicate the potential for damage to human health or to the environment.

2.3. INTENTIONAL POPs

With respect to intentionally produced POPs, different provisions are set for the various POPs, e.g., with entry into force, no Party will be allowed to further produce, use or trade endrin and toxaphene. Whereas PCB contained in safe equipment can stay in use until 2025, no new PCB production or placing in use is permitted. Presently, exemptions for the following seven chemicals can be registered by Parties in the Register of specific exemptions: aldrin, chlordane, DDT, dieldrin, heptachlor, hexachlorobenzene, and mirex. For aldrin, dieldrin, and heptachlor stockpiles can be used for defined purposes but no new production is permitted. For chlordane, mirex, and HCB, new productions according to permitted uses will be allowed; Parties have to register these uses as "exemptions"; the entries are maintained in a publicly accessible register (see http://www.pops.int/documents/registers/specexempt.htm) and are valid for 5 years. Production and use of DDT – but only for public heath purposes in vector control – can continue under the Convention and according to WHO guidelines; the entries are valid for 3 years in a DDT register.

2.4. UNINTENTIONAL POPs

Polychlorinated dibenzo-*p*-dioxins and polychlorinated dibenzofurans (PCDD/PCDF) are the two unintentionally produced POPs contained in the Stockholm Convention that should be eliminated where feasible or at a minimum being continuously reduced (Article 5 and Annex C). The same applies to PCB and PCB as unintentional POPs: They are thought to be formed in the same processes and under similar conditions as trace contaminants like PCDD/PCDF. In order to fulfill this obligation, Parties have to establish national dioxin and furan release inventories to identify and quantify their sources; Parties also have to update these inventories and make projections of future releases. The means to reduce the releases are the application of best available techniques (BAT) and best environmental practices (BEP). These may include process changes, use of alternative materials or products, etc. that would prevent the formation and release of unintentional POPs. An action plan has to be made and the strategies therein be reviewed every 5 years.

2.5. POPs IN STOCKPILES AND WASTE

Parties shall develop and implement strategies for identifying stockpiles, products, and articles containing POPs. Once identified, the stockpiles need to be managed in an environmentally sound manner. In POPs containing wastes, the POPs content needs to be destroyed; the Convention does not allow recovery, recycling, reclamation, direct reuse or alternative uses of the POPs. Transboundary transport is to be undertaken according to the provisions of the Basel Convention (SBC, no date). The Convention gives governments until 2025 to phase out PCB in equipment such as transformers and capacitors as long as the equipment is maintained in a way that prevents leaks. Efforts should be undertaken to destroy the PCB until 2028. Parties have to submit a progress report to the Secretariat every 5 years.

2.6. EFFECTIVENESS EVALUATION

The Convention includes a provision to evaluate the effectiveness of measures implemented. The first evaluation shall be undertaken by the Conference of the Parties 4 years after entry-into-force of the Convention (Article 16). Comparable data monitoring data should be made available. The details of the evaluation still need to be agreed upon.

3. Preliminary Results

3.1. NEW POPs

Presently, the POPs Review Committee agreed to evaluate the following ten additional POPs for consideration of inclusion into the Convention. They are at different stages of the evaluation and the first new POPs could be added at COP-4 in 2009 (chemicals in bold letters are in advanced stage): **Chlordecone, γ-hexachlorocyclohexane (γ-HCH, lindane), pentabromodiphenyl ether, hexabromobiphenyl, perfluorooctyl sulfonate (PFOS);** α- and β-hexachlorocyclohexane, octabromodiphenyl ether, pentachlorobenzene, and short-chained chlorinated paraffins.

3.2. INTENTIONAL POPs

By March 2007, the Secretariat for the Stockholm Convention has received notification for production and use of three POPs by the following Parties (the register of specific exemptions is updated by the Secretariat, see http://www.pops.int/documents/registers/specexempt.htm):

Chlordane	China – Production and use as termiticide
	Botswana – Use as termiticide in buildings and dams
DDT	China – Production (80% used as intermediate for dicofol)
	India – Production for use in dicofol manufacture
Mirex	China – Production and use as termiticide
	Australia – Use against giant termite

The DDT Register presently has entries from 12 Parties that may be in need to use DDT in the public health sector:

Since 2004 – Botswana, Marshall Islands, Mauritius, and South Africa
Since 2005 – China, Morocco, and Yemen
Since 2006 – Ethiopia, India, Myanmar, Senegal, and Swaziland

Production of DDT is listed for China, Ethiopia, and India. The actual DDT register is publicly accessible at http://www.pops.int/documents/ registers/ ddt.htm.

Other notifications on the presence or use of POPs are:

1. Listing of Closed-System Site-Limited Intermediates, which presently contains DDT for production of dicofol (by Brazil, China, and India) and HCB for production of sodium pentachlorophenate (http://www.pops.int/ documents/registers/closedsys.htm)
2. Notification of Articles in Use, where Japan has notified on the use of chlordane and heptachlor as termiticide in structures of houses (http://www.pops.int/documents/registers/artinuse.htm)

3.3. UNINTENTIONAL POPs

3.3.1. PCDD/PCDF Release Inventories

In support of the Stockholm Convention, UNEP Chemicals has developed a methodology – Standardized Toolkit – where without sampling and dioxin analysis, releases of PCDD/PCDF can be quantified according to performance criteria of the source (UNEP, 2007; Fiedler, 2006).

The basic aim of the Toolkit is to enable an estimate of average annual release to each vector (air, water, land, product, and residue) for each process identified as a potential dioxin source. In order to quantify the source strength, emission rates must be determined as annual mass flow rates of PCDD/PCDF expressed in grams of toxic equivalents (TEQ) of PCDD and PCDF released per year. The annual releases are calculated according to Equation (1):

$$\text{Source strength} = \text{emission factor} \times \text{“activity rate”} \qquad (1)$$

The Toolkit includes a set of default emission factors and each emission factor is the amount of PCDD/PCDF (in micrograms I-TEQ) that is released to any of the five vectors per unit of feed material processed or product produced (e.g., tons or liter). The activity rate for each individual source (plant or whole subcategory) will be taken from centrally available data, e.g., national statistics, or from responses obtained from individual plants via standardized questionnaires.

The Toolkit is being used by predominantly developing countries to develop their dioxin inventories within the National Implementation Plans. The Toolkit is a living document and is being updated and improved through an open and transparent process established by the 2nd Conference of the Parties and assisted by a Toolkit Expert Group.

From the inventories obtained so far, it can be concluded that the major sources in developing countries are different from those in developed countries. Whereas in developed countries the inventories are dominated by emissions from the ferrous and nonferrous metal industries and waste incineration, the predominant sources in developing countries are the releases from open burning processes such as burning waste dumps and individual fires to dispose of private waste as well as forest fires or agricultural fires, especially preharvest sugarcane burning. Also, power and steam generation or cooking with biomass is more prevalent.

An overview of PCDD/PCDF releases estimated with the Toolkit and according to country and release vector is shown in Table 2. The Table also provides – as a rough orientation and quality check for the results of the national estimates – the per capita releases in micrograms TEQ per year and person. It can be seen that on average 13 μg TEQ/a and person are released to air.

The Toolkit application has its special value in the identification of high emitters of PCDD/PCDF. The results may serve as a starting point in identifying PCDD/PCDF sources that are of special importance and in assigning priorities among sources, which have to be addressed as a priority by a Party in order to meet the Stockholm Convention's obligations. The Toolkit generates numeric results and these quantitative estimates of PCDD/PCDF releases can be used for priority setting and thus, pointing to potential areas to apply BAT and BEP to reduce the releases.

Unfortunately, there are presently not the necessary data available to publish Toolkits for PCB and HCB, so that these releases are hard to quantify.

3.4. EFFECTIVENESS EVALUATION

COP-2 has established the outline for the effectiveness evaluation and the amended global monitoring plan has been adopted by COP-3. Among others, the Global Monitoring Plan (GMP) will be established on a regional basis according to the five UN regions. The global monitoring plan aims for the evaluation of time trends and spatial trends of the POPs concentrations in the environment and in humans. The first global report would aim to determine the baseline for further evaluations, and that future evaluations would enhance global core comparable datasets. The first core data will include ambient air and mother's milk or human blood (as indicators for human exposure). A guidance document for the Global Monitoring of Plan (GMP) has been drafted and was found to be an appropriate basis for Parties to implement the GMP.

TABLE 2. PCDD/PCDF releases per country and year to air and all release vectors and releases per inhabitant and year – Toolkit application

Country		Annual release g TEQ/a		Annual release µg TEQ/person*a		Reference year
	Pop.*mio	Air	Total	Air	Total	
Argentina	37.4	874	2,111	23	56	2001
Armenia	2.976	5.49	52	1.8	17	2001
Australia	19.7	495	1,800	25	91	2002
Brunei	0.340	0.75	1.40	2.2	4.1	2001
Cambodia	13.4	273	607	20	45	Unknown
Chile	15.7	51.7	85.7	3.3	5.5	2003
Côte d'Ivoire	17.3	416	432	24	25	2002
Croatia	4.50	116	168	26	37	2001
Cuba	11.2	195	319	17	28	2000
Ecuador	13.7	65.5	98.5	4.8	7.2	2002
Estonia	1.42	14	29.2	9.9	21	Unknown
Fiji	0.81	11.2	19.2	14	24	2002
Jordan	5.30	64.3	81.6	12	15	2003
Latvia	3.40	22	54.5	6.5	16	2005
Lebanon	3.70	79.0	165.8	21	45	2004
Lithuania	3.60	37.44	56.91	10	16	2005
Mauritius	1.21	16.5	26.5	14	22	2003
Morocco	29.9	167	425	5.6	14	2003
Nicaragua	5.36	266	638	50	119	2004
Niue	0.002	0.392	0.563	181	259	2004
Paraguay	5.20	70.7	156	14	30	2002
Philippines	84.5	328	534	3.9	6.3	1999
Poland	38.6	490	1,039	13	27	Unknown
Seychelles	0.081	4.1	5.41	51	67	2003
Slovenia	2.01	6.194	30.5	3.1	15	2005
Sri Lanka	19.9	172	257	8.6	13	2002
Tanzania	36.8	517	947	14	26	Unknown
Thailand	62.4	286	1,070	4.6	17	Unknown
Uruguay	3.30	18.7	48.5	5.7	15	2003
Vietnam	78.4	16	69	0.20	0.88	2002
Zambia	10.3	290	483	28	47	2004
TOTAL		**5,368**	**11,811**	**13**	**22**	

To assist the effectiveness evaluation and other needs for POPs analysis, UNEP is implementing a GEF project on "Assessment of Existing Capacity and Capacity Building Needs to Analyze POPs." Needs for POPs analytical data arise in three areas:

1. Effectiveness evaluation: monitoring data for all POPs in ambient air, human blood, and mother's milk
2. BAT/BEP guidelines for BAT-associated PCDD/PCDF performance levels in stack emissions
3. Determination of the "low POP content in wastes"

A databank containing information from POPs laboratories is accessible at http://www.chem.unep.ch/databank/Home/Welcome.aspx. The information (from 182 laboratories) therein shows that least experience is with PCDD/PCDF but also mirex and toxaphene are hardly analyzed. The most popular matrices are water and soild/sediment. The ones of the first evaluation – ambient air, mother's milk, and human blood – are the least commonly analyzed matrices (UNEP/GEF, 2007). UNEP Chemicals will maintain th databank beyond the period of the project as a source of information.

3.5. GUIDELINES

3.5.1. BAT/BEP Guidelines

The means to reduce the formation and release of PCDD/PCDF are the application of best available techniques (BAT) and best environmental practices (BEP). Guidelines on BAT/BEP have been developed by an Expert Group on BAT/BEP established by the Conference of the Parties of the Stockholm Convention. The draft document is available at the Convention's WebPage and has been adopted by COP-3. The guidelines address the 20 source categories listed in Annex C of the Convention. The guidelines include "performance levels for PCDD/PCDF emissions to air that are associated with BAT" as follows (Table 3):

3.5.2. Basel Convention Technical Guidelines

With respect to stockpiles and wastes that consist, contain or are contaminated with POPs, there is a close cooperation between the Secretariats of the Stockholm and the Basel Conventions. In order to define the "low POP content," an open-ended working group established by the Conference of the Parties (COP) of the Basel Convention has provided a provisional definition in a technical guideline, which was adopted by the 7th COP of the Basel Convention in October 2004 and confirmed with the revised guidelines at COP-8 in November 2006 (SBC, no date). The provisional definitions of the low POP content are as follows:

- PCB: 50 mg/kg
- PCDD/PCDF: 15 μg TEQ/kg
- Aldrin, chlordane, DDT, dieldrin, endrin, heptachlor, hexachlorobenzene, mirex, toxaphene: 50 mg/kg for each POP

TABLE 3. Summary of performance levels for emissions of PCDD/PCDF associated with BAT (BAT/BEP Guidelines, 2007)

Source according to Annex C	Performance level (ng TEQ/Nm^3)
Part II categories	
(a) Waste incinerators for municipal solid waste, hazardous waste, medical waste, and sewage sludge	<0.1
(b) Cement kilns cofiring hazardous waste	<0.1
(d) Thermal processes in the metallurgical industry	
(i) Secondary copper production	<0.5
(ii) Sinter plants in iron and steel industry	<0.2
(iii) Secondary aluminium production	<0.5
(iv) Secondary zinc production	<0.5
Part III categories	
(b) Other sources in the metallurgical industry	
(i) Secondary lead smelting	<0.1
(ii) Primary aluminium production	<0.1
(iii) Secondary steel production	<0.1
(iv) Primary base metals smelting	<0.1
(d) Fossil fuel-fired utility and industrial boilers	<0.1
(e) Firing installations for wood and other biomass	<0.1
(g) Crematoria	<0.1
(i) Destruction of animal carcasses	<0.1

4. Context

As with all international environmental agreements, the Stockholm Convention on POPs should not be seen in isolation but in the context of other relevant agreements at the national and international level and a general chemicals' management strategy.

References

1. Fiedler (2006): National PCDD/PCDF Release Inventories under the Stockholm Convention on Persistent Organic Pollutants. Chemosphere **67**, S96-S108
2. SBC (no date): Secretariat of the Basel Convention www.basel.int
3. SSC (no date): Secretariat of the Stockholm Convention www.pops.int
4. UNEP (2007): Toolkit methodology and reports can be viewed at http://www.chem.unep.ch/pops/pcdd_activities/default.htm
5. UNEP/GEF (2007): Databank and reports accessible from http://www.chem.unep.ch/databank/

PERSISTENT ORGANIC POLLUTANTS (POPs) IN THE REPUBLIC OF MOLDOVA: PROBLEMS AND SOLUTIONS

GHEORGHE DUCA,[*] ION BARBARASA
Academy of Sciences of Moldova, 1 Stefan cel Mare bld. MD-2001, Chisinau, Republic of Moldova

Abstract: The aim of this article is to disseminate information about Moldova experience in the field of the Stockholm Convention on Persistent Organic Pollutants (POPs) and how Moldova managed to solve a part of its problems related to POP-contaminated pesticides and PCB-contaminated dielectric oil. This letter will not describe the technical issue related to POPs but only the achievements of Moldova in the field of POPs and how the Moldovan results could be used by other countries in solving their national POPs issues.

Keywords: Persistent Organic Pollutants (POPs); Global Environmental Facility (GEF), Stockholm Convention, National Implementation Plan

1. Introduction

Mounting evidence of health and environmental damage has focused the attention of the international community on a category of substances referred to as Persistent Organic Pollutants (POPs). Some of these are used as pesticides, while others are industrial chemicals. POPs are also generated unintentionally as by-products of combustion and industrial processes. POPs possess toxic characteristics which are persistent, accumulate in the fatty tissues of most living organisms, are prone to long-range transboundary transport, and are likely to cause significant adverse human health or environmental effects near to and distant from their sources.

[*] To whom correspondences should be addressed. E-mail: duca@mrda.md or duca@asm.md

E. Mehmetli and B. Koumanova (eds.),
The Fate of Persistent Organic Pollutants in the Environment, 13–20.

Due to POPs' persistence and propensity to cross-border movement, countries have sought multinational cooperation to address the challenge. The Stockholm Convention for POPs seeks to eliminate a group of 12 POPs.[1] Under the Convention, which entered into force on 17 May 2004 and was ratified by Moldova on 19 February 2004, parties are required to develop implementation plans to indicate how they will meet their obligations under the Convention. In 2001, Moldova requested from the Global Environment Facility (GEF) financial assistance in fulfilling its Stockholm Convention obligations and it received a GEF POPs Enabling Activity (EA) grant. The Moldova POPs NIP was one of the main outputs of the project and was submitted 20 October 2004. The NIP identified the POPs chemicals of concern in Moldova as being stockpiles of obsolete pesticides and Polychlorinated Biphenyls (PCBs) contained in electrical equipment, primarily capacitors.

2. Persistent Organic Pollutants Stockpiles in Moldova: POPs Obsolete Pesticides

Moldova has used extremely high amounts of pesticides in the past. Stockpiling of now banned and useless pesticides collectively referred to as "obsolete" pesticides is a significant POPs issue. The stockpiles of POPs obsolete pesticides pose a continuous threat to the environment and public health. In the absence of an obsolete pesticides management strategy, over the years, significant amounts of obsolete pesticides have been stockpiled in warehouses on an ad hoc basis. Often the warehouses are dilapidated; usually the passage of time and exposure has resulted in deterioration of the packaging material. Both increase possibilities for leakage of the toxic materials to the environment – the main pathway for contamination. When obsolete pesticides were placed in storage they were generally indiscriminately mixed with each other in bags and drums. This resulted in a mixture of POPs pesticides with non-POPs pesticides and there is no economically viable way of determining the compositions of all the resultant mixtures in the repackaged plastic and steel drums. Representative sampling analysis indicates that the average amount of POP pesticides out of the total stock of obsolete pesticides is about 20–30%. There are approximately 3,000 t of obsolete pesticides stored in warehouses in Moldova. Studies have shown conclusively that these materials have contaminated the sites and surrounding soils and nearby surface waters. Other 4,000 t, including 650 t of DDT and 1,300 t of HCH are stored in one warehouse situated in the south of the country.

Figure 1. Repackaging of POPs contaminated obsolete pesticides in Moldova

In November 2003, the Ministry of Defense (MOD) and the State Department of Emergency Situations (SDES) initiated the repackaging and transportation of the obsolete pesticide stocks from some 344 warehouses scattered across the country, to 37 centralized district storage facilities, one in each of the administrative districts. These warehouses were selected based on a number of criteria to ensure safety and security. The warehouses were each examined during the Environmental Assessment of the project to evaluate their integrity. While this system of centralizing the storage is an improvement it is not a long-term solution. Centralizing the materials allows for improved security and monitoring and will facilitate ultimate disposal. The repackaging, transport and storage activity has been, and will continue to be until completion, cofinanced by NATO. All of the obsolete pesticides will be relocated to the 37 warehouses and the works to be performed by the end of 2007.

2.1. PCBs AND PCB-CONTAINING ELECTRICAL EQUIPMENT

Moldova has an unusually high amount of PCBs requiring disposal because in former Soviet Union times it was the energy hub transmitting electricity to Bulgaria. Most of the PCBs in Moldova are concentrated in electrical power installations where the dielectric oils in capacitors and transformers contain PCBs. Most of this equipment is out of use but still in place. The main pathways of environmental pollution are PCB containing oil spills and leaks from electrical equipment no longer in use. The capacitors at power installations are situated outdoors and PCBs leak from corroded capacitors to the soil below the capacitor batteries. Twenty thousand PCB-containing capacitors, now unused and referred to as "discarded," are located in 20 electrical substations throughout the country but most (12,000) are at the Vulcanesti Power station where there are also two dumps with approximately 1,000 broken capacitors. The total PCB content in oils within the 20,000 capacitors (excluding the pits at Vulcanesti) is estimated at 380 t while the total weight of the capacitors is approximately 1,060 t. Since PCBs are mobile in the soil the contamination can spread. The PCB-containing capacitors in the electrical substations are in a condition where leakages, due to corrosion, are occurring and leakages are expected to increase in the coming years. Analysis has determined very high levels of PCB contamination at the capacitor sites primarily at Vulcanesti where the 10,000 m^2 of soil underneath this assemblage, to a depth of 60 cm, has been found to be contaminated with PCBs.

Figure 2. Vulcanesti power station

An inventory of capacitors held by users of electricity has not been undertaken as yet, but based on expert judgments the total PCB-content of those capacitors is roughly estimated at approximately 20–50 t. A detailed inventory of PCBs in transformers, switches and other electrical equipment has not been finalized but according to a preliminary inventory about 30,000 t of dielectric oils are used in electrical power installations.

3. Regulatory and Institutional Capacities

From a regulatory and institutional perspective Moldova does not have any specific requirements for the management and control of POPs, but to a certain extent some of the requirements of the Stockholm Convention are covered in a general sense under existing legislation and institutional arrangements. The Ministry of Ecology and Natural Resources (MENR) is the central national environmental authority which has been designated the Stockholm Convention competent authority and as such is responsible for coordinating the POPs related activities of the following government bodies involved in chemicals management issues: *Ministry of Health, Ministry of Agriculture and Food Industry, Ministry of Industry, Ministry of Energy, Ministry of Economy, Ministry of Internal Affairs, Ministry of Defense, Department of Customs, Department of Standardization and Metrology, and the Department for Emergency Situations*. Local authorities have responsibilities for environmental protection and management in the limits of their territory. A National Coordination Committee (NCC) for implementation of the Stockholm Convention, with MENR, as the lead agency has been established to provide overall guidance and coordination for POPs NIP development.

The MENR is responsible for developing the POPs National Implementation Plan (NIP) and for compliance and enforcement of national legal requirements and international obligations related to management of toxic and hazardous products and substances. Capacity for enforcement of provisions of the Articles covered under the Convention is currently nonexistent in Moldova. The NIP has identified the need for a gap analysis of existing legislative requirements for POPs to bring them in line with that of the Convention and comparable to those in EU countries for POPs and other toxic chemicals and wastes.

4. International Support in the Field of POPs Provided by the International Donors

The efforts done by the Moldovan Government in the filed of the implementation of the Stockholm Convention as well as the results of the first GEF project related to the POPs Enabling Activity had motivated GEF to allocate a new grant to assist the Moldovan Government in the preparation of the Sustainable Persistent Organic Pollutants Stockpiles Management, which will complement national ongoing activities in this field, including those on collection and environmentally safe storage of obsolete pesticides in agriculture. The main objective of the project is to prevent the POPs pollution of the environment and to protect human health by the environmentally safe packaging and storage of obsolete pesticides and PCBs; by improving the overall management of POPs; and by creating national capacities for the implementation of the Stockholm Convention requirements, and of those stipulated under other relevant international Conventions and Protocols ratified by Moldova.

The GEF Implementing Agency of the project is the World Bank. During last 10 years the World Bank has been very active in supporting Moldova in improving its environmental management capabilities and in incorporating environmental and social concerns into its sector operations. The proposed project is consistent with the Bank's Environment Strategy and 2005 Country Assistance Strategy (CAS) for Moldova. The key pillars of the strategy – improving people's quality of life, quality of economic growth, and quality of regional and global commons. The Bank's technical knowledge on POPs management and its experience in the design and implementation of GEF investment projects give it a comparative advantage among the GEF Implementing Agencies in providing this assistance.

5. Higher-Level Objectives to which the Project Contributes

The project will contribute to both national and global objectives. On the national level, it will support implementation of the country's *Economic Growth and Poverty Reduction Strategy Paper*, which has several references to POPs, and its action plan proposes interventions in the area of obsolete pesticides and PCBs, along with the recognition of the need to strengthen institutional capacity in this area. The implementation of proposed activities would provide many local benefits by reducing the impact of POPs on public health and environment. The project would address land degradation by preventing further soil pollution by various POPs.

The project will serve as a triggering mechanism for modernization of the current national chemicals management system steering it towards environmentally sound management of toxic, persistent, harmful and bioaccumulative substances in all spheres of human society. Minimization and final elimination of POPs-related pressures and impacts to the natural and human environment is an integral part of national environmental policy. It is considered that environmentally sound management of chemicals, if being adequately set up and functioning, is an important element which contributes to a well-being of the country, society sustainable development and poverty alleviation. Resolution of POPs pesticides problems, as well as sound management of other prohibited and unused agricultural chemicals, is considered to be helpful for promotion of Moldovan ecologically clean agricultural products world wide.

The project will also provide the following *global benefits*: (i) reduced threats on biological diversity – the elimination of POPs stockpiles and their sound management would decrease both the global pollutant burden and possible impacts on wildlife, domestic animals and humans; and (ii) improved transboundary water quality – by ensuring sustainable POPs management, the project will contribute to the prevention of future contamination and threats to the quality of the global hydrological regime. It will also contribute to the objectives of two other international environmental agreements – the Basel Convention on the Control of Transboundary Movements of Hazardous Wastes and Their Disposal and the Rotterdam Convention on the Prior Informed Consent Procedures for Certain Hazardous Chemicals.

6. Project Development Objective and Key Indicators

The *main development objective* of the project would be to protect the environment and human health by safely managing and disposing of stockpiles of POPs contaminated pesticides and PCBs. The *global project objective* of the project is sustainable POPs stockpiles management and strengthening of the regulatory and institutional arrangements for long-term control of POPs and other toxic substances in line with the requirements of the Stockholm Convention and other related conventions and protocols ratified by Moldova. To track the progress toward achieving this development objective, the project will use four key results indicators as summarized below:

- Reduced risks of POPs environmental pollution to human health by safely storing and disposing of stockpiles of POPs contaminated pesticides and PCBs
- Destruction of 2,210 t of PCBs and POPs containing and contaminated obsolete pesticides
- Modern regulatory system established within Moldova for the management and control of POPs and other toxic and harmful chemicals and wastes

Institutional and human capacities for enforcement of the POPs regulatory framework and for sustainable POPs stockpiles management strengthened.

7. Project Components

The project consists of three components: (1) Management and Destruction of POPs; (2) Development of a Regulatory Framework for PCB Management and Control; and (3) Institution Strengthening and Project Management.

Management and Destruction of POPs includes destruction of stockpiles of POPs containing and contaminated obsolete pesticides (approximately 1,150 t of stockpiled obsolete pesticides out of the total 3,000 t) and management of PCBs and destruction of obsolete capacitor stockpiles (19,300 capacitors – approximately 1,060 t) by incineration.

Strengthening the Regulatory Framework and Capacity Building for POPs Management includes actions for modernization of current legislation specifically related to the Stockholm Convention and incorporation of provisions for establishing a broader chemical safety approach in the country based on EU directives. Full transposition of all relevant EU legal acts shall be achieved.

Institutional Strengthening and Project Management Support foresees strengthening of the MENR's capacity for POPs management and includes developing of the Information Management and Reporting System for POPs; POPs Monitoring Network; Identification of POPs residuals and Mapping of Polluted Areas; and POPs Awareness and Educational Activities.

The Project is cofinanced by the Government of the Republic of Moldova in an amount of US$3.72 million, and envisages the implementation of a range of activities amounting to US$12.6 million, including with US$2.53 million from international donors for similar activities under some other projects implemented in Moldova.

References

1. National Implementation Plan for the Stockholm Convention on Persistent Organic Pollutants, Chisinau, Stiinta, 2004, www.moldovapops.md
2. National Inventory of Persistent Organic Pollutants of the Republic of Moldova, by Dr. Valentin Ciubotaru, Chisinau, July 2003
3. Project Appraisal Document on a Global Environmental Facility Trust Fund Grant to the Republic of Moldova for a Persistent Organic Pollutants (POPs) Stockpiles Management and Destruction Project, October 2005

STOCKHOLM CONVENTION AND CURRENT SITUATION ON POPs IN TURKEY

SÖNMEZ DAĞLI*
TUBITAK Marmara Research Center Chemistry and Environment Institute, Anibal Mevkii, Gebze, Kocaeli-Turkey

Abstract: POPs are mainly aromatic compounds with extreme high toxicity and received high publicity for their toxic characteristics because they are:

- Highly toxic to humans and environment
- Persistent in the environment, resisting biodegradation
- Taken up and bioaccumulated in terrestrial and aquatic ecosystems and capable of long-range transboundary atmospheric transport and deposition

The Stockholm Convention on Persistent Organic Polluters (POPs) was adopted by 125 countries including Turkey, on 22 and 23 May 2001 with the objective of protecting human health and environment, focusing on eliminating or reducing releases of 12 POPs, the "Dirty Dozen". These 12 chemicals include aldrin, chlordane, DDT, dieldrin, endrin, heptachlor, mirex, and toxaphene used principally as pesticides, two industrial chemicals polychlorinated biphenyls (PCBs) and hexachlorobenzene (HCB) used in industry but also produced unintentionally together with dioxins and furans. *In Turkey, on 15 Jan. 2004, the POPs project started and the main objective of this project was to prepare inventories by the responsible authorities that would agree upon the information and experience gathered national environmental priorities evolve.* Turkey has prepared National Implementation Plan (NIP) in 2005 and ratification process is underway. According to the first evaluation, Turkey has 10,930 kg of DDT and 2,700 t of HCB in stocks.

They have been used as insulators in almost every transformer for years and most of the transformers in Turkey contain PCBs. The numbers of transformers are about 250,000 in Turkey and only 10% of these were

---

* To whom correspondence should be addressed. E-mail:Sonmez.Dagli@mam.gov.tr

E. Mehmetli and B. Koumanova (eds.),
The Fate of Persistent Organic Pollutants in the Environment, 21–30.

examined yet. According to preliminary studies, country has 1,972 capacitor and 254 transformers contain PCB. Some of these products and equipments are exported to European Countries for destruction and very little of these are being destructed by Turkish Waste Incinerator Company Izaydas. According to first PCB analysis results in waste soils and treatment plant sludges, PCB levels in soils and treatment sludges are lower than 0.01 ppm. The most important POPs (dioxins and furans) introduced to air are discharged from ferrous and nonferrous metal production, production of mineral products, waste incineration, and power generation plants.

Keywords: POPs, Stockholm Convention, pollution

1. Introduction

In the last 50 years, around the world as a consequence of rapid development in modern production techniques, chemicals, and products, containing these chemicals were produced widely in order to increase agricultural production and as a result of their use, they became diffused into the environment. The main course of these chemicals is pesticides, herbicides, fungicides, polychlorinated biphenyl (PCBs), polychlorinated dibenzodioxin, and various synthetic organic halogen hydrocarbons. In the last 25 years, many of the developed countries banned especially the use of organic chlorinated hydrocarbon pesticides. However, persistent properties of these compounds in the environment, without decomposing, caused not only accumulation of these pesticides in various human and animal bodies, but also encounter these compounds even in the nonexposed areas. The reason is, due to persistent character and long-range effects of organic chlorinated hydrocarbon pesticides.

Ministry of Environment and Forestry of Turkish Republic signed Stockholm Convention, which has been carried out by United Nations Environment Program, Chemicals division and was opened for signature in 22 May 2001.

It has been agreed that use and production of 12 chemicals, Permanent Organic Pollutants (POPs), should be stopped. These 12 chemicals, classified under three categories, are nine pesticides (aldrin, chlordane, dieldrin, endrin, heptachlor, mirex, toxaphene, DDT, hexachlorobenzene), two industrial chemicals (hexachlorobenzene, PCBs), and two unintentional products (dioxin, furan). An action plan is aimed to be initiated after the ratification of the Convention to prepare the inventory and destruction of such chemicals that could have been still in stockpiles.

2. Polychlorinated Biphenyls

Polychlorinated biphenyl (PCBs) are organic compounds, which were used for electric machinery insulation. They have been used as insulators in almost every transformer production, for years. PCB's also appear as by-products of organochloro productions like PVC production and waste incineration.

Because of their aromatic structure, chloride content, and accumulation in the body, they are considered risky chemical substances. However, their use in completely closed systems causes no negative effects on environment and humans.

Studies on transformers containing PCBs, conducted by the Turkish Electricity Generation and Transmission Cooperation (EUAŞ) showed that in Turkey, the total number of 187, 28 are inactive, 3 of them eliminated and 97 of them would be eliminated later date and would be replaced with new, dry type transformers containing silicon oil, instead of the oil with PCBs. The proper types of transportation facilities for POPs are creating important problems in Turkey.

Some of the transformers are scattered to different locations. The majority are located in Kahraman Maraş, Kütahya, and Elazığ provinces. These areas need further studies to be able to identify how many of transformers cause contamination in those areas. In addition, training is given in many places on transformers to move them from one place to another for maintenance and change their oil with PCBs. For above mentioned transformers, total oil account is about 126,000 kg and approximate PCB content in the product is estimated 8.82 g TEQ[1]/year.

Due to the size and industrial population of Turkey also when limited financial human resources and time were added whole of the country could not be covered by a field search. But it is expected that PCB-containing equipments should be more than thousands in Turkey.

The number of transformers is about 250,000 in Turkey and only 10% of these has been examined yet. According to preliminary studies, only 2,500 transformer has been examined and country has 1,972 capacitor and 254 transformers that contain PCB or PCB-contaminated oil. Some of these products are exported to European Countries for destruction and very little of these equipments and products are being destructed by Turkish Waste Incinerator Company Izaydas. The main hazardous and medical waste incinerator (İzaydaş) licensed by Turkish Ministry of Environment and Forestry is located in İzmit. The incineration capacity of this plant is 35,000 t/year and total waste destructed by IZAYDAS is given in Table 1.

TABLE 1. Total disposed waste in İzaydaş Incinerator

Waste Type	2000 (t)	2001 (t)	2002 (t)	2003 (t)
Bunker	10,435	10,797	6,989,034	13,415
Tubes	722	1,508	485	552,9
Flammable liquid	2,542	1,141	1,923	1,8
Liquid with water	142	253,9	283,9	181,9
Special liquid	59	211,1	135,8	192,0
Medical	2.4	89,4	207,5	290,7
Total	13,903	14,002	10,024	16,449

Hexachlorobenzene (HCB) is used for seed disinfection as a fungicide in agriculture, as mentioned before. At the same time, they are by-products that are formed during productions of organic chloride substances, like chloride solvent and PVC. It is a product of volatile ashes, released from incinerator facilities and gases of the chlor alkaline facility, which together, form a pollution system. Their production and use are banned in 1985, by the Ministry of Agriculture and Rural Affairs.

2.1. PCB STOCKPILES AND CONTAMINATION

Approximately, 2,700 t of HCH (hexachlocyclohexane or α-HCB) and DDT in sacks and barrels were stored in the storage house of Merkim Industrial Products A.Ş in Şirintepe, Derince, Kocaeli.

Agricultural Protection Chemicals A.Ş produced the substance, which was in white powder form and preserved in 50 kg nylon bags and barrels to be used for agricultural protection in the past. In 1985, the General Directorate of Protection and Control of Ministry of Agriculture and Rural Affairs, according to the (Law for Agricultural Protection, Law no. 6968) forced the products to be stored in storage house.

Works, regarding the disposal of the waste, continue at different times, but still no concrete results have been achieved. TUBİTAK-MRC, Food Sciences and Technology Institute analyzed the product and DDT and HCB were identified.

The storage house of the Merkim Industrial Products A.Ş in Şirintepe, Derince. Kocaeli does not have the proper conditions for storing HCB, the entire building needs repair against rain. At present, this storage house is a potential contamination area in Kocaeli.

3. Pesticides

Pesticides have been controlled by Ministry of Agriculture and Rural Affairs (MARA) in Turkey. Registration Committee of MARA register agricultural pesticides after studying the chemical and physical properties, biological activity, residues, toxicological and ecotoxicological properties with internationally approved analysis methods and techniques. MARA also monitors registered pesticides from their production or import to their consumption. Pesticide retailers are constantly controlled by the Provincial and County Directorates of MARA responsible from market control. MARA has also performed some monitoring studies on pesticides mainly on food samples affected by pesticides.

A total of 1,231 commercial pesticides had been registered in 1995, though the picture changed in 2004 as 485 active ingredient and 3,006 commercial pesticides were registered in Turkey. However, due to various reasons only 252 active ingredients containing registered pesticides are commercially available. Total pesticides produced in Turkey are given in Table 2.

TABLE 2. Total pesticides produced in Turkey. (From State Planning Organisation, 2006.)

Year	Production (t)	Import (t)	Export (t)
1998	23,710	15,025	1,991
1999	24,343	20,686	1,675
2001	23,969	9,465	3,976
2002	23,807	–	1,616
2003	28,639	14,823	2,133
2004	26,480	16,513	2,234

Records from Undersecretariat of Foreign Trade and Undersecretariat of Customs show the fact that POPs substances have not been imported or exported after they had been banned. Nationwide residue monitoring of chloride containing pesticides has being performed on agricultural products, soil, and streams. No illegal use of such substances has been detected.

POPs pesticides ban was started in Turkey by 1970s, and their use, production, import, and export are prohibited by law.

Dieldrin was banned in 1971, aldrin, chlordane, heptachlor, endrin were banned in 1979 and toxaphene was banned in 1989. In addition, registrations of plant protection products containing these active ingredients were canceled. Mirex and its products have never been registered in Turkey. There is no record of production of POPs pesticides in Turkey since their ban. After registration, plant protection products were being prepared by using imported active ingredients. Production of those active substances in

Turkey has never been an issue. Records obtained from Undersecretariat of Foreign Trade and Undersecretariat of Customs show the fact that POPs substances have not been imported or exported after they had been banned.

Nationwide residue monitoring of chloride-containing pesticides has being performed on agricultural products, soil, and streams. No illegal use of such substances has been detected. As required by the Stockholm Convention, necessary arrangements are initiated for frequent inspection by MARA.

After the restriction and ban of use of POPs pesticides in some countries, necessary precautions were taken in Turkey. Starting from 1968 the use of aldrin, dieldrin, heptachlor, DDT, chlordane and toxaphene were restricted. Application of soil with aldrin and heptachlor was forbidden, but application on seed was allowed. There were no restrictions on HCB at those years. However, after some applications, HCB residues were found on weeds and DDT residues were found in straw (Güvener et al., 1974). Eighteen pesticides that were banned by MARA up to now (Burçak et al., 2005).

3.1. PESTICIDES STOCKPILES AND CONTAMINATION

After the ban on POPs pesticides the stock records of retailers and firms were collected as one of the precautions taken by MARA. Turkey has 10,930 kg of DDT in the stocks. DDT stock is located in the facilities of Ankara Central Supply Directorate of MARA.

Central Supply Directorate facilities are located in Yenimahalle/Ankara. DDT stocks are not stored in good conditions. The storage facility has broken windows and doors; the closeness of the storage facility to civil population may also create problems. These storage facilities should be improved until the disposal of pesticides.

Extensive quantities of hazardous waste, including POPs in barrels, which were dumped illegally has been found by the Turkish Authorities in the area between Sinop and Rize, along the Black Sea coast in 1988 and about 80% of them were in the area between Sinop and Samsun.

Turkey became a party to the Basel Convention on the Control of Transboundary Movements of Hazardous Wastes and their disposal in 1994. Turkey has a national regulation on Hazardous Waste Management and it became in to force in 1995.

At present, the storage conditions are very poor. Most of the metal drums have become heavily corroded and waste materials are contaminating the concrete and soil around the area. In order to protect human health and environment, disposal of the absolete stocks and the contamination studies are urgently needed (Ünlü and Avcı, 2004).

Although some samples exist higher pesticide concentrations (16 ppb endosulfan in mussels) there are no significant amounts of chlorinated pesticides and PCBs were found in the sea water and mussel samples are taken from the Black Sea, by Kurt and Özkoç (2004). Chlorinated pesticides and PCBs concentrations determined in mussel samples, taken from Turkish coasts of Central Black Sea.

4. Dioxin and Furanes

The Stockholm Convention requires all Parties endeavor to collect information and based on this information to develop a National Implementation Plan for meeting its obligations under the Convention. Related with this requirement this section intends to summarize the inventories of the emission of polychlorinated dibenzo-*p*-dioxins and dibenzofurans (PCDDs/PCDFs), hexachlorobenzene (HCB), and polychlorinated biphenils (PCBs) which are unintentionally formed in a wide range of thermal and industrial chemical processes. These are the chemical substances covered by Annex C of the Stockholm Convention.

Potential sources are:

- Thermal processes: waste incineration, uncontrolled waste burning, metal smelting, and refining processes, thermal power generation, cement kilns, wood and other biomass burning and transportation fuel combustion.
- Industrial chemical processes: production of pulp and paper when bleaching with elemental chlorine is used.

It detailed database of emission factors, which provides suitable default data were applied as representative of the class into which the processes are grouped. The main source categories for PCDDs/PCDFs emissions in the Toolkit are identified as:

- Waste incineration
- Ferrous and nonferrous metal production
- Power generation and heating
- Production of mineral products
- Transport
- Uncontrolled combustion processes
- Production of chemicals and consumer goods
- Miscellaneous
- Disposal
- Hot points

In Turkey, according to the results obtained among the emission into air, water and land, the most important POP's producer sectors into air, amount in order are ferrous and nonferrous metal production (113,854 g TEQ/year or gram toxic equivalent of dioxins per year), production of mineral products (25,088 g TEQ/year), waste incineration (19,883 g TEQ/year), power generation on heating (15,772 g TEQ/year), uncontrolled combustion process (5,000 g TEQ/year) and transport (3,063 g TEQ/year) (Dağlı et al., 2005).

The total emissions (air, water, soil, product) of unintentionally produced POPs (dioxin and furane) in Turkey are given in Table 3. According to this table, the most important POPs producer sectors are ferrous and nonferrous metal production, production of mineral products, waste incineration, and power generation.

TABLE 3. Dioxin and furan emissions (g TEQ/year) in Turkey

Category	Source categories	Annual releases (g TEQ/a)				
		Air	Water	Land	Products	Residue
1	Waste incineration	19.883	0.000	0.000	0.000	0.6
2	Ferrous and nonferrous metal production	113.854	0.000	0.000	0.000	195.7
3	Power generation and heating	15.772	0.000	0.000	0.000	0.0
4	Production of mineral products	25.088	0.000	0.000	0.000	3.3
5	Transportation	3.063	0.000	0.000	0.000	0.0
6	Uncontrolled combustion processes	5.000	0.000	4.000	0.000	0.0
7	Production of chemicals and consumer goods	0.000	0.000	0.000	21.589	0.0
8	Miscellaneous	0.010	0.000	0.000	0.000	0.0
9	Disposal/landfilling	0.000	0.000	0.000	0.001	0.0
10	Identification of potential hot spots					
1–9	Total	182.7	0.0	4.0	21.6	199.6

Ferrous and nonferrous metal production values are not surprising for Turkey, as most of the European Union Candidate Countries (e.g., Poland and Romania) have same emission values for metallurgical sector (Pulles, 2003). But for the validation of these results, detailed research and investigation are needed for this sector in Turkey.

There are no regular monitoring studies conducted in Turkey in the aspect of dioxin and furans. There is a recent published study carried out by

Karademir (2001), from Kocaeli University, Faculty of Environmental Engineering in collaboration with İzaydaş and University of Newcastle upon Thyne, Faculty of Environmental Engineering for the estimation of PCDD/F levels in an industrialized area in Turkey. Air, soil, grass, and milk samples for this study were collected in the İzaydaş incineration area in Kocaeli. PCDD/F concentrations are given in Table 4.

According to Table 4, there is no serious risk is valid on dioxin pollution but further studies should regularly be conducted.

TABLE 4. The concentrations of PCDD/F in air, soil grass, and milk

	Concentration range	Average
Air	23–563 fg TEQ/m^3	176 fg TEQ/m^3
Soil	0.5–4.1 ng TEQ/kg	1.2 ng TEQ/kg
Grass	0.3–5.0 ng TEQ/kg	1.8 ng TEQ/kg
Milk	33–308 pg TEQ/kg	105 pg TEQ/kg

Since the POPs pesticides were banned, other hazardous chemical were controlled and managed by law. In Turkey, there are no recently identified chemicals at present. This needs further studies by the responsible authorities in Turkey.

5. Conclusions

Because of Turkey's Candidature Process, future releases (production) of POPs will be started to decrease after 2005. Most of the regulations has been renewed and/or changed and a lot of new regulations are also produced in this process. For example, Ministry of Environment and Forestry has performed Turkish side necessities together with Dutch Coordinator, for the implementation of IPPC Directive in Turkey. So, it is possible to say that the legislation in Turkey is in compliance with EU Legislation. But, implementation of this legislation should need a long transition period to achieve compliance because of Turkey's local conditions.

References

Güvener, A., Çifter, F., Türker, O., and Körtimur, G., 1974, "Research on insecticide levels on foods", Plant Protection Bulletin, 14(3):229–235.

Burçak, A. and Dağli, S., 2005, Chapter Coordinator of the Report on the Prelımınary Inventories of the Pops Project in Turkey (Unıdo-Project No. Gf/Tur/03/008) by Professor Dr. Altan Acara, The National Project Coordinator, Ankara.

Professor Dr. Kahraman ÜNLÜ, METU, Environment Engineering Department & Professor Dr. Cem B. Avci, March 2004, Bosporus University, Merkim Industry A.Ş. Preliminary Feasibility Study at the store in İzmit for Elimination of Duration Time Ended Pesticides.

Kurt, P.B. and Özkoç, H.B., 2004, Marine Pollution Bulletin 48:1076–1083.

Pulles, T., 2004, Dioxin Emissions in Candidate Countries, Interim Report, TNO, Holland.

LESSONS LEARNED AND GOOD PRACTICE IN THE PREPARATION OF THE NATIONAL IMPLEMENTATION PLAN (NIP) FOR THE MANAGEMENT OF POPs IN THE REPUBLIC OF BULGARIA

TSVETANKA DIMCHEVA*
Hazardous Chemicals Control and Management Department, Ministry of Environment and Water, W. Gladstone 67, 1000 Sofia, Bulgaria

SVETLA KRAPTCHEVA
Ministry of Environment and Water, W. Gladstone 67, 1000 Sofia, Bulgaria

Abstract: The present paper presents the lessons learned and the good practice in the preparation of the National Implementation Plan (NIP) for the management of POPs in Bulgaria, developed under Stockholm Convention, Assessment of POPs issues in the country, based on preliminary National POPs Inventories and the Measures taken to implement the Bulgarian NIP for POPs.

Keywords: Bulgaria, Persistent Organic Pollutants (POPs), National Implementation Plan (NIP), POPs Action plans, National priorities, POPs management, POPs issue assessment, POPs inventory, POPs pesticides, PCB, HCB, dioxins, furans, lesson learned

1. Introduction

The negative effects of Persistent Organic Pollutants (POPs) for humans and for the environment became evident throughout the world, as well as in the Central and Eastern European countries.

*To whom correspondence should be addressed. E-mail: dimcheva@moew.government.bg

E. Mehmetli and B. Koumanova (eds.),
The Fate of Persistent Organic Pollutants in the Environment, 31–49.

Persistent Organic Pollutants (POPs) have been recognized by the international community as representing threats to human health and the environment, because they are toxic, resist degradation, bioaccumulate, and are transported through air, water, and migratory species across international boundaries and deposited far from their place of release, where they accumulate in terrestrial and aquatic ecosystems. The Stockholm Convention on POPs, agreed in 2001, identifies 12 chemicals of particular concern – aldrin, dieldrin, endrin, chlordane, heptachlor, DDT, mirex, hexachlorobenzene, toxaphene, polychlorinated biphenyls (PCBs), polychlorinated dibenzo-*p*-dioxins, and dibenzofurans (PCDD/PCDF).

The Republic of Bulgaria ratified the Stockholm Convention with a Law by the National Assembly (SG No. 89/12.10.2004). The Convention is effective for Bulgaria from 20 March 2005. The country is taking appropriate measures to implement its obligations under the Convention and has prepared and endorsed its National Implementation Plan (NIP) in March 2006.

The Republic of Bulgaria has completed and endorsed its National Implementation Plan (NIP)[1] for the management of POPs in March 2006 as required under Article 7 of the Stockholm Convention. The Bulgarian NIP was officially submitted to the Stockholm Convention Secretariat on 27 September 2006, although the deadline for NIP submission is 20 March 2007.

2. Process of NIP Development

The National Implementation Plan (NIP) under Stockholm Convention for POPs was developed by Ministry of Environment and Water (MoEW) of Bulgaria with GEF funds contribution and with the assistance of UNEP[2] and cofinancing provided from German Government as a part of Global GEF/UNEP project: GF/2732-02-4452 "Development of National Implementation Plans for the management of POPs in 12 pilot countries."

The NIP preparation took 4 years and all the activities were coordinated by a National Coordinating Committee (NCC). A number of experts, representatives of interested state institutions, ministries, universities, professional organizations, NGOs, etc., and the members of NCC took part actively within the process of formulating and development of Bulgarian NIP for POPs.

[1] National Implementation Plan for the management of POPs in the Republic of Bulgaria, Sofia, March 2006.

[2] Bulgaria is one of 12 countries that participates in the Global Project GEF/UNEP: GF/2732-02-4452 "Development Of National Implementation Plans for the management of Persistent Organic Pollutants (POPs)."

The NIP includes a set of measures which implementation shall allow safe storage of POPs and obsolete pesticides stockpiles, gradually phasing out of PCBs equipment and the reduction of unintentional production of POPs releases derived by different industrial sectors.

2.1. NIP EXPECTED OUTCOMES

- Assessment of national capacity to implement the Stockholm Convention
- Preliminary inventories of POPs
- National Implementation Plan, including specific Action Plans and strategies
- Strengthened POPs management infrastructure
- Raised public awareness on POPs

2.2. NIP OUTPUTS

- Final reporting and publication of assessment studies addressed to health and environmental impacts of POPs and actions to reduce or remove them
- Final preparation of the National Implementation Plan, development of three Specific POPs Action Plans and related Annexes and their endorsement
- Execution of a comprehensive information program to raise public and professional awareness of POPs issues, including the development, publication and wide dissemination of several popular POPs booklets and leaflets

2.3. NIP STRUCTURE

The National Implementation Plan for the management of POPs in the Republic of Bulgaria (NIP) comprises of two main parts. The first part summarizes:

1. Country baseline
2. Assessment of the POPs issue in the country based on the preliminary POPs inventories

The second part of the NIP includes:

3. Strategy and Action plan elements of NIP
4. Implementation strategy
5. Specific Action Plans

- Action plan for POPs Pesticides
- Action plan for PCBs in Equipment
- Action plan for POPs releases (Dioxin/Furans) from unintentional production

6. Development and capacity – building proposals and priorities
7. Timetable for plan implementation and measures of success

2.4. NATIONAL PRIORITIES

The Bulgarian NIP for POPs has formulated 10 major national priorities, aiming to achieve the strategic goal to reduce the POPs impacts on human health and the environment:

1. Development and enforcement of a Plan for the management of POPs and obsolete pesticides
2. Development of a Plan for identifying and remediation of POPs contaminated sites
3. Development of a Plan for identification, marking and phasing out PCB containing equipment in use
4. Development of Plan for safe storage and disposal of phased-out equipment and waste, containing PCBs
5. Development of Plan for reduction/elimination of releases from unintentional production (D/Fs, HCBs, and PCBs)
6. Evaluation of negative POPs impacts on human health and monitoring of POPs levels in humans and the environment
7. Encourage and support research on POPs effect on humans and the environment
8. Public awareness raising with regard to POPs issues
9. Endeavor to secure financial resources for implementation NIP measures by attracting investments from international finance institutions and donors
10. NIP integration within the existing National Environmental and Sectoral policies

3. Assessment of the POPs Issue in Bulgaria Based on the Preliminary POPs Inventories

3.1. ASSESSMENT WITH RESPECT TO POPs AND OBSOLETE PESTICIDES

In Bulgaria, the use of POPs pesticides has been largest in the 1960s of last century. Most POPs pesticides have been applied mainly as pest control preparations to treat termites and soil insects. Although never been manufactured in the country, they had been used on more than 2 million hectares of agricultural land, forests, etc. during this period.

POPs pesticides have never been produced in Bulgaria and their import and use have either not been approved or progressively banned (Table 1).

TABLE 1. Bulgaria's record of regulation concerning POPs plant protection chemicals

POPs chemicals	Import and use
Chlordane, hexachlorobenzene, mirex	Never permitted
Aldrin, dieldrin, endrin, DDT	Banned 1969
Toxaphene	Banned 1985
Heptachlor	Banned 1991

Government efforts to deal with its POPs and obsolete pesticides problems are longstanding and directly linked to national priorities, strategies, action plans and programs. In spite of the prohibitive and restrictive measures, that have reduced and discontinued the use of POPs pesticides, in Bulgaria there are large amounts of obsolete pesticides stocks available (Photos 1 and 2).

In former decades, plant protection chemicals, including POPs pesticides were accumulated and stored at a large number of sites around Bulgaria (Figure 1). Inventories prepared over the last 10 years, including during the NIP preparation work, have identified in 2006 a total inventory of obsolete pesticide wastes of an estimated 13,922 t out of which 52.3 t are assumed to be POPs pesticides or mixtures, containing or contaminated with POPs (Table 2).

Photo 1. Unsecured municipal storage sites for obsolete and POPs pesticides

Photo 2. Unsecured municipal storage sites for obsolete and POPs pesticides (Continue)

TABLE 2. Obsolete pesticides stockpiles in Bulgaria by years

Obsolete pesticides stockpiles	Amount (t)
Year 2003	
Total in warehouses and BB cubes	12,394
In 651 warehouses	8,835
In 72 centralized safeguarded warehouses	4,656
In 579 unrepaired nonsecured warehouses	4,179
In 957 BB cubes	3,559
Total OP stockpiles, stored in safe centralized warehouses and in BB cubes	8,215
Year 2004	
Total in warehouses and BB cubes	11,222
In 561 warehouses	7,011
In 84 centralized safeguarded warehouses	4,703
In 477 unrepaired nonsecured warehouses	2,308
In 1,255 BB cubes	4,211
Total OP stockpiles, stored in safe centralized warehouses and in BB cubes	8,914
Year 2005	
Total in warehouses and BB cubes	13,519
In 519 warehouses	7,492
In 105 centralized safeguarded warehouses	4,979
In 414 unrepaired nonsecured warehouses	2,513
In 1,612 BB cubes	6,027
Total OP stockpiles, stored in safe centralized warehouses and in BB cubes	11,006
Year 2006	
Total in warehouses and BB cubes	13,922
In 492 warehouses	7,605
In 84 centralized safeguarded warehouses	5,179
In 408 unrepaired nonsecured warehouses	2,426
In 1,588 BB cubes	6,317
Total OP stockpiles, stored in safe centralized warehouses and in BB cubes	11,496

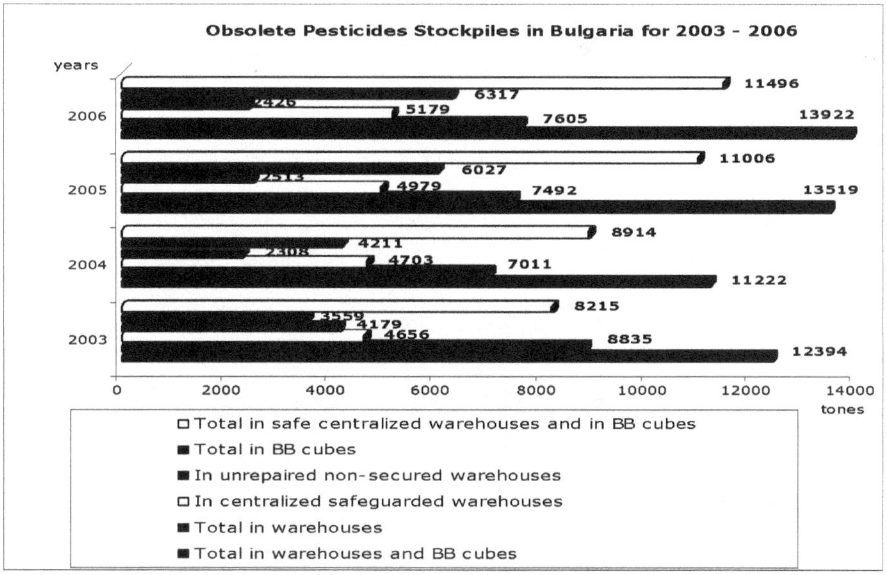

Figure 1. Obsolete pesticides stockpiles in warehouses and BB cubes in Bulgaria for 2006

Figure 2 clearly illustrates the steady upward trends toward the management of obsolete pesticides for safe storage in secured warehouses and BB cubes for the years 2003–2006.

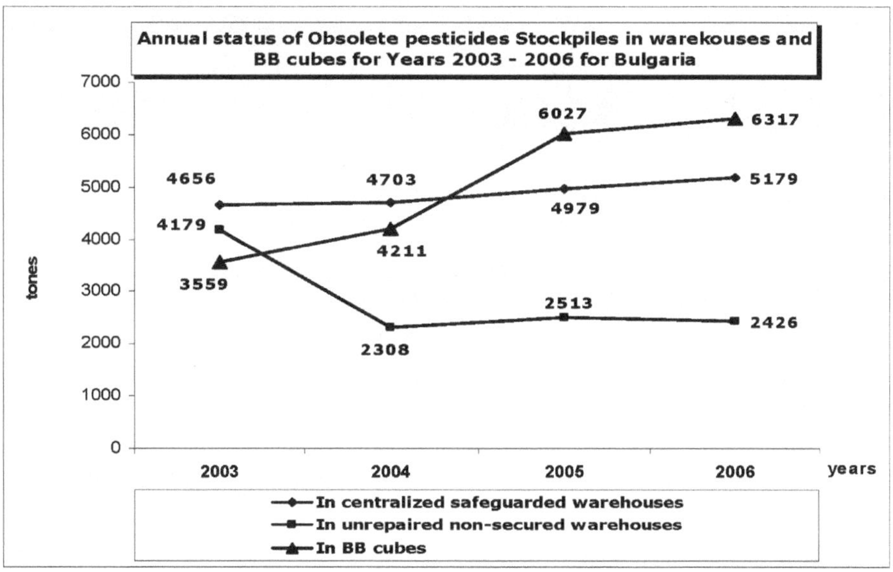

Figure 2. Annual status of obsolete pesticides in warehouses and BB cubes for 2003–2006

Figure 3 indicates the location of obsolete pesticide storage sites – centralized secured warehouses, municipal unsecured warehouses, and encapsulated in BB cubes.

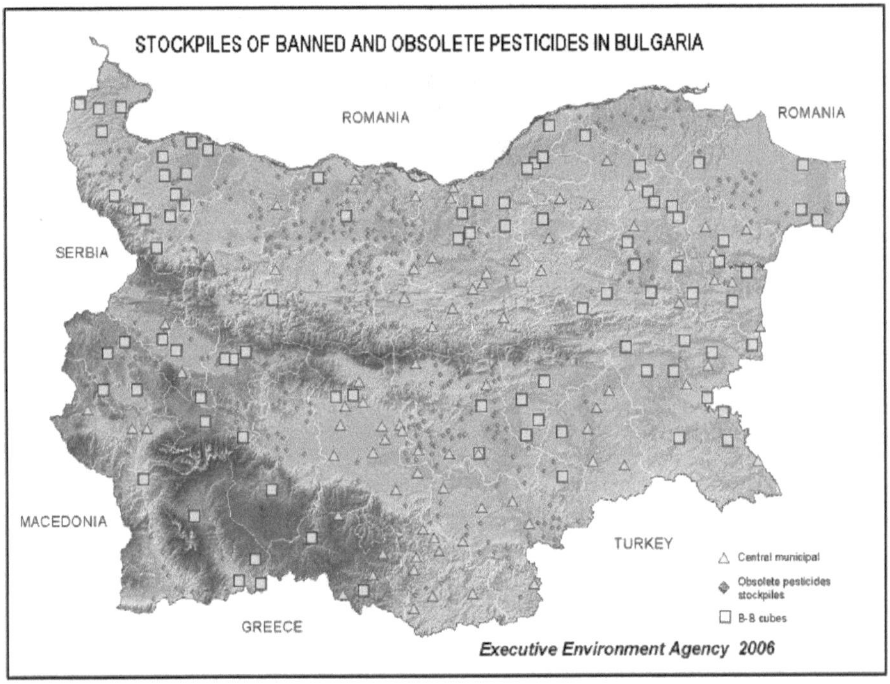

Figure 3. Storage sites map of obsolete pesticides stored in centralized secured and municipal unsecured warehouses and encapsulated in BB cubes for 2006 in Bulgaria

3.1.1. Management of POPs and Obsolete Pesticides

The management of POPs and obsolete pesticides in Bulgaria is implemented in conformity to the effective national regulations, mechanisms, and procedures. Their enforcement guarantees prevention to the maximum extends the negative impact of POPs and obsolete pesticides on human health and the environment.

An interagency expert committee was formed in 1998 between Ministry of Environment and Water and Ministry of Agriculture and Forestry to solve the problem of safe storage of obsolete pesticides. The Committee had developed a Program at national level, implementation mechanism and criteria for assessment and allocation of grand financial funds for projects related to save storage of obsolete pesticides. All municipalities, which apply with such projects, receive necessary funds from the Ministry of Environment and Water, promptly and as a grant. As a result of the implementation of this program, many of warehouses, storing obsolete pesticides have been repaired and transformed into storages with safeguard provided.

The storage facilities for obsolete pesticides are one of the sources for local environmental pollution and pose health and environmental risks.

The construction of centralized municipal warehouses and BB cubes conforming to the legislative requirements for safe disposal, liable storage of available obsolete pesticides stockpiles and cleaning up of emptied warehouses are activities that illustrate consistency in environmental protection policy and sustainable management of obsolete pesticides (Photos 3 and 4).

Photo 3. Repaired centralized storages for obsolete and POPs pesticides

Photo 4. Hermetically sealed steel-concrete containers "BB cubes" for disposal of obsolete and POPs pesticides

The state authorities, responsible for the inventory of obsolete pesticides in Bulgaria are the Ministry of Agriculture and Forestry (MAF) through its National Plant Protection Service (NPPS), the Ministry of Environment and Water (MOEW) though the Executive Environmental Agency (EEA), and their regional structures.

From 2001 the Ministry of Environment and Water (MoEW) through its Executive Environmental Agency (EEA) and the regional structures (RIEWs) implement the monitoring of warehouses status and the obsolete pesticides, stored there on annual bases. Using information cards, the RIEWs collect annually and submit to the EEA information about the status of warehouses and record the amount of obsolete plant protection chemicals, stored there.

The data are compared with the records given by NPPS, summarized and verified by EEA. The database is updated annually.

The Interagency Expert Committee of Experts from MAF, and MOEW, considers and decides on all activities regarding the repair of facilities for storage of obsolete pesticides.

During the period 2001–2006, steady positive trends toward the management of warehouse for banned and obsolete pesticides and the surrounding areas are observed as a result of:

- Restriction of existing and prevention of future pollution in cosequence of the effective enforcement of national legislation.
- Restriction of negative impact of warehouses and obsolete pesticide on the environment and human health by repacking, and shifting in centralized municipal warehouses and cleaning up of emptied premises.
- Permanent disposal of obsolete pesticides in BB cubes (steel-concrete containers with size 195 × 195 × 195 cm, hermetically sealed with useful volume 5 m^3) with upto 300 years waste storage period.
- Awareness rising and providing public access to the available information – a number of popular brochures and leaflets (3) has been published and widely disseminated. All of them can be found also on MoEW's web page (www.moew.government.bg).
- Several Training Workshops and Seminars on POPs and obsolete pesticides issues have been carried out with regional environmental authorities and general public.
- In April 2006, Bulgaria was a host country of an international workshop for Central and Eastern Europe and Asia (CEECA) countries on "Lessons learned and good practice on NIPs' preparation."
- In March 2007, Bulgaria was a host country of TAIEX 23981 Multi-Country Workshop on Obsolete Pesticides in Central and Eastern European countries, organized in cooperation with the Ministry of Environment and Water, Bulgaria.
- Maintaining and annual updating of obsolete pesticides stockpiles database, stored in warehouse and BB cubes on national in Executive Environmental Agency (EEA) and on regional level in Regional Inspectorates for environment and water (RIEWs).
- Financing of project proposals/programs for improvement of storage facilities status and safe and environmentally sound storage of obsolete pesticides by the Enterprise for the management of environmental protection activities (EMEPA).

Bulgaria has made considerable efforts to consolidate these wastes, properly repackaged, into secure storages in order to protect human health and prevent environmental releases. Till 2006, a total of 5,179 t has been repacked and moved to 84 safe centralized warehouses. A further 6,317 t has been encapsulated in 1,588 reinforced concrete “BB cubes.”

Apart from the activities executed within the framework of the Stockholm Convention, Bulgaria has made efforts at an early stage to involve international donors to help. Due to these efforts the Dutch government has offered help to eliminate DDT in Bulgaria. In August 2000, under the project “Destruction of Risk Pesticides from Bulgaria in the Netherlands”, 27,680 kg of POP pesticides such as DDT, aldrin, and dieldrin from four regions of Bulgaria have been exported to the Netherlands, and destroyed in an incinerator in Rotterdam.

In 2006, there remains an estimated ~2,426 t of obsolete pesticide wastes, including POPs, stored in 408 unrepaired and unsecured warehouses, representing ~17% of total obsolete pesticides stockpiles. These wastes are thus representing a significant risk to the environment and the population and are a high priority for action.

Nevertheless, the steady positive trends observed during recent years and constantly increased funds allocated by Bulgarian state (for the period 1998–2006: more than €5 million) for the management of POPs and obsolete pesticides, Bulgaria can not cope alone with final elimination of POPs and obsolete pesticides stockpiles without international financial support, due to limited national funding available.

For the final disposal of obsolete and POPs pesticides stocks Bulgaria shall seek to provide international funding from Global Environmental Facility (GEF) and multilateral sources and funds.

3.2. ASSESSMENT WITH RESPECT TO POLYCHLORINATED BIPHEYLS (PCBs) IN ELECTRICAL EQUIPMENT

Regulation (EC) No850/2004[3] of the European Parliament and Directive 96/59/EC[4] and their amendments has been transposed into Bulgarian legislation by CM Decree 50/09.03.2006, promulgated SG 24/21.03.2006.

This decree entered into force immediately and serves to regulate the use and disposal of PCBs. The regulation limits the use of PCBs to closed-system applications such as transformers and capacitors and requires that decontamination or disposal of equipment with PCBs concentrations greater than 0.05% by weight (500 mg/kg) and volumes greater than 5 dm^3 should be accomplished in an appropriate manner no later than the end of 2010.

---

[3] On persistent organic pollutants and amending Directive 79/117/EC.

[4] On the disposal of polychlorinated biphenyls and polychlorinated terphenyls (PCBs/PCTs).

The regulation also provides for the establishment of detailed inventories, register of equipment and their proper labeling.

In the NIP, PCB management, phase-out, and disposal are identified as amongst the top priorities, requiring immediate attention and action to address problems identified during the country assessment and inventory process.

The preliminary inventory carried out in the Republic of Bulgaria as part of the enabling activities recorded only very restricted data. This is because the inventory was based on voluntary returns from industry as there was, at that time, no legislation in force obliging companies to present information about PCBs containing equipment in their ownership. Only a relatively low rate of return was achieved to survey questionnaires and many of the returns are incomplete. Furthermore, this initial inventory concerned only transformers and capacitors and did not consider other equipment and articles in use.

3.2.1. Inventory of PCB-containing Transformers

PCBs and equipment containing PCBs have never been produced in Bulgaria. None of the manufacturers of transformers, capacitors and their dielectric oils have ever produced equipment and oils containing PCBs. A total of 1954 transformers were imported by Bulgaria between 1970 and 1990. The inventory recorded that from a total of over 43,000 transformers 158 were confirmed to contain PCB oil; a further 41 were confirmed to be contaminated with PCBs; and 3,082 are assumed to contain or be contaminated with PCBs and require further testing (Table 3).

TABLE 3. Inventory results for transformers in Bulgaria

Transformers	>500 ppm	>50 ppm	Assumed	Not PCB
Pieces of equipment	158	41	3,082	40,363
Recorded oil wt. (t)	310.5	1,642	2,531	
Assumed total wt.[5] (t)	1,000	6,500–8,000	10,000–12,000	
Used oil held (t)	9.88		10.24	
Oil in stock (t)	6.8		36.84	
Total[6] (t)	1,016	6,500–8,000	10,050–12,050	

[5] Assumed weight of equipment calculated from oil weight on the basis that PCB oil ≈30% of total weight of PCB transformers (>500 ppm) and 20–25% of total weight of contaminated (>50 ppm) and "assumed" transformers.

[6] Total tonnage is the sum of the assumed total weight of equipment (including oil), waste, and fresh oils.

These results suggest that ~7,500–9,000 t of transformer equipment recorded in the inventory require attention while a further 3,000 items of equipment, with an estimated total weight of 10,000–12,000 t, require further testing.

3.2.2. Inventory of PCB-containing Capacitors

A total of over 3,000 PCB-containing capacitors and a similar number assumed to be PCB containing was reported in an overall inventory of over 11,000 items (Table 4). While the equipment weights were not recorded, the estimated weight of PCBs suggests that 100–150 t of capacitors may contain PCBs and a similar weight requires further investigation.

TABLE 4. Inventory results for capacitors in Bulgaria

Capacitors	PCB containing	Assumed PCB	Non-PCB
Pieces of equipment in service	2,415	2,634	
Pieces of equipment in storage	646	475	
Total	3,061	3,109	11,519
Estimated oil weight (t)	40	3,264.2	

Taking into consideration the low returns on inventory questionnaires, the incomplete nature of many returns, and the fact that questionnaires were not sent to all possible holders of PCB equipment, national inventory totals for both transformers and capacitors are likely to be higher.

3.2.3. Intermediate Results of 2006 Inventory of PCB-containing Equipment

The intermediate inventory results of the commenced in September 2006 detailed Inventory of PCB in equipment show much higher number of PCB capacitors and transformers than the data of the preliminary inventory, done in 2003.

Data from the detailed PCB inventory till December 2006 recorded 249 transformers and 14,738 capacitors, containing PCB above 500 mg/kg in ownership of 128 companies. As per EU and Bulgarian legislation this equipment should be phased out and exported for destruction abroad till December 2010 due to the absence of appropriate disposal facility in Bulgaria.

The detailed PCB inventory is not yet finalized and it is expected the numbers to be much higher than these found at the moment.

Similarly to implement its obligations under Stockholm Convention, Regulation (EC) No 850/2004 on POPs and Directive 96/59/EC on the disposal of PCBs and PCTs Bulgaria need to be supported financially by international sponsor programs and funds.

3.3. ASSESSMENT WITH RESPECT TO POPS RELEASES (DIOX/FURANS, PCBs, AND HCB) FROM UNINTENTIONAL PRODUCTION, 2003

The emissions are calculated under National CORINAIR – 94 methodology, developed by adapting the emission inventory Guide – CORINAIR-94, SNAP-94 for the Bulgarian conditions.

3.3.1. National annual POPs Emissions (dioxins/furans, PCBs, and HCBs)

- In comparison to base year 1990, the annual emissions of dioxin/furans for 2003 show a downward trend of 53, 9% or 2, two times had been observed, following the European trend (Table 5). According to official data for PCDDs/PCDF emissions in Europe within the period 1990–2003, the decrease is 2, seven times (63%) (Figure 4).
- The annual PCBs emissions for the period 1990–2003 are almost the same.

TABLE 5. National annual POPs emissions (DIOX/Fs, PCBs and HCB) in Bulgaria by years

Year	1990	1995	1996	1997	1998	1999	2000	2001	2002	2003
PCDDs/Fs (g I-TEQ/year)	554.2	456	340.9	309.7	288.3	245.2	232.5	200.9	218.5	255
PCB (kg/year)	258.5	382.3	261.7	226.9	252.8	234.3	228.5	211.9	250.1	260.7
HCB (kg/year)	544	79	87	47	76	46	54	42.5	38	45

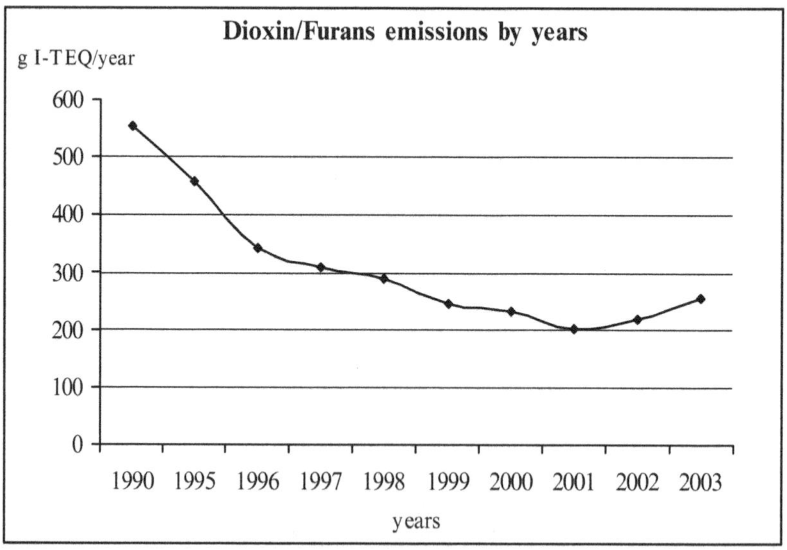

Figure 4. Annual PCDDs/PCDFs releases in the atmosphere by years

- For the period 1990–2003, the HCB emissions in the atmosphere show a significant downward trend. Compared to the base year 1990, for the HCB emission in 2003, a sharp decrease with 91.7% or 12.1 times has been registered due to the decline of industrial production (Figure 5).

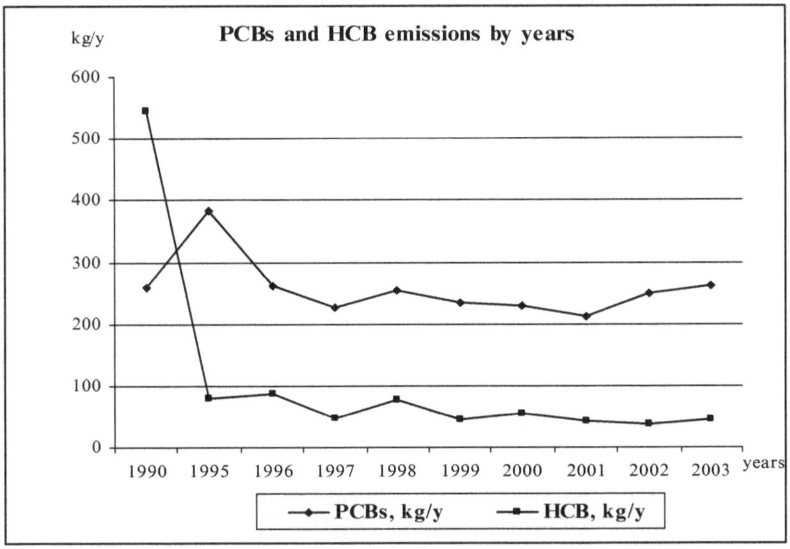

Figure 5. Annual PCBs and HCB releases in the atmosphere by years

3.3.2. National Annual Sector POPs Emissions by Category Sources

POPs releases, generated in Bulgaria in the last 5 years are within the range as follows:

- Dioxins/Furans: 200–255 g I-TEQ/year, and in 2003 have reached 254.9 g I-TEQ/year
- PCBs: 212–261 kg/year, and for 2003 have reached 260.7 kg
- HCB: 38–54 kg/year, and for 2003 being 45 kg

Compared to the base year 1990, PCDDs/PCDFs and HCB note a sharp decline, respectively with 53.9% or 2.2 times and with 91.7% or 12.1 times. The annual PCBs emissions for the same period are almost the same, which could be explained with upward or downward change of the PCBs emissions formed by various category sources.

4. NIP Postimplementation Activities

Bulgaria has already started to implement or has finished some of the activities included in our NIP:

- Bulgarian NIP was published as a book in English and Bulgarian language. An NIP executive Summary has been also published in English.
- The NIP and its Annexes, PCB Guidelines, Leaflets and Brochures are available also in the MoEW's web page: www.moew.government.bg
- POPs and obsolete pesticides inventory is updated annually.
- The Stockholm Convention was translated into Bulgarian language, published and widely disseminated to the public.
- National Profile for the management of Chemicals in Bulgaria was updated in 2003.
- For public awareness rising on POPs issues – a number of popular brochures and leaflets (2) has been published and widely disseminated.
- Under Small Grant Component was implemented a Project, titled "Programme for Public Awareness – POPs impacts on Human Health and the Environment" with funds granted from Global GEF/UNEP Project Umbrella Budget. Several roundtable discussions with students from Chemical Universities and Colleges, general public and professionals has been carried out in Sofia, Plovdiv, Burgas, Pleven, etc. Three brochures and two leaflets has been published and disseminated to the academia and to the public.
- In April 2006, Bulgaria was a host country of an international workshop for Central and Eastern Europe and Asia (CEECA) countries on "Lessons learned and good practice on NIPs' preparation."
- In March 2007, Bulgaria was a host country of TAIEX 23981 Multi-Country Workshop on Obsolete Pesticides in Central and Eastern European countries, organized in cooperation with the Ministry of Environment and Water, Bulgaria.
- Guidelines for the Inventory, Labeling and Management of equipment, containing PCBs was published and disseminated among the holders of PCB electrical equipment.
- Several training workshops and seminars on NIP implementation and POPs issues has been carried out with regional environmental authorities, industry and general public.
- The preliminary PCBs inventories have been completed and a detailed inventory in PCB equipment is going on in the moment.
- A number of training workshops (8) on the inventory of PCB-containing equipment for the industry, PCB equipment holders and experts of the regional structures of MoEW have been carried out.

- For appropriate record keeping, a PCB software program is in place and PCB-containing equipment register is dully maintained by MoEW on national and regional bases.
- POPs emission inventory is updated annually and reported to the European Environmental Agency.

5. Lessons Learned in the Preparation of the NIP for of POPs

- Toolkits, Guidelines, and other support materials must remain flexibly and adjusted in accordance with lessons learned and delivered in time.
- The process should be designed so that the potentially impacted sectors in the country can precisely define the industry conditions (process, emissions, etc.) up front, so that discussions can begin on a factually correct basis. A lot of time was reportedly lost in Bulgaria for correcting information.
- It should be developed financial mechanisms for providing funds to implement the measures set up in POPs Action plans through GEF financing of project proposals, promotion of multiple-source funding approaches and arrangements, twinning programs, and other bilateral, regional and multilateral sources, and channels.
- Moreover, financing modalities should not be limited only to grants. Other options such as state budget funds, commercial loans and private investments should also be considered.
- In prioritizing activities, included in POPs Action plans, it should be important the information related to the costs and benefits of certain interventions that have been successfully implemented in other countries to be made available to key decision-makers.
- For the successful implementation of NIP and POPs Action plans, adequate information pertaining to the environmental impact of POPs should be made available to Bulgarian government. Information related to availability and efficacy of various commercially available technologies as well as the latest information on the development of new emerging technologies should be disseminated to the competent expert bodies in charge and key decision-makers.

6. Conclusions

The successful implementation of NIP for POPs and specific Action plans should employ a realistic approach. Investment and noninvestment activities as well as policies and regulatory measures have to be interrelated and integrated in the existing overall National Environmental and Sectoral policies to ensure implementation of the NIP for POPs.

Various financing modalities, including GEF grants and donor funding programs, should be explored in order to enhance the effective use of limited country resources.

Bulgaria has acted responsibly to reduce environmental and health threats posed by obsolete and POPs pesticides by placing significant quantities in secure and long-term storages. For the final elimination of obsolete pesticides waste a definitive solution can be achieved only with the common actions of all EU member states and other countries within the region of Central and Eastern Europe.The Republic of Bulgaria can not cope alone with the full incremental costs of meeting the obligations under Stockholm Convention and EU POPs legislation and requires international financial support.

To reduce the risk to human health and the environment of POPs stockpiles urgent measures should be taken for safe storage and/or environmentally sound disposal abroad, due to absence of appropriate disposal facility in the country.

To implement the most urgent activities, the Ministry of Environment and Water of Bulgaria has prepared two project proposals for GEF funding:

- POPs and obsolete pesticides risk reduction by environmentally sound disposal abroad and elimination of POPs and obsolete pesticides waste from most affected storage sites.
- PCB risk reduction by environmentally sound management of PCBs and disposal planning measures to eliminate PCBs waste from selected industrial sectors by export abroad.

For the implementation of the NIP for the management of POPs, Bulgaria shall promote taking measures by the enterprises' operators, unintentionally producing and/or using POPs to reduce the risks for the environment and human health by introducing the best available techniques (BAT) and the best environmental practices (BEP), etc.

All these activities, if properly and timely done, shall guarantee the successful implementation of measures, envisaged in the Bulgarian NIP for the management of POPs and the specific Action plans, and shall help Bulgaria to meet its obligations under Stockholm convention and EU legislation.

References

1. National Implementation Plan for the management of POPs in the Republic of Bulgaria, Ministry of Environment and Water, Sofia, Bulgaria, March 2006.
2. Inventory of POPs and Obsolete Pesticides in the Republic of Bulgaria, Ministry of Environment and Water, Sofia , Bulgaria, February 2007.
3. Inventory of PCBs in Equipment in the Republic of Bulgaria, Ministry of Environment and Water, Sofia, Bulgaria, December 2006.
4. Inventory of Dioxin and Furan, PCBs and HCB in the Republic of Bulgaria, Ministry of Environment and Water, Sofia, Bulgaria, July 2004.

PROBLEM OF POPs MANAGEMENT IN UKRAINE: THE LACK OF INTERDEPARTMENTAL COOPERATION AND PUBLIC INVOLVEMENT

VOLODYMYR KUZNYETSOV*
Ukrainian Scientific Research Institute of Ecological Problems, 6. Bakulin Street, Office 237, 61166, Kharkiv, Ukraine

Abstract: Reviewed are problems of POPs management, application and safe disposal in Ukraine and historical roots of this problem as well as legal, institutional, economic and technological approaches to its solution. Existing constraints to efficient management of POPs in Ukraine such as lack of interdepartmental coordination, insufficient funding, poor public involvement, education, and awareness of the problem are discussed.

Keywords: POPs, obsolete pesticides, inventory, pesticides disposal, human and environmental impacts of POPS, risk assessment

1. Background and General Aspects of POPs Problem

When we are talking about persistent organic pollutants (POPs) we mainly mean a number of primary and secondary by-products resulting from industrial and agricultural activities. Currently, 12 substances generally known as "the dirty dozen" and including, among others, aldrin, dieldrin, dioxins, furans, DDT, and PCB are a curse to nature and human health. All of them are highly toxic substances, resistant to decomposition and enhanced capacity for bioaccumulation which makes them dangerous for human health and environment. Generally speaking they can be divided into three broad groups, such as: (1) chlorine organic pesticides; (2) polychlorinated diphenyls; and (3) dioxins and furans. They are not only highly toxic, but

* To whom correspondence should be addressed. Tel.: +380972372128; e-mail: vladkuz@ukr.net

E. Mehmetli and B. Koumanova (eds.),
The Fate of Persistent Organic Pollutants in the Environment, 51–64.

generally are carcinogenic and mutagenic substances difficult to be rid-off from the environment and safely disposed. Due to economic stagnation after collapse of the USSR the level of environmental pollution during the recent years has somehow decreased in Ukraine. The volume of pesticides used in agriculture has decreased also. For example, the pesticide application load at the territories surveyed after the Chernobyl accident (1986) decreased 13.7 times and currently makes up about 0.16 kg/ha. However, the use of POPs-containing pesticides started to grow up again in recent years. Their presence in soils, water and air causes pollution of the environment and a threat to human health. The studies showed that the historic accumulation of considerable amounts of banned pesticides and pesticides that cannot be used for specific purposes in the Ukraine is estimated at 22,000 t, though these estimations are not final.

1.1. ACCUMULATION OF POPs STOCKPILES

Alongside with industrial waste materials, agricultural pesticides are main source of POPs in Ukraine. Several factors have contributed to the accumulation of pesticide stockpiles. Centralized supply systems often did not take into account actual need; pesticides were overordered and oversupplied. In many cases, pest management strategies for migratory pests required large stocks of pesticides to be amassed for pest outbreaks that never came. Often, foreign companies coming to the Ukrainian market supplied pesticides of the wrong type, in the wrong place, or in inappropriate packaging or formulations. There were and there continue to be many cases of unscrupulous behavior such as dumping of obsolete chemicals, intentional oversupply to boost sales, and provision of products with labels in foreign languages or with no labels at all.

In Ukraine, chemicals are often stored in unsuitable conditions and rapidly lose their effectiveness as pest control agents. Modern pesticides have a shelf life of about 2 years when properly stored. But certain components in pesticide formulations deteriorate more quickly if exposed to extreme heat or cold, or if their containers are damaged and thus expose the chemicals to air. The combined result of these multiple failings in pesticide supply and management has been accumulation of large quantities of obsolete pesticides. In many cases these chemicals are escaping into the environment, contaminating soil, water, and air, and threatening wildlife and human health. Elsewhere it is only a matter of time before the pesticides corrode their drums or before paper sacks of DDT, lindane, chlordane, and other pesticide powders are punctured.

1.2. CURRENT DISPOSAL METHODS

Solving the existing problem of obsolete pesticide stockpiles is both technically challenging and expensive. The location of the stocks in some of the poorest regions of the world adds to the complexity and cost. The most commonly used methods for the disposal of hazardous chemical waste are incineration, chemical processing, or landfilling. Unfortunately, virtually none of these existing technologies is able to eliminate the harmful materials without generating potentially dangerous by-products. Incineration of POP pesticides leads to the formation of dioxins and furans, which are, in themselves, extremely hazardous POP chemicals. Chemical treatment generally produces a larger volume of less hazardous chemicals, which then require further treatment and disposal. Landfill, no matter how carefully engineered, which is not the case in Ukraine, can never be a permanent solution, since seals and linings eventually degrade and release the contained toxic waste into the environment.

1.3. POPs DISPOSAL OPTIONS

At present, developing countries and transitional economies seeking to dispose of hazardous materials have limited options. Some have buried their old chemical stocks, but this accelerates the corrosion of containers and release of the chemicals, in some cases, into ground water. Such burial has also been applied in Ukraine. A recent survey of some burial sites revealed that it led to heavy soil contamination with unidentified pesticides, which have leaked from their containers and are contaminating ground water. As to incineration of obsolete pesticides and other hazardous wastes it has been considered as a cheap solution to the problem of POPs disposal. Cement kilns, iron smelters, and open fires are used for destroying waste locally. However, judging by international and the Ukrainian experience such facilities, despite the fact that they helped to reduce the amount of dangerous stocks of POPs, unfortunately also contributed to serious environmental contamination and associated health hazards, through the release of both the waste chemicals and their toxic by-products, such as dioxins.

1.4. INVENTORY ACTIVITIES

Since 1994, the FAO has been working with developing countries to quantify and dispose of accumulated obsolete pesticide stocks. Ukraine was a part of this exercise. Which turned out to be expensive and complicated. Disposal costs average to US$5,000 per t which makes it nearly impossible for this country to afford. UNEP has also initiated similar efforts to address

POPs stockpiles, as part of a Global Environmental Facility (GEF) project entitled "Assessing National Management Needs of Persistent Toxic Substances." Ukraine also takes part at this project. It involves evaluation of national sources and imports of POPs and other chemicals, and identification of stockpiles and/or contaminated environmental sites requiring interventions to reduce risks to human health and the environment.

1.5. RISK ASSESSMENT

While making inventories and identifying the stored pesticides, the internationally applied WHO classification provides a basis for risk assessment. According to the classification, difference between pesticide toxicity and potential hazards is taken into consideration and commercial formulations are evaluated by the active ingredient concentration and physical state of a compound (solids, liquids). While working in warehouses the possibility of skin reabsorption and skin irritation effects are taken into account. The risk assessment for population living at the territories near the pesticide storage sites is based on the hygienic classification. The fact that the criteria of pesticide persistence and migration have been added to the classification is quite important. The risk of potential unpredictable negative impacts resulting from treatment and disposal of various types of obsolete pesticides using the same disposal process is also very high.

While evaluating the risk of obsolete pesticides it is important to take into account not only their composition and toxicity, but also the characteristics of the territory where they are stored, length of the contact and their physical and chemical properties and temperature at which an exposure took place. Thus, the approaches to the problem of risk assessment of obsolete or abandoned pesticides require the use of a methodology taking into consideration of environmental and toxicological effects and the results of epidemiological studies. The risk assessment of obsolete pesticides should serve as a first step and it is essential for the development of action plans and selection of safe method of pesticides disposal (detoxification, landfilling, incineration, composting, etc.) as well as development of preventive measures on control of possible emergencies that may happen due to substantial accumulation of POPs [1].

1.6. MONITORING

One of the serious problems in organization of the efficient management of POPs is lack of reliable monitoring information on volumes of production, application rates, storage sites, and disposal facilities and infrastructure. Lack of measuring equipment for monitoring infrastructure and organizations results in uncontrolled introduction of some POPs (e.g., dioxins) into human food chain which is responsible for delivery of up to 90% of human dioxin unintentional intake. Lack of means for systematic monitoring of POPs causes a tendency among experts to distinguish not POPs pollution zones, but mainly the sources of pollution with POPs such as chemical industry, pulp and paper factories, steel mills, oil refineries, and pesticides application and storage sites. Principal source of dioxin input into environment are waste incinerators. In general, dioxin pollution zone around even the best waste incinerators is at least 1.5 km. or more. Concentration of dioxins in air emissions of an incinerator changes not much within the distance of 24 km from the incinerator stack. Luckily, in Ukraine, number of waste incinerators historically was very limited and many of them are out of operation now [2].

2. Legal Aspects of POPs Management in Ukraine

The legal basis of environmental policy in Ukraine is Constitution (1996) which guarantees the right of citizens to safe and healthy environment and to compensation for damages resulting from the violation of this right. Every citizen is also guaranteed the right to free access to information on environment condition, quality of food and consumer goods as well as the right to disseminate this information. Main legal tool is the Law of Ukraine "On Environmental protection" (1991). Most relevant laws and codes dealing with management of POPs are the Law of Ukraine "On Ambient Air protection" (1992), Water Code of Ukraine (1995), the Law of Ukraine "On Wastes" (1998), Law of Ukraine "On Ecological Expertise" (1996) regulates environmental impact assessment and approval of measures proposed for the management of POPs. Legal and regulatory basis dealing with POPS is rather well developed in Ukraine, but its practical enforcement is poor. Among specific pieces of such regulations mentioned should be the following ones:

- Law of Ukraine "On pesticides and Agrochemicals"
- Sanitary and Epidemiological Department Regulation: "List of substances, products, technological processes, home and nature factors carcinogenic for man"; approved by the Ministry of Health of Ukraine on 7 February 1997, Order No 25
- Decree of the Cabinet of Ministers on 27 March 1996, No 354: "On approving of the order of the removal and utilization of unfit or banned for use pesticides and agricultural chemicals and their packaging"

- Decree of the Cabinet of Ministers of Ukraine No 440 dated 20 June 1995: "Order of receiving permission on the production, storage, transportation, use, destruction and utilization of poisonous substances including toxic industrial waste, biotechnology products and other biological agents"
- List of pesticides banned for use in agriculture, which may not be registered or reregistered in Ukraine "approved by the State Commission on Testing and Registration of Plant Protection Products, Growth Regulators and Fertilizers," dated 5 August 1997, agreed with the Ministry of Health of Ukraine
- List of pesticides and agricultural chemicals allowed to use in Ukraine" (1999)
- Specific bans on the use of Annex A to Stockholm Convention (2001) POP-containing pesticides: Aldrin, Dieldrin, Chloradane, Endrin, Heptachlor, and Hexachlorobenzene, and Annex B pesticide

DDT was intensively applied in the 1970s and 1980s in the Soviet Union but now it is formally banned in the "List of pesticides banned for use in agriculture, which may not be registered or re-registered in Ukraine," dated 5 August 1997.

Annex A pesticide: Mirex is not explicitly banned but it is not registered as required under the Law of Ukraine "On Pesticides and Agricultural Chemicals." At present no explicit regulatory controls are applicable to PCBs or Annex C POPs.

The adoption the National Action Plan on POPs, in line with the requirement of the Stockholm Convention (2001) involves a formal review and endorsement by the Cabinet of Ministers of Ukraine and approval of the Parliament of Ukraine. Two existing National Programs such as National Program of Hazardous Waste Management and National Program of Improvement of Occupational Safety, Occupational Health and Working Environment, adopted by the Parliament of Ukraine in 2000 and 2001, provide conceptual, institutional, and regulatory background and specific action plans for the management of chemical pollutants, including POPs.

2.1. ENVIRONMENTAL MANAGEMENT REGULATION

The Ministry of Environment and Natural Resources of Ukraine is national executive agency and state authority responsible for environmental protection, and is the lead authority responsible for national action and control measures related to POPs. Some executive responsibility in respect to the agrochemicals application issue also lies with the Ministry of Agrarian Policy of Ukraine, as well as with the Sanitary and Epidemiological Department of the Ministry of Health of Ukraine which imposes ambient standards for air, soils, drinking water and is in charge of monitoring public health. Important role is also played by Ministry of Industry – producer of the most of POPs containing materials and equipment for their disposal and treatment. Ukraine still does not have a fully developed regulatory and assessment scheme for pesticides and industrial chemicals but has been working toward such a system for a number of years [3].

Under the National Program of Hazardous Waste Management a National Plan for the Elimination of Risks Related to Stockpiled Obsolete Pesticides in Ukraine is under implementation through NCHWM. Ukraine has regularly collected data on chemical production and usage, and participates in the voluntary reporting to UNEP on chemicals from the PIC List.

2.2. INTERNATIONAL ASPECTS

Ukraine is an active participant in most major global and regional environmental Conventions and the processes involved in their development. The following summarizes the status of Ukraine's participation in those Conventions the may be related to the POPs issue:

- Stockholm Convention on POPs
- Basel Convention on Transboundary Movements of Hazardous Wastes and their Disposal
- Rotterdam Convention on the Prior Informed Consent Procedure for Certain Hazardous Chemicals and Pesticides in International Trade
- UNECE LRTAP Convention
- UNECE LRTAP Convention – (Aarhus) POPs Protocol
- UNECE LRTAP Convention – (Aarhus) Heavy Metals Protocol
- Aarhus Convention on Access to Information, Public Participation in Decision-making and Access to Justice in Environmental Matters

2.3. POLLUTANTS RELEASE AND TRANSFER REGISTRY

In collaboration with UNEP, Ukraine has initiated the development of a national Pollutant Release and Transfer Register System. Under the framework of this project, pilot regions were selected for more detailed monitoring: Cherkassy, Donetsk, and Dnipropetrovs'k. Under the direction of the Ministry of the Environment and Natural Resources, an Interagency Executive Committee to supervise the development of a PRTR was established and a Working Group has been created. This has representation from relevant ministries and organizations as well as NGOs. The first working meeting was held in August 2001 resulting in the identification of immediate tasks and development of an initial implementation plan.

Adoption and implementation of the Protocol on PRTRs within the framework of the Aarhus Convention on Access to Information, Public Participation in Decision-Making and Access to Justice in Environmental Matters is currently considered.

3. POPs Usage, Stockpiles, and Disposal

Pesticides of Annex A and B of the Stockholm Convention of 2001 were never produced in Ukraine (but mainly in Russia) and application pesticides, included into these annexes was prohibited for some years or they were never applied. However, as in most countries, Ukraine has accumulated stocks of obsolete pesticides. There safe disposal became a priority environmental problem for the country and its environmental authorities at all levels. The term "obsolete pesticides" covers a group of chemical products used in agriculture, property, and health protection applications. As a result of their prolonged storage, they can transform into a hazardous waste due to the fact that:

- These products are now considered dangerous to the environment, and prohibited (e.g., DDT).
- Due to their age these products lost their original properties and/or acquired new or unknown properties.
- These products are of unknown composition due to lost documentation or labeling.
- Unintentional or accidental or intentional mixture of several other products, and consequently resulted in unknown properties.

Common characteristic of these materials is lack of data as to their chemical composition. Under Ukrainian regulations such materials are generally considered Class 1 or 2 hazardous wastes, which are the highest environmental hazard classes assigned to waste.

3.1. POPs STOCKPILING INVENTORY ISSUE

The accumulation of the large quantities of obsolete pesticides in Ukraine is a result of extensive and, in some cases, excessive use of pesticides in the agricultural areas of the former Soviet Union as well the centrally managed practices applied to their distribution. Inefficiency of the management system existing in the USSR led to accumulation of considerable quantities of outdated chemicals throughout Ukraine, which used to be a main agricultural supplier in the USSR. However, in the 1960s, demand for pesticides in the agricultural sector also began to fall as a result of bans for some pesticides (e.g., DDT) or stricter controls on their use.

Changes in some agricultural practices and techniques such as introduction of crop rotation also played a role in reduction of pesticides application. In the 1970s, a program for reformulating and repacking the outdated pesticides was initiated and the obsolete pesticides began to be collected from the farms and re-stored at the central stores with the intention that they be returned to producers. The program lasted for about 20 years, but was never fully implemented. By 1993, more than 52% of the banned and outdated pesticides were collected in the central storage sites, originally used as pesticides distribution points, while the rest remains scattered throughout the country usually in substandard storage facilities [4]. Despite original intentions, no considerable amount of the outdated chemicals was in fact returned to the producers for reformulation or treatment.

The general condition of both the central and end-use storage facilities is poor by current environmental standards. In many cases, these facilities are in poor condition and do not have fencing, warning signs or other security arrangements. Actual storage of the pesticides in the buildings is typically in metal containers or bags placed on the floors, which are also likely in poor condition. Little or no segregation or organization of this material has been undertaken. In the areas around these buildings, bulk storage in steel tanks, originally those used for transport of liquids, have been modified, and are now used for "long-term" storage of mixed types of obsolete pesticides. This is a product of the general lack of resources to maintain them, change of ownership and, as a result, uncertainty in responsibility for storage facilities and their content management. Besides, Ukraine has not fully developed integrated systems of state standards regulating pesticides management practices. Though conditions of storage facilities vary, they typically lack appropriate containment, security and supervision. This results in risks of unauthorized use and leakage of material into the environment and direct exposure of local population and wildlife to contaminants. A project jointly financed by the MENR and the Danish Environmental Protection Agency was begun in 1998 to systematically define the extent of the obsolete pesticide problem and develop a long-term plan to address it. The first phase of this program initiated a national inventory of obsolete pesticides, undertook representative analysis of storage sites and began evaluation of domestic treatment and disposal option. The principle product of this work was a National Plan for the Elimination of Risks Related to Stockpiled Obsolete Pesticides in Ukraine. This has been included in the National Program on Hazardous Waste Management, which was approved by the Parliament of Ukraine in 2001. The National Centre for Hazardous Waste Management (NCHWM) has been designated by the Ministry of Environment to carry-out this Plan. Currently, the NCHWM are undertaking Phase 2 of the work jointly with Danish waste management experts.

3.2. SCALE OF THE POPS STORAGE PROBLEM

The inventory of accumulated obsolete pesticides performed recently has shown that there are about 15,000 t of them stored in Ukraine, including 6,500 t that were collected from the farms and stored in the centralized storage facilities formerly owned by agricultural chemicals maintenance organizations. Altogether there are 147 centralized storage facilities containing 7,900,000 t of obsolete pesticides. Similarly, 4,967 smaller storage sites (which belong to various legal entities or have no owner at all) have been identified with an estimated 7,000 t of obsolete pesticides. Substances that are banned make up an estimated 24% of this volume while 10% is classed as out of date material. The remainder (60%) is not identified due to mixing or loss of labeling. Screening level analytical work on a sample basis suggests that up to 70% of the volume will be or contain some form of chlorinated organic compounds and 4% are mercury based. The MENR and the Ministry of Agricultural Policy of Ukraine have also made preliminary estimates of Annex A and B POPs pesticides that indicate that relatively small quantities of endrin, heptachlor and HCB may exist (less than 15 t total) but relatively large quantities of DDT in various concentrations exist. These estimates range from approximately 1,500,000 t to 1,800,000 t [5]. In addition, it is estimated that approximately 550 t of lindane in various concentrations is in storage.

It should be noted that all of the above estimates apply only to the agricultural sector and additional stocks that are still unidentified may be found in other sectors. The only unofficially reported stocks reported to date are 23 t of DDT held by the Ministry of Defense. However, it would be anticipated that agencies and enterprises in the transportation, utility and forestry sectors will also hold stocks of obsolete pesticides, potentially including Annex A and B POPs pesticides.

4. PCBs Management in Ukraine

PCB's were never produced in Ukraine but like in other industrialized countries were used in a number of applications, mainly transformers. Assuming that 50% of the PCB-containing material used in large capacitors and transformers, manufactured in the USSR is still in service or stored in Ukraine, an overall first order estimate of PCB volumes that might remain in Ukraine would be approximately 10,000 t. This estimate recognizes that the above figure for directly contaminated materials and equipment will be much greater and it does not account for cross contamination caused by mixing dielectric fluids which may have occurred during servicing operations. PCB-containing equipment is mainly used in power generation and distribution systems while some use in various industrial sectors (ferrous and nonferrous metallurgy, food, chemical, and coal) is also recorded. It is estimated that 17,700 pieces of PCB-containing equipment (transformers and capacitors) remain in operation in the country and 1,600 pieces of equipment containing PCBs are stored [6].

Data on dioxin, furan, HCB, and PCBs emissions show that an increasing trend in their releases, which has been attributed to increasing utilization of municipal waste incineration. In accordance with the Stockholm Convention, there are a number of requirements regulating the releases of dioxins and furans by incinerators. Unfortunately, Ukraine does not have the monitoring capability and, which is the most important, there is no laboratory that is capable of performing a qualified chemical analysis of dioxins and furans. The results of the experts' work demonstrate the need to address urgently the personnel training problem in Ukraine and to provide the laboratories with modern analytical equipment meeting international standards.

5. Contaminated Sites

No inventories or evaluations of contaminated sites generally or specifically those that may involve contamination by POPs have been undertaken in Ukraine so far. However, based on recent assessments of the condition of pesticide storage facilities noted above it can be anticipated that a significant number of these sites may be contaminated with pesticides, including POPs pesticides. Similarly, it would also be anticipated that the experience of other countries relating to PCB contamination in and around facilities where electrical equipment is stored, operated and serviced would apply. The general low quality of municipal and industrial waste management facilities which is based almost exclusively on land disposal without adequate controls on or segregation of incoming waste streams would also suggest that POPs contamination may be found in such locations. A priority for the National Implementation Plan required under the Stockholm Convention (2001) will be development of a capacity to carry out inventory, evaluation, prioritization and proper containment and environmental remediation of all contaminated sites.

6. Waste Management Infrastructure, and Monitoring and Analytical Capability

While Ukraine posses strong technical and scientific capacity in most if not all the disciplines required to address the POPs issue, it generally lacks modern infrastructure for storage, handling, treatment and disposal of POPs wastes, and capability for release monitoring and the necessary chemical analysis. As reported above, a pilot initiative is underway to upgrade at least one central obsolete pesticide storage facility to Western European standards and to undertake nation wide training in improved handling storage and evaluation of obsolete pesticide storage. Research and development activities are active related to various treatment and disposal methods for POPs and technologies that should provide for the reduction in their releases. These include consideration of various thermal destruction technologies. Those include utilization of stationary and mobile facilities for plasma pyrolysis, and the development of biological techniques for contaminated site remediation. In the recent years, a number of new technologies were tried on a pilot basis to be introduced in this country. These include the following technologies:

- Gas-phase hydrogenation
- Electrochemical oxidation
- Plasma arc
- Molten metal

They offer different ways of dealing with chemical waste. But these systems tend to be highly specific with regard to the chemicals they can decompose, however, and often the process is slow or extremely costly. Unfortunately, Ukraine lacks the infrastructure necessary to support resource intensive technologies mentioned above mainly not due to lack of appropriate know-how, but due to lack of funding.

7. POPs, Environmental and Public Health

Monitoring and assessment of potential human health impacts associated with chlorinated organic pesticides has been active in Ukraine as early as the 1950s. Investigation of impacts of DDT and other chemicals on workers using these materials was initiated in the 1950s. This addressed cardiovascular, endocrine, liver and kidney, and respiratory impacts. The accumulation of DDT and its metabolites in a human organism as well as in the breast milk of Ukrainian women, who had no previous direct contacts with chlorinated organic chemicals, were found during the 1960s. In the 1970s, the impacts of lindane and HCB accumulation have been studied and the presence of PCBs in the human organism was first recorded in the 1980s. National Program on "Development and implementation of measures for the prophylactic of peritoneal and infant morbidity and mortality into the heath care practice" initiated a biological monitoring of chlorinated organic pesticides in breast milk in different Ukrainian regions during 1992–1997. These results showed that contamination of breast milk results in transfer to nursing children in quantities exceeding permissible levels. As a general conclusion, these studies show that accumulation of chlorinated organic chemicals is generally occurring in any Ukrainian citizen irrespective of the region under scrutiny [7].

8. Public Awareness

While a relatively high level of awareness regarding the POPs issue exists with the Government and professional community, awareness among the general public is not high. This is seen as a priority in the development of effective responses to the issue of POPs reduction or elimination in Ukraine.

9. Conclusions

Despite of its threat to human and environmental health the problem of POPs management, application and disposal so far has not become a focus of concentrated activities at national, regional, or local level.

Funding allocated to solve the issue is insufficient, poorly managed, and disbursed. There is no well-functioning and efficient system of POPs management in Ukraine. There are no specialized organizations which would be capable and properly equipped to deal with problem of obsolete POPs collection, transportation and final utilization. Ukraine has not fully developed integrated systems of state standards regulating pesticides management practices Works dealing with POPs are not concentrated in one hand. Many different institutions and organizations deal with it and all of them have differing views and approaches as to what and how should be done in respect to manufacturing, application and disposal of POPs and coordination of their activities is poor. General level of awareness of the authorities and general public about potential threats of POPs misuse and neglect remains quite low and needs to be enhanced.

References

1. Relevant Problems in the field of PCBs management in Ukraine, Sukhorebra S., Chetverykov V., National Center for Hazardous Waste Management, Ministry of Ecology and Natural Resources of Ukraine, Kyiv, Ukraine, 2002
2. POPs survey and situation analysis in Ukraine, development of recommendations for the National Implementation Plan of the Stockholm Convention on POPs (Country situation report), The Ecological Charitable Fund (Ukraine), Kyiv, Ukraine, 2002
3. Ministry of Environment and Natural Resources of Ukraine, Danish Environmental Protection Agency, "Elimination of Risks Associated with Stockpiles of Obsolete Pesticides in Ukraine: National Plan," June 2000
4. The Problem of POPs Regulating in the Ukraine: Characteristic Features and Priorities, Belous A. Proceedings of the Sub-regional Meeting on National Strategy and Action Plan for Reducing/Eliminating POPs Releases Moscow, Russian Federation, July 1998
5. The Problem of Persistent Organic Pollutants in the Ukraine, Dobrovolsky L. Institute of Labour Medicine, Academy of Medical Sciences of the Ukraine, Kiev, Ukraine, 2004
6. Arctic Monitoring and Assessment Programme (AMAP) and Centre for International Projects, "PCB in the Russian Federation: Inventory and Proposal for Priority Remedial Actions. Executive Summary," 2000. Internet
7. Partnerships Between NGOs and Research Facilities for Capacity Building to Reduce Adverse Health and Environmental Impacts of POPs, Tsyguleva O., MAMA-86 NGO Kharkiv, 2004

PERSISTENT ORGANIC POLLUTANTS (POPs) MANAGEMENT IN GEORGIA

IRAKLI LEGASHVILI*
Department of Waste and Chemical Substances Management, Ministry of Environment Protection and Natural Resources of Georgia, 6 Gulua street, Tbilisi 0114, Georgia

Abstract: Quite a quantity of old pesticides has been accumulated in the country since the Soviet times. They are kept in broken-down and obsolete storehouses and their adjacent territories. Due to the precipitation, they get into the environment (soil, surface waters, and ground waters). Certain quantity of these pesticides are Persistent Organic Pollutants (POPs), characteristic by the ability to be accumulated in the environment and all kinds of living beings creating threat to the human health and other living organisms. The situation is aggravated by the fact that there is no special burial sites in the country for the toxic wastes, where they would be safely disposed. It is worth mentioning that near the city of Rustavi on the mountain of Iagluji there has been a burial place since the Soviet times, which, however, does not meet any environmental requirements and needs urgent relevant measures to be carried out. The aim of Stockholm convention is to take measures for reducing or eliminating POPs releases from intentional and unintentional production and use, also the releases from existing stockpiles and wastes. Georgia is currently in the final stages of preparing its National Implementation Plan (NIP) for POPs. The NIP priority setting exercise conducted during the NIP process gave the first priority among the various POPs categories to obsolete POPs pesticides.

Keywords: pesticide, PCBs, dioxin, furan, inventory, public awareness

*To whom correspondence should be addressed. Tel.: (995 93) 200085; E-mail: chem_ira@yahoo.com

E. Mehmetli and B. Koumanova (eds.),
The Fate of Persistent Organic Pollutants in the Environment, 65–79.

1. Introduction

It is widely known that Georgia is characterized by diverse climatic conditions, agriculture herbs and vegetative cover, providing for extensive spread of entophytes, plant diseases and weeds. Hence, the pesticides of various chemical groups are applied. Pesticides are also widely used in municipal economy and veterinary for struggling against rodents and various parasites.

According to the existing information, till the 1990s, annual consumption of pesticides made 30,000–35,000 t. They were applied in a very wide range of fields. Part of the pesticides was produced locally and part imported from outside of the republic.

Pesticide storage, carriage and use norms were never strictly observed in the country and, as a result, the environmental pollution was quite high. Particularly, as it can be observed from the monitoring data, concentration of some pesticides (DDT, HCH, hexachloran, treflan, etc.) in the environment (soil, water) 4–12, an in some case 30–50 times exceeds MACs. Unfortunately, since 1990, due to a heavy economical situation existing in the country, no systematic environmental monitoring, other than single trials under various programs, has been performed.

Heaviest socioeconomical and political conditions of 1990–1995 destroyed state pesticide management at all stages of their use (production, import, export, transit, transportation, storage, consumption, and disposal). During the same period, the import of pesticides has significantly reduced, and mainly the old, often expired soviet pesticides remaining in the warehouses were used. Rules and requirements for pesticide consumption were violated. The state registration was not performed.

In 1996, the Law of Georgia on Protection of Environment has been passed. The law covered chemical (including pesticides) management issues. In 1998, the law on Pesticides and Agrichemicals increased state control and relatively stabilized the situation, although, the requirements of the law are not always observed.

2. Existing Situation

According to official data for 2000–2004, annual import and consumption of pesticides makes 1,000–2,500 t. Sharp reduction of their use (in comparison to the 1970s–1980s) can be explained by the following: introduction of modern pesticides and low dosages of their use; import of not registered illegal pesticides; low purchase capacities of the farmers. According to the existing information, only 30–35% of the total agricultural lands of Georgia are treated with pesticides, and this reduces load per hectare. For instance, the load for vine makes 5 kg, for grain 1.25 kg, maize 0.61 kg, vegetables

6.6 kg, and citrus fruits 15.8 kg (in this case, the main load falls on use of mineral oils, consumption norm for which is considerably high). However, in the future we shall expect rise in use of pesticides in Georgia.

Despite the fact that Georgia has respective legislation concerning chemical (including pesticides) management, the provisions of these laws are frequently violated [1]. Particularly:

- Pesticides are illegally imported.
- Pesticide trading network is not duly organized.
- Pesticides are often falsified.
- Low-quality pesticides are used.
- In many cases technical equipment (sprayers, pulverizers, etc.) do not comply with the requirements.
- Practically none of the farming economies has specially arranged areas for preparation of working blend of pesticides and repair of technical equipment.
- Storages do not comply with modern requirements.
- Destroyed and worn warehouses of Soviet times still keep the obsolete and expired pesticides.
- Special equipment (apparatus, tare, and special ware) are not properly treated.
- None of the farming economies performs due registration of volumes and type of the used pesticides.
- Public awareness on safe and effective use of the pesticides is very low.
- The environment (soil, water, and air) is not monitored for pesticide contents.

Due to the above reasons, the pesticides, including persistent organic pollutants (POPs) [2], get to soil, ground and surface waters, air, cause environmental pollution and impose hazard to human health and safety of living organisms.

3. Stockholm Convention

Stockholm Convention on Persistent Organic Pollutants defines international legislation on regulation of POPs. The problem these substances pose on environment is of transboundary character because POPs have toxic properties, resist degradation, bioaccumulate and are transported through air, water and migratory species, across international boundaries. They can be deposited far from their place of release, where they accumulate in terrestrial and aquatic ecosystems. Georgia signed Stockholm convention on 23 May 2003, and ratified on 16 October 2006.

The Convention takes into account the technical as well as financial support for the developing countries and the countries with transition economies, aiming to increase their potential and capacity for dealing with POPs problem according to the requirements of Stockholm Convention. The important components of the corresponding projects should be information exchange, public awareness building and creation of adequate legal bases at the country level.

The main objective of the project was to create sustainable capacity and ownership in Georgia to meet its obligations under the Stockholm Convention. The project covered inventory [3] of the POPs in Georgia, informing public regarding POPs problem, and preparation of a POPs National Implementation Plan. The National Implementation Plan must describe how Georgia will meet its obligations under the Stockholm Convention to phase-out POPs sources and manage POPs contaminated sites.

In 2003, within the frame of Stockholm convention, the program on "Preparation of the Persistent Organic Pollutants (POPs) National Implementation Plan under the Stockholm Convention" was developed and approved in Georgia. The project is executed by the Ministry of Environment and Natural resources of Georgia.

4. Global Significance and Problem Statement

Persistent Organic Pollutants (POPs) pose severe global risks to human health and the environment due to their toxicity, their persistence, their ability to travel long distances on air and water currents, and their propensity to bioaccumulate in food chains. They include some of the world's most harmful pesticides such as DDT; industrial chemicals such as PCBs; and unintended by-products of industrial processes and incineration such as dioxins and furans.

POPs, when released into the environment, can travel by air and water far from their origins. This typically consists of a number of "hops," each consisting of three stages: (i) evaporation; (ii) transport in the atmosphere; and (iii) condensation at lower temperatures – the "grasshopper effect." In this way, POPs can travel long distances and become widely dispersed in a matter of days or weeks on air currents, and more slowly in rivers and by ocean currents.

The Stockholm Convention works towards the reduction and elimination of the use of POPs and environmentally sound disposal of existing stockpiles.

The major part of the obsolete pesticide stores have already been identified by the Government of Georgia as part of the NIP preparation and earlier inventory programs. The data was collected in accordance with the

FAO methodology including collection of information on location of obsolete and expired pesticide storages, volumes of such chemicals, their trade names, chemical composition, identification labels, packaging, producer and a visual assessment of area potentially contaminated.

Out of 214 storages containing obsolete and expired pesticides were found at 46 facilities (Table 1). In aggregate, 3,057 t of chemicals were found, 2,700 t being stored at Ialghuji Pesticide polygon, and the rest 357 t is distributed in other chemical storages of the country (Table 2). Most of these pesticides are POPs containing.

TABLE 1. Number of registered sites and taken samples by the regions

No.	Region	Number of registered sites	Number of sites with pesticides	Number of taken samples	
				Pesticide	Soil
1	Achara	2	2	6	2
2	Imereti	23	9	8	
3	Samtskhe-Javakheti	10	2	7	1
4	Shida Kartli	11	8	13	–
5	Khashuri	19	10	8	–
6	Kakheti	8	3	17	6
7	Guria	22	2	1	
8	Mtskheta-Mtianeti	8	5	8	2
9	Samegrelo	87	–	–	–
10	Kvemo Svaneti-Lechkhumi	16	–	–	–
11	Racha	1	–	–	–
12	Kvemo Kartli	7	5	3	
13	Tbilisi	–	–	–	–
	Total	**214**	**46**	**71**	**11**

The stockpiles remain in Georgia from Soviet times, and are accumulated at worn and destroyed warehouses of former Soplkimia ("rural chemistry"). These pesticides are carried with precipitation to surrounding soil, ground and surface waters, thus creating serious hazard to human health and the environment.

There are no processing/destruction centers for obsolete POPs pesticides in Georgia. The aim of the project was to repack and dispose POPs pesticides found in the regions as well as at the Iaghuja pesticide dump.

Most of the stores are disassembled and used for construction materials. In many cases the obsolete pesticides are stored under open sky and mixed with each other. Much of the pesticides are without packaging. Precipitation washes them making any labels or other identification markings illegible. It is very difficult to classify them by the chemical groups and exactly identify their composition. Therefore, samples were taken to establish possible POPs contents in the found pesticides.

TABLE 2. Aggregate volume of expired pesticides in Georgia by the regions

No.	Region	Volume of obsolete pesticides	
		kg	L
1	Achara	15,000	
2	Shida Kartli	18,500	100
3	Imereti	10,000	100
4	Mtskheta-Mtianeti	23,000	500
5	Samtskhe-Javakheti	4,200	
6	Kakheti	230,000	
7	Khashuri District	50,500	
8	Kvemo Kartli	4,500	500
9	Guria	–	–
10	Samegrelo	–	–
11	Kvemo Svaneti-Lechkhumi	–	–
12	Racha	–	–
13	Tbilisi	–	–
	Sub total	357,000	1,200
	Ialghuji	2,700,000	
	Total	**3,057,000**	**1,200**

Altogether 71 unknown chemical and 11 soil samples were taken for identification of POPs contents in them. Laboratory analyses showed that 47 out of 71 unknown chemical samples and all soil samples contained POPs. Particularly, in Kakheti (Achinebuli, Tsnori) heptachlor and α, γ-hexachlorohexane was discovered, in Shida Kartli (Kaspi, Kareli, Gori) – heptachlor, α, γ-hexachlorohexane, DDT, and DDE, in Khashuri – heptachlor, α, γ-hexachlorohexane, and DDT, in Achara (Batumi, Kobuleti) – heptachlor, α, γ-hexachlorohexane, DDT, and DDE, in Imereti (Zestaponi, Samtredia) – heptachlor, α, γ-hexachlorohexane and DDT.

Extrapolating from the laboratory analyses, it is estimated that about 66% of total volume of obsolete pesticides are POPs containing, bringing the volume of POPs pesticides in the regional stores to 235 t.

5. Iaghuji Landfill

In Marneuli district of eastern Georgia, at Iaghuja Mountain there is the polygon (landfill) for obsolete and expired pesticides, which, according to the existing information, was functioning between 1976 and 1985. During this period, some 2,700 of various pesticides were buried at the polygon. Since 1986, the polygon is closed and currently it is in a very bad condition. Territory has no fencing, the burial is not protected and domestic animals have a free access to it. There are many cases of theft of pesticides and their metallic containers. In many places the buried pesticides and their packaging (bags, drums) are exposed to the surface.

Unfortunately, it is impossible to obtain the design of the polygon and, hence, to clarify the disposal scheme used for pesticides. The samples taken from the site showed high content of POPs. Both samples analyzed contained 10% of α-hexachlorohexane and 10% of heptachlor.

It should be noted that the polygon was constructed for disposal of organochlorine pesticides which were banned shortly before the polygon was built. Based on this and the laboratory analysis, it is probable that the majority of the obsolete pesticides buried at Iaghuja are organochlorine. The total amount of POPs containing pesticides is approximately 2,000 t or more.

The Iaghuja pesticide dump poses serious threat to the nearby environment and communities. It can be safely assumed that if the pesticides are not removed, there will be a severe contamination of the nearby environment as well as drinking water sources wider in the region. This will result in environmental damage and negative health consequences to the population.

6. Expected Goal, Objectives and Outcomes of Final Project

The outcome of the project is to avoid releases of POPs pesticides stored under unsafe conditions. The project will repack and remove POPs pesticides from unsafe storage sites, transport the pesticide stockpiles to centralized location for transportation to a disposal facility. If the Iaghuja site will be covered by the project it may be more economical to transport the POPs pesticide waste directly for disposal.

Also the strengthening of the policy and regulatory framework of environmentally sound management of POPs and other chemicals are among the outcomes of the project.

7. Description of Preparatory Inception Stage

The major part of the obsolete pesticide stores in the regions have already been identified by the Government of Georgia as part of the NIP preparation and earlier inventory programs.

Analyses have shown that obsolete pesticide stock piles are to large extent POPs pesticides. In this regard it should be noted that most of the pesticides are mixed with each other making it very difficult to identify and separate the pesticides.

8. Laboratory

Selection of the suitable laboratory was required for identifying the trace of POPs into suspected samples, gathered in the regions. The requirements to the lab to be contracted included existence of the necessary equipment (gas chromatographer and mass spectrometer) and qualified staff.

9. Pesticides

Total of 214 sites have been checked in 12 regions of Georgia, out of which at 46 sites the contamination by pesticides was reported. Roughly estimated total amount of pesticides discovered is ~357 t. Seventy-one samples of unknown chemicals and 11 samples of contaminated soils have been taken by the regional inspectors and were transferred to the laboratory. The lab has performed qualitative and quantitative analysis of the samples on POPs.

The summary of the sample analysis looks as follows: out of 71 samples of unknown chemicals, 47 contain at least one of the POPs (~66% of the total number). All 11 samples of contaminated soils bare the trace of few POPs. The main contaminants are hexachlorocyclohexane (hexachlorobenzene), heptachlor, DDT and its derivatives.

Registration-inventory of the pesticides shall assist the country in resolution of issues concerned with obsolete pesticides. Pesticide inventory envisages observation of all the chemicals at their storage places, as well as identification of all unlabeled or mixed chemicals.

This issue becomes very problematic, when it concerns not entire pesticide group, but isolation of those belonging to POPs.

The pesticide inventory and identification shall be carried out by unified method, which will allow receiving objective and easy to compare information on range and volume of the obsolete pesticides.

Results of the inventory provide for possibility of making decision on: (1) possible use of the pesticides in accordance with recommended norms and (2) disposal of pesticides (liquidation, burial), which belong to banned and restricted groups. While carrying out the inventory, it was necessary to observe the following basic conditions:

- Compilation of reliable information on places of storage and burial of the pesticides
- Observation of unified methodics and methodical approach
- Identification of potential hazard of pesticides for the personnel and development of protective measures
- Equipment of the personnel by the protection means
- Sampling, analyzing, and identification of unlabeled pesticides

- Possible secondary packaging and labelling of the chemicals, their collection and preparation for transporting
- Separation of POPs pesticide inventory from general pesticide inventory

Samples of unidentified pesticides were passed to the laboratories for the analyses and determination of their type.

10. PCBs

Investigation of PCB-containing equipment was focused on transformers, capacitors and high-voltage converters – the equipment potentially containing substantial amount of synthetic oils [4]. The sectors covered by the initial inventory were: power generation; high-voltage distribution; low-voltage distribution; railways and metro; milling industry; main chemical industry; metallurgy; paper industry; mining; and oil industry.

10.1. FINDINGS

10.1.1. Transformers

Up to date, total of 15,757 transformers have been registered by experts and regional inspectors. The brief description of the size and company belonging of most of these follows:

1. Power generation: 90 large size transformers, out of which 80 are of USSR make and remaining 10 originating from various countries. Some of the transformers originate from the first half of the 20th century.
2. High-voltage power transmission network: The national high-voltage network contains 458 substations with 4–5 transformers each. The total number is hence around 2,000.
3. The low-voltage distribution system is managed by four entities: The biggest is Georgian Distribution Company, covering all regions of Georgia apart from Kakheti, Adjara and Tbilisi. This company has 9,411 transformers of different effect levels in its distribution network. The same applies for the around 2,000 transformers in the Tbilisi (Telasi) network, ~1,200 transformers in Kakheti, and 2,000 in Adjara.
4. Railways: Five regional departments have 58 substations with 4–5 transformers totaling in 300 transformers. These have been inspected at site and detailed data is available for these as well as for the transformers in the Tbilisi metro system.
5. Mills: 15 including the major units involved in milling different agricultural crops have been included in the survey. Here, some 60 transformers have been accounted for.

The survey has relied on questionnaires and inspections for obtaining the presented numbers. Altogether, the survey on PCB-containing equipment has obtained data from some 400 big and medium-sized companies out of a total of 840 functioning industries. It is estimated that this amounts for some 80% of the total industrial activity in the country.

10.1.2. Capacitors and Converters

According to the survey covering both energy sector and the main industries in 12 regions of Georgia (excluding Racha and Kvemo Svaneti/lechxumi, where the data on synthetic oil-containing equipment are close to zero) total of 5,460 capacitors of 25 types have been registered up to date. Out of these, ~1,100 are still functioning, while ~4,400 are out of order. The total mass of the equipment (including oils and casing) is estimated to be ~295 t, and the estimated mass of oils only ~95 t. The number of converters registered is ~3,200.

In the Soviet period, three types of PCBs were produced: Sovol, Sovtol, and Trichlordiphenyl. They were produced in two plants located in Russia: Orgmina (Dzerzhinsk, Gorki Oblast) and Orgsintez (Novomoskovsk, Tula Oblast). Within 1939–1993 periods, these plants produced ~180,000 t of PCBs.

Sovol, being insulation liquid, was sent to Chirchik (Uzbekistan) Transformer Plant and was used in TH3 type transformers produced there. Sovtol and Trichlordiphenyl were sent to Serpukhov (Russia), Ust-Kamenogorsk (Kazakhstan) and Kamair (Armenia) capacitor plants, where the produced capacitors were filled by the mentioned liquids. Therefore, when the decision on carrying out PCB inventory in Georgia has been made, we have planned and selected mainly the spheres, which are using big capacity powers. Let's consider each of them separately.

Over 2,000 UNEP forms were filled out. Using the CLOR-N-OIL kits, the local, field tests of PCBs in transformer oils were carried out. Total number of tests carried out exceeded 200. In 45 cases we have received positive results. Besides this, 550 specimens of suspicious oil samples received from the regions were sent to laboratories for density analyses.

In the working process, in enterprises located on the territory of Georgia, has been found thousands of all type PCB-containing capacitors produced by Serpukhov JSC KBAR and Ust-Kamenogorsk Capacitor Plant. Also there is registered tens of PCB-containing capacitors manufactured in German Democratic Republic by VEB in 1977–1991.

Also special attention is paid to Convention requirements on assessment of the condition of capacitors existing at food industry sites (mills, reefer facilities, bread bakeries, etc.).

It also should be mentioned that inventory did not cover hundreds of sites belonging to Ministries of Defense and Interior, as well as sites located in Abkhazia and South Osetia. Therefore, the presented data can be considered orientational.

For the present stage, the information that is beyond any doubt can be formulated as follows: the biggest share (77%) of PCB-containing equipment (capacitors) in energy, chemical, oil refinery, ferrous and nonferrous metallurgy, wood processing, railway, food production sectors falls on five regions (Imereti, Achara, Tbilisi, Kvemo Kartli, and Shida Kartli). In three regions (Racha, Kvemo Svaneti-Lechkhumi, and Samegrelo-Zemo Svaneti) share is insignificant (0.3%), and for the remaining six regions (Khashuri, Mtskheta-Mtianeti, Poti, Guria, and Kakheti), these indicators vary between 21% and 23%.

11. Unintentional Products (Dioxins/Furans)

The investigation was carried out in two directions [5]: first approach was based on questionnaires directly distributed to industrial enterprises, and the second – through research into statistical information available at central and regional levels. The data obtained were further compared and verified against each other.

The questionnaires were sent to 379 enterprises and entities. The response was unfortunately very low – 41 forms filled [6].

Based on statistical information obtained from the Central Statistical Department, total annual emissions throughout the country (main point sources being waste incineration, noncontrolled combustion of forests and landfill fires) is gradually diminishing from 359.08 g TEQ/a in 1990 to 121.0 g TEQ/a in 2004. The tendency seems to be logical taking into attention the economic downturn of Georgia since 1990. The findings are summarized in Table 3.

Annual emissions according to the regional data: sum of the emissions in 12 key regions of Georgia – 118.57 g TEQ/a in 2004 – is practically in accordance with the calculations based on information from the central office. See the table summarizing the regional data below.

The inventory of (i) polychlorinated dibenzo-*para*-dioxins – PCDD and (ii) polychlorinated dibenzodifurans PCDF commenced in Georgia in June 2004, in accordance with the *Inventory Plan-Schedule*. The inventory was performed with active participation of Inventory Groups created on the basis of all 14 regional services of Ministry of Natural Resources and Protection of Environment. This process was preceded by the preparation works (organizational and informational–methodological provision of pre-inventory period, respective training workshops).

Inventory of the mentioned composites was carried out on the basis of methodology developed by UNEP (Standardized Toolkit for Identification and Quantification of Dioxin and Furan Releases, 1st edition, Prepared by UNEP Chemicals, Geneva, May 2003), by using proposed coefficients of specific emissions, and filling out respective questionnaires for all *large and medium capacity* enterprises out of total 3,181 enterprises registered in Georgia for 2003 [7].

Information collected by the regional inventory groups was presented to main inventory group and was summarized. At the same time, in order to improve accuracy of the mentioned data, emission of dioxins and furans has also been calculated on basis initial state statistical data for entire country (in consideration of regional specifics) for 1990–2004 period.

TABLE 3. Emissions of dioxin–furans (g TEQ/a) in regions of Georgia from waste incineration and uncontrolled processes categories

No.	Region	Hazardous waste incineration	Medical waste incineration	Uncontrolled combustion process (Landfill fires)	Total emissions g TEQ/a	%
1	Tbilisi	8.99	6.34	22.1	37.43	30.9
2	Shida Kartli	1.32	1.126	5.2	7.646	6.3
3	Kvemo Kartli	2.04	1.739	8.5	12.27	10.1
4	Imereti	2.8	2.301	11.14	16.241	13.4
5	Mtskheta-Mtianeti	0.57	0.482	2.23	3.282	2.7
6	Samtskhe-Javakheti	0.88	0.754	3.48	5.11	4.2
7	Guria	0.63	0.533	2.4	3.563	2.9
8	Kakheti	1.73	1.407	6.6	9.737	8.0
9	Samegrelo-Zemo Ssvaneti	1.98	1.688	7.7	11.368	9.4
10	Kvemo Svaneti-Lechkhumi	0.38	1.15	0.53	2.06	1.7
11	Achara	1.63	1.387	6.3	9.317	7.7
12	Racha	0.1	0.08	0.37	0.55	0.5
	Sum	**23.05**	**18.987**	**76.55**	**118.574**	**97.8**

12. Public Awareness

The Preparation of National Action Plan under Stockholm Convention on Persistent Organic Pollutants Project (Global Environmental Facility, United Nations Development Programme, Ministry of Natural Resources and Protection of Environment of Georgia) initiated activities directed towards rising public awareness [8].

The list of the activities carried out up to the date in accordance with the obligations of Consultant on Public Awareness Issues determined in the respective Terms of Reference is as follows:

Just before the start of soccer CE, we have placed the supplement – banner – dedicated to Euro 2004 in two Georgian newspapers (Sakartvelos

Respublika daily and sports paper Sarbieli). The banners also contained general information on POPs.

According to the plan developed by the Consultant on Public Awareness Issues has been conducted workshop for media representatives. Workshop aimed towards rising awareness on POPs in media representatives. For this purposes, we have invited 18 journalists from various organizations (TV and radio stations, daily and weekly newspapers, information agencies, and Internet edition). The participants had opportunity to share their vision of POPs management and approaches towards rising public awareness. Later there were incorporated recommendations in strategy for rising public awareness.

Several newspaper published articles on POPs issues. Particularly, there were four publications in 24 Saati daily and several publications in Mtavari Gazeti, Rezonansi, and Akhali Taoba. One of these publications (24 Saati) was announced winner in the media contest for "the Best Environmental Article" conducted by Regional Environmental Center for Caucasus. Twenty-four Saati also published large interview with Project consultant. Informational agency Interpress disseminated information on POPs situation existing in Akhaltsikhe.

TV program Perekriostok, Internews broadcasted 5 min plot on POPs, particularly, on the situation in Achara. Besides Georgia, this program is received by Armenia and Azerbaijan.

One of the most popular TV Channels of Georgia – Rustavi 2 – placed in its informational program Courier several plots on POPs, particularly, on the situation existing in Ali village of Khashuri district with this regard.

TV company 1st Channel (in new format) placed large plot on POPs existing in Samtredia and problems concerned with them in its informational program Mtavari.

Program Mtavari broadcasted special reportage on POPs problems existing in Achinebuli settlement, Telavi district. This plot was shot in the presence of director of International Pesticide Association, FAO and European Parliament expert, which have assessed the existing situation. This problem and plot will be presented to FAO and Special Committee of European Parliament, as well as to the International Pesticide Association.

One of the most prestigious Georgian galleries hosted photo-exhibition on POPs issues, which was fully dedicated to the situation existing in Georgian regions, and reflected activities implemented under the Project. The POPs exhibition lasted for 3 days. Opening session was attended by representatives of international organizations and diplomatic corps, government and parliament of Georgia, cultural and public workers, famous persons, and journalists. The event was covered by:

- Broadcasting Company Rustavi 2 – live morning program, 15 min plot;
- TV Company 1st Channel – live morning program, 10 min plot
- Broadcasting Company Rustavi 2 – informational program Courier
- TV Company Imedi – informational program Kronika
- TV Company 1st Channel – informational program Mtavari
- TV Company Maestro – live evening program, 10 min plot

The exhibition was also covered by *24 Saati*, *Rezonansi*, and *Akhali Taoba* newspapers.

The event was touched in a large article (journalistic investigation on situation existing in Achinebuli settlement, Telavi district) published in Sunday supplement of *Alia* daily.

Radio stations Fortuna+ and Liberty also broadcasted information on the exhibition.

Prior to opening of the exhibition, newspaper *24 Saati* published two articles with comments of the Project Management concerning POPs issues.

Information on the exhibition was disseminated through Internet editions of local NGOs REC and CENN. The event was also covered by TV Company Mir.

Accordingly, the event caused big response, which resulted in various consequences. Particularly, the Ministry of Health of Georgia arranged a special briefing and made comments on POPs and the exhibition. This briefing was also broadcasted by various TV stations. Along with the mentioned coverage, informational program of TV Company 1st Channel Mtavari broadcasted programs on POPs issues. Tbilisi sanitary service presented its comments on the exhibition. The exhibition was invited to Bulgaria, to participate in 8th Sophia International Forum on Pesticides.

13. Measures to be Taken/Ways for Problem Resolution

In order to resolve the above-mentioned problem, it is necessary to apply the complex approach and to introduce and implement modern systems of pesticide management. Particularly:

- Use of pesticides rapidly dissociating in the environment, not causing its pollution and damage to human health
- Decrease of volumes and frequencies of pesticide use
- Use of integrated methods in combating with hazardous parasites, including biological, microbiological, mechanical and other environmentally friendly means
- Consulting/training farmers on issues of safe and effective use of pesticides and production of ecologically pure commodities

- Implementation of chemically safe and effective environmental programs
- Implementation of state control of pesticide import, storage, trade, quality and norms of use
- Harmonization of legal framework with international standards
- Disposal of obsolete and expired pesticides with environmentally safe methods and technologies (collecting them and packaging into plastic drums, temporary disposal at the safe sites, decontamination of storage facilities by the respective means, incineration of the collected pesticides in respective devices or their final disposal at the burials complying with the modern requirements, removal of the contaminated soil layers and burning in the respective incinerators or their reclaiming/rehabilitation by other chemical/biological means).

References

1. L. Ritter, K.R. Solomon, J. Forget, M. Stemeroff, and C.O'Leary, A review of selected persistent organic pollutants, PCS/95.39, 1995.
2. UNEP Chemicals, Guidelines for the Identification of PCBs and Materials Containing PCBs, Prepared by UNEP Chemicals, First Issue, 1999.
3. Dioxin and Furan Inventories National and Regional Emissions of PCDD/PCDF, Prepared by UNEP Chemicals, Geneva, Switzerland, IOMC, 1999.
4. Chemical Inventories – Chemical Management Instrument – A Series of Resource Documents for Countries Making Choices, Produced by UNITAR and UNEP within the framework of IOMC, 1999.
5. Inventory of Information Sources on Chemicals – Persistent Organic Pollutants, Prepared by UNEP Chemicals, 1999.
6. Guidance on Chemicals Legislation – An Overview, Prepared by UNEP Chemicals and UNEP Environmental Law and Institutions – Programme Activity Center, 1995.
7. The Inventory of Sources of Dioxin in the United States, Review Draft, EPA/600/P, 98002Aa (NTIS PB98-137037), National Centre for Environmental Assessment, Washington, DC, April 1998.
8. Combustion Emissions Technical Resource Document (CETRED), Report No. EPA 530-R-94-014, Washington, DC, 1994.

Part II

MONITORING ACTIVITIES AND CURRENT SITUATION

MONITORING OF DIOXINS AND DIOXIN-LIKE PCBs IN FOOD, FEED, AND BIOLOGICAL SAMPLES IN GREECE

LEONDIOS LEONDIADIS*, DANAE COSTOPOULOU, IRENE VASSILIADOU, ATHANASIOS PAPADOPOULOS
Mass Spectrometry and Dioxin Analysis Laboratory, National Center for Scientific Research "Demokritos," Aghia Paraskevi Attikis, 15310, Greece

Abstract: In an attempt to evaluate levels of polychlorinated dibenzo-*p*-dioxins, polychlorinated dibenzofurans and polychlorinated biphenyls in Greece, blood serum, breast milk, food, and feed samples were analyzed. Analytical results reported in this study include blood samples and milk from people living in an urban (Athens) and a rural (Kozani) area of Greece, dairy products, meat, fish, olive oil, eggs, fruit, vegetable, and feed. Contamination levels in blood and human milk from Greece reported here are low compared to the previously reported dioxin data from other European countries. Besides, all food and feed samples had a dioxin content far below the EC Regulation 1881/2006/EC.

Keywords: polychlorinated dibenzo-*p*-dioxins, polychlorinated dibenzofurans, polychlorinated biphenyls, human serum, human milk, Greek food, chemical analysis

1. Introduction

Persistent organic pollutants are widely distributed contaminants that accumulate in the environment and cause many toxic actions. Among these contaminants, dioxins and related compounds, such as polychlorinated dibenzo-*p*-dioxins (PCDDs), polychlorinated dibenzofurans (PCDFs), and polychlorinated biphenyls (PCBs) have caused great concern because of their severe health effects in humans following chronic exposure. All these

* To whom correspondence should be addressed. E-mail: leondi@rrp.demokritos.gr

E. Mehmetli and B. Koumanova (eds.),
The Fate of Persistent Organic Pollutants in the Environment, 83–98.

contaminants are highly lipophilic. Due to their lipophilicity and their long biological half-lives, PCDDs/Fs and dioxin-like PCBs accumulate in predators at the top of the food chain (Froescheis et al., 2000; Startin and Rose, 2003). Toxic actions in humans include reproductive and developmental effects, neurological and behavioral effects, dermal toxicity, immunomodulatory, and carcinogenic effects (Van den Berg et al., 1998; WHO, 1998; Bleeker et al., 1999; ATSDR Tox Profiles, 2002; Bencko, 2003).

Incomplete combustion of organic materials in the presence of chlorine in the feedstock or in the air supply (common in most combustion systems) is responsible for the formation of PCDDs/Fs. This is the case in waste incineration (municipal solid waste, sewage sludge, and hazardous waste), burning of fuels such as coal or wood, metallurgical processing, manufacturing of some herbicides, pesticides, and petroleum products (Schramm et al., 1998; McKay, 2002). Other high temperature sources are cement kilns and poorly controlled combustion sources such as building fires (Ruokojarvi et al., 2000). In the near future, nonindustrial emission sources will probably become more important and may soon dominate the overall annual emissions of PCDDs/Fs in Europe. Domestic solid fuel combustion (wood and coal) contributes more than 60% of all nonindustrial PCDDs/Fs emissions. Other significant emissions originate from accidental fires, pentaclorophenol-treated wood products and illegal incineration of household waste (Quass et al., 2004). A major source of PCDDs/Fs pollution is the incineration of medical waste, due to PVC products (Walker and Cooper, 1992; Lerner, 1997; Vesilind et al., 2002). The European Council has set air emission limit values in EC Directive 2000/76/EC, in an attempt to restrict the emission of toxic substances including dioxins from the incineration of waste into air, soil, surface water, and ground water.

In Greece, the dominant source of PCDDs/Fs is considered to be the uncontrolled combustion of municipal solid waste in open landfills, since industrial emission sources are very limited (Martens et al., 1998).

PCBs have been produced since 1930 and were used in the past mainly in electronic appliances, heat transfer systems and hydraulic fluids, as coolant fluids in transformers and dielectric fluids in capacitors. Other applications include coatings, inks, flame retardants, and paints. PCBs were produced as mixtures; individual congeners were seldom synthesized. PCBs production was banned in the late 1970s, because of their health effects that were reported. Human exposure still occurs through leakage of old capacitors and transformers and disposal of contaminated materials.

Food is generally recognized as the main source for human intake of PCDDs/Fs and dioxin-like PCBs. Over 90% of human exposure to dioxin-like compounds has been attributed to food intake (Liem et al., 2000). In evaluating the risk for humans, surveys in various countries have determined the intake of PCDDs/Fs and dioxin-like PCBs from total diet study (TDS) samples and individual foodstuffs (Liem et al., 2000). The European Union has set maximum limits for PCDDs/Fs and dioxin-like PCBs in food products which are specified in EC Regulation 1881/2006. Results presented in this study include the analyses of different kinds of dairy products, olive oil, meat, fish, vegetable, and feed obtained from public markets in different locations in Greece.

Development of specific and sensitive analytical techniques in the recent years has facilitated the determination of PCDDs/Fs and dioxin-like PCBs in blood. It is therefore possible to monitor human exposure to PCDDs/Fs and PCBs through measurement in blood plasma and blood serum (Covaci et al., 2002; Koppen et al., 2002; Link et al., 2005). In addition, monitoring of PCDDs/Fs and dioxin-like PCBs in human milk is of critical importance, since lactational exposure to dioxin and related compounds may adversely affect brain development and the immune system of infants and children (Lanting et al., 1998; Weisglas-Kuperus et al., 2000). This study presents analyses of PCDD/Fs, PCBs (non-ortho, mono-ortho, and the six indicator PCBs accounting for 50% of the total non-dioxin-like PCB concentration) in human blood samples collected from people living in an urban (Athens) and a rural (Kozani) area of Greece. Additionally, eight human milk samples were analyzed for PCDDs/Fs, dioxin-like PCBs and indicator PCBs.

2. Materials and Methods

2.1. COLLECTION OF SAMPLES

2.1.1. Collection of Human Blood Samples

Blood samples were collected at the blood donor facilities of the Athens Regional General Hospital "G. Gennimatas" and the General Hospital of Kozani between November 2002 and February 2004. Approximately 50–300 ml of blood was collected from each individual. Blood samples were collected in triple bag blood collection systems. Immediately after sampling, blood samples were processed for serum separation, frozen right after separation and transported to the laboratory for measurement of PCDDs/Fs and PCBs.

Milk samples were collected in glass vessels from mothers living in different areas in Athens, between the third and eighth week after delivery, in 2003.

The blood serum samples, as well as the collected breast milk samples remained frozen at a temperature of –40°C until they were analyzed.

2.1.2. Collection of Foodstuffs

All food items were collected through the services of the Hellenic Food Authority and were appropriately transported to the laboratory between

August and December 2002. The samples were kept at –20°C until they were processed.

Foodstuffs were sampled according to the EC Regulation laying down methods of sampling and analysis for the official control of levels of dioxins and dioxin-like PCBs in foodstuffs (EC Regulation 1883/2006/EC), and the number of collected samples is defined in the EC Recommendation 2004/705/EC which takes into account the production of each country. Foodstuffs sampled were typical of the Greek diet.

2.2. MATERIALS

All solvents used were residue analysis picograde and were purchased from Promochem (Germany). The isomers for the preparation of the $^{13}C_{12}$ internal standard solutions were purchased from Wellington Laboratories (Canada) and Cambridge Isotope Laboratories (USA). Carbosphere 80/100 mesh was purchased from Alltech (it was cleaned by elution with methanol and consequently with toluene for several weeks before usage). Alumina was Basic activity Super 1 for dioxin analysis, MP Biochemicals GmbH. The sulfuric acid impregnated silica gel was prepared as follows: Silica gel (100 g, 60–200 mesh, Merck) was activated in an oven at 200°C for at least 2 days and then mixed with concentrated sulfuric acid (44 g).

2.3. EXTRACTION OF LIPIDS

Milk and serum samples were subjected to a liquid–liquid extraction procedure consisting of mixing with sodium oxalate and methanol, followed by extraction steps with a combination of diethyl ether–petroleum ether (AOAC, 1990). The organic layer of this extraction was evaporated and the lipid residue was dried and weighed for calculation of lipid percentage in serum and milk.

Fats and oils are generally assumed to be homogeneous, and normally do not require any extraction procedures. Aliquots of such samples were dissolved in dichloromethane to the desired fat concentration.

Meat and fish samples were initially blended and homogenized. Next, a representative test sample was ground with anhydrous sodium sulphate, until a free flowing powder was obtained. This mixture was then extracted using Soxhlet extraction with dichloromethane (De Boer, 1998).

Labeled quantification standards were added to each sample before extraction. These were toluene solutions containing a mixture of $^{13}C_{12}$ isomers of all the 17 PCDDs/Fs congeners except OCDF, the four $^{13}C_{12}$ non-ortho PCBs, the eight $^{13}C_{12}$ mono-ortho PCBs, as well as $^{13}C_{12}$PCB 52 and 101 for indicator PCBs.

2.4. CLEANUP

Sample cleanup was based on the method initially described by Liem et al. (1990), that we routinely use for sample analysis (Papadopoulos et al., 2004; Costopoulou et al., 2006). A brief description follows.

2.4.1. Dioxins and Non-ortho PCBs

(a) Carbon chromatography

The sample residue was dissolved in dichloromethane and brought onto the top of a Carbosphere column, which was placed in a reflux unit and refluxed for 2 h with dichloromethane. This fraction, including residual fat amounts was discarded. Next, the column was rinsed with toluene and refluxed with toluene for 1 h. This fraction, containing the non-ortho PCBs, was carefully evaporated to dryness. The Carbosphere column was then inverted in the reflux unit and the PCDD/F fraction was eluted from the column by refluxing with of toluene for 16 h. The PCDD/F fraction was evaporated to dryness.

(b) Alumina chromatography

The residue containing the non-ortho PCBs, was dissolved in hexane and the mixture was placed onto a column containing 0.5 g of 44% H_2SO_4-silica gel and 5 g of alumina. The alumina column was rinsed twice with hexane and then non-ortho PCBs were eluted with a hexane/dichloromethane mixture (1:1 v/v). Finally, the eluate was evaporated to dryness and redissolved in toluene containing injection standard $^{13}C_{12}$ PCB-80.

The residue containing PCDD/Fs was dissolved in hexane and the mixture was brought onto a column containing 0.5 g of 44% H_2SO_4-silicagel, and 5 g of alumina. The alumina column was rinsed twice with hexane and then washed with a hexane/dichloromethane mixture (93:7 v/v). This eluate was discarded. PCDDs and PCDFs were eluted with a hexane/dichloromethane mixture (60:40 v/v). Finally, the eluate was evaporated to dryness and redissolved in toluene containing injection standard $^{13}C_{12}$ 1,2,3,4-TCDD.

2.4.2. Mono-ortho and Indicator PCBs

The residue obtained by the extraction step was dissolved in 5 ml of hexane and brought onto a column (length 30 cm, 8 mm ID) plugged with glass wool and filled with 10 g 44% H_2SO_4-silica gel. The column was eluted with 100 ml of hexane. The eluate was concentrated to a volume of about 5 ml and brought onto an alumina chromatography column as described above and eluted with 50 ml of a hexane/dichloromethane mixture (1:1 v/v).

Finally, the eluate was evaporated to dryness and redissolved in 50 μl of *n*-nonane containing 20 ng/ml of performance standard ($^{13}C_{12}$ PCB 80).

2.5. INSTRUMENTAL ANALYSIS

The quantification of PCDDs/Fs, non-ortho, mono-ortho, and indicator PCBs was performed by HRGC-HRMS (EI) in MID mode on a Trace GC gas chromatograph (ThermoFinnigan) coupled to a MAT-95 XP mass spectrometer (ThermoFinnigan) equipped with a CTC A 200S autosampler at 10,000 resolving power (10% valley definition). Instrumental conditions and purity control criteria were according to the EPA 1613 and EPA 1668A methods (US Environmental Protection Agency, 1994, 1999). The quantification of concentration and recovery was carried out by the isotopic dilution method. The limit of detection (LOD) for each congener was determined as the concentration in the extract which produced an instrumental response at two different ions to be monitored with a signal to noise ratio of 3:1 for the less sensitive signal (EC Directive 2004/44/EC). For TEQ calculations the WHO-98 toxicity equivalent factors (TEF) were used (Van den Berg et al., 1998).

2.6. QUALITY CONTROL

A method blank and a quantitative control sample (reference) were included for every eight samples. As reference sample, olive oil spiked with PCDDs/Fs (TEQ value 3.0 pg/g), non-ortho PCBs (TEQ value 2.2 pg/g) and mono-ortho PCBs (TEQ value 2.0 pg/g) was used. The reference sample was measured 40 times during the study period. PCDDs/Fs TEQ value showed an average of 3.26 pg/g with a confidence interval of 2.74–3.78 pg/g (confidence level 95%). PCBs TEQ value showed an average of 4.31 pg/g with a confidence interval of 3.62–5.00 pg/g (confidence level 95%).

The LOD was determined for each congener to be between 0.1 and 0.15 pg/g fat. Recovery ranged between 70% and 120%.

The laboratory participated successfully in the following interlaboratory studies:

- For the years 2003–2006 in Interlaboratory Comparison of Dioxins in Food, Division of Environmental Medicine, Norwegian Institute of Public Health (Folkehelse) Norway
- For 2003 and 2006 in Interlaboratory Study on Fish Tissue Reference Material, Cambridge Isotope Laboratories/Cerilliant, USA
- For 2005 and 2006 in FAPAS Proficiency Testing for PCBs and Dioxins/UK. In 2005, the laboratory achieved a z-score equal to zero for the determination of WHO-dioxin-TEQ upper

2.7. STATISTICAL ANALYSIS

Statistical analysis was conducted using the statistical software program for SAS Institute (SAS Inc.). Mean values were compared by *t*-test. Significance probabilities (*p** values) were calculated for the respective number of samples analyzed.

3. Results and Discussion

3.1. FOOD SAMPLES

Results from PCDDs/Fs and dioxin-like PCBs measurements in food samples collected in Greece are presented in Table 1. Seventy-seven samples were analyzed for PCDDs, PCDFs, and dioxin-like PCBs. As might be expected, contamination appears to differ among the foodstuffs.

The contamination was calculated as the TEQ value by multiplying the concentrations with the corresponding WHO-TEFs for each congener (Van den Berg et al., 1998).

Concentrations and TEQ values of all compounds are reported on a fat basis (pg/g fat) except for fish, vegetable and feed, which are expressed on a whole weight basis (pg/g product).

TABLE 1. Food and feed average total concentrations of PCDD/Fs, non-ortho PCBs and mono-ortho PCBs, along with minimum and maximum values for each sample category

	Concentration (pg WHO-TEQ/g lipid)			
Foodstuff	PCDD/F	non-ortho PCB	mono-ortho PCB	Total TEQ
Butter	0.79 (0.29–1.33)	0.32 (0.01–0.66)	0.05 (0.02–0.06)	1.16 (0.36–1.82)
Cheese	1.03 (0.46–1.88)	0.13 (0.01–0.19)	0.04 (0.03–0.06)	1.20 (0.56–1.89)
Yoghurt	0.44 (0.38–0.51)	0.41 (0.14–0.67)	0.06 (0.06–0.06)	0.91 (0.65–1.10)
Cow milk	0.39 (0.34–0.47)	0.18 (0.07–0.34)	0.11 (0.02–0.27)	0.68 (0.48–0.96)
Milk powder	0.40 (0.35–0.48)	0.04 (0.01–0.08)	0.01 (0.01–0.02)	0.45 (0.42–0.52)
Beef	0.55 (0.52–0.59)	0.41 (0.01–0.69)	0.05 (0.05–0.06)	1.01 (0.65–1.19)
Lamb	0.46 (0.22–0.71)	0.38 (0.20–0.74)	0.14(0.03–0.50)	0.98 (0.51–1.73)
Pork	0.39 (0.32–0.42)	0.69 (0.27–1.22)	0.03 (0.01–0.05)	1.11 (0.68–1.50)
Poultry	0.30 (0.18–0.40)	0.06 (0.04–0.11)	0.12 (0.02–0.31)	0.48 (0.27–0.73)
Bovine liver	1.64 (1.17–2.47)	0.02 (0.01–0.04)	0.24 (0.22–0.27)	1.90 (1.58–2.47)
Fish, aquaculture*	0.47 (0.22–1.12)	1.19 (0.56–2.90)	0.24 (0.04–0.34)	1.90 (0.92–3.88)
Fish, wild*	0.12 (0.00–0.39)	0.33 (0.01–0.98)	0.09 (0.06–0.12)	0.54 (0.08–1.32)
Fish oil	1.01 (0.30–1.83)	3.76 (0.96–6.18)	2.90 (2.01–3.80)	7.67(3.68–10.50)
Eggs	0.37 (0.26–0.45)	0.13 (0.06–0.33)	0.04 (0.01–0.12)	0.54 (0.37–0.80)
Olive oil	0.30 (0.23–0.36)	0.04 (0.01–0.09)	0.03 (0.01–0.05)	0.37 (0.28–0.44)
Vegetable*	0.01 (0.00–0.01)	0.01 (0.01–0.01)	0.00 (0.00–0.00)	0.02 (0.01–0.02)
Feed*	0.02 (0.01–0.03)	0.01 (0.01–0.01)	0.00 (0.00–0.00)	0.03 (0.02–0.04)

*Concentrations in fish, vegetable, and feed are expressed in pg WHO-TEQ/g product, as established in European regulations

Upperbound TEQ values are calculated, assuming that nondetected individual congener concentrations are equal to their corresponding limits of detection.

The contribution of dioxin-like PCBs to total TEQ values is variable, as shown in Figure 1. For most samples this contribution is in the range of 20–50%. However, fish show much higher levels of dioxin-like PCBs, whose contribution exceeds 75%.

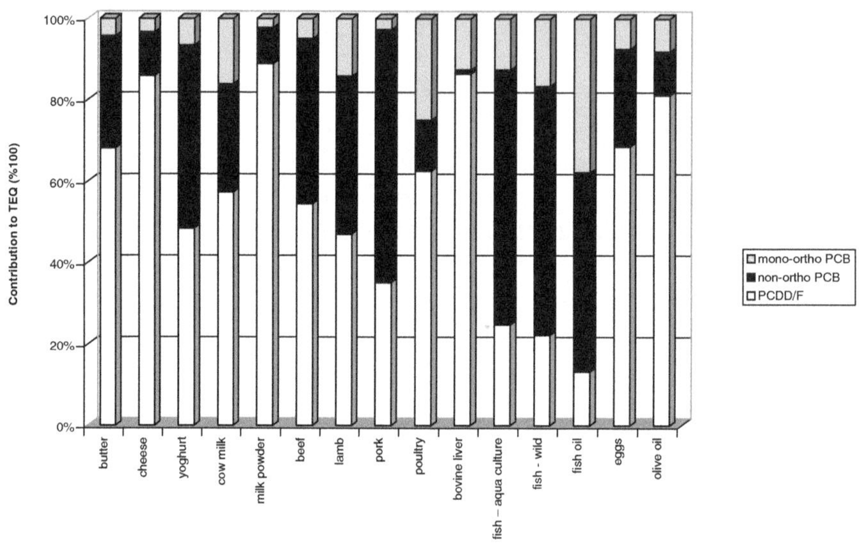

Figure 1. Contribution (%) of PCDD/Fs, non-ortho PCBs and mono-ortho PCBs to total TEQ in foodstuffs

3.1.1. Milk and Dairy Products

Cow milk, milk powder, yogurt, butter, yellow cheese, and feta cheese samples were analyzed. The average values for lipid content were 24.85% for cheese, 8.32% for yogurt, 4.10% for cow milk, and 24.73% for milk powder. PCDDs, PCDFs, non-ortho PCBs, and mono-ortho PCBs concentration values were at low levels. PCDD/F upperbound TEQ values for cow milk ranged from 0.34 to 0.47 with a mean value of 0.39 pg/g fat and from 0.29 to 1.88 with a mean value of 0.81 pg/g fat for dairy products. These values are lower than those monitored in other European and Mediterranean countries, ranging from 0.28 to 1.33 pg/g fat for cow milk and from 0.69 to 2.26 pg/g fat for dairy products (Bascompta et al., 2002; Bocio et al., 2002; Bordajandi et al., 2002; Hosseinpour et al., 2002).

3.1.2. Meat and Poultry

The average values for lipid content were 2.69% for beef, 6.20% for pork, 13.06% for lamb, 9.84% for poultry, and 4.05% for bovine liver. PCDD/F and dioxin-like PCB TEQ values of all meat samples were lower than those reported in other European and Mediterranean countries (PCDD/F TEQ 0.89–2.26 pg/g fat for beef and pork, 1.36–8.92 pg/g fat for poultry; non-ortho PCB TEQ 0.15–0.44 pg/g fat for beef and pork, 0.56–3.74 pg/g fat for poultry) (Bascompta et al., 2002; Bocio et al., 2002; Bordajandi et al., 2002; Hosseinpour et al., 2002).

For a specific pork sample, the non-ortho PCB concentration was high and the non-ortho PCB toxicity equivalent was greater than that of PCDD/Fs, dominated by PCB-126 and PCB-169. This indicates a possible PCB-contamination of the feed used for raising the pig.

3.1.3. Fish and Fish Oil

TEQ values were expressed in pg/g or ng/kg product, in accordance with the European Council Regulation. The average values for lipid content were 2.47% for wild fish and 11.78% for farmed fish, respectively.

The present study showed a relatively wide range of PCDDs, PCDFs, non-ortho PCBs, and mono-ortho PCBs concentrations in saltwater fish. Upperbound TEQ values were at low levels. Generally, the toxicity pattern of fish is dominated by 2,3,7,8-TCDF, 2,3,4,7,8-PeCDF, 1,2,3,7,8-PeCDD and to a smaller extent by 2,3,7,8-TCDD. The pattern is consistent with that reported in previous studies (Fiedler et al., 1997; Karl et al., 2002).

3.1.4. Olive Oil and Eggs

For the olive oil sample, almost all PCDD/F and dioxin-like PCB congeners were below the detection limit. Free range egg samples (average values lipid content 30.44%), contained very few congeners detected at low concentrations. PCDD/F, non-ortho PCB and mono-ortho PCB upperbound TEQ values were lower than those typically monitored in other European countries (PCDD/F TEQ 0.55–0.99 pg/g fat (Bascompta et al., 2002).

3.1.5. Fruits and Vegetables

Concentrations of congeners and TEQ values were expressed in pg/g or ng/kg product. Fruits (apples, grapes, oranges, peaches, and pears), and vegetables (carrots, garlic, onion, and potatoes), showed almost all PCDD, PCDF non-ortho PCB and mono-ortho PCB congeners below the detection limit, as expected due to very low fat content.

3.1.6. Feed

Concentrations of congeners and TEQ values were expressed as pg/g or ng/kg product, in accordance with the European Council Recommendation 13/2006/EC. Very low PCDD/F and dioxin-like PCB concentrations were measured in cereals used in animal feed.

3.2. HUMAN SAMPLES

3.2.1. Blood Serum Samples

From the area of Athens, 105 blood samples were collected from the general population (average age 48.5 years, range 27–66 years). All samples were analyzed for mono-ortho and indicator PCBs content. Indicator PCBs have been defined by the European Food Safety Authority as the major congeners that are almost always present in various sample matrices in high concentrations (European Food Safety Authority, 2005). In addition, 10 of the blood samples were of sufficient volume and were analyzed for non-ortho PCBs and the 17 WHO-TEQ PCDDs/F congeners.

Twenty-two blood samples were collected in Kozani, a rural area (general population, average age 43.5 years, range 28–65 years) and analyzed for mono-ortho, non-ortho, and indicator PCBs.

Results are presented in Table 2. The PCDDs/F and dioxin-like PCB levels were calculated as TEQ values by multiplying with the corresponding WHO-TEF for each congener (Van den Berg et al., 1998). Upperbound total TEQ values were calculated for each category of congeners assuming that non-detected individual congener concentrations are equal to their corresponding limit of detection. Indicator PCBs were reported as concentrations. Concentrations and TEQ values of all compounds were reported on a fat basis.

Regional differences were examined using the formula of comparison of two groups of normally distributed values. Indicator, mono-ortho and non-ortho PCBs population total means in Athens (156.65 ng/g fat, 4.03 pg TEQ/g fat and 3.20 pg TEQ/g fat, respectively) were significantly higher than the respective population total means in Kozani (56.99 ng/g fat, 1.84 pg TEQ/g fat, and 1.24 pg TEQ/g fat, respectively). A relatively wide range of concentrations of PCDDs/Fs in blood samples of people belonging to the same population was also shown in the present study.

Literature references concerning PCDD/F concentration in blood samples are limited and are summarized in Table 3 (Päpke et al., 1997; EU Dioxin Exposure and Health Data, 1999; Schuhmacher et al., 1999; Buckland et al., 2001; Koppen et al., 2002; Harden et al., 2004a, b; Sampaio et al., 2004).

For PCBs, it has proved impossible to identify comparable sets of data on which to base a regional comparison.

Overall, the PCDD/F and PCB values of all blood samples in the Greek population are very low by international standards and comparable with, although lower than those typically monitored in other European and Mediterranean countries.

TABLE 2. Blood serum and human milk analyzed lipid content (%) and average total concentrations of PCDD/Fs, non-ortho PCBs, mono-ortho PCBs (pg/g fat WHO-TEQ) and indicator-PCBs (ng/g fat), along with minimum and maximum values (in brackets) for each sample category

	Athens serum	Kozani serum	Athens human milk
Percentage lipid values	0.47 (0.27–0.76)	0.31 (0.18–0.48)	4.18 (2.30–7.76)
	Concentration (pg WHO-TEQ/g fat)		
PCDD/F	6.82 (1.86–11.01)		7.27 (3.43–11.28)
Non-ortho PCB	3.20 (1.46–5.50)	1.24 (0.22–7.22)	3.08 (1.29–5.92)
Mono-ortho PCB	4.03 (0.74–16.96)	1.84 (0.59–4.64)	3.48 (1.76–8.47)
	Concentration (ng/g fat)		
Indicator PCB	156.65 (38.39–660.24)	56.99 (24.06–122.55)	94.4 (50.62–231.42)

TABLE 3. Mean concentrations of PCDDs/Fs in human serum from different countries

Country	PCDDs/Fs (pg/g fat WHO-TEQ)
Finland	37.0
Spain (Tarragona)	27.0
USA	22.1
Norway	21.1
Germany	16.5
Portugal	15.3
Spain (Madrid)	14.4
New Zealand	12.4
Australia	9.33
Greece (Athens)	**6.82**
Greece (Kozani)	

3.2.2. Human Milk Samples

Eight human milk samples from the area of Athens (general population, average age 33.5 years, range 28–44 years) were analyzed for PCDDs/Fs, non-ortho, mono-ortho, and indicator PCBs contamination. Results are reported in Table 2. The total TEQ PCDD/PCDF level was in the lowest end of the range measured in Europe (EU Dioxin Exposure and Health Data, 1999; Malisch and van Leeuven, 2003). This is clearly demonstrated in Table 4, where results of the fourth round of WHO-coordinated exposure study on the levels of PCBs, PCDDs, and PCDFs in human milk are presented.

TABLE 4. Mean concentrations of PCDD/Fs, dioxin-like PCBs, and indicator PCBs in human milk of different countries

Country	PCDDs/PCDFs pg/g fat WHO-TEQ	PCBs pg/g fat WHO-TEQ	Indicator PCBs ng/g fat
Egypt	22.33	5.48	106
The Netherlands	18.27	11.57	192
Belgium	16.92	12.60	191
Luxembourg	14.97	13.67	217
Italy	12.66	16.29	253
Germany	12.53	13.67	220
Spain	11.56	9.42	241
Ukraine	10.04	19.95	136
Sweden	9.58	9.71	146
Finland	9.44	5.85	91
Russia	9.36	13.45	126
Slovak Republic	9.07	12.60	443
Romania	8.86	8.06	173
Hong Kong SAR	8.69	4.73	45
Czech Republic	7.78	15.24	502
Ireland	7.72	4.57	60
Norway	7.30	8.08	119
Greece (Athens)	**7.27**	**6.56**	**94**
USA	7.18	4.61	54
New Zealand	6.86	3.92	37
Hungary	6.79	2.87	34
Croatia	6.40	7.17	135
Bulgaria	6.14	4.21	42
Australia	5.57	2.89	30
Philippines	3.94	2.38	26
Brazil	3.92	1.77	16
Fiji	3.34	1.75	17

The contribution of dioxin and dioxin-like PCBs in total TEQ values in human samples is shown in Figure 2. Blood and breast milk distribution is similar. Dioxin-like PCBs contribution to total TEQ is about 50%.

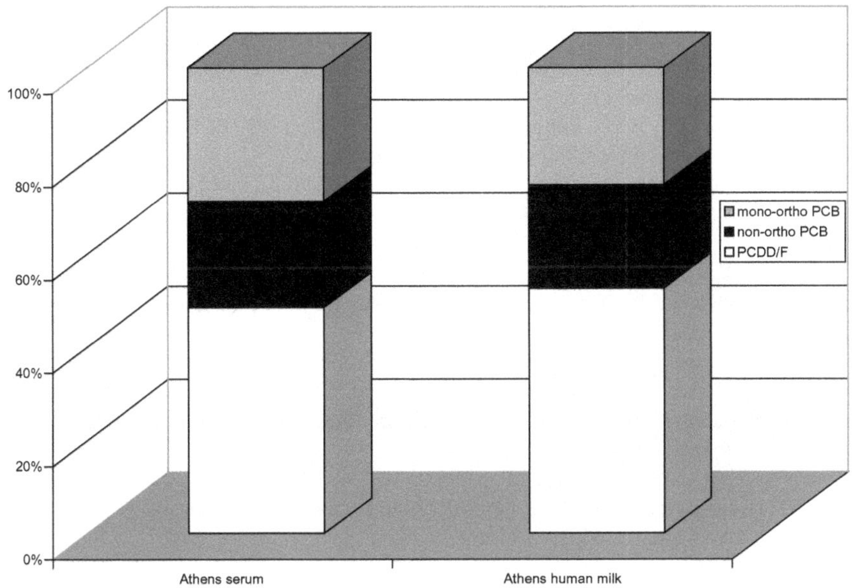

Figure 2. Contribution (%) of PCDD/Fs, non-ortho PCBs and mono-ortho PCBs to total TEQ in serum and breast milk from Athens

4. Conclusions

In conclusion, dioxin and PCB contamination in Greek biological samples and food and feed samples reported here is low compared to the previously reported dioxin data from other European countries. All concentrations in food and feed were lower than the EU action levels (EC Recommendation 88/2006/EC and 13/2006/EC, respectively) and in the lowest end of the EU SCOOP data (EC, SCF/CS/CNTM/DIOXIN/8). Blood and human milk results are amongst the lowest reported in other European and Mediterranean countries. These results are in agreement with the fact that Greece is not highly industrialized and has never produced PCBs.

References

1. AOAC (Association of Official Analytical Chemists), 1990. Official Methods of the Association of Official Analytical Chemists, 15th edn. AOAC, Arlington, VA.
2. ATSDR, 2002. ATSDR Tox Profiles, Atlanta, 2002, http://www.atsdr.cdc.gov
3. Bascompta, O., Montaña, M.J., Martí, R., Broto-Puig, F., Comellas, L., Díaz-Ferrero, J., and Rodríguez-Larena, M.C., 2002. Levels of persistent organic pollutants (PCDD/F and dioxin-like PCB) in food from the Mediterranean diet, Organohalogen Compounds 57, 149.

4. Bencko, V., 2003. Risk assessment and human exposure to endocrine disrupters. In: Jedrychovski, W.A., Petera, F.P., and Maugeri, U., Editors, Molecular Epidemiology in Preventive Medicine. International Center for Studies and Research in Biomedicine, Luxembourg, pp. 315–327.
5. Bleeker, I., Fischer, A.B., Tilkes, F., and Eikmann, T., 1999. PCB Konzentrationen im menschlichen Blut. Umwelt Forsch Prax 4, 84–96.
6. Bocio, A., Llobet, J.M., Domingo, J.L., Casas, C., Teixidó, A., and Müller L., 2002. Levels of PCDD/PCDFs in food samples from Catalonia, Spain, Organohalogen Compounds 57, 105.
7. Bordajandi, L.R., Herrero, L., Abad, E., Rivera, J., and González, M.J., 2002. Presence of PCDDs, PCDFs and PCBs in selected food samples from Huelva (Spain), Organohalogen Compounds 57, 69.
8. Buckland, S.J., Bates, M.N., Garrett, N., Ellis, H. and Kvan Maanen, T., 2001. Concentrations of selected organochlorines in the serum of the non-occupationally exposed New Zealand population. In Ministry for the Environment, Wellington. www.mfe.govt.nz/publications/hazardous/serum-study-may01.pdf
9. Costopoulou, D., Vassiliadou, I., Papadopoulos, A., Makropoulos, V., and Leondiadis, L., 2006. Levels of dioxins, furans and PCBs in human serum and milk of people living in Greece. Chemosphere 65, 1462–1469.
10. Covaci, A., Koppen, G., Van Cleuvenbergen, R., Schepens, P., Winneke, G., van Larebeke, N., Nelen, V., Vlietinck, R., and Schoeters, G., 2002. Persistent organochlorine pollutants in human serum of 50–65 years old women in the Flanders Environmental and Health Study (FLEHS). Part 2: correlations among PCBs, PCDD/DFs and the use of predictive markers. Chemosphere 48, 827–832.
11. De Boer, J., 1998. Trends in chlorobiphenyl contents in livers of Atlantic cod (Gadus morhua) from the North Sea. Chemosphere 17(9), 1811.
12. EC Directive 2000/76/EC of 4 December 2000 on the incineration of waste.
13. EC Directive 2004/44/EC of 13 April 2004 amending Directive 2002/69/EC laying down the sampling methods and the methods of analysis for the official control of dioxins and the determination of dioxin-like PCBs in foodstuffs. OJEC L113/17.
14. EC Recommendation 2002/201/EC of 4 March 2002 on the reduction of the presence of dioxins, furans and PCBs in feeding stuffs and foodstuffs.
15. EC Recommendation 2004/705/EC of 11 October 2004 on the monitoring of background levels of dioxins and dioxin-like PCBs in foodstuffs.
16. EC Recommendation 13/2006/EC of 3 February 2006 amending Annexes I and II to Directive 2002/32/EC of the European Parliament and of the Council on undesirable substances in animal feed as regards dioxins and dioxin-like PCBs.
17. EC Recommendation 88/2006/EC of 6 February 2006 on the reduction of the presence of dioxins, furans and PCBs in feedingstuffs and foodstuffs.
18. EC Regulation 1881/2006/EC of 19 December 2006 setting maximum levels for certain contaminants in foodstuffs.
19. EC Regulation 1883/2006/EC of 19 December 2006 laying down the sampling methods and the methods of analysis for the official control of dioxins and the determination of dioxin-like PCBs in foodstuffs.
20. EC, SCF/CS/CNTM/DIOXIN/8 Final, 23 November 2000. Opinion of the SCF on the Risk Assessment of Dioxins and Dioxin-like PCBs in Food, Adopted on 22 November 2000.
21. EU Dioxin Exposure and Health Data, 1999. Report produced for European Commission Environment, UK Department of the Environment, Transport and the Regions (DETR). Available at: <http://europa.eu.int/comm/environment/dioxin>

22. European Food Safety Authority, 2005. Opinion of the scientific panel on contaminants in the food chain on a request from the commission related to the presence of non dioxin-like PCBs in feed and food. EFSA Journal 284, 1–137.
23. Fiedler, H., Cooper, K.R., Bergek, S., Hjelt, M., and Rappe, C., 1997. Polychlorinated dibenzo-*p*-dioxins and polychlorinated dibenzofurans ((P)CDD/PCDF) in food samples collected in Southern Mississippi, USA. Chemosphere, 34(5–7), 1411.
24. Froescheis, O., Looser, R., Cailliet, G.M., Jarman, W.M., and Ballschmiter, K., 2000. The deep-sea as a final global sink of semivolatile persistent organic pollutants? Part 1. PCBs in surface and deep-sea dwelling fish of North and South Atlantic and the Monterey Bay Canyon (California). Chemosphere 40, 651–660.
25. Geyer, H.J., Schramm, K.W., Feicht, E.A., Behechti, A., Steinberg, C., Bruggemann, R., Poiger, H., Henkelmann, B., and Kettrup, A., 2002. Half-lives of tetra-, penta-, hexa-, hepta-, and octachlorodibenzo-*p*-dioxin in rats, monkeys, and humans – a critical review. Chemosphere 48, 631–644.
26. Harden, F., Müller, J., and Toms, L., 2004a. Dioxins in the Australian population: levels in blood, technical report No. 9, Department of the Environment and Heritage, Canberra.
27. Harden F., Mueller, J.F., Toms, L.-M.L., Gaus, C., Moore, M., Paepke, O., et al., 2004b. Determination of the levels of dioxin in the Australian population by analysis of blood serum. Organohalogen Compounds 66, 2853–2858.
28. Hosseinpour, J., Rottler, H., Joas, R., Potrykus, A., and Schott, R., 2002. Dioxin-like PCB levels in feed and food from across the European Union, Organohalogen Compounds 57, 77.
29. Karl, H., Ruoffb, U., and Blüthgenb, A., 2002. Levels of dioxins in fish and fishery products on the German market. Chemosphere 49(7), 765.
30. Koppen, G., Covaci, A., Van Cleuvenbergen, R., Schepens, P., Winneke, G., Nelen, V., Van Larebeke, N., Vlietinck, R., and Schoeters, G., 2002. Persistent organochlorine pollutants in human serum of 50–65 years old women in the Flanders Environmental and Health Study (FLEHS). Part 1: concentrations and regional differences. Chemosphere 48(8), 811–825.
31. Lanting, C.I., Patandin, S., Fidler, V., Weisglas-Kuperus, N., Sauer, P.J.J., Boersma, E.R., and Touwen, E.R., 1998. Neurological condition in 42-month-old children in relation to pre- and postnatal exposure to polychlorinated biphenyls and dioxins. Early Human Development 50(3), 27, 283–292.
32. Lerner, B., 1997. Prevention of Dioxin Formation in Medical Waste Incineration, 90th Annual Meeting and Exhibition, Air & Waste Manage. Assoc., Paper No. 97-FA 166.01, Toronto, Ontario, Canada, June 8–13.
33. Liem, A.K.D., De Jong, A.P.J.M., Marshman, J.A., Den Boer, A.C., Groenemeijier, G.S., Den Hartog, R.S., De Korte, G.A.L., Hoogerbrugge, R., Kootstra, P.R., and Van 't Klooster, H.A., 1990. A rapid clean up procedure for the analysis of polychlorinated dibenzo-p-dioxins and dibenzofurans in milk samples. Chemosphere 20, 843.
34. Liem, A.K.D., Fürst, P., and Rappe, C., 2000. Exposure of populations to dioxins and related compounds. Food Additives and Contaminants 17, 241.
35. Link, B., Gabrio, T., Zoellner, I., Piechotowski, I., Paepke, O., Herrmann, T., Felder-Kennel, A., Maisner, V., Schick, K.-H., Schrimpf, M., et al., 2005. Biomonitoring of persistent organochlorine pesticides, PCDD/PCDFs and dioxin-like PCBs in blood of children from South West Germany (Baden-Wuerttemberg) from 1993 to 2003. Chemosphere 58, 1185–1201.
36. Malisch, R. and Van Leeuwen, R., 2003. PCDD/F Results of the WHO-coordinated exposure study on the levels of PCBs, PCDDs and PCDFs in human milk. Organohalogen Compounds 64, 140.

37. Martens, D., Balta-Brouma, K., Brotsack, R., Michalke, B., Schramel, P., Klimm, C., Henkelmann, B., Oxynos, K., Schramm, K.-W., Diamadopoulos, E., and Kettrup, A., 1998. Chemical impact of uncontrolled solid waste combustion to the vicinity of the kouroupitos ravine, Crete, Greece. Chemosphere 36, 2855–2866.
38. McKay, G., 2002. Dioxin characterization, formation and minimization during municipal solid waste (MSW) incineration: review. Chemical Engineering Journal 86, 343–368.
39. Papadopoulos, A., Vassiliadou, I., Costopoulou, D., Papanicolaou, C., and Leondiadis, L., 2004. Levels of dioxins and dioxin-like PCBs in food samples on the Greek market. Chemosphere 57, 413–419.
40. Päpke, O., Herrmann, T., and Ball, M., 1997. PCDD/PCDFs in humans, follow up of background data for Germany, 1996. Organohalogen Compounds 33, 530–534.
41. Quass, U., Fermann, M., and Bröker, G., 2004. The European Dioxin Air Emission Inventory Project-Final Results. Chemosphere 54, 1319–1327.
42. Ruokojarvi, P., Aatamila, M., and Ruushanen, J., 2000. Toxic chlorinated and polyaromatic hydrocarbons in simulated house fires. Chemosphere 41, 825–828.
43. Sampaio, C., Reis, M.F., Miguel, J.P., and Aguiar, P., 2004. Levels and trends of PCDD/Fs in human blood and milk of residents in the vicinity of a modern municipal solid waste incinerator near to Lisbon. Organohalogen Compounds 66, 2813–2817.
44. Schramm, K.-W., Kaune, A., Lehnardt, R., Hofmaier, A., Henkelmann, B., and Kettrup, A., 1998. Isokinetic sampling of PCFF/F response in low and high volatile fractions of a wood incinerator. Organohalogen Compounds 36, 289–292.
45. Schuhmacher, M., Domingo, J.L., Llobet, J.M., Lindström, G., and Wingfors, H. 1999. Dioxin and dibenzofuran concentrations in blood of a general population from Tarragona, Spain. Chemosphere 38, 1123–1133.
46. Startin, J.R. and Rose, M.D., 2003. Dioxins and dioxin-like PCBs in food. In: Schecter, A. and Gasiewicz, T.A., Editors, Dioxins and Health, Wiley, New York, pp. 89–136.
47. US Environmental Protection Agency, 1994. Tetra- through Octa-chlorinated Dioxins and Furans by Isotope Dilution HRGC/HRMS, Method 1613.
48. US Environmental Protection Agency, 1999. Chlorinated biphenyl congeners in water, soil, sediment and tissue by HRGC/HRMS, Method 1668, Revision A.
49. Van den Berg, M., Birnbaum, L., Bosveld, B.T.C., Brunstrom, B., Cook, P., Feeley, M., Giesy, J.P., Hanberg, A., Hasegawa, R., Kennedy, S.W., Kubiak, T., Larsen, J.C., van Leeuwen, F.X.R., Liem, A.K.D., Nolt, C., Peterson, R.E., Poellinger, L., Safe, S., Schrenk, D., Tillitt, D., Tysklind, M., Younes, M., Waern, F., and Zacharewski, T., 1998. Toxic equivalency factors (TEFs) for PCBs, PCDDs, PCDFs for humans and wildlife. Environmental Health Perspectives 106, 775.
50. Vesilind, P., Worrell, W., and Reinhart, D., 2002. Solid Waste Engineering, Thompson Learning, London.
51. Walker, B. and Cooper, C., 1992. Air pollution emission factors for medical waste incinerators. Journal of the Air and Waste Management Association 42, 784–789.
52. Weisglas-Kuperus, N., Patandin, S., Berbers, G., Sas, T.C.J., Mulder, T.C.J., Sauer, T.C.J., and Hooijkaas, H., 2000. Immunologic effects of background exposure to polychlorinated biphenyls and dioxins in Dutch preschool children. Environmental Health Perspectives 108, 1203–1207.
53. WHO, 1998. Consultation on assessment of the health risk of dioxins; re-evaluation of the tolerable daily intake (TDI). Food Additives and Contaminants 17(4), 223–240.

ENVIRONMENTAL FATE OF LEGACY CHIRAL PESTICIDES IN BACKGROUND SOILS

PERIHAN B. KURT-KARAKUS, TERRY F. BIDLEMAN*, LIISA M. JANTUNEN
Centre for Atmospheric Research Experiments (CARE), Environment Canada, 6248 Eighth Line, Egbert, L0L 1N0, ON, Canada
KEVIN C. JONES
Department of Environmental Science and Centre for Chemical Management, Lancaster Environment Centre, Lancaster University, Lancaster, LA1 4YQ, UK

Abstract: Variability in the enantioselective degradation of chiral organochlorine pesticides (OCPs: α-HCH, *cis*- and *trans*-chlordane (CC and TC), chlordane MC5 and *o,p'*-DDT) in background soils was investigated on a global scale and in small-scale transect studies. Background soils collected from 32 countries displayed a wide range of enantiomer profiles for a particular compound, showing varying degradation preferences. In some locations, grassland and woodland soils which were collected only a few hundred meters apart showed differences in the enantiomer profiles and even reversals in degradation preferences. Further studies at small plots in Scotland indicated that such differences could be found within a few meters horizontally and a few centimeters in depth. Soil organic matter content and pH exerted a minor influence on the variability. For chlordanes, the enantiomer profiles in background soils were much more variable than those in ambient air, suggesting that degradation occurred following atmospheric deposition. Thus, background soils play a role in sequestering and degrading OCPs, although enantiomer profiles are not a quantitative measure of overall degradation. The results of this study show that caution is needed when considering the enantiomer signatures in air as a marker of volatilization of weathered soil-derived OCPs.

* To whom correspondence should be addressed. E-mail: Terry.Bidleman@ec.gc.ca

E. Mehmetli and B. Koumanova (eds.),
The Fate of Persistent Organic Pollutants in the Environment, 99–112.

Keywords: organochlorine pesticides, chiral, enantiomer, background soils

1. Introduction

Like other chemicals, organochlorine pesticides (OCPs) in soil are subjected to microbial degradation. Chiral OCPs such as α-hexachlorocyclohexane (α-HCH), *o,p′*-DDT, and some components of technical chlordane are produced as racemates, but undergo enantioselective degradation in soil to yield nonracemic residues. Nonracemic enantiomer distributions in soils provide evidence of microbial degradation and serve as a basis for interpreting bioaccumulation by plants (Mattina et al., 2002, 2006). Since enantiomers have the same volatilities, their profiles in air can be linked to those in the underlying soil (Bidleman and Falconer, 1999; Bidleman and Leone, 2004; Eitzer et al., 2003; Leone et al., 2001). However, enantioselective degradation in soils can occur over horizontal distances of 1 m or so (and possibly less) and a few centimeters in depth. This supports the hypothesis that the soil microorganisms which are responsible for the preferential degradation of enantiomers vary at the local level, probably depending on differences in soil properties. Therefore, in a given area, background soils receiving the same enantiomer "signature" of a compound from atmosphere can display varying profiles over a small spatial scale.

Identification of the origin of OCPs is important to understand their observed distribution in different environmental compartments. A key issue is whether the OCPs found in background regions are atmospherically transported from where they currently applied or volatilized residues from past applications. The focus of previous studies has been on the characterization of chiral signatures in treated agricultural soils, using them to track soil–air exchange and to infer sources of OCPs in ambient air (Bidleman and Falconer, 1999; Bidleman and Leone, 2004; Eitzer et al., 2003; Leone et al., 2001). A similar approach has been used to distinguish soil emissions vs primary sources for chiral PCBs (Jamshidi et al., 2007; Robson and Harrad, 2004). This study was undertaken to investigate enantioselective degradation of chiral OCPs and its variability in background soils, where the source was most likely atmospheric deposition.

2. Chirality and Expression of Chiral Data

The term "chiral" which is known in chemistry since the 1870s, originates from Greek word "kehir" meaning "hand". Chiral chemicals consist of two or more configurations called "enantiomers" which are nonsuperimposable mirror images of each other (Figure 1). Enantiomers possess the same physical and chemical properties but often have different rates of metabolism and toxicological properties. A racemate, or a racemic mixture, contains equal proportions of the enantiomers. (Ali et al., 2003; Kallenborn and Huhnerfuss, 2001).

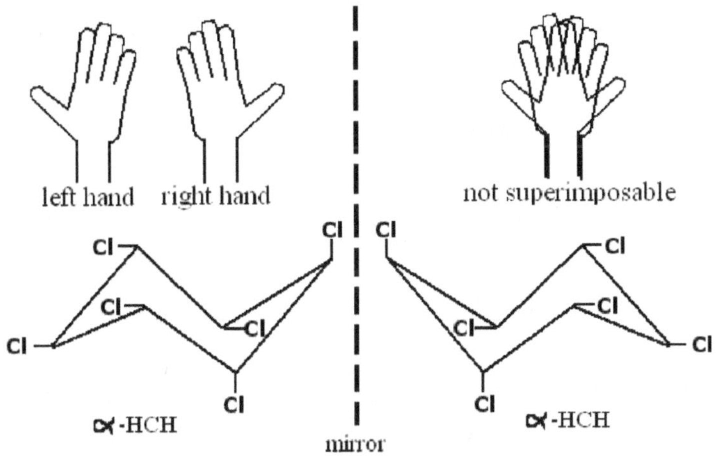

Figure 1. Illustration of chirality and (below) enantiomers of the common chiral pesticide α-hexachlorocyclohexane (α-HCH)

Figure 2 shows the chromatograms of a chiral compound that has been subjected to preferential degradation of one enantiomer in the environment. If the optical signs are known (e.g., as in Figure 2), it is common to calculate the enantiomer ratio (ER) from peak areas as:

$$ER = \frac{\text{Area }(+)}{\text{Area }(-)} \tag{1}$$

or in cases where the enantiomer designations are R and S:

$$ER = \frac{\text{Area }(R)}{\text{Area }(S)} \tag{2}$$

Note that since there is no correspondence between R, S, and (+), (–), it is important to define whether Eq. 1 or 2 is being used to calculate ER.

If the enantiomer designations are not known, ER is calculated from:

$$ER = \frac{\text{Area }(1)}{\text{Area }(2)} \tag{3}$$

where Eqs. 1 and 2 are the first- and second-eluting peaks on a specified chromatographic column. As it is seen in Eqs. 1–3, the ER can go to very high values, or to an indefinite value, in cases of the enantiomer peak in the denominator being small or below detection. Therefore the enantiomer fraction (EF), which varies only between 0 and 1, is preferred to the ER for statistical purposes as well as its handiness in graphing (de Geus et al., 2000; Harner et al., 2000; Ulrich et al., 2003). If enantiomers are designated as (+) and (–), the EF is defined:

$$EF = \frac{\text{Area }(+)}{\left[\text{Area }(+) + \text{Area }(-)\right]} \tag{4}$$

$$EF = \frac{ER}{ER + 1} \tag{5}$$

and similarly for Eqs. 2 and 3 formats. Enantioselective degradation often involves preferential loss of either enantiomer, resulting in EF values greater or less than 0.500. To express enantioselective degradation, irrespective of which enantiomer is depleted, the concept of "deviation from racemic" (DEVRac) was introduced, where DEVRac = absolute value of (0.500 – EF) (Kurt-Karakus et al., 2005).

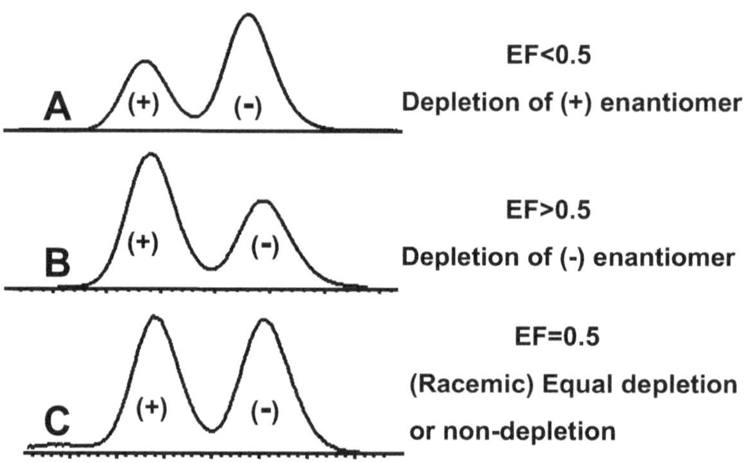

Figure 2. Possible enantiomer fractions (EFs), designated according to Eq. 4, resulting from enantioselective degradation of a racemic chiral pesticide in the environment

3. Case Studies of Chiral OCPs in Background Soils

3.1. BACKGROUND TO THE STUDIES

In 1998, researchers at Environmental Science Department of Lancaster University conducted a global background soil sampling campaign and over 200 soil samples were collected from different countries (Meijer et al., 2003a). In addition, a background surface soil sampling (41 soil samples) campaign was carried out on a UK–Norway transect in the same year (Meijer et al., 2002). Sampling locations were chosen at remote locations which were far from any potential sources of persistent organic pollutants (POPs) so that they represented the background levels of the compounds of interest. These soil samples were analyzed for POPs such as PCBs, hexachlorobenzene (HCB) (Meijer et al., 2002, 2003a) and DDT (Kurt-Karakus et al., 2007b).

3.2. EXPERIMENTAL APPROACH

A subset of 65 representative samples from 32 countries was analyzed to investigate the enantioselective degradation, as indicated by EFs, of α-HCH, *trans*- and *cis*-chlordane (TC and CC), chlordane MC5 and *o,p*′-DDT. Soils were collected from different ecosystems (i.e., grasslands, forests) to investigate the variations in EFs due to the system variables annual mean temperature and soil organic matter. A map of the locations and analytical methodology are given in Kurt-Karakus et al. (2005). Prior studies to characterize chiral signatures (EFs) have focused on pesticides applied to agricultural land and tracking signatures in the atmosphere for source apportionment purposes (Leone et al., 2000, 2001; Mattina et al., 2002; Wiberg et al., 2001). This was the first large-scale study of EFs in background (untreated) soils.

The differences in enantioselective degradation of OCPs found in the background soils study (Section 3.3.1) brought to our minds the question of small-scale variability of EFs in soils. Further investigations were carried out to assess this at a site in the Kintyre Peninsula of the UK. This site was chosen because coniferous woodland and grassland background soils, collected only a few hundred meters apart, displayed reversals in their OCP enantiomer degradation preference (Kurt-Karakus et al., 2005). Details of the experimental setup and sample collection are given in Kurt-Karakus et al. (2007a). A study was conducted in the area to make the following observations:

(a) EF variability was examined within nine grids of 2 × 2 m plots in grassland and woodland areas (Figure 3)
(b) EF time trends were determined for soil-spiked chiral compounds under field conditions in grassland and woodland areas. For this purpose, a second set of nine soil cores was collected to obtain a composite sample at each site in woodland and grassland area. The composite sample was divided into four subsamples; three of these were spiked in the field using an acetone solution containing the compounds of interest, transferred into the aluminum tins (11 cm diameter and 13 cm depth) and placed back in situ. Three tins were buried up to their tops at each sampling site and covered with grass. The sites were revisited to collect soil samples from the tins after 209 and 445 days. The fourth subsample was used to conduct controlled laboratory degradation experiments whose results are reported elsewhere (Kurt-Karakus et al., 2007a).

3.3. RESULTS AND DISCUSSION

3.3.1. EFs of OCPs in Global Background Soils

The distributions of EFs in background soil samples are summarized in Table 1. A preferred degradation direction occurred for some compounds; e.g., 72% of soils showed (+)TC loss, 10.3% (–)TC loss and racemic TC was found in 17.3%. Preferential degradation of (–)CC (57.4%) was favored over loss of (+)CC (27.8%) and occurrence of racemic CC (14.8%). Residues of *o,p′*-DDT were split nearly equally among the categories: EF < 0.5 (32.5%), EF > 0.5 (37.5%) and EF = 0.5 (30%). The lowest EF (0.080) for this study, as well as in the literature, was found for CC in a UK woodland soil sample; whereas the highest EF value (0.846) occurred also for CC in a Switzerland woodland soil sample.

Results in Table 1 show the nonspecific nature of microbial degradation processes in soils. Most studies of agricultural soils in the USA reported preferential degradation of (+)TC and (–)CC (Aigner et al., 1998; Eitzer et al., 2003; Leone et al., 2001; Wiberg et al., 2001), although racemic residues were occasionally found (Falconer et al., 1997). As in the background soils, degradation of *o,p′*-DDT in US agricultural soils (Bidleman and Leone, 2004; Wiberg et al., 2001) was ambivalent with EF distributions similar to those in our background soils. Similar degradation preferences were found in archived agricultural soils in the UK (Meijer et al., 2001). A study of soils in the Pearl River Delta, China, found preferential degradation of either enantiomer for α-HCH, TC, CC and *o,p′*-DDT, resulting in EFs above and below 0.5. Similar ranges of EFs were found in crop, paddy and natural soils (Li et al., 2006). Residues of α-HCH in crop and paddy soils tended to be close to racemic, whereas depletion of (−)α-HCH was found in most natural soils. In our background soils survey, (−)α-HCH was depleted in 37.5%, (+)α-HCH in 17.9% and racemic α-HCH was found in 44.6% (Table 1).

TABLE 1. Distribution of EFs of OCPs in global background soils

	TC	CC	MC5	α-HCH	*o,p'*-DDT
n[a]	58	54	52	57	40
Min. EF[b]	0.341	0.080	0.284	0.399	0.393
Max. EF[b]	0.616	0.846	0.763	0.655	0.535
Mean	0.457	0.507	0.469	0.514	0.501
Standard deviation	0.062	0.094	0.077	0.041	0.041
			Samples with EF < 0.5		
% of total	72.4	27.8	67.3	17.9	32.5
N	42	15	35	10	13
			Samples with EF > 0.5		
% of total	10.3	57.4	25.0	37.5	37.5
N	6	31	13	22	15
			Samples with EF = 0.5 (racemic)		
% of total	17.3	14.8	7.7	44.6	30.0
n	10	8	4	25	12

[a]Number of samples in which the compound was detected
[b]EFs designated according to Eq. 4, except for MC5, for which EFs < 0.5 and > 0.5 indicate preferential loss of the first- and second-eluting peaks from a Betadex-120 capillary column (Kurt-Karakus et al., 2005)

During the background soils survey, soils which were collected from locations only hundreds of meters apart but in different ecosystems sometimes showed remarkable differences in degradation preferences for chiral OCPs. For example, EFs in adjacent UK woodland and grassland soils were 0.375 and 0.577 for TC, 0.478 and 0.575 for α-HCH, and 0.439 and 0.606 for MC5. However, no significant differences ($p > 0.05$) were found between grassland and woodland soil samples for the entire data set.

3.3.2. EFs of OCPs on a Small-scale Transect

The variability in EFs of chlordanes at the Woodland 2 (W2) site in the Kintyre Peninsula is shown in Figure 3. EFs for adjacent woodland (W1) and grassland (G1) sites can be found in Kurt-Karakus et al. (2007a). EFs in composite soil samples, collected by pooling cores from the nine grids, ranged from 0.453 to 0.496 for TC and 0.406 to 0.456 for CC. Although the EFs of TC and CC in surface soils of the nine grids were all <0.5, they varied by up to 0.068 for TC and 0.222 for CC. Surface (0–5 cm) and underlying (5–10 cm) soil samples were collected in central grid 5. These displayed different EFs for TC, whereas those for CC were the same. Moreover, in the grassland plot G1, the EF of TC was 0.426 in the surface layer and 0.511 in the bottom layer, indicating a reversal in the degradation preference. In summary, there was considerable variability of EFs and reversal of degradation preference over spatial range of meters among the grids for chlordanes, especially for CC in W2 (0.272–0.494), TC in W1 (0.358–0.487) and chlordane MC5 in W1 (0.454–0.668).

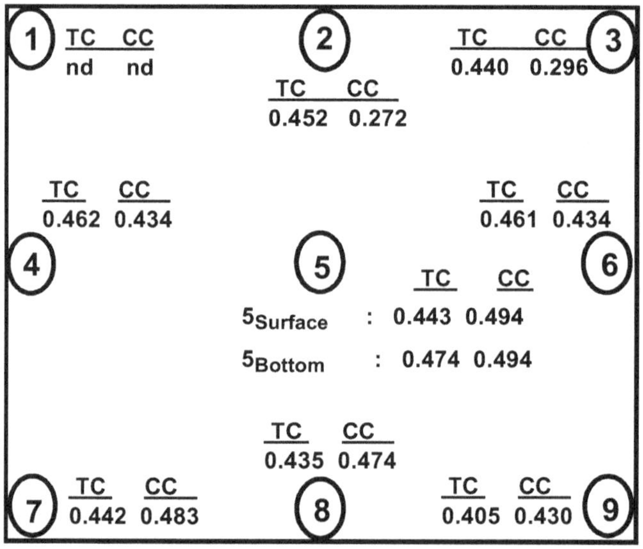

Figure 3. EFs of TC and CC in nine grid points on a woodland site (W2)

3.3.3. EFs of Chlordanes in Ambient Air and Soil

Figure 4 shows the ranges of average EFs in ambient air samples collected from Europe, North America, and the Arctic as well as EF ranges in global background soils and Kintyre plot soils (G1, W1, and W2 soils). Air in the vicinity of the Kintyre plots, monitored by passive samplers (Kurt-Karakus et al., 2007a), displayed chlordane EF profiles similar to those in ambient air reported in other studies (Bidleman et al., 2004; Shen et al., 2005). This is not surprising since air masses originating from a wide area would affect the chlordane concentrations and EFs in the passive air samples, which were integrated over 8 months. The ranges of EFs in global background soils (Table 1) and Kintyre transect samples are greater than those in ambient air, suggesting that degradation of TC and CC occurred in these soils following atmospheric deposition.

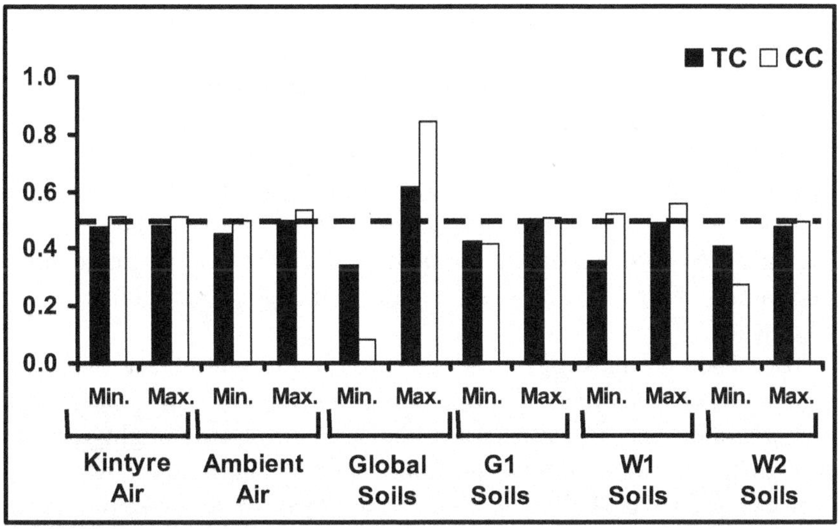

Figure 4. EFs of TC and CC in ambient air and soil. Kintyre air: EFs for ambient air in the vicinity of the Kintyre plots (Kurt-Karakus et al., 2007a), Ambient air: EFs for ambient air reported by Bidleman et al., 2004 in arctic and Sweden locations and by Shen et al., 2005 for urban, rural and background North American locations. Global soils reported by Kurt-Karakus et al. (2005)

Although it is shown in several studies that EFs of chiral OCPs in the air above agricultural soils are mainly driven by the emission from these treated soils (Bidleman and Falconer, 1999; Bidleman and Leone, 2004; Eitzer et al., 2003; Leone et al., 2001; Mattina et al., 2002; Meijer et al., 2003b), it is still unknown whether OCPs deposited to background soils are subject to reemission. The small-scale variability in EFs in Figure 3 shows why making such a determination is difficult.

3.3.4. EF–concentration Relationships

Quantitative determination of OCPs was not carried out in our studies because losses occurred when the extracts were blown down to the low volumes necessary to obtain good enantiomer chromatograms. Tests using solvent concentration procedures in our laboratory showed that recoveries of HCHs and chlordanes were only ~50% when sample extracts were concentrated to ~25 μL, however enantiomer proportions was not altered since enantiomers possess the same volatility.

Lacking concentration information, ratios of TC and CC to the more stable *trans*-nonachlor (TN) were to estimate the relative concentrations (Kurt-Karakus et al., 2005, 2007a). Correlations between the deviation of EFs from racemic (DEVRac, Section 2) and relative concentrations to TN for TC and CC were not statistically significant, neither in the agricultural soils ($n = 22$, $r^2 = 0.13$, $p = 0.11$ for TC and $n = 23$, $r^2 = 0.12$, $p = 0.10$ for CC; data set from Aigner et al. (1998), nor in background soils ($n = 55$, $r^2 = 0.018$, $p = 0.33$ for TC and $n = 37$, $r^2 = 0.024$, $p = 0.36$ for CC; data set from Kurt-Karakus et al. (2005). Li et al. (2006) examined relationships between EFs and concentrations for chiral OCPs in the Pearl River Delta. No relationship was apparent for α-HCH. For *o,p′*-DDT and TC, nonracemic EFs with values above and below 0.500 were found for soil concentrations <1 ng g^{-1}, whereas residues at higher concentrations were mainly racemic.

These observations suggest that EFs are not a good measure of overall degradation rates for organochlorine compounds in soils. It is possible that both enantiomers are degraded but at the same or at only slightly different rates. Figure 5 shows results of the spiked soils experiments at Kintyre plot W1 for CC and α-HCH. Residues of CC at days 209 and 445 had racemic EFs in the three spiked tins, ranging from 0.499 and 0.510. These were not significantly different from the EF of a racemic CC standard (Figure 5A). However, relative concentrations (CC/TN, average of three tins buried in the site) show losses at days 209 and 445 compared to day 0 (Figure 5B). For α-HCH, relative concentrations (α-HCH/TN, average of spiked three tins buried in the site) showed losses at days 209 and 445 compared to day 0 (Figure 5D) and EFs also showed increasing enantioselective degradation with time (Figure 5C). EFs at days 209 (0.453 average) and 445 (0.402 average) were significantly different from a racemic α-HCH standard.

3.3.5. Relationship of EFs to Soil Properties from a Global and Small-scale Perspective

Relationships between EFs, annual average temperature and soil organic matter content (SOM%) were investigated for the global soil data set, and also between EFs and soil pH in the Kintyre plots.

As seen in Table 2, there is a weak but statistically significant positive correlation between DEVRac and SOM% for TC in global background soil samples. The global soil set and W2 site small-scale transect samples showed a positive correlation between DEVRac and SOM% for CC. Annual mean temperature data were available only for the global soil set and there was no significant correlation with DEVRac for any of the chiral OCPs. Buerge et al. (2003) reported a correlation between soil pH and enantioselective degradation of metalaxyl, however, our data showed a positive correlation between soil pH and DEVRac only in W2 site.

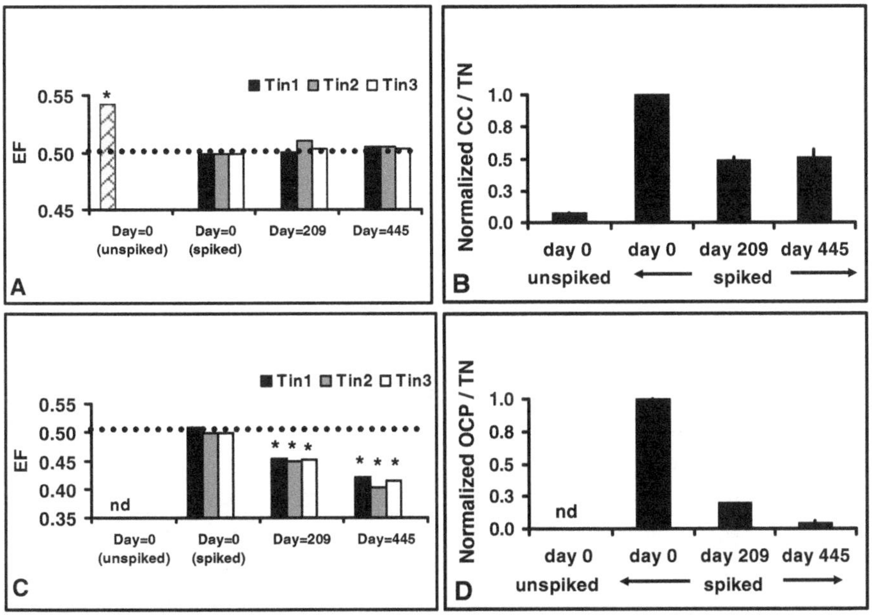

Figure 5. Enantiomer fractions (EFs) and normalized OCP/TN ratios for CC (A and B) and α-HCH (C and D) in three tins of unspiked W1 soil and soil spiked with CC and α-HCH under field conditions. *EFs significantly different from those of racemic standards; nd: not available due to nondetection or interference problems.

TABLE 2. Correlations between DEVRac and soil properties for global soils[a] and Kintyre Plot soils[b]

	TC		CC		MC5		α-HCH		*o,p′*-DDT	
	r^2	p	r^2	p	r^2	p	r^2	p	r^2	p
				DEVRac–SOM%						
Global[a]	0.15	**0.003**	0.17	**0.002**	0.004	0.650	0.06	0.080	0.02	0.360
W1[b]	0.60	**0.013**	0.30	0.140						
W2[b]	0.01	0.750	0.60	**0.012**						
G1[b]	0.03	0.610	0.01	0.840						
				DEVRac–temperature						
Global	0.04	0.130	0.004	0.620	0.01	0.400	0.04	0.160	0.02	0.350
				DEVRac–soil pH						
W1	0.18	0.250	0.50	0.05						
W2	0.16	0.260	0.50	**0.02**						
G1	0.02	0.700	0.04	0.70						

[a]Kurt-Karakus et al. (2005)
[b]Kurt-Karakus et al. (2007a); bolded numbers show the statistically significant correlations

4. Conclusions

Enantioselective degradation of OCPs and its environmental consequences have been under investigation for several years. Most studies have focused on characterizing EFs in agricultural soils and making links to the chiral signatures found in the atmosphere. Studies on background soils are also of importance due to differences in levels of OCPs between background and agricultural soil types as well as differences in nutrient levels, SOM%, microbial communities, and the degree of soil mixing.

EFs of chiral OCPs in background soils are much more variable than in background air. The delivery of chiral OCPs to background soils is probably influenced by patterns/timing of the initial atmospheric deposition; however, due to diverse enantioselective microbial degradation processes in soil, the EFs display different postdepositional patterns.

The variability of EFs in background soils brings to mind many issues related to mechanisms of enantioselective degradation by microbial communities. Results of the Kintyre plot studies (e.g., Figure 3) show that enantioselective degradation operates over spatial scales of meters and probably less. Distances between soil bacteria microcolonies in soil are on the order of a few hundred micrometers (Bosma et al., 1997) and transport along "fungal highways" enables bacteria to bridge air gaps in the soil (Wick, 2007). Similar small-scale variations in EFs have not been investigated in agricultural soils which receive mixing due to tillage. Tracking enantioselective degradation of chiral chemicals in soils may provide a useful tool to examine the extent/scale of mixing among soil microbial communities.

Complex and interacting factors, such as SOM type and burden, pH, nutrient and redox status, type of vegetation, temperature, moisture and, for agricultural soils, tillage affect the capability of a soil to degrade pesticides. (Andersson and Nilsson, 2001; Boul et al., 1994; Buerge et al., 2003; Grayston et al., 2001; Lewis et al., 1999; Spencer et al., 1996). Lewis et al. (1999) hypothesized that enantioselective sorption to structurally similar chemical species in the SOM could affect the availability of enantiomers for microbial transformation. These factors are likely to operate in combination rather than individually and this is suggested by the weak correlations of DEVRac to soil properties (Table 2).

References

Aigner, E.J., Leone, A.D. and Falconer, R.L., 1998, Concentrations and enantiomeric ratios of organochlorine pesticides in soils from the US Corn Belt, *Environ. Sci. Technol.* **32**:1162–1168.

Ali, I., Gupta, V.K. and Aboul-Enein, H.Y., 2003, Chirality: a challenge for the environmental scientists, *Current Sci.* **84**(2):152–156.

Andersson, S. and Nilsson, S.I., 2001, Influence of pH and temperature on microbial activity, substrate availability of soil-solution bacteria and leaching of dissolved organic carbon in a moor humus, *Soil Biol. Biochem.* **33**:1181–1191.

Bidleman, T.F. and Falconer, R.L., 1999, Using enantiomers to trace pesticide emission, *Environ. Sci. Technol.* **33**:206A–209A.

Bidleman, T.F. and Leone, A.D., 2004, Soil-air exchange of organochlorine pesticides in the Southern United States, *Environ. Pollut.* **128**:49–57.

Bidleman, T.F., Wong, F., Backe, C., Södergren, A., Brorström-Lundén, E., Helm, P.A. and Stern, G.A., 2004, Chiral signatures of chlordanes indicate changing sources to the atmosphere the past 30 years, *Atmos. Environ.* **38**:5963–5970.

Bosma, T.N.P., Middeldorp, P.J.M., Schraa, G., and Zehnder, A.J.B., 1997, Mass transfer limitation of biotransformation: quantifying bioavailability, *Environ. Sci. Technol.* **31**(1):248–252.

Boul, H.L., Garnham, M.L., Hucker, D., Baird, D., and Aislable, J., 1994, Influence of agricultural practices on the levels of DDT and its residues in soils, *Environ. Sci. Technol.* **28**(8):1397–1402.

Buerge, I.J., Poigner, T., Müller, M.D., and Buser, H.-R., 2003, Enantioselective degradation of metalaxyl in soils: chiral preference changes with soil pH, *Environ. Sci. Technol.* **37**:2668–2674.

de Geus, J.H., Wester, P.G., De Boer, J., Brinkman, U.A., 2000, Enantiomer fractions instead of enantiomer ratios, *Chemosphere* **41**(5):725–727.

Eitzer, B., Ianucci-Berger, W., and Mattina, M.J.I., 2003, Volatilization of weathered chrial and achiral chlordane residues from soil, *Environ. Sci. Technol.* **37**:4887–4893.

Falconer, R.L., Bidleman, T.F., and Szeto, S.Y., 1997, Chiral pesticides on soils of the Fraser Valley, British Columbia, *J. Agric. Food Chem.* **45**(5):1946–1951.

Grayston, S.J., Griffith, G.S., Mawdsley, J.L., Campbell, C.D., and Bardgett, R.D., 2001, Accounting for variability in soil microbial communities of temperate upland grassland ecosystems, *Soil Biol. Biochem.* **33**:533–551.

Harner, T., Wiberg, K., and Norstrom, R., 2000, Enantiomer fractions are preferred to enantiomer ratios for describing chiral signatures in environmental analysis, *Environ. Sci. Technol.* **34**(1):218–220.

Jamshidi, A., Hunter, S., Hazrati, S., and Harrad, S., 2007, Concentrations and chiral signatures of polychlorinated biphenyls in outdoor and indoor air and soil in a major UK conurbation, *Environ. Sci. Technol.* **41**(7):2153–2158.

Kallenborn, R. and Hühnerfuss, H., 2001, *Chiral Environmental Pollutants*, 1st edn. Springer, Berlin/Germany, pp. 1–209.

Kurt-Karakus, P.B., Bidleman, T.F., and Jones, K.C., 2005, Chiral organochlorine pesticide signatures in global background soils, *Environ. Sci. Technol.* **39**(22):8671–8677.

Kurt-Karakus, P.B., Stroud, J.L., Bidleman, T.F., Semple, K.T., Jantunen, L., and Jones, K.C., 2007a, Enantioselective degradation of organochlorine pesticides in background soils: variability in field and laboratory studies, *Environ. Sci. Technol.* **41**(14):4965–4971.

Kurt-Karakus, P.B., Bidleman, T.F., Jones, K.C., Meijer, S.N., and Staebler, R.M., 2007b, Persistent organic pollutants (POPs) and air-soil exchange: case studies for DDTs, in *The Fate of Persistent Organic Pollutants in the Environment*, NATO Science Series, Ref.# ARW 982400, E. Mehmetli and B. Koumanova (Eds), Springer, The Netherlands.

Leone, A.D., Ulrich, E.M., Bodnar, C.E., Falconer, R.L., and Hites, R.A., 2000, Organochlorine pesticide concentrations and enantiomer fractions for chlordane in indoor air from the US Cornbelt, *Atmos. Environ.* **34**:4131–4138.

Leone, A., Amato, S., and Falconer, R., 2001, Emission of chiral pesticides from agricultural soils in the cornbelt region of the US, *Environ. Sci. Technol.* **35**:4592–4596.

Lewis, D.L., Garrison, A.W., Wommack, K.E., Whittemore, A., Steudler, P., and Melillo, J., 1999, Influence of environmental changes on degradation of chiral pollutants in soils, *Nature* **401**:898–901.

Li, J., Zhang, G., Qi, S., Li, X., and Peng, X., 2006, Concentrations, enantiomeric compositions, and sources of HCH, DDT and chlordane in soils from the Pearl River Delta, South China, *Sci. Total Environ.* **372**(1):215–224.

Mattina, M.I., White, J., Eitzer, B., and Lannucci-Berger, W., 2002, Cycling of weathered chlordane residues in the environment: Compositional and chiral profiles in contiguous soil, vegetation and air compartments, *Environ. Toxicol. Chem.* **21**(2):281–288.

Mattina, M.J.I., Isleyen, M., Eitzer, B.D., Iannucci-Berger, W., and White, J.C., 2006, Uptake by *Cucurbitaceae* of soil-borne contaminants depends upon plant genotype and pollutant properties, *Environ. Sci. Technol.* **40**(6):1814–1821.

Meijer, S.N., Halsall, C.J., Harner, T., Peters, A.J., Ockenden, W.A., and Jones, K.C., 2001, Organochlorine pesticides in archived UK soil, *Environ. Sci. Technol.* **35**:1989–1995.

Meijer, S.N., Steinnes, E., Ockenden, W.A., and Jones, K.C., 2002, Influence of environmental variables on the spatial distribution of PCBs in Norwegian and UK soils: implications for global cycling, *Environ. Sci. Technol.* **36**(10):2146–2153.

Meijer, S.N., Ockenden, W.A., Sweetman, A., Breivik, K., Grimalt, J.O., and Jones, K.C., 2003a, Global distribution and budget of PCBs and HCB in background surface soils: Implications for sources and environmental processes, *Environ. Sci. Technol.* **37**(4):667–672.

Meijer, S.N., Shoeib, M., Jantunen, L.M.M., Jones, K.C., and Harner, T., 2003b, Air-soil exchange of organochlorine pesticides in agricultural soils. 1. Field measurements using a novel in situ sampling device, *Environ. Sci. Technol.* **37**(7):1292–1299.

Robson, M. and Harrad, S., 2004, Chiral PCB signatures in air and soil: implications for atmospheric source apportionment, *Environ. Sci. Technol.* **38**(6):1662–1666.

Shen, L., Wania, F., Lei, Y.D. Teixeira, C., Muir, D.C.G., and Bidleman, T.F., 2005, Atmospheric distribution and long-range transport behavior of organochlorine pesticides in North America, *Environ. Sci. Technol.* **39**(2):409–420.

Spencer, W.F., Singh, G., Taylor, C.D., LeMert, R.A., Cliath, M.M., and Farmer, W.J., 1996, DDT persistence and volatility as affected by management practices after 23 years, *J. Environ. Qual.* **25**:815–821.

Ulrich, E.M., Helsel, D.R., and Foreman, W.T., 2003, Complications with using ratios for environmental data: comparing enantiomeric ratios (ERs) and enantiomer fractions (EFs), *Chemosphere* **53**(5):531–538.

Wiberg, K., Harner, T., Wideman, J., and Bidleman, T.F., 2001, Chiral analysis of organochlorine pesticides in Alabama soils, *Chemosphere* **45**:843–848.

Wick, L.Y., Remer, R., Wurz, B., Reichenbach, J., Braun, S., Schafer, F., and Harms, H., 2007, Effect of fungal hyphae on the access of bacteria to phenanthrene in soil, *Environ. Sci. Technol.* **41**(2):500–505.

PERSISTENT ORGANIC POLLUTANTS (POPs) HOT SPOTS IN RUSSIA

B. REVICH*
Institute of Forecasting of Russian Academy of Sciences, Moscow

A. SHELEPCHIKOV
A. N. Severtsov Institute of Problems of Ecology and Evolution of Russian Academy of Sciences, Moscow

Abstract: There are five hot spots of dioxins and PCBs pollution in Russia. These hot spots are situated in various regions of the country, where DDT, PCB, and other chlorinated substances had been produced earlier. Despite phasing-out of these production facilities, high levels of dioxins, PCBs and DDT are routinely detected in the environment. These substances represent chlorinated organic substances that are most frequently found in the environment, food products, or breast milk. The levels of dioxins in breast milk of the residents of Chapaevsk and Ufa are among the highest in the world. These residents have been shown to have health effects, typically caused by persistent organic pollutants (POPs).

Keywords: POPs, PCDD/PCDF, PCB, DDT, DDE, environmental pollution, breast milk, blood, biomonitoring, Russian Federation, breast cancer, sexual development of children

1. Introduction

Russian Federation has many chemical factories, which once produced chlorinated organic pesticides and other chlorinated organic substances. There are also hundreds of industrial facilities and installation which used

*To whom correspondence should be addressed: Institute of Forecasting, Nakhimovsky Prospect 47, Moscow 117418, Russia, revich@ecfor.ru

E. Mehmetli and B. Koumanova (eds.),
The Fate of Persistent Organic Pollutants in the Environment, 113–126.

PCBs in the past. Five laboratories in different regions of the country have been certified to detect polychlorinated dibenzo-*p*-dioxins (PCDD), dibenzofurans (PCDF), biphenyls (PCB), and other POPs. Three of these laboratories utilize high-resolution mass-spectrometry technique, recommended by USEPA and EU Directives for PCDD/PCDF analysis in foodstuffs. Over 100 labs of Public Health Ministry test food products on content of DDT and its metabolites. These labs have been actively modernized recently.

Considerable fraction of country's rural population experiences the consequences of exposure to chlorinated organic pesticides, especially in south agricultural regions. There are five cities which have the most severe problems with dioxin pollution and PCBs. These cities hosted chemical factories which produced chlorinated organic pesticides, other chlorinated organic substances, or molecular chlorine in the past. There are still very few municipal waste incinerators (MWI) in Russia. MWI have not become the priority sources of emissions of dioxin-like compounds in Russia, like they have in most developed industrial countries. But construction of many new MWIs is being actively discussed today. At the same time, industrial waste incinerators have become widespread. Such incinerators often do not have efficient pollution control equipment. Nevertheless, they are used, inter alia, for incineration of chlorinated organic waste.

POP inventory projects have been carried out during the past decade, but such projects have local and one-time character, because of lack of financing. These projects are not carried out in a systemic way, and many results remain unpublished. Most regions do not have any information on POPs content in their environment, or there exist only data about some isolated measurements. Epidemiological research projects have been carried out in very few Russian regions. Unlike other countries, Russia does not have a federal or nationwide program for control and reduction of POP emissions. Neither it has national biomonitoring system for control of POPs in breast milk and blood of various population groups.

The most extensive projects have been completed in the framework of international projects (AMAP and ACAP projects in Arctic), or by individual cities. Several projects have been carried out in the framework of IPEN program.

2. Pesticides

Of the nine pesticides listed in Stockholm Convention on POPs, Russia continues to produce only DDT, hexachlorobenzene (HCB), heptachlor, and toxaphene on industrial scale. Hexachlorocyclohexane (HCH) is also widespread in Russia, despite its being a likely candidate to be included in the extended list of persistent organic pollutants. According to expert estimates,

USSR produced up to 100,000–150.000 t of DDT and 11,000–16,000 t of toxaphene between 1957 and 1988. Toxaphene production was stopped in 1988, and application of DDT was banned in 1969. Nevertheless, DDT was periodically used in Russian agriculture for 15 years after its official ban. Currently, up to 2,000–3,000 t of this substance are still kept in stock. Although, on national scale, DDT and its metabolites are rarely detected in soil and food in the amounts which exceed national standards, several regions, which intensely used this pesticide in the past, still have high levels of DDT and its metabolites. For example, average concentration of DDT + DDE in chicken eggs from such regions reach 420 nanograms per gram of lipid (ng/g lipid), whereas corresponding value for "clean" regions is only 21–101 ng/g lipid. Russian standard for maximum allowable concentration of DDT and its metabolites in chicken eggs is 100 ng/g. Hotspots of local pollution may be found in various regions. For example, concentration of DDT + DDE in chicken eggs was 12 900 ng/g lipid in Novomoskovsk, where PCB had been produced in the past, and 101–505 ng/g lipid in Chapaevsk, near the former HCH factory. Concentrations of hexachlorobenzene in Chapaevsk eggs was 66.4 ng/g lipid (Revich et al., 2007). One-third of soil samples, taken in Moscow, had elevated concentrations of DDT and its metabolites (DDE, DDD), above national standard of 100 ng/g; the mean concentration was 150 ng/g, while maximum concentration reached 14,400 ng/g (Shelepchikov et al., 2007).

Concentrations of DDT and other POPs in breast milk may be used as indicators of exposure of population to these pollutants. The first measurements of chlorinated organic pesticides in breast milk were conducted in the Former USSR Republics in the 1980s. These measurements showed high concentrations of DDT and DDE in breast milk: from 1 to 67 mkg/L (Sofina et al., 1995), or from 0.025 to 1.67 mkg/g lipid. These concentrations were higher than in Central Asia states. Average content of DDT/DDE in breast milk of Moscow women was lower: 7.7 or 0.19 mkg/L lipid (Bobovnikova et al., 1987). Twenty years later, new measurements showed that concentrations of DDT and DDE in breast milk of women from several cities of Moscow region have diminished by several times (Konoplev et al., 2006). Unfortunately, new data on DDT in breast milk of women who live in south regions of Russia (where DDT had been applied most intensely) are absent.

Other pesticides, listed in Stockholm Convention, have not been widely used in Russia. Nevertheless, these substances persist in the environment and may come to Russia via transboundary transport, or with imported foods. Monitoring of soil pollution in Moscow city showed that chlordane was found in one-third of all samples, and sum its isomers varied from 14.4 to 399.3 pg/g, while target concentration of chlordane in Holland is only 30 pg/g. The concentrations of chlordane metabolites, *cis*- and *trans*-nonachlor, varied from 21.3 to 62.2 pg/g. Mirex, which was used in USA and some other countries as pesticide and as fire retardant, was found in 3 out of 40 soil samples. In two of these samples, its concentrations were below 15 pg/g, while in the third sample its concentration was 594 pg/g. This fact deserves attention, because it cannot be explained only by transboundary transfer and confirms the possibility of existence territories, which are polluted by "exotic" for Russia substances.

3. Polychlorinated Dibenzo-*p*-Dioxins and Furans

In the framework of dioxin emission inventory project, PCDD/Fs were analyzed in stack gas samples taken at more than 20 chlorine chemical plants, metallurgical plants, and incinerators. Annual emission of dioxins in Russia may be between 6.9 and 10.8 kg T-TEQ, which is more than in most European countries (Kluev et al., 2001). Dioxins in soil were analyzed in 25 regions. The highest levels of dioxins were detected near chemical factories which earlier produced chlorinated organic pesticides: in Ufa (Republic Bashkortostan) and Middle Volga Region (Chapaevsk). Dioxin content of soil near chemical plants varied from ND to 24.6 ng/kg in Ufa; from 54 to 86 pg I-TEQ/kg in Chapaevsk in 1993 and from 6.8 to 51 pg I-TEQ/kg in 2006; from 13.4 to 52.7 pg I-TEQ/kg near Vladivostok incinerator. Thus, even after 18 years chlorine production was closed concentrations of dioxins in Chapaevsk soils exceeded German standard for allowable content of dioxins in agricultural soils (5–40 pg I-TEQ/kg) by almost two times.

Dioxins and other organic pollutants, accumulated in sludge tanks of Chimprom chemical factories may pose considerable environmental risks in Ufa, Sterlitamak, Novocheboksarsk, Dzerzhinsk, Novomoskovsk, and other cities. Incinerators of industrial chlorinated organic waste and electrolytic processes at nonferrous metallurgy plants may also be important sources of dioxin contamination. Titanium and magnesium production in Berezniaki is an example of such pollution source (Shelepchikov et al., 2006). The exact amounts of POPs, which still exist in reservoir sources, have not been assessed yet, but the risks of their evaporation and infiltration in ground waters and river waters is apparent. Dioxin concentration in carp fish caught in a pond near sludge tanks reached 93.9 pg/g WHO-TEQ (f.w.).

The levels of PCDD/PCDF in Russian cities are not unique; similar levels have been observed in Europe and USA. However, Russian residents often use contaminated land plots for vegetable gardens, cattle breeding or poultry farming, such practices greatly increase health risks for local population (Sotskov et al., 1999, Revich et al., 2000b). This problem has been studied most extensively in Chapaevsk. PCDD/PCDF profile is characterized by domination of low-chlorinated PCDFs, which formed during chlorine production. There are also high concentrations of OCDD/HpCDD/OCDF, typical for pentachlorophenol of HCB production. Similar profile of PCDD/PCDF was observed in chicken eggs (corrected for diminishing bioavailability of highly chlorinated compounds). In 1998, dioxin content in cow milk in Chapaevsk varied from 8.6 to 29.6 pg I-TEQ/g fat. In 2006, PCDD/Fs content in chicken eggs in Chapaevsk were 8.3–279.7 pg WHO-TEQ/g fat. This value greatly exceeds EU standard (3.0 pg/g WHO-$TEQ_{PCDD/F}$/6.0 pg/g WHO-$TEQ_{PCDD/F,PCB}$ in lipid base).

The women of Chapaevsk have the highest concentrations of dioxins in breast milk: 43.3 pg WHO-TEQ/g lipid (Revich et al., 1999, 2001). The second highest concentrations were registered in Usolie-Sibirskoye: between 23.7 and 37.0 pg WHO-TEQ/g lipid (Schecter, 1999, Mamontova et al., 1999). Table 1 lists some other results of measurements of dioxins in breast milk in Russia.

TABLE 1. PCDD/PCDF in breast milk, pg WHO-TEQ/g lipid

City and year	N	WHO-$TEQ_{PCDD/Fs}$	Reference
Chlorine production sites			
Chapaevsk, 1998	40	43.3	Revich et al. (2001)
Usolie-Sibirskoye, 1998	11	23.7	Schecter (1999)
	6	37.0	Mamontova et al. (1999)
Dzerzhinsk	10	10.7	Traag and Yufit (1997)
Volgograd	10	9.1	Ibid.
Bashkiria			Amirova et al. (1999)
Eight cities	43	15.9	
Rural areas	23	12.8	
Large cities without chlorine production – Irkutsk, Novosibirsk, Nizhni Tagil, Murmansk, Angarsk,	62	8–17	Schecter et al. (1990); Schecter (1999); Traag and Yufit (1997); Amirova et al. (1997); Mamontova et al.
Cheremchovo			Mamontova et al. (1999)
Cities of Kola Peninsula, 1993	30	15.5	Polder et al. (1998)
Magnitogorsk	25	7.16	Determination (2004)

The highest concentrations of PCDD/Fs in human blood have been registered during survey of workers of former pesticide plant in Chapaevsk (412.4 pg WHO-TEQ/g lipid) and Ufa (490 pg WHO-TEQ/g lipid). Somewhat lower concentrations have been measured in the blood samples of firemen who extinguished the major fire at Shelekhov cable factory in Irkutsk Region in 1992. Twelve years after the event, mean total content of dioxins and PCBs in blood of these firemen reached 152 pg WHO-TEQ/g lipid, while median value was 123 pg WHO-TEQ/g lipid, 95% CI 96–210 pg WHO-TEQ/g lipid (Chernyak et al., 2004).

Dioxins in blood are continuously monitored only in one city – Chapaevsk. Such monitoring was initiated in 1997, and continues today in the framework of joint Russian–US project in collaboration with Harvard School of Public Health. Pesticide plant workers used to have higher dioxin content in blood, than local residents. Concentrations of dioxins in blood of local residents also differed from one location to another. Women who lived next to the plant had higher dioxin content (75.7 pg/p lipid WHO-TEQ), while women who lived farther away from the plant had lower concentrations in their blood (44.1 pg WHO-TEQ/g lipid, $p = 0.04$) (Akhmedkhanov et al., 2002) (see Table 2).

Dioxin content in blood of boys is much lower. This is explained by bioaccumulation of dioxins with age. Arithmetic means and 25th; 75th percentiles of PCDD and PCDF in blood of children were 95.8 (40.9; 144) and 33.9 (20.4; 61.8) pg/g lipid correspondingly. Recalculated with WHO dioxin toxicity weights, these concentrations will be equal to 0.29 (0.1; 9.14) and 7.98 (5.27; 12.3) correspondingly. Higher concentrations of dioxin-like compounds in blood serum generally were associated with age and dietary

TABLE 2. PCDD/PCDF in blood of residents of different regions of the Russian Federation, pg/p lipid WHO-TEQ

City and year	Number of samples	Mean dioxin content	Source
Chapaevsk			
1997, within 3 km from the plant	6	75.2	Revich et al. (2001)
5–8 km	4	24.1	Ibid.
1998, within 5 km from the plant		75.7	Akhmedkhanov et al. (2002)
More than 5 km		44.1	Ibid.
Boys, 2002	30	8.27	Revich et al. (2006)
Bashkiria			Amirova et al. (1999)
Industrial cities	124	39.8	Ibid.
Small towns	138	25.0	Ibid.
Rural areas	102	24.8	Ibid.
Irkutsk region:			Schecter (1999)
Sayansk	1	37.3	Ibid.
Baikalsk	8	20.0	Ibid.
Angarsk	5	14.8	Ibid.
Urals – towns with metallurgical works	?	21.7–64.4	Amirova and Kruglov (1998)
Chuvashia, Novo-Cheboksarsk	14	12.0	Ibid.
Saint-Petersburg	60	29.8	Ibid.
Komi Republic		44.5	Amirova et al. (2002)
Arctic region	27	0.34–9.0	Konoplev et al. (2006)

habits: consumption of fish and local meat products (except poultry) (Hauser et al., 2005; Revich et al., 2006). Several studies of reproductive health of population have been conducted in Chapaevsk and Ufa. Former pesticide plant workers had statistically significant reduction of spermatozoid count, lower specific gravity of normal sperm and other disturbances of spermatogenesis (Britvin, 2000; Goncharov, 1999). The residents of these cities, who lived near the chemical plants, also had significantly higher rates of spontaneous abortions and complicated pregnancy (Basharova, 1996; Revich et al., 2001, 2006). These regions are characterized by abnormally high proportion of newborn girls (Basharova, 1996; Revich et al., 2001).

Dioxins affect sexual and physical development of children. The first large-scale epidemiological study in Chapaevsk analyzed cohort of adolescents aged from 10 to 16. Among the boys of the cohort, 45 boys (1.7%) had true cryptorchism, which is higher than average incidence for Russia (0.7–1%); and 18 boys (0.7%) had hypospadias – a rare development defect. Average incidence of hypospadias in Russia is 0.2–0.33% (Sergeyev et al., 2002; Lee et al., 2003).

Dioxins in serum were analyzed to investigate possible relationship between exposition of mothers and their children to dioxins, and disturbances of sexual development of boys (cryptorchism, hypospadias, retardation of sexual development). Dioxins, furans, and PCBs were analyzed by CDC lab in 30 samples of blood serum of boys aged 14–16. Total concentrations of dioxins increased with age and consumption of meat and fish products (odds ratios were correspondingly 1.31, 1.75, and 1.62) (Hauser et al., 2005).

Epidemiological cross-sectional study has been conducted to establish the influence of dioxins on probability of development of breast cancer. The women who worked at Chapaevsk chemical plant had higher incidence of breast cancer (relative risk 2.1). Average time of service at the plant was 20.4 years for women with breast cancer, versus 12.1 years for women from the control group. In comparison with the control group, the women with breast cancer, and their parents were more frequently employed at chemical production (27.7% vs 19.7%). The survey of women identified their dietary habits. Local residents of Chapaevsk (much like in other small Russian towns) raise cattle and pigs, grow vegetables and fruit in their vegetable gardens. The women who used this locally grown foods had higher incidence of breast cancer (OR = 5.7 for local pork and OR = 2.3 for local fish). Women in case group used lard for cooking more frequently than women in control group (Table 3). This finding is important for planning of preventive measures to reduce incidence of breast cancer (Revich et al., 2002).

TABLE 3. Odds ratios (ORs) for breast cancer risk factors in Chapaevsk. (From Revich et al., 2002.)

Risk factor	OR*	p
Nulliparity	3.8 [1.1–13.5]	0.038
Oral contraception	2.4 [1.03–5.5]	0.042
Breast cancers in relatives	9.0 [1.85–43.6]	0.006
Usage more than 50% of pork from farms in the Chapaevsk region	5.7 [1.3–25.5]	0.021
Usage more than 50% of fish from the nearest lakes or rivers	2.3 [1.13–4.80]	0.022
Parents worked at Chemical plant	1.8 [0.80–4.00]	0.16
Occupation at chemical plant	2.1 [0.95–4.68]	0.07
Age at menarche <13 years	0.82 [0.34–1.97]	0.64

*Crude odds ratios

4. Polychlorinated Biphenyls

PCBs were produced in two towns in European Part of Russia: Novomoskovsk and Dzerzhinsk. Transformers were produced in Serpukhov. Between 1939 and 1990, 180 t of PCBs were produced. These substances were used as dielectrics in transformers and capacitors, and as additives to paints and lubricants. According to the inventory, there is about 30,000 t of PCB in Russia (20,000 t in transformers, 10,000 t in condensers) (Treger and Rozanov, 2000). Used PCB-containing oils are incinerated in various furnaces, including the blast furnace of Novolipetsk metallurgical combine. Although long-term monitoring did not reveal elevated concentrations of dioxins in blast furnace gas there, PCDD/PCDF content in indoor air at the combine approaches European standard for flue gases of incinerators.

PCB-producing towns have high levels of dioxin-like substances in the environment and local food products. For example, locally produced chicken eggs in Dzerzhinsk have elevated concentrations of seven indicator PCBs (63.5–167 pg/g lipid). This is equivalent to 9.08–18.37 pg WHO-TEQ_{PCB}/g lipid, and to 21.8–63.1 pg/g lipid, if PCDD/PCDFs are taken in account (IPEN, 2005).

Locally produced chicken eggs in Novomoskovsk and adjacent districts have very high concentrations of dioxin-like PCBs. Chicken eggs from the farms next to the plant, where Sovol/Sovtol was produced, contain 324.2 pg WHO-TEQ_{PCB}/g lipid, while chicken eggs, collected at the farms 2 km away from the plant, have 69.6 pg WHO-TEQ_{PCB}/g lipid. The profile of concentrations of dioxin-like PCBs is fairly stable and specific to the products of the plant. PCDD/PCDF were analyzed only in one joint sample (6.05 pg WHO-TEQ/g lipid), while concentration of non-ortho-PCB was 53.8 pg WHO-TEQ/g lipid.

In Serpukhov, PCBs concentration in soil near the plant where PCB-containing electric equipment had been produced varied from 13.6 to 172 mkg/g (Bobovnikova et al., 2000a). Women, who worked at the plant or lived near it, had high concentrations of PCB in breast milk (Bobovnikova et al., 2000b), see Table 4.

PCB levels in blood of 15 workers of Serpukhov capacitor plant were measured in 1988, when PCB use was terminated. PCB levels in blood serum of individual workers varied from 120 to 1080 mkg/L, with the mean value 460 ± 360 mkg/L. Cross-sectional study of influence of PCB on infertility has been conducted in the same city. After correction for age of women and their sexual behavior, regression analysis showed that each 1 ng/g increase of PCB content in blood was associated with 1.7 OR of infertility, and each 1 ng/g lipid increase of PCB content in blood was associated with 1.1 OR of infertility. These results proved that risk of infertility increased with PCB content in blood of women (Korrick and Altshul, 1998; Revich et al., 2000a).

TABLE 4. PCB in beast milk of women in different regions of the Russian Federation (mkg/g lipid)

Region/city and year	Number of samples	Arithmetic mean (standard deviation)	Min/Max	Source
Serpukhov, women-workers of capacitor plant, 1986	4	42.2 (26.5)	17.0/79.7	Bobovnikova et al. (1993)
Serpukhov, women residents	10	4.8 (6.6)	0.7/22.0	Ibid.
Moscow, Moscow region, Baikalsk	88	1.4 (n/a)	0.3/3.7	Bobovnikova et al. (1987)
Chapaevsk, 1998	6 samples, combined from 40 individual samples	0.5 (0.2)	0.4/0.6	Revich and Bobovnikova (unpublished)
Baikal region			0.091/5.0	PCB in Baikal Region (2006)
Kola Peninsula	30	0.46		Polder et al. (1998)
Astrakhan Region, 2000	9	0.283	0.236/1.028	Egorov et al. (2003)
Magnitogorsk	29	0.011	0/0.004	Analysis of content... (2004)

Wide distribution of PCB-containing equipment caused development of local hotspots in different regions. One sample of chicken eggs from Chapaevsk had PCB content of 663.2 pg/g WHO-TEQ lipid. Further investigation showed that chicken eggs, collected near the chemical plant, had TEQ 3.5 times higher than that of chicken eggs collected farther away (61.8 vs 17.6 pg/g lipid WHO-TEQ, correspondingly) (Revich et al., 2007). Monitoring of soil pollution in Moscow showed that 90% of soil samples met environmental standards, while one sample had very high concentration of seven indicator congeners (4.0 mkg/g) and total toxicity equivalent (331.2 pg/g WHO-TEQ) (Shelepchikov et al., 2007).

A study of PCB in the environment has been carried out in Baikal region, where two sources of PCB pollution have been identified: industrial waste of chlorine production and (possibly) transformer plant in Usolie-Sibirskoye. Breast milk of residents of nine cities and villages was monitored between 1998 and 2004. The sum of 28 analyzed PCB congeners varied from 91 to 5,300 ng/g lipid. The women from Irkutsk (regional capital) and Usolie-Sibirskoye had the highest concentrations of PCB in breast milk. PCB levels in breast milk also depended upon dietary habits (fish consumption) (PCB in Baikal Region, 2005).

Wide utilization of PCB-containing equipment and disregard of industrial safety norms and operating instructions led to infiltration of large quantities of PCB in the environment. This source of dioxin-like substances may be as important from public health perspective, as PCDD/PCDF emissions from high-temperature sources.

Selective measurements of PCDD/PCDF and dioxin-like PCB in fish and seafood, caught in Russian zone of Baltic Sea, showed that pollution levels of most fish and sea products closely approach EC standard for maximum TEQ content. The share of PCB in total WHO-TEQ varied from 45% to 95%. Hence, if PCBs are included in total WHO-TEQ, then TEQ of all studied fish and sea products may approach or exceed the EC standard, established for fish and fishery products – 4 ng/kg $\text{WHO-TEQ}_{\text{PCDD/F}}$ and 8 ng/kg $\text{WHO-TEQ}_{\text{PCDD/F,PCB}}$ (12 ng/kg $\text{WHO-TEQ}_{\text{PCDD/F,PCB}}$ for eel). If dioxin-like PCBs are not considered, then only TEQ of cod liver would violate Russian standard (11 ng/kg I-TEQ), and TEQ of Atlantic salmon would slightly exceed EC limit (Shelepchikov et al., 2005).

5. Conclusion

Between 1995 and 2006, RF completed inventory of pesticides and PCB-containing equipment. Utilization of pesticides has been completed in several regions of the country. These pesticides were transported to special waste management facilities or destroyed.

Nevertheless, Russia lacks concerted public policy in the area of control and prevention of toxic pollution of the environment by persistent organic pollutants. Several studies of this problem have been initiated by regional authorities of NGOs, but this is not enough for countrywide health risk assessment. Some important major hotspots and POP emission sources are yet to be found. There may be even more local hotspots of environmental pollution to be discovered.

Several problematic territories have been studied most thoroughly, where the levels of dioxins and PCBs in the environment approach those of the most polluted sites in Vietnam and Sevezo. Chlorine chemical plants in the most polluted Russian cities (Ufa, Chapaevsk, Dzerzhinsk, and Novomoskovsk) were closed many years ago, but the levels of pollution of the environment and food products are still high. Contaminated soil near pesticide plants, industrial facilities and waste are very important secondary sources of dioxin and PCB pollution. For example, dioxin content of stucco in Chapaevsk buildings varied from 3 to 99 pg/g WHO-TEQ in 2006. Russia does not implement complex programs of environmental rehabilitation of polluted territories.

The residents of small towns grow vegetables, raise cattle and poultry on polluted lands. This is also an important source of POPs intake for local population. The relationship between PCB in breast milk and consumption of local fish was established in Volga and Baikal regions (Egorov et al., 2003, Polychlorinated biphenyls 2005). The relationship between consumption of locally grown food products, incidence of breast cancer, and dioxin content in blood of boys has been established in Chapaevsk.

POPs in breast milk or blood of local residents may serve as a reliable indicator of effectiveness of environmental policy on polluted territories. For example, dioxins in blood of German people fell from 46 pg TEQ-WHO/g lipid in 1988 to 12–13 pg TEQ-WHO/g lipid in 1998 (Papke, 2000). Similar trends have been observed in the Netherlands, Japan and other countries. Very few studies confirmed downward trend of dioxin content in breast milk of women of Moscow region. The dynamics of PCBs and dioxins in breast milk is difficult to monitor, because the levels of these pollutants vary greatly among population, and only one city implements a long-term program of monitoring of POPs in biological media.

Russian researchers who studied relationship between POP and public health have reported several important findings. Health effects among general population of Chapaevsk and Ufa were quite similar to those among the Vietnamese exposed to Agent Orange. Reproductive disorders among Chapaevsk men, increased incidence of infertility, disturbance of disperma-togenesis among local residents of Chapaevsk do look like reproductive defects observed among Vietnam War veterans in the USA. Chapaevsk was the first city where the researchers observed the relationship between consumption of locally grown foods, incidence of breast cancer, and disturbances of sexual development among adolescents.

The results of studies of dioxins and PCBs have led to closure or modernization of environmentally hazardous industrial facilities, and implementation of environmental protection measures. For example, open dialogue among the public, municipal authorities, and industrial establishment of Archangelsk pulp and paper mill has helped to modernize this factory and reduce emissions of dioxins by an order of magnitude. Special federal and regional program of environmental reclamation and replacement of contaminated soils in Chapaevsk greatly reduced soil pollution. In addition, a new hospital was constructed in Chapaevsk, and public health care centers received new equipment. There are future plans of massive rehabilitation of the most polluted territories.

References

Akhmedkhanov, A., Revich, B., Adibi, J. J., et al., 2002, Characterization of dioxin exposure in residents of Chapaevsk, Russia, *Journal of Exposure Analysis and Environmental Epidemiology*, **12**:409–417.

Amirova, Z. and Kruglov, T. A., 2002, Second AMAP International Symposium on Environmental Pollution of Arctic, Rovaniemi, Finland, P-H13.

Amirova, Z., et al., 1999, PCDD/PCDF levels in blood and breast milk from urban and rural areas of Bashkortostan, Russia, *Organohalogen Compounds*, **44**:75–78.

Amirova, Z. K. and Kruglov, E. A., 1998, *Situation with Dioxins in Republic Baskortostan*, Ufa.

Basharova, G. R., 1996, Medical and biological consequences of exposure to dioxins, Dr. Med. thesis, Moscow.

Bobovnikova, T. I., et al., 1987, Pesticides in breast milk USSR woman, Proceeding Ins. Experimental Meteorology, **4**:101–110.

Bobovnikova, T. I., et al., 2000a, The influence of capacitor plant in Serpukhov on contamination of vegetables by PCB, *The Science of the Total Environmental*, **246**:51–60.

Bobovnikova, C. I., et al., 2000b, Influence of capacitor plant on pollution of the environment by polychlorinated biphenyls near Serpukhov, *Polychlorinated Biphenyls*, Vol. 5, Nauka, Moscow, pp. 87–103.

Britvin, A. A., 2000, Reproductive status of males, who lived in territories contaminated by polychlorinated hydrocarbons, Dr. Med. thesis, Samara.

Chernyak, Y., et al., 2004, Assessment of PCDD, PCDF and PCB levels in serum of firemen exposed to combustion products during the 1992 fire at Irkutscable factory fire in Shelekhov, Russia, *Organohalogen Compounds*, **66**:2481–2487.

Contamination of the environment by dioxins and related compounds in Irkutsk region (hygienic aspects of the problem), Irkutsk, 2000.

Determination of content of PCDD/PCDF and PCB in breast milk of Magnitogorsk women, 2004, Report of International Project for Liquidation of POPs, Chelyabinsk; http://www.ipen.org.

Egorov, A., Mikhailov, and G., Altshul, L., 2003, Fish consumption and levels of PCB and chlorinated pesticides in breast milk PCB in Astrakhan region, Russia, *Organohalogen Compounds*, **64**:91–94.

Goncharov, N. P., et al., 1999, Influence of dioxins on reproductive system of males in Samara Region of Russia, *Organohalogen Compounds*, **42**:61–65.

Hauser, R., Williams, P., Altshul, L., et al., 2005, Predictors of serum dioxin levels among adolescent boys in Chapaevsk, Russia: a cross-sectional pilot study, *Environmental Health*, **4**(1):8.

IPEN, 2005, Dioxin, PCBs and hexachlorobenzene contamination of chicken eggs from Dzerzhinsk region, Russia, http://www.ipen.org.

Kluev, N. A., Kurliandsky, B. A., Revich, B. A., and Filatov, B. N., 2001, *Dioxins in Moscow*, Moscow.

Konoplev, A., et al., 2006, POPs in human milk from Moscow region, *Organohalogen Compounds*, **68**:1611–1614.

Korrick, S. A. and Altshul, L., 1998, *Environmental Health Perspectives*, **106**(8):513–518.

Lee, M. M., Sergeyev, O., Williams, P., et al., 2003, Physical growth and sexual maturation of male adolescents in Chapaevsk, Russia, *Journal of Pediatric Endocrinology and Metabolism*, **16**:169–178.

Mamontova, E. A., et al., 1999, PCDD/PCDF in Human milk and blood samples from a contaminated region near Lake Baikal, *Organohalogen Compounds*, **44**:37–40.

Papke O., 2000, Lessons learned on performing more than 15,000 dioxin analyses *Organohalogen Compounds,* **48**:115–119.

PCB in Baikal Region: Sources, Long-range Transport and Risk Assessment (Project Report under Grant INTAS 2000-00140), Institute of Geography of Siberian Department of Russian Academy of Sciences, Irkutsk, 2005.

Polder, A., et al., 1998, Dioxins, PCBs and some chlorinated pesticides in human breast milk from the Kola Peninsula, Russia, *Chemosphere*, **37**:1795–1806.

Revich, B., Brodsky, E., and Sotskov, Y., 1999, Dioxin in the environment, blood, breast milk, and cow milk in Chapaevsk, *Organohalogen Compounds*, **44**:229–232.

Revich, B. A., Korrick, S., Altshul, L. et al., 2000a, Polychlorinated biphenyls and disorders of reproductive health – a study in Serpukhov, in: *Polychlorinated Biphenyls: Super-Toxicants of the 20th Century*, Vol. 5, VINITI, Moscow, pp. 104–116.

Revich, B., Aksel, E., Ushakova, T., Ivanova, J., Zhuchenko, N., Klyuyev, N., Brodsky, E., and Sotskov, Y., 2000b, Dioxin exposure and public health in Chapaevsk, Russia, *Chemosphere*, **43**:951–966.

Revich, B., Ushakova, T., Sergeev, O., and Zeilert, V., 2002, Dioxin pollution as a risk factor for development of female breast cancer: a study in Chapaevsk, Russia, *Organohalogen Compounds*, **59**:361–364.

Revich, B. A., Sotskov, Y. P., Kluev, N. A., Brodsky, E. S., Lipchenko, Y. N., Muzurov, V., and Zeilert, V. Y., 2001, Bioxins in the environment and biological media of population of Chapaevsk, *Hygiene and Sanitary*, **6**:6–10.

Revich, B. A., Sergeev, O. V., and Hauser, R., 2006, Dioxins, furans, and PCB in blood of adolescents in Chapaevsk – preliminary results of prospective epidemiological study, *Toxicology Vestnik*, **5**:2–8.

Revich, B. A., Shelepchikov, A. A., and Sergeev, O. V., 2007, PCB and chlorinated organic pesticides in chicken eggs collected in different regions of Russia, *Voprosy Pitanija*, **4**.

Sofina, L. I., Kolycheva, S. S., Nefiodov, P V., Zaria, N. Y., and Kutumova, S. L, 1995, Content of trace quantities of pesticides in biological media of humans, in: *Actual Environmental and Hygienic Problems of North Caucasus*, Krasnodar, pp. 239–241.

Sotskov, Y. P., Lipchenko, Y. N., Muzurov, I. V., Revich, B. A., et al., 1999, *Ecology of Chapaevsk – Environment and Public Health*, Chapaevsk, Moscow.

Schecter, A., 1999, Elevated levels of dioxins in human tissues after occupational exposure to dioxins and related chemicals, *Organohalogen Compounds*, **44**:23–26.

Sergeyev, O., Revich, B., Williams, P., et al., 2002, Case-cohort study of cryptorchism, hypospadias and delayed sexual maturation in a region with dioxin contamination: Chapaevsk, Russia, *Organohalogen Compounds*, **59**:385–388.

Shelepchikov, A. A., Shenderyuk, V. V., Brodsky, E. S., and Baholdina, L. P., 2005, Contamination of Russian Baltic fish by polychlorinated dibenzo-p-dioxins, dibenzofurans and dioxin-like biphenyls, *Organohalogen Compounds*, **67**:1502–1505.

Shelepchikov, A. A., Brodsky, E. S., Feshin, D. B., and Orenbah, D., 2006, PCDD/PCDFs in soils around of titanium-magnesium industrial complex, *Organohalogen Compounds*, **67**:2302–2305.

Shelepchikov, A. A., Brodsky, E. S., Balashova, S. P., and Feshin, D. B., 2007, Contamination of Moscow soils by POPs and others organic contaminants, *Organohalogen Compounds*, **69**.

Traag, W. A. and Yufit, S., 1997, Contamination of breast milk with PCDD, PCDF and PCB in Russia, *Organohalogen Compounds*, **33**:524–529.

Treger, Y. A. and Rozanov, V. N., 2000, Production and consumption of polychlorinated biphenyls in Russia, *Polychlorinated Biphenyls*, Vol. 5, Moscow, pp. 64–70.

PCB POLLUTION OF IZMIT BAY (MARMARA SEA) MUSSELS AFTER THE EARTHQUAKE

LEYLA TOLUN,* FATMA TELLI-KARAKOC
TUBITAK-MRC Chemistry and Environment Institute, P.O. Box 21, 41470 Gebze, Kocaeli, Turkey

BERNHARD HENKELMANN, KARL-WERNER SCHRAMM
GSF – National Research Center for Environment and Health, Institute of Ecological Chemistry, Ingolstaedter Landstrasse 1, D-85764 Neuherberg, Germany

OYA S. OKAY
Istanbul Technical University, Faculty of Naval Architecture and Ocean Engineering, 34469, Maslak, Istanbul, Turkey

Abstract: On 17 August 1999 a powerful earthquake struck the eastern part of Marmara Region and Izmit Bay, the most industrialized and populated area of the region. Environmental effects of the earthquake and subsequent refinery fire were partly investigated by several scientists (Okay et al., 2001, 2003; Tolun et al., 2001, 2006; Balkıs, 2003; Unlu and Alpar, 2004; Karakaş and Pekey, 2005). As a part of the previous efforts, in this study, the levels of total PCBs, sum of the marker PCBs and TEqs in mussels collected immediately after and 1 year later the earthquake were presented. Total PCB (# 1-209) levels in mussels from different coastal locations were found between 5.47 and 14 ng/g in September 1999 and between 11.2 and 36.0 ng/g in June 2000. June 2000 samples from throughout the Bay contained higher quantities of total, marker PCBs and TEq values when compared with the before earthquake values. The elevated concentrations of PCBs in mussels collected after the earthquake implies an input of PCBs to the water column form different sources such as underlying sediments and/or land-based sources such as industrial inputs. Although WHO-TEq values found in this study for Izmit Bay ($0.4 \pm 0.1 - 1.4 \pm 0.7$ pg/g) are lower than the proposed values for fisheries products (4 pg/g) by EC

* To whom correspondence should be addressed. E-mail: Leyla.Tolun@mam.gov.tr

E. Mehmetli and B. Koumanova (eds.),
The Fate of Persistent Organic Pollutants in the Environment, 127–139.

Regulation 199/2006, considering the TDI value as 2 pg WHO-TEq/kg bw (COT, 2001) it was concluded that consumable mussels from Izmit Bay pose a risk for health.

Keywords: polychlorinated biphenyls, Izmit Bay, mussels, TEq values, environmental effects of earthquake

1. Introduction

The Marmara region of Turkey has a long history of damaging earthquakes, which was tragically recorded on 17 August 1999 when an Mw 7.4 earthquake occurred in the heavily industrialized province of Kocaeli, centered on the ancient city of Izmit (Scawthron and Jhonson, 2000).

The Region especially around the semienclosed Izmit Bay (30 km) has experienced very rapid growth in the last 20 years. This growth is due to rapid industrialization in the triangle bounded by Istanbul, Izmit and Bursa. A widespread growth of large spectrum industrial activities ranging from pulp and paper to petrochemical complexes, from food to metal processing and fertilizer industries have increased the pollution problems due to uncontrolled discharges for many years. Industrial loads have been reduced by treatment and waste minimization within the last 10 years, but domestic wastes have doubled, due to the increasing population in the Bay. Therefore, the total (domestic + industrial) discharge load into the Bay during the last 10 years has not changed significantly (Morkoç et al., 2001). The dissolved oxygen content of Izmit Bay decreased dramatically from 1984 to 1999 and reached a minimum value at 20 m throughout the Bay (Okay et al., 2001).

Previous studies carried out at Izmit Bay have provided important information about the oceanographic characteristics (Tugrul et al., 1985) and the pollution levels of the bay before the earthquake. Recent investigations showed that the subsequent fire after the earthquake caused an increase in the total PAH concentrations of the local mussels and sediments (Okay et al., 2003; Tolun et al., 2006; Telli-Karakoç et al., 2002; Karakaş et al., 2005) and the dissolved oxygen content of the lower layer decreased below the detection limit (Okay et al., 2001; Balkis, 2003).

Polychlorinated biphenyls (PCBs) are a family of man-made chemicals that contains 209 individual compounds (known as congeners) with varying harmful effects. EPA considers all PCB mixtures to be toxic. A number of studies show that PCBs cause cancer in animals. They are claimed as probable human carcinogens (USEPA, 1998). Due to their useful physical and chemical properties, PCBs have been used in a variety of applications, e.g., dielectric fluids in transformers, hydraulic system and paints. One of the major sources of PCBs are burning processes (waste incineration and backyard burning), metal industries, contaminated soil and sediments and landfill sites with contaminated sediments (Kivirianta, 2005). Furthermore, PCBs are environmentally stable and biologically persistent. These characteristics together with high lipophilicity result in accumulation of PCBs in food chain.

Although, environmental effects of the earthquake and subsequent refinery fire were partly investigated by several scientists, information on PCB contamination are limited with the studies carried out at Marmara Sea (Coelhan et al., 2006) and Izmit Bay, before earthquake (Telli Karakoç et al., 2002). In this study, PCB contents of mussels collected from several coastal locations of Izmit Bay after the earthquake were presented.

2. Materials and Methods

2.1. STUDY SITE AND SAMPLE COLLECTION

Mussels (*Mytilus galloprovincialis*) were collected from the Izmit Bay in September 1999 and June 2000 after the Kocaeli earthquake (17 August 1999). The sampling locations selected to be the nearest to the land based discharges are shown in Figure 1 and the characteristics are given in Table 1. All sampling sites are located in the north and northeastern part of the bay (most industrialized part). 15–25 mussels of each sampling station were divided in two or three groups, including 3–5 individuals in each group, dissected and homogenized. The homogenized samples were stored in glass jars at 20°C until the analysis.

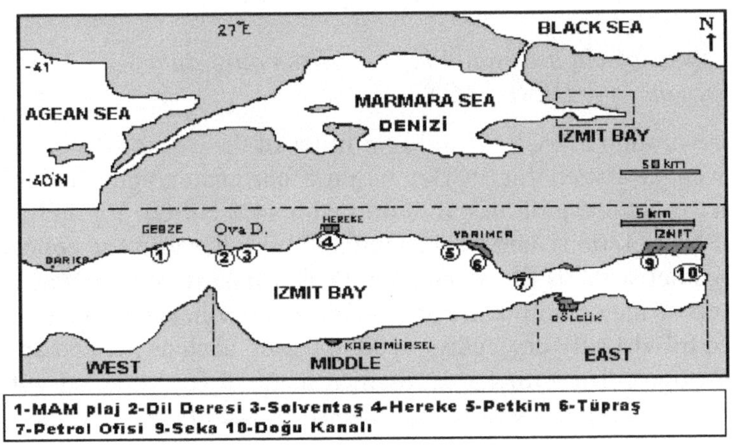

Figure 1. Location of sampling stations

TABLE 1. Description of the sampling locations

Sampling place	Characteristics
1	The coast near MRC. There is no direct discharge but suspected diffuse pollution source from heavy ship traffics
2	The river which carries industrial and domestic wastewater and discharge to the Bay
3	Near solvent production industry, there is no direct wastewater discharge
4	Domestic discharge but it is not a big source
5	The petrochemical industry
6	Petroleum refinery. Fire happened subsequent to the earthquake
7	Petroleum refinery + storage
9	Pulp and paper factory
10	A big channel, carrying more than 30 industrial wastewater discharges and domestic wastes

2.2. CHEMICAL ANALYSIS

2.2.1. Sample Preparation for Mussels

10–15 g of the homogenized mussel sample were mixed with pure anhydrous sodium sulphate in a clean porcelain mortar and then extracted with 250 ml *n*-hexane in a Soxhlet apparatus fitted with precleaned thimble. One blank (consisting of all chemicals except the sample) for each set of extraction were also run. After 7 h extraction (4–5 cycles per hour), *n*-hexane phase was removed at 40°C in rotary evaporator. The residue was dissolved in 1 ml of *n*-hexane before clean up process.

2.2.2. Analysis by High-Resolution Gas Chromatography with Mass Spectrometry (HR GC/MS)

Sample preparation of PCBs: Half a milliliter of the extracted samples from mussels were analyzed for PCBs. A glass chromatographic column was filled with 2 g SiO_2, 5 g of SiO_2 (bottom) with 44% H_2SO_4, 1 g SiO_2, and 5 g anhydrous Na_2SO_4 (top), and washed with 30 ml *n*-hexane. The concentrated extract together with 10 µl ^{13}C-labelled PCB standard mixture was injected onto the top of the column and eluted with 90 ml *n*-hexane. The eluate was evaporated till virtually dry, redissolved in 0.2 ml acetonitrile and applied to a SPE-cartridge filled with 1 g C18-silica material (conditioned with 4 ml acetonitrile). The sample was eluted with 4 ml acetonitrile and the volume of the eluate was reduced to 0.1 ml by a gentle stream of N_2. Ten microliters of recovery standard ^{13}C-1,2,3,4-TCDD in *n*-nonane was added. The total sample volume was reduced to 10 μl and analyzed using a HRGC/HRMS (High-Resolution Gas Chromatography/High-Resolution Mass Spectrometry).

An HP 5890 Series II gas chromatograph with a 60 m DB-5MS capillary column was interfaced to a Finnigan MAT95 magnetic sector field mass spectrometer. Chromatographic separation was achieved by splitless injection of 0.5 μl (cold injection system CIS3, Gerstel, Germany). The GC oven was programmed as follows: initial temperature 90°C held for 1.5 min, increased to 170°C at 20°C/min, held for 7.5 min, increased to 280°C at 3°C/min and held for 10 min. The carrier gas was helium. The MS was operated in SIM mode at a resolution of 8,000 and the two most intense ions of the molecular ion cluster were monitored for the unlabeled and labeled isomers (tri- to decachlorinated PCBs; Wu et al., 1997; Martens et al., 1998). The methods used in HRGC/HRMS are validated by inter-laboratory calibration studies and in between are checked by internal reference materials.

2.2.3. Calculation of the Toxicity Equivalent Concentrations (TEq)

TEFs (Toxic Equivalency Factor) indicate a value to estimate the toxicity of a compound relative to TCDD (2,3,7,8-tetrachlorodibenzo-*p*-dioxin). TEF values, in combination with chemical residue data can be used to calculate toxic equivalent concentrations in various environmental samples including sediments and mussels (Ahlborg et al., 1994). In this study TEq concentrations were calculated using the equation and TEF values of World Health Organization as given in Van den Berg et al. (1998).

3. Results and Discussion

3.1. CONTAMINATION OF MUSSELS

PCBs were detected in mussels collected from the different locations of Izmit Bay (upper layer) after the earthquake. Their values were represented as total PCBs (1–209), non-ortho, mono-ortho, sum of six marker congeners (28, 52, 101, 138, 153, and 180) and toxic equivalent quantities (TEq) in Table 2a (for September 1999) and Table 2b (for June 2000) comparing with the before earthquake (June 1999) results previously published by Telli Karakoç et al. in 2002; spatial distribution was shown in Figure 2a–c.

In general, two different trends were observed in the total PCB and sum of the marker PCB contents of the samples (Figure 2a and c). The first trend shows a small decrease after the earthquake (September 1999 data) and then a sharp increase approximately 1 year later (June 2000 data) for deeper stations such as 1, 2, and 3 located in the western part of the Bay (Figure 1). Previous studies showed that the suspended particulate matter values less in upper waters and display highest values towards the east (Algan et al., 1999). Remote sensing data obtained after the earthquake confirmed the increase in turbidity and showed highest values in northeastern region (Tufekci and Akman, 2005) near our sampling stations 4, 5, 6, 7, 9, and 10. It is also known that the Secchi disk depths decreased after the earthquake in the western part of the Bay (Morkoç et al., 2007). Furthermore, the fishermen observed more transparent water column, after the earthquake for a few weeks. Those observations may imply lesser amount of suspended solids for the filer feeders such as mussels and may explain the decrease in PCB pollution in the mussels collected immediately after the earthquake in those stations (1, 2, and 3). The other trend, belonging to the shallower stations (located in the northeastern part of the Bay) shows significant increases in PCB contents of mussels collected following the earthquake (September 1999 and June 2000 data). These stations are close to the refinery area and incineration plant.

In June 2000, samples from throughout the Bay contained the highest quantities of PCBs when compared with the previous collections. The highest value of total PCBs was found as 36 ng/g (ww), approximately ten times higher then the pre-earthquake value (3.1 ng/g), in mussels collected from the station 7. The minimum value was detected as 11.21 ng/g belonging to the station 4 in the same sampling period. The sum of the marker PCBs (28, 52, 101, 138, 153, and 180) shows a similar trend as the total PCBs. The highest value was found as 12.4 ng/g (ww) at station 7 in June 2000 and the lowest value was 1.1 ng/g (approximately ten times lower) for the same station before the earthquake.

The elevated concentrations of total PCBs in mussels collected after the earthquake implies an input of PCBs to the water column from different sources. The underlaying contaminated sediment may be one of the possible PCB sources of the water column. Previous studies related with PAH pollution of Izmit Bay sediments shows that the chemical composition of the sediments changed after the earthquake due to resuspension especially in the shallowest eastern part of the Bay (Tolun et al., 2006.). The other possible sources, like atmospheric inputs from incineration plants and land based contamination, especially from constructions like transformers and tanks damaged after the earthquake, may cause increases in the PCB contents of mussels after the earthquake.

Significant increments in PCB pollution of mussels were also observed for the TEq values at all stations. The minimum and maximum TEQ values were obtained as 0.1 and 1.40 pg/g before and 1 year after the earthquake, respectively, for the station 7. In the June 2000 samples, the lowest TEq value (0.43 pg/g) was obtained at station 4 (a small port located in the middle part of the Bay).

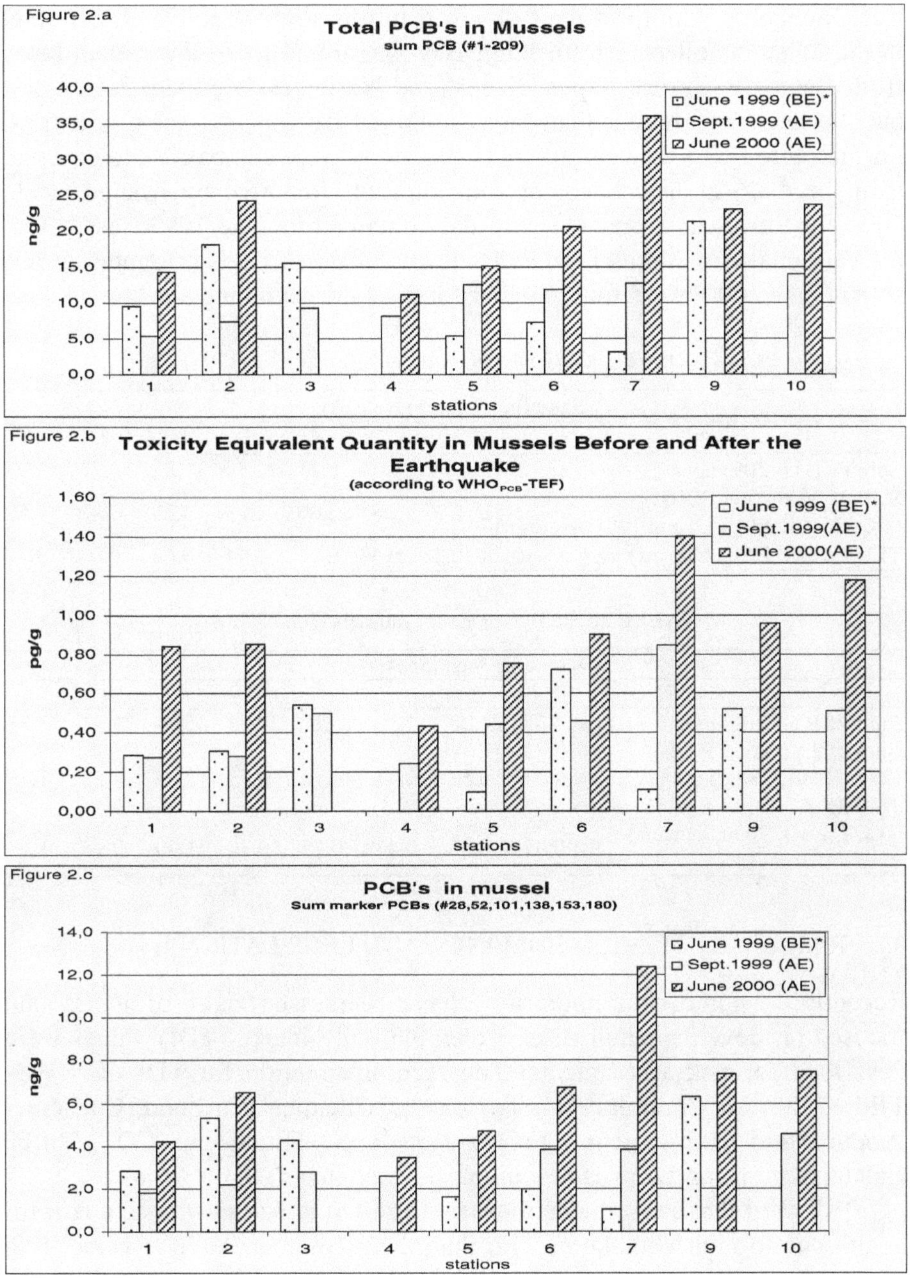

Figure 2. (a–c) Contamination of PCBs in mussels after the earthquake

Mussels have been used worldwide for the rapid assessment of bioaccumulation for large number of pollutants and offer wide geographic distributions. Mussel contents of PCBs from different seas and countries were compared with the values obtained in this study (Table 3). PCB level in the mussels collected from Izmit Bay was found to be lower than those from Denmark, France, Japan, Korea, and Northwest Mediterranean Coast but the level was higher than those obtained from many Asian countries. Furthermore PCBs concentrations in this study were comparable to or lesser than the 4–66 ng/g wet weight concentrations reported by Khaled et al., 2004 for Red Sea. Compared with data obtained in June 1999 (just before the earthquake in August) the PCBs levels of the mussels presented in this study (June 2000) were found higher for the same sampling sites.

TABLE 2a. PCB contents of mussels (September 1999)

	Sampling locations (st. no)							
	1	2	3	4	5	6	7	10
Sum PCB (1–209) ng/g	5.47	7.40	9.26	8.21	12.54	11.82	12.73	13.98
Sum of the marker PCBs (28, 52, 101, 138,153, and 180) ng/g	1.81	2.20	2.81	2.60	4.28	3.86	3.87	4.53
TEQ (WHO, 2000, humans) pg/g	0.28	0.28	0.50	0.24	0.44	0.46	0.85	0.51

TABLE 2b. PCB contents of mussels (June 2000)

	Sampling locations (st. no)							
	1	2	4	5	6	7	9	10
Sum PCB (1–209) ng/g	14.35	24.23	11.21	15.14	20.56	36.04	23.03	23.73
Sum of the marker PCBs (28, 52, 101, 138,153, and 180) ng/g	4.24	6.52	3.44	4.70	6.75	12.39	7.35	7.43
TEQ (WHO, 2000, humans) pg/g	0.84	0.85	0.43	0.75	0.90	1.40	0.96	1.18

3.2. RISK ASSESSMENT, GUIDELINES, AND LEGISLATION

Recently, a number of authorities have reassessed risks of PCBs and focused on developmental risks. Tolerable daily intake (TDI) values were described for risk assessments. The recommendation for TDI of WHO-TEq of the UK Committee on Toxicity of Chemicals in Food, Consumer Products and Environment (COT) is 2 pg WHO-TEq/kg bw (COT, 2001). Calculations using Izmit Bay mussels TEq contents ($0.4 \pm 0.1 - 1.4 \pm 0.7$ pg/g) showed that consumable mussels from Izmit Bay may pose a risk for health due to higher intake of TEq than the recommended TDI value. It is very clear that if a person (50 kg body weight) eats 100 g mussel from the contaminated site of Izmit Bay, he will intake higher than the recommended TDI value (2 pg WHO-TEq/kg bw). Such as:

TABLE 3. Concentrations of PCB's in mussels collected worldwide

Location	Mussel	Range ng/g ww	References
Red Sea	*Brachiodontes*	7–66	Khaled et al. (2004) (sum of 7 congeners)
Perth (Australia)	*Mytilus edulis*	<10	Burt and Ebell (1995)
Denmark	*M. edulis*	3–328	Granby and Spliid (1995)
USA	*M. edulis*	10–3,800	Sericano et al. (1995)
Korea	*M. edulis*	6–100	Khim et al. (2000)
Ariake Sea-Japan	*M. edulis*	Av. 590	Nakata et al. (2002)
Hong Kong (China)	*Perna viridis*	1–152	Liu and Kueh (2005)
Cambodia	*P. viridis*	0.5–5.1	Monirith et al. (2003)
Mainland (China)	*P. viridis*	0.3–13	"
India	*P. viridis*	0.2–11	"
Indonesia	*P. viridis*	0.1–2.7	"
Japan	*P. viridis*	7.4–84	"
South Korea	*P. viridis*	0.8–7.2	"
Malaysia	*P. viridis*	0.05–5.1	"
Philippines	*P. viridis*	0.4–14	"
Singapore	*P. viridis*	2.4	"
Vietnam	*P. viridis*	1.4	"
Thailand	*P. viridis*	0.01–20	Tanabe et al. (2000)
NW Mediterranean Coast	*Mytilus galloprovincialis*	10–700	Villenevue et al. (1999)
Izmit Bay (Turkey)	*M. galloprovincialis*	3–21	Telli Karakoç et al. (2002) (before EQ, June 1999)
Izmit Bay (Turkey)	*M. galloprovincialis*	5–14 11–36	This study (after EQ, September 1999) This study (after EQ, June 2000)

The TDI level for 50 kg person is 2 pg WHO-TEq/kg bw × 50 kg = 100 pg
Izmit Bay mussel content (highly contaminated site) is 1.4 TEq pg/g

100 g mussel contains 140 pg TEq

140 > 100 so the recommended TDI level is exceeded.

The Council of European Union proposed maximum levels for dioxin like PCBs in food stuffs (EC Regulation 199/2006). According to this regulation the fisheries products exceeding the proposed values will not be exported to other EU countries (Kivirianta 2005). The proposed WHO-TEq value for fisheries products is 4 pg/g (ww). Although this value is higher than those found in this study for Izmit Bay (0.4±0.1 – 1.4±0.7 pg/g), the amount of daily or weekly consumptions should be considered for risk assessments.

4. Conclusions

This study represents the first findings on PCB pollution in Izmit Bay after the earthquake (17 August 1999). Total PCBs (1–209), sum of six marker congeners (28, 52, 101, 138,153, and 180) and TEq values of mussels were compared with the before earthquake values. June 2000 samples from throughout the Bay contained higher quantities of PCBs when compared with the before earthquake values. The highest contamination of total PCBs (36.0 ng/g) was observed at station 7 in the eastern part of the bay. Despite this value is lower than those found for other industrialized countries, considering the TDI value as 2 pg WHO-TEq/kg bw (COT, 2001) it was concluded that consumable mussels from Izmit Bay pose a risk for health. Although WHO-TEq values found in this study for Izmit Bay (0.4–1.4 pg/g) are below the proposed values for fisheries products (4 pg/g) by EC Regulation 199/2006, the amount of daily or weekly consumptions should be considered for risk assessments. The elevated concentrations of total PCBs in mussels collected after the earthquake imply an input of PCBs to the water column from different sources such as underlying sediments and/or land based sources such as industrial inputs. Further studies should be carried out to estimate those sources and monitoring studies should be continued especially on fisheries products.

Acknowledgments

This study was supported by the Research Centre Julich, Germany, the GSF National Research Center for Environment and Health, Germany, and Turkish Scientific and Technological Research Council (TUBITAK) bilateral cooperation. The Authors wish to thank Enis Morkoç, Vildan Tufekci, and Huseyin Tufekci for their positive criticism; Gulumser Yakupoglu, and Mehtap Celemen for their help in sampling and also Jarmila Kotalik and Torsten Wottgen from GSF for their assistance in the laboratory studies.

References

Ahlborg, U.G., Becking, G.C., Birnbaum, L.S., Brouwer, A., Derks, H.J.G.M., Freely, M., Golor, G., Hanberg, A., Larsen, J.C., Liem, A.K.D., et al. (1994) Toxic equivalency factors for dioxin like PCB's; report on WHO-ECEH and IPCS consultation, Chemosphere 28, 1049–1067.

Algan, O., Altıok, H., Yüce, H., 1999. Seasonal variation of suspended particulate matter in two-layered Izmit Bay, Turkey, Estuarine Coastal Shelf Science 49, 235–250.

Balkis, N., 2003. The effect of Marmara (Izmit) Earthquake on the chemical oceanography of Izmit Bay, Turkey. Marine Pollution Bulletin 46, 865–878.

Burt, J.S. and Ebell, G.F., 1995. Organic pollutants in mussels and sediments of coastal waters of Perth, Western Australia. Marine Pollution Bulletin 30, 723–732.

Coelhan, M., Strohmeier, J., and Barlas, H., 2006. Organochlorine levels in edible fish from the Marmara Sea. Turkey. Environment International 32, 775–780.

COT, 2001. The recommendation for TDI of WHO-TEq of the UK Committee on Toxicity of Chemicals in Food. Consumer Products and Environment.

Granby, K. and Spliid, N., 1995. Hydrocarbon and organochlorines in common mussels from the Kattegat and the belts and their relation to condition indices. Marine Pollution Bulletin 30, 74–82.

Khim, J.S., Villeneuve, D.L., Kannan, K., Hu, W.Y., Giesy, J.P., Kang, S.G., Song, K.J, and Koh, C.H., 2000. Instrumental and bioanalytical measures of persistent organochlorines in blue mussel (*Mytilus edilus*) from Korean coastal waters. Archives of Environmental Contamination and Toxicology 39, 360–368.

Kivirianta, H., 2005. Exposure and Human PCDD/F and PCB body burden in Finland, National Public Health Institute Publications KTL A14, Helsinki-Finland, pp. 17–19.

Karakaş, D. and Pekey, B., 2005. Source apportionment of polycyclic aromatic hydrocarbons in surface sediments of Izmit Bay (Turkey). International Journal of Environmental and Analytical Chemistry 85(7), 433–442.

Khaled, A., Nemr, A.E., Said, T.O., Sikaily, S.A., and Abd-Alla, A.M.A., 2004. Polychlorinated biphenyls and chlorinated pesticides in mussels from the Egyptian Red Sea coast. Chemosphere 54, 1407–1412.

Liu, J.H. and Kueh, C.S.W., 2005. Biomonitoring of heavy metals and trace organics using the intertidal mussel Perna viridis in Hong Kong coastal waters. Marine Pollution Bulletin 51, 857–875.

Martens, D., Balta-Brouma, K., Brotsack, R., Michalke, B., Schramel, P., Klimm, C., et al., 1998. Chemical impact of uncontrolled solid waste combustion to the vicinity of the Kouroupitos Ravine, Crete, Greece. Chemosphere 36, 2855–2866.

Morkoç E., Okay, O.S., Tolun, L., Tufekç V., Tufekç H., and Legoviç T., 2001. Towards a clean Izmit Bay. Environmental International 26, 157–161.

Morkoç E., Legoviç T., Okay, Tufekci, H., Tufekci, V., Tolun, L., and Karakoc, F., 2007. Changes of oceanographic characteristics and the state of pollution in the Izmit bay following the earthquake of 1999. Environmental Geology (in publ.).

Monirith, I., Ueno, D., Tkahashi, S., Nakata, H., Sudaryanto, A., Subramanian, A., Karuppiah, S., Ismail, A., Muchar, M., Zeng, J.S., Tka lin, A.V., and Tanabe, S., 2003. Asia-Pacific mussel watch: monitoring contamination of persistent organochlorine compounds in coastal waters of Asian countries. Marine Pollution Bulletin 46, 281–300.

Nakata, H., Sakai, Y., and Miyawaki, T., 2002. Growth-dependent and species-specific accumulation of polychlorinated biphenyls (PCBs) in tidal flat organisms collected from the Ariake Sea, Japan. Archives of Environmental Contamination and Toxicology 42, 222–228.

Okay, O.S., Tolun, L, Telli-Karakoç, F., Tüfekçi, V., Tüfekçi, H., and Morkoç, E., 2001. İzmit Bay ecosystem after Marmara earthquake and subsequent fire: the long-term data. Marine Pollution Bulletin 42, 361–369.

Okay, O.S., Tolun, L., Telli-Karakoç, F., Tüfekçi, V., Tüfekçi, H., Olgun, A., and Morkoç, E., 2003. The changes of T-PAH levels and health status of mussels in İzmit bay (Turkey) after Marmara earthquake and subsequent refinery fire. Environment International 965, 1–5.

Scawthron, C. and Johnson, G.S., 2000. Preliminary report Kocaeli (Izmit) Earthquake of 17 August 1999. Engineering Structures 22, 727–745.

Sericano, J.L., Wade, T.L., Jackson, T.J., Brooks, J.M., Tripp, B.W., Farrington, J.W., Mee, L.D., Readmann, J.W., Villeneuuve, J.P., and Goldberg, E.D., 1995. Trace organic contamination in the Americas: an overview of the US national status and trends and the international mussel watch programmes. Marine Pollution Bulletin 31, 214–225.

Tanabe, S., Purudente, M.S., Kan-atirekalp, S., and Subramanian, A., 2000. Mussel watch: marine pollution monitoring of butyltins and organochlorides in coastal waters of Thailand, Philippines and India. Ocean and Coastal Management 43, 819–839.

Telli-Karakoc, F., Tolun, L., Henkelmann, B., Klimm, C., Okay, O., and Schramm, K.-W., 2002. Polycyclic aromatic hydrocarbons (PAH) and polychlorinated biphenyls (PCB) distribution in the bay of Marmara Sea: İzmit Bay. Environmental Pollution 119, 383–397.

Tolun, L., Okay, O.S., Gaines, A.F., Tolay, M., Tüfekçi, H., and Kıratlı, N., 2001. The pollution status and toxicity of surface sediments in Izmit Bay (Marmara Sea), Turkey. Environmental International 26, 163–168.

Tolun, L., Okay, O.S., Martens, D., and Schramm, K.W., 2006. Polycyclic aromatic hydrocarbon contamination in coastal sediments of the Izmit Bay (Marmara Sea): case studies before and after the Izmit Earthquake. Environment International 32, 758–765.

Tufekci, K. and Akman, U., 2005. Monitoring the turbidity and surface temperature changes and effects of the 17 August 1999 earthquake in the Izmit Gulf, Turkey By The Landsat Tm/Etm Data, Environmental Monitoring and Assessment, 108, 45–57.

Tugrul, S., Morkoc, E., Okay, O., and Balkaş, T. 1985. Determination of Oceanographic Characteristics and Assimilation Capacity of Izmit Bay, Final Report, NATO TU-WATERS project, TUBITAK, MRI publicatins, pp. 193–259.

Unlu, S. and Alpar, B., 2004. Hydrocarbon balance of surface sediments in Izmit Bay (Marmara Sea) Turkey. Bulletin of Environemntal Contamination and Toxicology 73, 85–92.

USEPA, 1998. Carol Browner's testimony before the Committee on Environmental Conservation of New York State Assembly, 9 July 1998.

Van den Berg, M., Birnbaum, L., Bosveld, Albertus, T.C., Brunström, B., Cook, P., Feeley, M., Giesy, John, P., Hanberg, A., Hasegawa, R., Kennedy, Sean, W., Kubiak, T., Larsen, J.C., Rolaf van Leeuwen, F.X., Djien Lien, A.K., Nolt, C., Peterson, Richard, E., Poellinger, L., Safe, S., Schrenck, D., Tillitt, D., Tysklind, M., Younes, M., Waern, F., and Zacharewski, T., 1998. Toxic equivalency factor (TEFs) for PCBs, PCDDs, PCDFs for Humans and for Wildlife. Environmental Health Perspectives 106, 775–792.

Villenevue, J.P., Carvalho, S.W., and Cattini, F.C., 1999. Levels and trends of PCB's chlorinated pesticides and petroleum hydrocarbons in mussels from the NW Mediterranean Coast: comparison of concentrations in 1973/1974 and 1988/1989. The Science of Total Environment 237–238, 57–65.

Wu, W.Z., Schramm, K.-W., Henkelmann, B., Xu, Y., Yediler, A., and Ketrupp, A., 1997. PCDD/Fs. PCB's and HCB in sediments and soils of Ya-Er Lake area in China: results on residual levels and correlation to the organic carbon and the particle size. Chemosphere 34, 191–202.

WHO, 2000. World Health Organization, Assessment of the health risk of dioxins: re-evaluation of the tolerable daily index(TDI), Food Additives and Contaminants 71 (14), 237.

INVENTORY OF UNINTENTIONAL RELEASES OF POPs IN BELARUS

SERGEY KAKAREKA,* ANNA MALCHYKHINA
Institute for Problems of Natural Resources Use & Ecology, National Academy of Sciences of Belarus, Skoriny 10, 220114, Minsk, Belarus

Abstract: The paper is devoted to results of the latest inventory of unintentional releases of POPs in Belarus. Methodology, input data, and results of inventory are discussed. Releases into the atmospheric air, water, land/soil, products and wastes were estimated. In accordance with the estimates the total annual releases of dioxins/furans in 2004 were 141.8 g I-TEQ. The principle sources of dioxin/furan releases to the environment of Belarus are waste incineration, ferrous and nonferrous metal production, power generation and heating. About 37 g I-TEQ of dioxins/furans are annually released to air, 103.3 g I-TEQ – into residues. Releases to water, to ground, and in products constitute 1.97 g I-TEQ. Releases of HCB to air are estimated at 0.75 kg. Releases of PCB to air make up 9.8 kg. Results of POPs inventory were used for the preparation of National Implementation Plan of the Stockholm Convention of the Republic of Belarus.

Keywords: inventory, POPs, dioxins/furans, HCB, PCB, releases, Belarus

1. Introduction

The inventory of POPs emission (releases) is the key element of POPs sources ecologically sound management, elaboration of emission abatement strategies and emission reduction costs estimation. From the quality and completeness of POPs emission inventory depends on results of all further emission management actions. That's why emission of POPs is regulated by

* To whom correspondence should be addressed. E-mail: kakareka@ns.ecology.ac.by

E. Mehmetli and B. Koumanova (eds.),
The Fate of Persistent Organic Pollutants in the Environment, 141–150.

a few international agreements. Thus unintentional POPs releases – one of obligations of the Parties to the Stockholm Convention. Article 5 of the Stockholm Convention require from parties to reduce or eliminate releases from unintentional production.

The inventory of POPs (dioxins/furans, HCB, PCB, and indicator polyaromatic hydrocarbons) emissions has been taken in Belarus since 2000 in the framework of preparation of the national emission data on pollutant releases for the Co-operative Program for Monitoring and Evaluation of the Long-Range Transmission of Air Pollutants in Europe (EMEP) (Kakareka, 2002; Kakareka et al., 2003, 2004). During implementation of this task main sources of POPs emission in Belarus were revealed, certain experiments on POPs measurement in flue gases were conducted. That works allowed to elaborate emission factors on POPs emission and to estimate in general PCDD/PCDF, PCB, HCB, PAH annual emission fluxes.

However a lot of problems connected with POPs remission remain. Thus only POPs releases to air were taken into consideration; estimates were made for the whole sources categories while detection of key point sources and hot spots are necessary etc. Level of POPs sources knowledge was not enough for their management. In connection with the above mentioned in the framework of the GEF project TF 053865 "Enabling activities related to the implementation of the Stockholm Convention on persistent organic pollutants (POPs) in the Republic of Belarus" the further inventory of unintentional releases of POPs was taken and the register of releases of PCB, PCDD/PCDF and HCB was elaborated. POPs releases not only to air but also to other environmental compartments – water, ground, food products, wastes and residues were identified.

In the course of this inventory the following tasks were implemented:

- Update of the methodology of the establishment of POPs unintentional releases inventory
- Collection of the required statistical, manufacturing and technological data for POPs releases assessment
- Processing of information, estimation of unintentional releases of PCDD/PCDF, HCB and PCB on the territory of the Republic of Belarus to the environment
- Elaboration of the register of unintentional releases of PCDD/PCDF, HCB, and PCB on the territory of Belarus

The year 2004 was assumed as the base year.

The paper includes the summary of the inventory methodology, the procedures, results, and discussion.

2. Methodology

2.1. METHODOLOGICAL BASIS OF THE INVENTORY OF UNINTENTIONAL POPS RELEASES

According to existing rules, enterprises and organizations of the Republic of Belarus are supposed to maintain the initial recording of the composition and magnitude of the releases of hazardous substances; to estimate, the composition and magnitude of the releases of hazardous substances to ambient air and other environment media. The data on the releases of any substances can be presumably obtained from the enterprise reports on releases and the respective summaries. However, in practice the enterprises record only the releases of substances for which the limits of maximum permissible concentration in the air and other media have been established and the methodologies of measuring concentration in releases have been developed and endorsed. Those are not available for the majority of POPs. Therefore, there are almost no statistical data on POPs releases. Apart from that most of the POPs emissions sources are not covered by the state statistical reporting. So international approaches should be used for POPs inventory. There are two basic guidance documents for the establishment of the inventory of PCDD/PCDF:

- Standardized Toolkit for Identification and Quantification of Dioxin and Furan Releases, UNEP Chemicals (2005)
- Atmospheric Emission Inventory Guidebook. A joint EMEP/ CORINAIR Production Prepared by the EMEP Task Force on Emission Inventories (4th edition – 2006)

In the framework of international programs and agreements a number of other methodological guidelines have been elaborated. It is worth mentioning "Technical Paper to the OSPARCOM-HELCOM-UNECE Emission Inventory" (1995).

In Belarus, considerable efforts have been made to improve the methodological framework of the inventory of POPs unintentional releases. In particular, "Methodological recommendations for identification and quantification of the sources of POPs releases" (2004) have been published.

The EMEP/CORINAIR Guidebook contains also chapters relating to the assessment of the PCB and HCB emission to air. Mentioned above "Methodological recommendations" (2004) also were used for the inventory of unintentional releases of PCB.

2.2. PROCEDURE

The POPs inventory is usually based on statistical data describing the scale of the process and emission factors describing the intensity of POPs releases for a unit of activity. The POPs releases inventory (by each group of compounds) has involved the following stages:

- Update of the list of POPs release sources
- Collection of the statistical data on the source activity
- Collection of the technological data, characterizing the processes (parameters of fuels and raw materials, abatement systems, etc.)
- Development (selection) of POPs emission factors
- Calculation of releases and validation of results
- Characteristics of uncertainty of POPs releases estimates
- Identification of hot spots
- Estimation of the spatial structure of releases, identification of areas with the highest level of releases
- Comparison of the outcomes of the inventory in the Republic of Belarus with the POPs inventory outcomes in other countries

The principle informational source was the Ministry of Statistic and Analysis of Belarus. To obtain the production statistics, the key ministries and agencies including the Ministry of Statistics and Analysis, main branch ministries, concerns and departments (Ministry of Housing and Communal Services, the Ministry of Architecture and Construction, the Concern on Oil and Chemistry and others) were also queried.

As the source of POPs emission factors the above mentioned UNEP Standardized Toolkit (2005), EMEP/CORINAIR Guidebook (Atmospheric Emission Inventory Guidebook, 2006) and other guideline documents were used, taking into account technological specificity of Belarus; some HCB and PCB emission factors were obtained by results of sources testing.

To specify POPs emission factors, enterprises relating to the following release source categories were queried: waste incineration; ferrous and nonferrous metal production; production of minerals; production and use of chemicals and consumer goods. For these purposes a special query form was elaborated. It comprised several entries: enterprise profile; data on production volume and consumption of fuels and raw materials; information on emission sources (type of installation, temperature of the flue gases, abatement system, and its efficiency); information on solvents use and on wastes incineration. About 100 responses were obtained from the enterprises which allowed to specify emission factors used for releases estimation.

3. Results

3.1. MAIN POPs EMISSION SOURCE SECTORS

In accordance with the estimates the total annual releases of PCDD/PCDF in 2004 were 141.8 g I-TEQ (the range is 25.2–396.4 g I-TEQ). Distribution of PCDD/PCDF releases by environment media is shown in Figure 1.

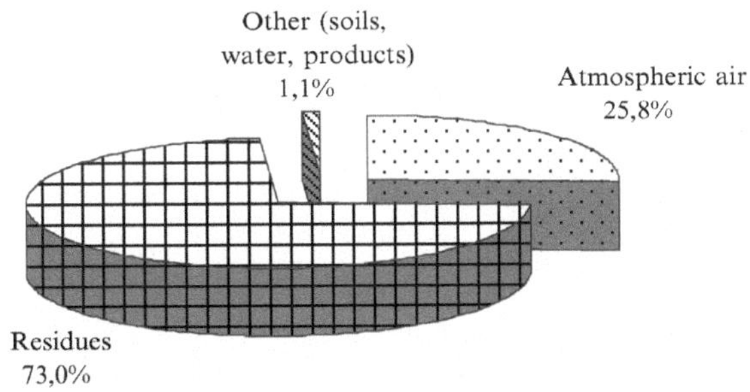

Figure 1. Distribution of dioxin/furan releases into environment in Belarus by types of media

The main sources of dioxin/furan releases to the environment of Belarus are categories Waste Incineration – 47.8%, Ferrous and Non-Ferrous Metal Production – 27.7%, Power Generation and Heating – 11.7%, Disposal/ Landfill – 9.9%. Amount of dioxins/furans annual releases comprised by estimates 36.6 g I-TEQ (the range is 9.1–169.7 g I-TEQ); Waste Incineration generate 46.2% of the total amount of air releases, Ferrous and Non-Ferrous Metal Production produce 22.8% of dioxin/furan releases, and Power Generation and Heating processes make up 23% of the total amount of the dioxin/furan air releases (Figure 2).

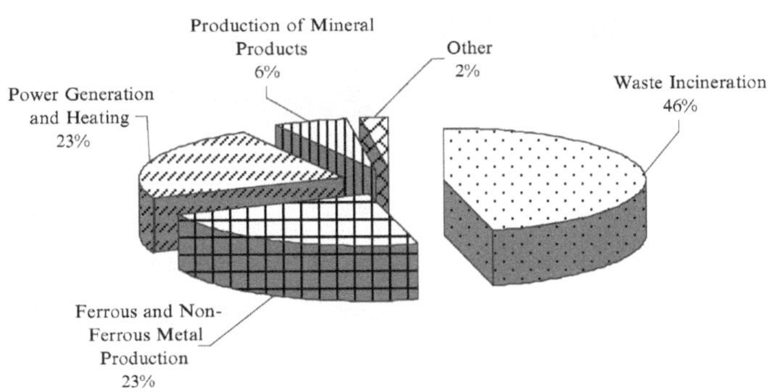

Figure 2. Contribution of sources sectors in total dioxin/furan releases into ambient air in Belarus

Dioxins/furans releases in residues are estimated at 103.3 g I-TEQ (the range is 14.5–220.3 g I-TEQ). The considerable part of releases to residues is produced by source category Waste Incineration – 49.1%, the source category Ferrous and Non-Ferrous Metal Production produces 29.9% of the total amount of releases in residues, 13.1% of releases in residues are produced by source category Disposal/Landfill (Figure 3).

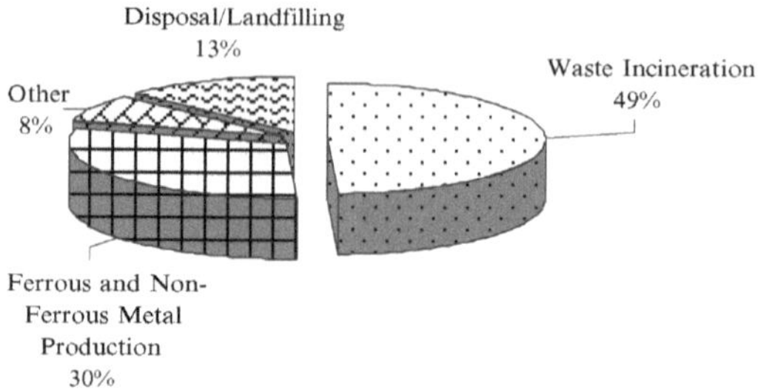

Figure 3. Contribution of sources sectors in total dioxin/furan releases into residues in Belarus

Releases to water, to soils, and in products constitute 1.87 g I-TEQ, at the same time the major part of releases from all these categories is produced by sources of one subcategory (releases to water – from category Disposal/Landfill, to soils – from Uncontrolled Combustion Processes, in products – from Production of Chemicals and Consumer Goods) (Figure 4).

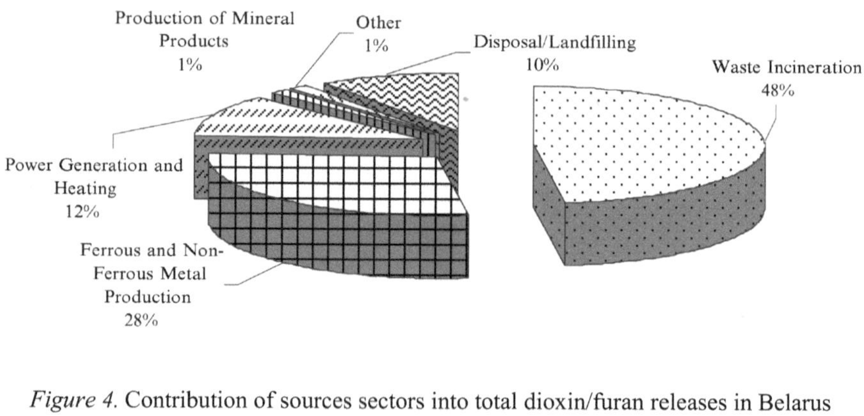

Figure 4. Contribution of sources sectors into total dioxin/furan releases in Belarus

Releases of HCB to air are estimated at 0.75 kg. The major sources are categories Waste Incineration and Transport.

Releases of PCB to air make up 9.8 kg. Principle sources of releases to air are Ferrous and Non-Ferrous Production, Waste Incineration, Power Generation and Heating. However, the amount of unintentional releases of

PCB to air is insignificant if compared to releases from PCB-filled electrical equipment (which considered as emissions from stockpiles). According to preliminary estimates about 1.6 t of PCBs are released annually (mostly to ground).

3.2. STRUCTURE OF THE REGISTER OF UNINTENTIONAL RELEASES OF POPs

Register of unintentional releases of PCDD/PCDF, HCB and PCB is a detailed list of process categories identified as sources of releases of these substances, with quantification of the releases of POPs into different compartments. The register of unintentional releases of POPs is compiled in the format MS Excel on the basis of the calculation module for releases of dioxins and furans (prepared by UNEP Chemicals). It includes the sheets for 10 source categories and a summary sheet. There is an additional sheet for calculation of releases of HCB and PCB.

Every calculation sheet contains production statistical data for the given source category, emission factors, and the obtained estimates of releases.

3.3. IDENTIFICATION OF MAJOR POINT SOURCES (HOT SPOTS) OF POPs RELEASES

The identification of the hot spots of POPs emission is based on the criteria set forth in the UNEP, EMEP, and other guidelines. The release threshold, activity level, the capacity threshold or the use of a certain process can be taken as the criteria of point source. In terms of contribution to the global releases, there are virtually no large point sources of POPs releases in Belarus in a "traditional" sense; chemical plants producing herbicides, domestic waste incineration plants, sinter plants that may release tens and hundreds grams of PCDD/PCDF per year (Fedorov, 1993; Dioxins in Russia, 2001). Given the production and technological specifics of Belarus, we have used a threshold of a plant's potential contribution to the gross releases of POPs of 1 g I-TEQ per year as the key criterion for classifying a plant as a major point source, 0.1 g I-TEQ per year for classifying a plant as a point source.

As the potential major point sources of dioxins/furans releases in category Waste Incineration are considered 12 enterprises. Thirty-four enterprises are considered as point sources. The list of point sources in this category is not stable and varies from year to year depending on incineration volumes. A special attention should be paid to medical waste incineration despite the relatively small volumes of incineration. The main point source of dioxin/furan releases in the category Ferrous and Non-Ferrous Metal

Production is Belarusian Metallurgic Plant, with the total volume of releases of 30–35 g I-TEQ. Six enterprises are classified as point sources.

The most widely used fuel at power plants of Belarus is natural gas, thus according to the preliminary estimates in Belarus there are no power generation plants with PCDD/PCDF releases more than 0.1 g I-TEQ per year. Point sources in this category were not identified.

Major point sources in category Production of Minerals have not been identified. Three enterprises have been classified as point sources. In category Miscellaneous the only source of dioxin/furan releases is cremation. The only point source in this category in Belarus is the Minsk crematorium.

On the whole in all sectors 16 major point sources and about 40 point sources of dioxins/furans releases have been identified.

Level of dioxins/furans releases by regions have been estimated: the highest level of unintentional releases of dioxins/furans is typical for the Gomel region.

For PCB and HCB no point sources of air emission have been allocated: Most of PCB hot spots are connected with places of PCB-containing equipment installations which were considered as emissions from stockpiles).

4. Discussion

Results of POPs inventory were compared with results of POPs inventories in other countries. EMEP database as most complete POPs emission database was used (Webdab, 2006). As basis official data reported for 2003 was considered.

For 2003 information on dioxin/furan emission was submitted by 26 countries. Annual PCDD/PCDF air emissions amounted in Europe 2,148.3 g I-I-TEQ. Main sources of emission – Fuel combustion in Industry (together with related contact processes) – 42% and Waste Combustion – 14.4%.

Greatest input among countries into total European emissions are of Great Britain, France, and Bulgaria (common – 36.7%); contribution of Belarus – is about 1.5%.

Information on HCB emission was reported by 15 countries. Total emission in 2003 amounted 8,372.4 kg. Main sources of HCB emission – Waste Combustion (71%), Industrial combustion (14.6%). The greatest contribution are from France – 73.4% and Spain – 21%. Contribution of Belarus into European air emissions is less than 0.01%.

PCB emission data was reported in 2003 by 13 European countries. Total PCB emission amounted 2,391.6 kg. Main sources: Industrial Processes (49.7%) and Combustion in Industry (15%). Greatest contributor

– Great Britain (57.6%). Contribution of Belarus into European PCB emission – 0.4%.

On the whole, analysis have shown that state of POPs inventory in countries is rather different. Especially this is characteristic for PCB and HCB inventories.

5. Conclusion

In accordance with the estimates the total annual releases of PCDD/PCDF in 2004 were 141.8 g I-TEQ. The principle sources of dioxin/furan releases to the environment of Belarus are such categories as Waste Incineration, Ferrous and Non-Ferrous Metal Production, Power Generation and Heating, Disposal/Landfill.

The major sources of HCB releases to air are waste incineration and transport. Principle sources of PCB releases to air are ferrous and nonferrous production, waste incineration, power generation, and heating. However, the amount of unintentional releases of PCB to air is insignificant if compared to releases from PCB-filled electrical equipment: according to preliminary estimates about 1.6 t of PCBs are released annually (mostly to ground).

The input of Belarus to European releases of dioxins/furans to air makes up about 1.5%, releases of HCB – about 0.01%, PCB – about 0.4%.

One of the outcomes of the inventory of unintentional releases of POPs is the identification of the point sources of PCDD/PCDF releases.

The established inventory of unintentional releases of POPs forms the basis for elaboration of the section on unintentional releases of POPs of the National Implementation Plan of the Stockholm Convention.

Acknowledgments

Authors express their gratitude to GEF and World Bank for funding of the inventory in Belarus, to the leaders and participants of the GEF project TF 053865, to all ministries, agencies and experts for support of this work and provision of information.

References

Atmospheric Emission Inventory Guidebook, 2006, A joint EMEP/CORINAIR Production Prepared by the EMEP Task Force on Emission Inventories.

Dioxin and Furan Inventories, 1999, National and Regional Emissions of PCDD/PCDF. UNEP Chemicals. Geneva.

Dioxins in Russia, 2001, ed. by Kurljandsky, B.A. Moscow: Neftjnnik (in Russian).

Fedorov, L.A., 1993, Dioxins as fundamental factor of technogenic pollution of environment, *Magazine of Ecological Chemistry*, 3, 169–187 (in Russian).

Kakareka, S., 2002, Sources of persistent organic pollutants emission on the territory of Belarus, *Atmospheric Environment*, 36, 1407–1419.

Kakareka, S. and Kukharchyk, T., 2004, Assessment of dioxins/furans emission on the territory of Belarus, *Natural Resources*, 1, 79–86 (in Russian).

Kakareka, S., Kukharchyk, T., and Khomich, V., 2003, *Persistent Organic Pollutants: Sources and Emission Assessment*. Minsk: RUE Minsktipproekt (in Russian).

Methodological recommendations for identification and quantification of the sources of POPs releases, 2004, Compilation of legislative documents on environmental protection. Minsk, 47, pp. 3–63 (in Russian).

Standardized Toolkit for Identification and Quantification of Dioxin and Furan Releases, 2005, UNEP Chemicals.

Stockholm Convention on Persistent Organic Pollutants, 2001, UNEP, Geneva.

Technical Paper to the OSPARCOM-HELCOM-UNECE Emission Inventory, 1995, TNO-report, TNO-MEPR95/ 247.

The National plan of the Republic of Belarus for the Implementation of its Obligations under the Stockholm Convention on Persistent Organic pollutants for the Period of 2007–2010 and until 2028, 2006, Minsk, p. 183.

WebDab EMEP activity data and emission database, 2006, http://webdab.emep.int

POLYCHLORINATED BIPHENYLS (PCBs) IN BELARUS: INVENTORY AND PRIORITIES FOR ECOLOGICALLY SOUND MANAGEMENT

TAMARA KUKHARCHYK*
Institute for Problems of Natural Resources Use and Ecology of National Academy of Sciences of Belarus, Skoriny Street 10, Minsk, 220114, Belarus

Abstract: In the paper the results of polychlorinated biphenyls (PCBs) inventory in Belarus are discussed. The approaches and methods used for PCB-containing equipment identification are given. Volumes of PCBs and types of PCB-containing equipment than are still in and out of operation are assessed. Distribution of PCBs in Belarus among ministries, economy sectors and regions is shown. Assessment of PCB-containing equipment condition and peculiarities of their storage are done. The problems of PCB-containing equipment storage and replacement, as well as measures for ecologically sound management in Belarus are given.

Keywords: polychlorinated biphenyls, inventory, PCB-containing equipment, ecologically sound management

1. Introduction

Polychlorinated biphenyls (PCBs) are one of the groups of persistent organic pollutants (POPs) which are belong to the most dangerous chemicals in the world and are regulated by several international agreements. The Stockholm Convention on POPs is the most important among them. To protect human health and the environment, Parties of the Convention shall eliminate the usage of polychlorinated biphenyls in equipment by 2025, and dispose of PCBs in an ecologically sound manner as soon as possible but no later than

*To whom correspondence should be addressed. E-mail: kukharchyk@mail.ru

E. Mehmetli and B. Koumanova (eds.),
The Fate of Persistent Organic Pollutants in the Environment, 151–160.

2028 (The Stockholm Convention, 2001). Belarus acceded to the Stockholm Convention on POPs in February 2004 in accordance with Decree of the President of the Republic of Belarus of 23 December 2003 No. 594.

PCB inventory is one of the obligations of Parties and the key element of efficient PCB management. In accordance with the priorities identified in Part 2 of Annex A to the Stockholm Convention on POPs, determined efforts should be made to identify, label and remove from use equipment containing greater than 10% polychlorinated biphenyls and volumes greater than 5 l, equipment containing greater than 0.05% polychlorinated biphenyls and volumes greater than 5 l (The Stockholm Convention, 2001). Endeavor should be also made to identify and remove from use equipment containing greater than 0.005% polychlorinated biphenyls and volumes greater than 0.05 l.

At present PCB inventory is fulfilled in many countries of the world. It should be noticed that this process has been actively developed in the end of 1990s and the beginning 2000s (AMAP, 2000; ADEME, 2003; Republic of Moldova, 2004; Republic of Croatia, 2004; Enabling activities, 2005; CLEEN, 2005; Persistent Organic, 2005). In most cases questionnaires for PCB accounting are used. In spite of the long history of PCB management (restrictions of use, incineration, etc.) only in some countries (e.g., in the USA and Canada) strict rules on PCB recordkeeping and reporting were adapted (National Inventory, 1995; 40 CFR, 2002).

PCB inventories in Belarus were undertaken in 2003–2004 in the framework of the State Scientific Technical Program "Environmental Safety" and in 2005 in the framework of the project GEF TF 053865. Some findings of the PCB inventory in Belarus, as well as priority measures for PCB management are discussed in this paper.

2. Methods and Objects

The inventory was carried out based on the following principles:

- The most complete identification of power capacitors and transformers taking into account their types, ways of use, location, and condition
- The most complete coverage of economy sectors for the purpose of identification of potential owners of PCB-containing equipment
- Filling in of preliminary reporting forms by the enterprises–owners of PCB-containing equipment

The PCB inventory was carried out by means of sending out inquiries, accompanied by a set of documents, including "Methodological guidance for PCB inventory taking," "PCB report form" and "Guidelines on filling the report form."

Inventory form were designed to account of PCB-containing equipment (pieces) with determination of its trademark, volume of PCBs, year of production, condition of equipment, level of operation, description of the place of installation, or storage.

To facilitate the identification of PCB-containing equipment the lists with the brands of power capacitors and transformers, produced at the enterprises of the former USSR and in other countries were compiled. This was done on the basis of reference guides on electrical equipment, state standards and technical specification for this equipment, literature sources (PCB Guidelines, 1999; PCB Transformers and Capacitors, 2002; UK Guidance, 2002; AMAP, 2000; Kondensatory, w ktorych, 2002, etc.), as well as full-scale studies of electrical equipment at a number of enterprises of Belarus.

More than 2,000 enterprises were covered by the inventory process. They present the main part of industry and energy sectors, a railway, some part of a communal sector and agriculture. A special attention has been given to the enterprises of food industry and the provender milling enterprises, since there is a risk of contamination of the products with PCBs.

Methodological and information provision of the inventory was performed at all stages. Reports from enterprises were collected and carefully analyzed for detection of uncertainties and errors.

In the course of projects implementation 762 enterprises – owners of PCB-containing equipment were revealed (Table 1).

TABLE 1. General data on PCB inventory in Belarus

Indicator	Number
Enterprises that reported the absence of PCB-containing equipment	1,351
Enterprises identified as owners of PCB-containing equipment including	762
Enterprises that have power capacitors with PCBs	748
Enterprises that have power transformers with PCBs	44
Enterprises that have barrels with PCBs	4

3. Results and Discussion

3.1. PCB VOLUMES IN BELARUS

According to the results of the inventory, in Belarus there are: 380 power transformers filled up with Sovtol-10 or similar liquids, about 47,000 PCB-filled power capacitors, 29 containers with PCB-containing dielectric liquids, about 40,000 small-size capacitors (Kukharchyk, 2006; The National plan, 2006). The total amount of the identified polychlorinated biphenyls in liquid form is estimated to be 1,564 t. The obtained data shows, that 55% of the total amount of PCBs is contained in power transformers and 44% – in power capacitors.

The most widely spread type of a transformer in Belarus (89% of the total amount) was manufactured at Chirchik Transformer Plant (Uzbekistan). 80% of power capacitors identified in Belarus were produced at Serpukhov Capacitor Plant (Russia) and Ust-Kamenogorsk Capacitor Plant (Kazakhstan). 20% of capacitors are of foreign origin.

PCB-containing equipment is installed at enterprises of various types: machine-building and metal processing, power industry, chemical and petroleum industry, light and food industries, housing and communal services, and many others (Table 2).

TABLE 2. PCB volumes at enterprises of different economy sectors, tons

Economy sector	In power capacitors	In power transformers	In barrels	Total
Machine-building and metal processing	294.9	323.5	3.4	621.8
Chemical and petroleum industry	75.4	306.2	4.4	386.0
Power industry	120.0	0.8		120.9
Light industry	51.6	46.9		98.5
Microbiological industry	2.2	91.2		93.4
Food industry	34.7	14.8		49.6
Wood-working, pulp, and paper industry	23.0	16.1		39.1
Ferrous industry	2.8	30.8		33.5
Building material industry	19.8	5.5		25.3
Transport and communication	20.2	2.8		23.1
Agriculture	1.3	18.6		19.9
Housing and communal services	15.1	2.0		17.1
Others	32.2	3.5		35.5
Total	693.2	862.7	7.8	1,563.7

Enterprises of the machine-building complex as well as chemical and petroleum industry have the largest amount of PCBs: 621.8 t or 40% of the totals and 386 t or 25%, respectively. The share of food industry makes up to about 3 per cent of the total amount of PCBs (49.6 t).

Enterprises which have the PCB-containing equipment belong to different ministries and agencies (about 30 altogether), among which Ministry of Industry and Oil and Chemistry Agency are the largest owners (Figure 1).

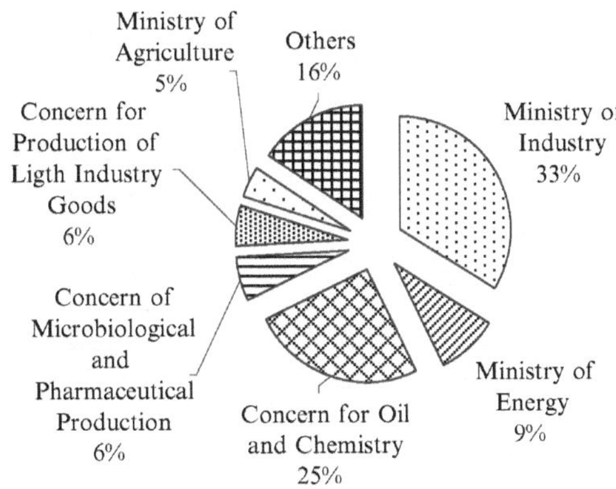

Figure 1. Distribution of PCB-containing equipment by ministries and agencies in Belarus

It was established that enterprises that have PCB-containing equipment are dispersed on the whole territory of the country. There are about 140 cities and towns where PCB-containing equipment is used. The largest volumes of PCBs are concentrated in three cities: Bobruysk, Minsk and Novopolotsk, where 44.4% of the total volumes of PCBs are found. There are 16 settlements with PCB volumes more than 20 t (Figure 2).

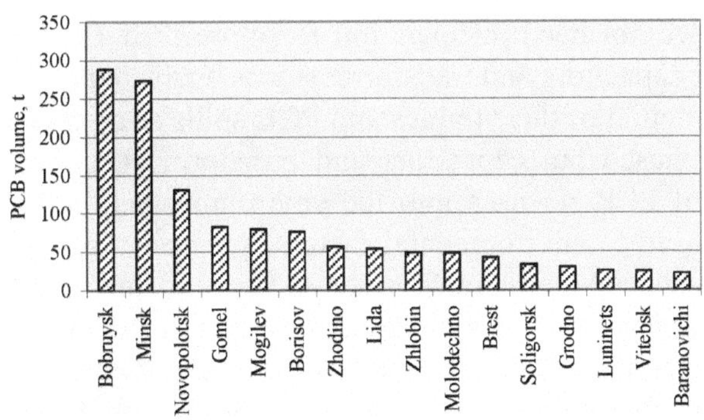

Figure 2. PCB volumes in cities and towns of Belarus

3.2. ASSESSMENT OF PCB-CONTAINING EQUIPMENT CONDITION AND PECUILARITIES OF THEIR STORAGE

About 43% of capacitors and 54% of transformers used in Belarus were manufactured before 1980, 44% and 20%, correspondingly – in the period between 1981 and 1986 (Figure 3). It means that significant part of PCB-containing equipment has already exhausted its operational capacity and should be removed from operation and replacement by non-PCB equipment.

a b

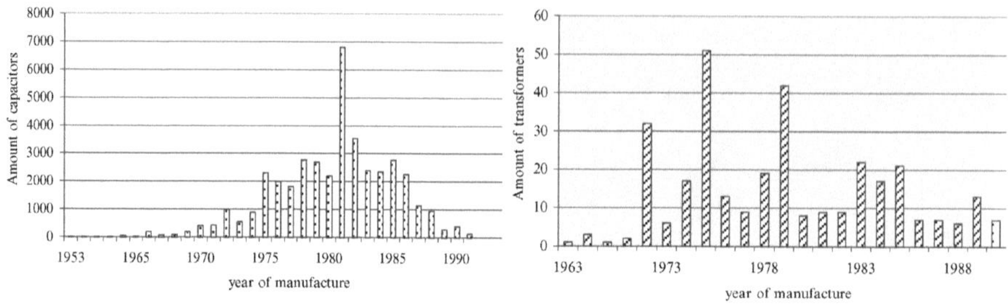

Figure 3. Distribution of PCB-containing capacitors (a) and PCB-containing transformers (b) by year of manufacture

According to the inventory, by now 27% of capacitors and 9% of transformers have been removed from service. The total amount of PCBs in the phased out equipment is estimated at 270 t. Data obtained shows, that in most cases PCB-containing equipment removed from operation was often stored in the open air (at substations and territory of enterprises) and in some cases – in storehouses. Quite often, removal of equipment from operation did not mean its dismantling. Even now in many cases capacitors are still kept within the banks of capacitors and capacitor installations, though they have not been in operation for more than 10 years. The condition of power capacitors and transformers was evaluated as unsatisfactory, due to the destruction of their frames and PCB spills as a consequence.

When the phased out capacitors and transformers are stored at open sites, the risk of PCB releases into the environment is the highest. Being constantly exposed to environmental factors (precipitation, air temperature perturbation), capacitor and transformer shells tend to corrode and cause PCB leakages (capacitor shells are most prone to corrosion). Investigation shows that PCB content in soils of the territories and impact zones of places for electrical equipment use and storage makes up to milligrams or sometime grams per kilogram. Maximum concentrations of PCBs were tracked in soils near destroyed capacitors and transformers as a consequence of PCB leakages (Kukharchyk, 2006). This underscores the need for soil clean up and remediation. Leakage of polychlorinated biphenyls from damaged equipment poses a risk of subsequent distribution of polychlorinated biphenyls in the broader environment, as well as the risk of water and air pollution and accumulation of polychlorinated biphenyls in biota and food products.

For the last 2 years, considerable effort has been made to improve the organization of PCB-containing equipment storage: PCB-containing capacitors stored in the open air were transported to storehouses; more that 3,000 damaged capacitors were packed into containers. The containers are very different in construction, size, capacity, etc. (Figure 4). Some of them were previously used for other purposes (e.g., as transformer tank, or electricity distribution plant) and now are not suitable for damaged capacitors storage and thus do not prevent PCB leakages.

Figure 4. Types of containers for PCB-containing capacitors storage

4. Priority Measures for Ecologically Sound Management of PCBs in Belarus

PCB management in Belarus and in other former Soviet Union countries is a very complicated problem. It can be explained by the following reasons: PCBs were not taken into consideration as dangerous chemicals until recent times; registration of PCB-containing equipment is not organized properly; there are no legislative documents on PCB management; there are no requirements and standards regulating arrangements for special storehouses for the damaged equipment and PCB-containing wastes; there is no control of compliance with the rules of environmentally sound management of such equipment.

Meanwhile, the necessity of measures for prevention of PCB spills and its releases into environment is evident. For these purposes the Recommendation for Prevention of Environment Pollution by Polychlorinated Biphenyls was developed (Recommendation for Prevention, 2006). This recommendation is based on experiences and achievements of other countries in the sphere of PCB management (40 CFR, 2002; Regulation (EC) 850/2004, 2004; Polychlorinated Biphenyls, 2001; Polychlorinated Biphenyls, 1999; Management of Polychlorinated, 2001; UK Guidance, 2002; Safe Management, 2005, etc.). There are: the procedure of PCB-containing equipment identification, labeling and accounting; inspection of PCB-containing equipment; priority activities during damaged equipment revelation and PCB spills; requirements for temporary PCB-containing equipment storage.

In the period of 2004–2006 the Ministry of Natural Resources and Environmental Protection in cooperation with the World Bank undertook a development of a formal National Plan, aimed at addressing the POPs issue in Belarus (The National Plan, 2006). Taking into account the current situation with PCBs in Belarus and obligation of Stockholm Convention on POPs, the following measures on PCB management in Belarus are suggested:

- Elaboration of methodological documents regulating PCB management
- Organization of environmentally sound storage of the phased out electrical equipment in order to prevent PCB leakages
- Labeling of PCB-containing equipment and organization of its primary registration at enterprises
- Ongoing inventory of PCBs
- Elaboration of a plan on PCB-containing equipment replacement
- Choice of the environmentally sound and economically efficient technology of PCB disposal

5. Conclusion

PCB investigations, which started in the late 1990s, are now developing very intensive, that can be explained the fact that Belarus acceded to the Stockholm Convention on POPs. The findings of the PCB inventory and the database of PCBs represent information base for PCB management in Belarus. First steps to improve the condition of damaged PCB-containing equipment storage are being done. However, the implementation of ecologically sound management of PCBs, and the reduction of adverse effect of PCBs on human health remain a difficult and enduring task, which requires improvement of Belarusian legal system, creation of the necessary infrastructure, development of chemical analytical investigation and PCB monitoring in the environment, increasing public awareness of the dangerous chemicals and elaboration of researches.

Acknowledgments

This work would not be possible without the participation of many specialists from the Institute for Problems of Natural Resources Use and Ecology of the National Academy of Sciences of Belarus. The author is very grateful to all colleagues who were involved in all stages of this investigation. The great appreciation to GEF and World Bank for funding this work as well as to all experts from the Ministry of Natural Resources and Environmental Protection of the Republic of Belarus, Ministry of Industry, Ministry of Energy and others one for their support and assistance.

References

40 CFR. Part 761 – Polychlorinated biphenyls (PCBs): manufacturing, processing, distribution in commerce, and use prohibitions, 2002. Environmental Protection Agency. USA; http:// www.epa.gov

ADEME, 2003. Plan national de decontamination et d'elimination des appareils contenant des PCB et PCT. Approuve par arrete du 26 Fevrier 2003. Ministère de l'écologie et du développement durable – ADEME (in France).

AMAP, 2000. PCB in the Russian Federation: inventory and proposals for priority remedial actions, AMAP report. ISBN 82-7971-008-6. Oslo. Centre for International Projects (CIP), Moscow.

CLEEN (Chemical Legislation Enforcement Network), 2005. EuroPCB: inventory PCB enforcement in member states. Part 1, final.

Enabling activities to facilitate early action on the implementation of the Stockholm Convention on persistent organic pollutants (POPs), 2004. Poland National implementation plan for the Stockholm Convention, 2004. Project N GF/POL/01/004. Warsaw.

Guidelines for the Identification of PCBs and Materials Containing PCBs, 1999. First issue. Prepared by UNEP Chemicals. United Nations, UNEP.

Kondensatory, w ktorych mogl y byc wykorzystane PCB, 2002 / Załacznik do Rozporzadzenia Ministra Gospodarki z dnia 26 wrzesnia 2002r. (poz.1416) // Dziennik Ustaw, Warszawa, dnia 17 pazdziernika 2002 r. – № 173, Poz. 1416.

Kukharchyk, T. *Polychlorinated Biphenyls in Belarus*. Minsk: Minsktipproekt, 2006. p. 264 (in Russian).

Management of polychlorinated biphenyls, 2001. Document 14.14. Environmental, Safety and Health Manual. Revision Definitions//www.llnl.gov.

National inventory of PCBs in use and PCB wastes in storage in Canada, 1995. Annual report. Commercial Chemicals Evaluation Branch Environmental Protection Service Environment Canada.

PCB transformers and capacitors, 2002. From management to reclassification and disposal. United Nations Environment Programme. First Issue. Prepared by UNEP Chemicals, Geneva, Switzerland.

Persistent organic pollutants: national implementation plan for the Republic of Latvia 2005–2020, 2005. Adopted by the cabinet of ministers of the Republic of Latvia on 31 March 2005, Riga

Polychlorinated biphenyls (PCBs), 2001. HELSINKI COMMISSION Baltic Marine Environment Protection Commission. Submitted by the Co-ordinator of the HELCOM project team on hazardous substances Christine Full, p. 24.

Polychlorinated biphenyls management plan, 1999. Australian and New Zealand Environment and Conservation Council, 1996. Revised 1999. Waste Management Secretariat Environment Australia, Canberra, Australia, p. 19.

Recommendation for Prevention of Environment Pollution by Polychlorinated Biphenyls, 2006/Kukharchyk, T., Kakareka, S., and Khomich, V.; Ministry of Natural Resources and Environmental Protection. Minsk: RUP Bel NIC ECOLOGY, p. 24 (In Russian).

Regulation (EC) No 850/2004 of the European Parliament and of the Council of 29 April 2004 on Persistent Organic Pollutants and Amending Directive 79/117/EEC, *Official Journal of the European Union*. L 158 of 30 April 2004, pp. 229/5–229/22.

Republic of Croatia, 2004. National Implementation Plan for the Stockholm Convention / Croatian Cleaner Production Centre.

Republic of Moldova, 2004. National implementation plan for the Stockholm Convention on persistent organic pollutants, Chisinau.

Safe management of PCBs, 2005, byFiedler, H.//www.unep.chemical/ pcbs/ regulation

Technical guidelines on wastes comprising or containing PCBs, PCTs and PBBs. Prepared by the Technical Working Group of the Basel Convention and Adapted by the Second Meeting of the Conference of the Parties to the Basel Convention in March 1994, Geneva. Basel Convention Series/SBC N: 97/009, Geneva, February 1997.

The National plan of the Republic of Belarus for the implementation of its obligations under the Stockholm Convention on persistent organic pollutants for the period of 2007–2010 and until 2028, 2006. Ministry of natural resources and environmental protection of the Republic of Belarus, Global Environment Facility, World Bank. Minsk, Belarus. p. 183.

The Stockholm convention on persistent organic pollutants, 2001. UNEP, Geneva.

UK Guidance, 2002. Collection and disposal of equipment containing small amounts of PCBs. Scottish Executive Environment Group. Paper 2002/7.

THE FATE OF PERSISTENT ORGANIC POLLUTANTS (POPs) IN THE ENVIRONMENT

ANNA CUMANOVA*, GAVRIL GILCA, NADIA ORLOVA
State Hydrometeorological Service, 193 Grenoble Street,
2043 Chisinau, Republic of Moldova

Abstract: In this paper is presented Monitoring Data of Persistent Organic Pollutants for period more than 25 years performed in the Monitoring Department on Environmental Quality that is part of State Hydrometeorological Service.

Keywords: dioxins/furans, polychlorinated biphenyls (PCBs), organochlorinated pesticides (OCPs), maximum allowable concentration (MAC), oriented allowable concentration (OAC), State Hydrometeorological Service (SHS), Republic of Moldova

1. Introduction

Concerning the Article 1 the objective of the Stockholm Convention is to protect human health and the environment from Persistent Organic Pollutants (12 POPs listed in Annex A, B, C – Aldrin, Chlordane, Dieldrin, Endrin, Heptachlor, Hexachlorobenzene (HCB), Mirex, Toxaphene, polychlorinated biphenyls (PCBs), 1,1,1-trichloro-2,2-bis(4-chlorophenyl)ethane DDT, polychlorinated-*p*-dioxins and dibenzofurans, and hexachlorobenzene).[1]

The Republic of Moldova has recognized and is aware of the problem of chemical hazards, and considers the health and environmental concerns that they pose as a high priority for actions, in particular the importance of POPs. The Republic of Moldova signed the Stockholm Convention on 23 May 2001 and ratified it on 19 February 2004.

* To whom correspondence should be addressed. E-mail: cumanova@yahoo.com

E. Mehmetli and B. Koumanova (eds.),
The Fate of Persistent Organic Pollutants in the Environment, 161–171.

2. National Policy

The national policy regarding POPs chemicals is driven by understanding that **national chemical safety management system** needs to be created to prevent degradation and societal, particularly health, impact. Minimization and final elimination of POPs related pressure and impact to the natural and human environment is an integral part of national environmental policy.

The policy has **two key** management objectives:

- Remediation and virtual elimination from the environment of POPs that result predominantly from human activity and that are persistent and bioaccumulative
- Management of other toxic substances and chemicals, throughout their entire life cycles, to avoid, prevent, or minimize their release into the environment

Research and development strategy will have to be target and phased, showing a high degree of flexibility, in order to be easily adapted as new data will be gathered. Develop **environmental and health-oriented monitoring strategies** and step-by-step implementation since 2005, beginning with the priority areas, zones, and concerns.[2]

3. Background

3.1. COUNTRY CONTEXT

The Republic of Moldova is a small landlocked country of approximately 33,840 km^2, geographically situated near the northwest section of Black Sea between the Ukraine and Romania (Figure 1). It is strategically located at the crossroads between Central, Southeast, and Eastern Europe. Moldova is one of newly independent European States. The biggest rivers in Moldova are the Nistru and Prut. The capital of Moldova is Chisinau with population about 700,000 people. The total population of Moldova is about 4.3 million people. Moldova has no significant mineral deposits, major are: cement components and gypsum, limestone, sand and raw materials for bricks and tiles. Natural oil and brown coal deposits have been discovered in South. However, the Republic's greatest natural recourse is its fertile "chernozem" soil (the world's most fertile soil) which covers nearly 75% of its territory, 86% of which is being utilized by agriculture. The major industries in

Moldova are agriculture, viniculture and food processing (fruits and vegetables). The country is an important regional producer of grapes and grape products, and its orchards produce significant amounts of fruit. Also sugar beets, sunflowers, and cereals, including wheat is grown in Moldova.[3] As Moldova is agriculture country it follows that, to protect crops in agriculture are used the different chemical plants protectants (pesticides).

Figure 1. Map of Republic of Moldova[4]

3.2. THE PESTICIDE HISTORY

The Republic of Moldova has never had, and does not currently have, pesticide-producing enterprises or factories, all agrochemicals for plant protection permitted for use in the country have been and are imported from abroad. Pesticides, including POPs, were used in large quantities in the past in Moldova (Figure 2). During 1950–1990 an estimated total amount of **560,000 t of pesticides** were used in Moldova including **22,000 t of**

organochlorinated pesticides (OCPs). Pesticides use registered a peak in 1975–1985 but reduced dramatically over the last 10–12 years (from 38,300 t in 1984 to some 2,800 t in 2,000 as active ingredient).

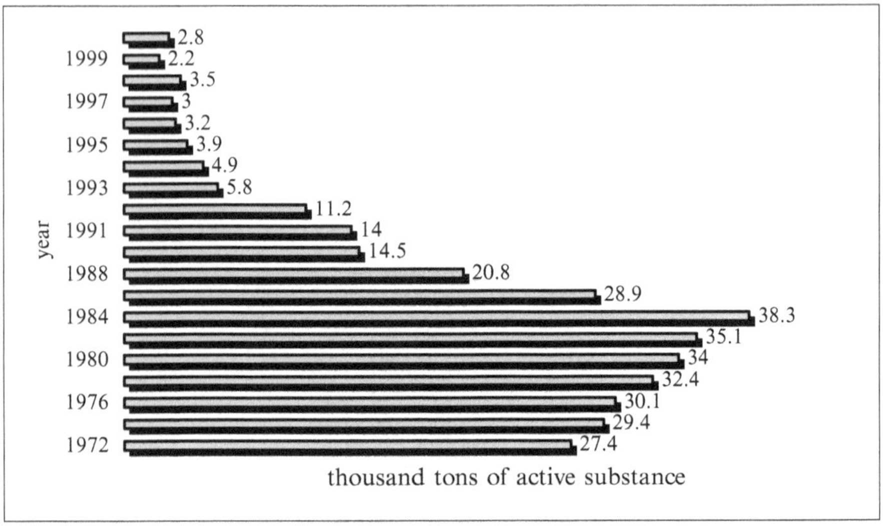

Figure 2. Pesticide use in Republic of Moldova

The absence in the past of controls on pesticides imports, storage and use have resulted in the stockpiling of banned and useless pesticides, and the pesticide dump was built in 1978 in 10 km to the south from the city of Vulkanesti nearly the village Cismichioi in the South of Moldova. Over a period of 10 years (1978–1988) 3,940 t of pesticides were buried there, including 654.1 t of DDT. Also 1,712 t of pesticides were stored in 344 warehouses in country site.

3.3. THE PCBs STORY

PCBs in Moldova are primarily used in the energy sector as dielectric fluids in power installations, especially transformers and capacitors. The resent POPs inventory, undertaken in the context of development of national implementation plan (NIP) for POPs, identified approximately 26,300 transformers, 17,000 capacitors, and other electro-energy equipment containing a total amount of dielectric oils of approximately 23,900 t much of which is expected to contain PCBs. The approximate breakdown reported is: capacitors from the electric industry – 380 t; other capacitors – 20–50 t; electrical transformers – 18,000 t; and, switches – 5,400 t for a total of approximately 23,900 t. Much of this equipment is no longer in use. Most of capacitors are concentrated at the Vulcanesti substation in the south of Republic of Moldova.[5]

4. POPs Monitoring in Republic of Moldova

According to Monitoring Program the determination of OCPs such as DDTs and HCHs are performed in soil from agricultural areas more than 25 years. Pesticides from stockpiles with obsolete pesticides and oil in energy installation were identified as sources of pollution of soil and other media. Investigation of soil around the stockpiles with obsolete pesticides, soil around pesticide dumps and soil around the electric power installations and sediments from main water bodies was included in Monitoring Program of State Hydrometeorological Service. There is no investigation regarding dioxins/furans because of lacking the HR/GC HR/MS.

4.1. OCPs CONTENTS

4.1.1. OCPs Contents in Agricultural Soil

Organochlorinated pesticides such as DDTs, HCHs, and lindane have been widely used in the 1960s and 1970s. High persistency, accumulation, and global migration to the long distances are the reason for their research in environment. Norm (MAC and OAC) of DDTs and HCHs were established and is 0.1 mg/kg (Figures 3 and 4). The application of DDT was banned in 1973–1974. Activities for measuring of contents of OCPs include DDTs (*p-p*'DDT, *p-p*'DDE, *p-p*'DDD) and HCHs (alpha-, beta-, and gamma-HCH). Largest mean value of HCHs was detected in soil in 1981 with concentration of 0.683 mg/kg (6.83 exceed norm), and DDTs – in 1983 with concentration of 1.324 mg/kg (13.24 exceed norm). Since the 1990s the results demonstrate the significant decrease of OCPs in soil. During 1992–2005 findings have shown that the mean contents of DDTs are not high and do not exceed the norm and varies from 0.009 to 0.088 mg/kg, except some particular and isolated cases.

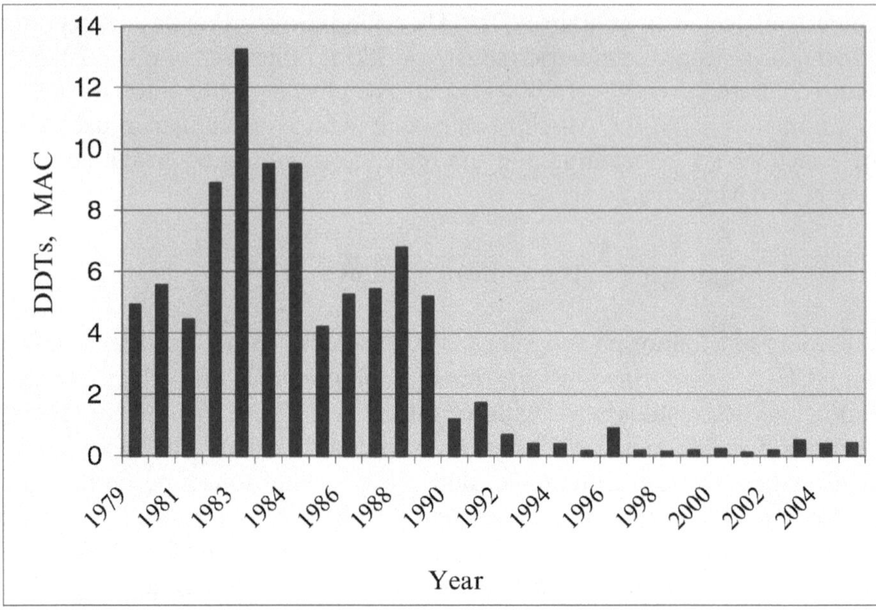

Figure 3. Mean contents of DDTs in agricultural soil

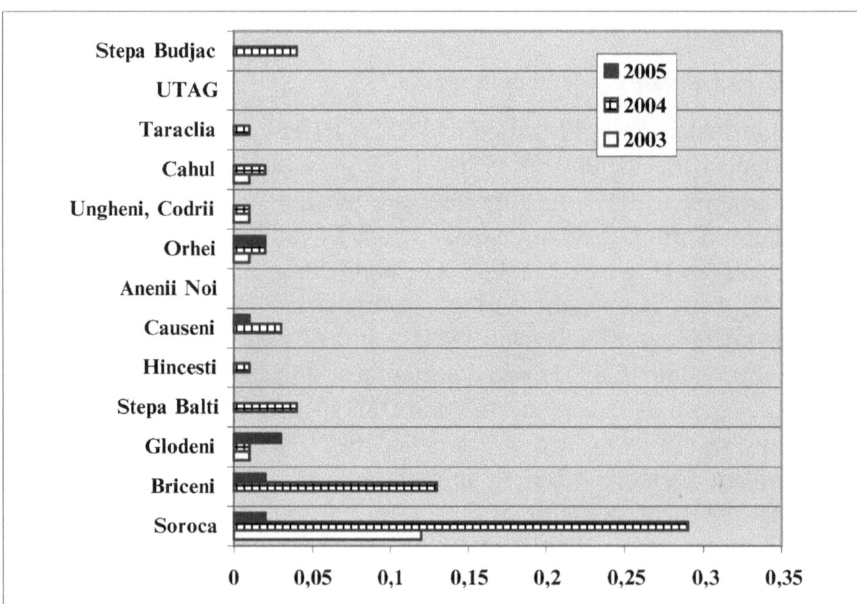

Figure 4. Mean contents of HCHs in agricultural soil, MAC

4.1.2. OCPs Contents Around the Warehouse with Obsolete Pesticides

During 2003–2005 according to the activities of SHS were analyzed 112 soil samples taken around the 14 warehouses with obsolete pesticides. In 16% of samples were revealed exceeding of MAC of HCHs and in 20% of soil samples were reviled exceeding MAC of DDTs with maximum concentration of HCH – 5,408 mg/kg and DDTs – 1,202 mg/kg (Table 1).

TABLE 1. Maximum level of HCHs and DDTs in soil around the stockpiles with obsolete pesticides, CMA

District	Location (mayoralty)	MAC, HCHs	MAC, DDTs
Briceni	Balasinesti	0.02	0.20
Glodeni	Petrunea	0.01	0.54
Hincesti	Mereseni	0.01	4.03
Taraclia	Albota de Jos	0.02	0.80
Hincesti	Mereseni	1.30	11.52
Taraclia	Albota de Jos	0.39	1.70
Glodeni	Petrunea	7.34	8.36
Briceni	Balasinesti	12.56	2.20
Causeni	Gradinita	1.54	0.09
Stefan-Voda	Tudora	17.97	0.40
Telenesti	Ratus	0.12	0.15
Cimislia	Cimislia	54.08	7.22
Floresti	Temeleuti	5.23	12.02
Briceni	Grimancauti	5.68	5.41

4.1.3. OCPs Contents in Soil Around the Pesticide Dump in Cismicioi

Fourteen cells are located there which are well visible from a surface. In general, in territory of a pesticide dump there is vegetation though naked sites are visible also (in warm seasons). Examination of pesticide dump was carried out twice – in 1997 (within TACIS ENVREG 9705 Project) and in 2004. Fifty-two samples of soil and ground were analyzed in 2004. Organochlorine pesticides, DDTs, HCHs, and HCB, were determined in soil samples. The most pollution area was revealed in the territory of Dump between cells up to 178.6 MAC of DDTs, up to 21.2 MAC of HCHs and up to 2.4 OAC HCB. In samples taken from the west, slope and valley out of pesticide dump the values of HCHs not exceed MAC, but were revealed the exceeding of DDTs in abovementioned samples up to 5.6 MAC. In total from 52 analyzed samples – 5% samples exceed HCHs MAC and 15% – DDTs MAC. In one sample taken from territory of Dump was revealed exceeding of OAC of HCB with value 2.4. The data shows that no leakage revealed from pesticide dump.

4.1.4. OCPs Contents in Sediments

Survey of contents of OCPs in sediments was included in Monitoring (Figures 5 and 6).

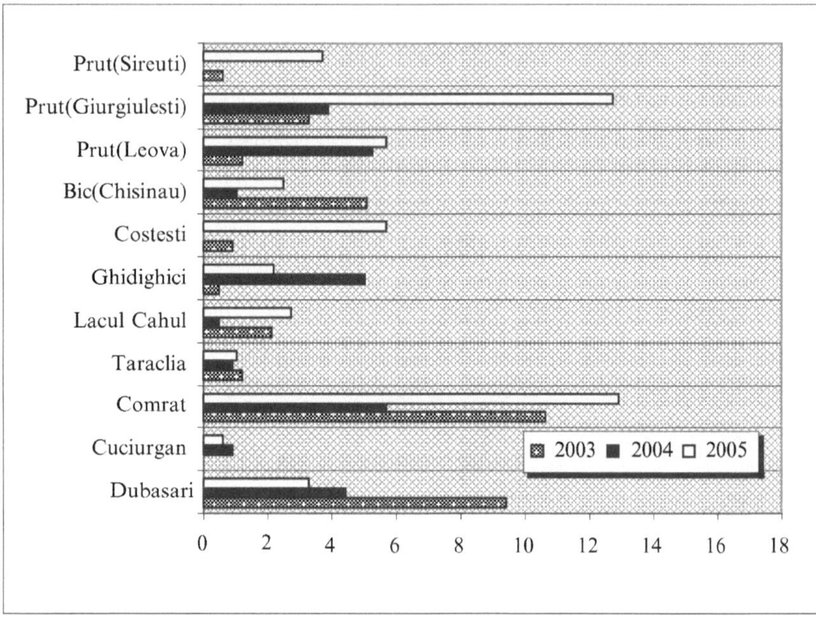

Figure 5. Contents of DDTs in sediments (mkg/kg)

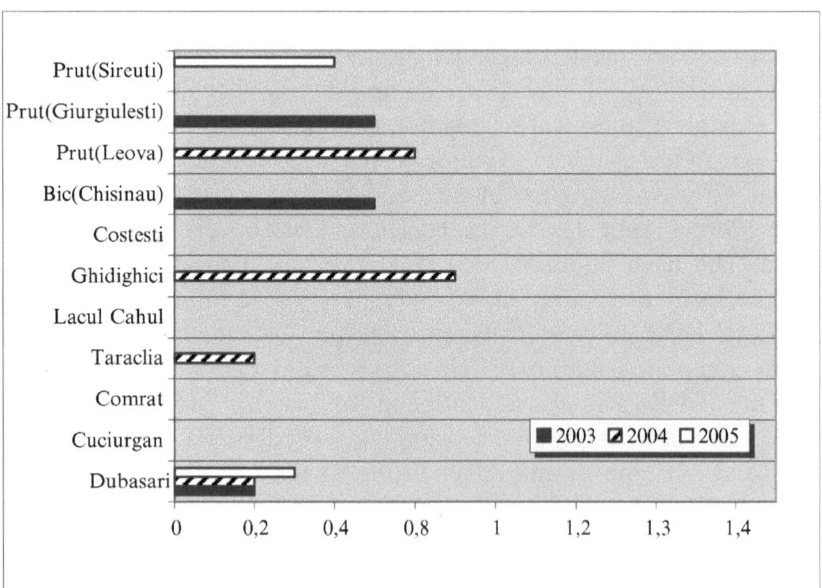

Figure 6. Contents of HCHs in sediments (mkg/kg)

Program of SHS in 2003. OCPs have been measured in samples of sediments taken from main water bodies of Republic of Moldova (six reservoirs, one lake, and two rivers). DDTs concentration varied from 0.5 to 12.9 mkg/kg, HCHs concentration – from 0.2 to 0.6 mkg/kg.

4.2. PCBs CONTENTS

4.2.1. PCBs Contents in Sediments

Surveys of contents of $PCBs_7$ (PCB28/PCB31, PCB52, PCB101, PCB118, PCB138, PCB153, and PCB180) in sediments were included in Monitoring Network in 2003. PCBs contents varied from 0.3 to 15.8 mkg/kg (Figure 7).

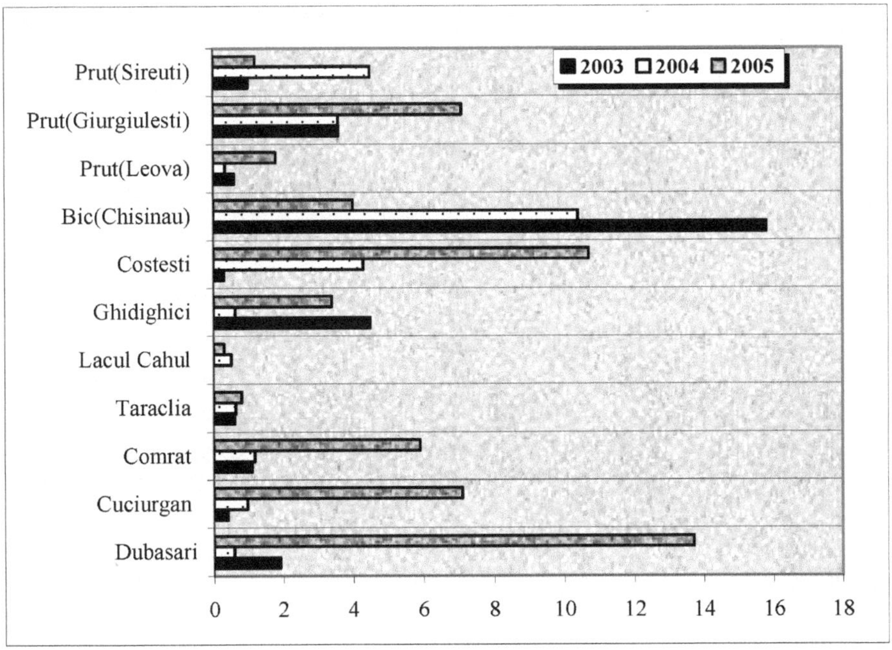

Figure 7. Contents of PCBs in sediments (mkg/kg)

4.2.2. PCBs Contents Around the Electro Energetic Installations

The norm of PCBs in soil (MAC) is 0.06 mg/kg. In year 2005, four substations were investigated. The PCBs contents were determined in 24 soil samples taken in depth 0–10 cm around the PCB-containing capacitors in the electrical substations that are in conditions where leakages due to corrosion are occurring. Data show that all investigated area is polluted, the exceeding of norm of PCBs in soil was revealed, and the most contaminated area was revealed in district Donduseni. The results are presented in Table 2.

TABLE 2. Contents of PCBs in soil near the substations, MAC

Substation	Max. value
Straseni	1.67
Soroca	42.5
Drochia	18.33
Donduseni	15.90

4.2.3. Contents of PCBs in soil in the Vulcanesti substations

In spring 2005 was carried out the examination of PCBs in soil taken from Vulcanesti substation (Table 3). The measurements demonstrate that the soil is contaminated with PCBs. The PCBs in the upper 10 cm varied from 1.3 to 7,098 mg/kg. In depth 50–60 cm the concentration ranged from 0.1 to 177 mg/kg.

TABLE 3. Results of prefeasibility analysis program spring 2005 in Vulcanesti substation (mg/kg[6])

Sample location	PCBs content 0–10 cm	PCBs content 50–60 cm
Between batteries underneath the assembly construction	1.3	0.1
Underneath capacitor battery	331.7	0.4
Underneath capacitor battery	95.1	0.2
Underneath capacitor battery	638.2	50.3
Between batteries underneath the assembly construction	7,097.7	44.8
Underneath capacitor battery	5,300.4	177.2
Underneath capacitor battery	5.1	3.3
Beside the assembly	0.7	n/d
Uphill of the assembly	0.5	n/d
By waste storage	33.4	n/d
By storage pit	0.03	n/d

5. Conclusion

- Contents of DDTs and HCHs in soil continue to decrease
- Soils around the stockpiles with obsolete pesticides are polluted and need the application of remediation technologies
- Soil around the power energy installations contaminated with PCBs has different pollution and needs various approach: remediation technologies or soil excavation and destruction

- Abovementioned contaminated sites and many others places are danger to the environment that is to be taken into consideration in the course of further investigations

References

1. Stockholm Convention on Persistent Organic Pollutants (POPs), pp. 4, 33–40.
2. National Implementation Plan for the Stockholm Convention on Persistent Organic Pollutants Republic of Moldova CHISINAU, Stiinta, 2004, pp. 44, 49.
3. G. Victor Buxton, M.Sc., P. Eng. Hazatdous Chemicals International Consultant Ottawa, Ontario, Canada, STAR Roster Reviewer Comments, 21 May 2005, Moldova Persistent Organic Pollutants (POPs) Stockpiles Management and Destruction Project, Project Document, October 2005, Report No. 33862-MD.
4. www.theodora.com/maps
5. Persistent Organic Pollutants (POPs) Stockpiles Management and Destruction Project, Project Document, October 2005, Report No. 33862-MD, pp. 32–33.

PERSISTENT ORGANIC POLLUTANTS (POPs): FOOD SAFETY CONTROL IN ESTONIA

OTT ROOTS[1,2*]
[1]*Estonian Environmental Research Centre, 10 617, Tallinn, Marja 4D, Estonia*
[2]*Estonian Marine Institute, University of Tartu, 12618 Tallinn, Mäealuse 10a, Estonia*

MART SIMM
Estonian Marine Institute, University of Tartu, 12618 Tallinn, Mäealuse 10a, Estonia

HANNU KIVIRANTA
National Public Health Institute (KTL), Department of Environmental Health, Laboratory of Chemistry, FI-70701Kuopio, Finland

PANU RANTAKOKKO
National Public Health Institute (KTL), Department of Environmental Health, Laboratory of Chemistry, FI-70701 Kuopio, Finland

Abstract: The objective of article is to harmonize and integrate the activities of the New European Union (EU) Member State – Estonia in the field of chemical food safety with those of Old Member States. This will be achieved by the activities planned in two paths: one aimed at gathering knowledge on sources of dietary data and another focussing on strategies/ approaches and capabilities of monitoring food chemical contamination, like polychlorinated dibenzo-*p*-dioxin (PCDDs), polychlorinated dibenzo-furans (PCDFs), dioxin-like polychlorinated biphenyls (DL-PCBs) and other polychlorinated biphenyls (other-PCBs). All studied Baltic Sea wild fish herring, sprat, perch, eel, pikeperch, flounder, and aquaculture rainbow

* To whom correspondence should be addressed: Estonian Environmental Research Centre, 10 617 Tallinn, Marja 4D, Estonia; E-mail: Ott.Roots@klab.ee

E. Mehmetli and B. Koumanova (eds.),
The Fate of Persistent Organic Pollutants in the Environment, 173–185.

trout, eel samples had WHO-PCDD/F-TEQ and also combined WHO-PCDD/F-PCB-TEQ upperbound values below the maximum limits set in Council Regulation (EC) No 2375/2001 and (EC) No199/2006. The same for all pork, butter, milk and fish oil samples. PCDD/PCDF and DL-PCB content in river lamprey and wild salmon exceeded the established limits in both cases. Results were calculated as upperbound, mediumbound, and lowerbound values per gram fresh weight, dry weight and fat.

Keywords: Estonia, Baltic Sea, PCDD/PCDF, DL-PCB, Estonian food, wild fish, aquaculture, fish oil, meat, butter, milk

1. Introduction

Chemicals are nowadays produced and marketed in great numbers. So far, the dangerous compounds have been found mainly as result of chemical monitoring, carried out after the appearance of problems relating to human health. Chemical residues may affect all major organs of the body, causing serious health outcomes like cancer, birth defects, and brain damage. The contamination of food, including drinking water, by potentially hazardous substances is a worldwide public health concern. National authorities have the responsibility and obligation to ensure that toxic chemicals, such as pesticides, heavy metals, environmental contaminants and naturally occurring toxins, are not present in food at levels that may adversely affect the health of consumers.

The objective of article is to harmonize and integrate the activities of the New European Union (EU) Member State – Estonia in the field of chemical food safety with those of Old Member States – like Finland. On the basis of this information, recommendations how to improve chemical food safety management in Estonia and also other EU countries will be prepared. This will be achieved by the activities planned in two paths: one aimed at gathering knowledge on sources of dietary data and another focusing on strategies/approaches and capabilities of monitoring food chemical contamination, like polychlorinated dibenzo-*p*-dioxins (PCDDs), polychlorinated dibenzofurans (PCDFs), dioxin-like polychlorinated biphenyls (DL-PCBs) and other polychlorinated biphenyls (other-PCBs). Dioxins as referred to in the EU Commission Regulation (EC) No 199/2006 cover a group of 75 PCDDs congeners and 135 PCDFs congeners, of which 17 are of toxicological concern. PCBs are a group of 209 different congeners of which a small number (12 congeners) exhibit toxicological properties similar to dioxins and are therefore termed "dioxin-like PCBs" (DL-PCBs).

On 30 May 2001, the Scientific Committee for Food (SCF) adopted an Opinion on the Risk Assessment of Dioxins and Dioxin-like PCBs in Food, updating its Opinion of 22 November 2000 on the basis of new scientific information that had become available since the latter adoption. The SCF fixed a tolerable weekly intake (TWI) to 14 pg WHO-TEQ/kg body weight for dioxins and dioxin-like PCBs.

2. Methods

The determination of PCDD/PCDFs, DL-PCBs, and other-PCBs was done according to standard operation procedure in the laboratory of Chemistry in the National Public Health Institute (KTL) (SOP called "Determination of PCDD/PCDFs, PCBs, DL-PCBs, and other POPs in tissue samples, KEM MO3"). The method is in part described in a previously published paper (Kiviranta et al., 2004).

Analytes were extracted in a 33 ml stainless steel extraction cells from freeze dried sample with an ASE 300 Accelerated Solvent Extractor (Dionex Sunnyvale, California). Extraction solvent was 50% acetone-hexane using 3 × 5 min extraction cycles, 120°C temperature, 1,500 psi pressure, and 80% flush volume. Solvent was evaporated and fat percentage was determined gravimetrically. Fat was decomposed with ASE from the sample in a 100 ml stainless steel clean-up extraction cells filled with sulphuric acid impregnated silica gel. Hexane was the clean-up extraction solvent using 5 × 1 min extraction cycles, 40°C temperature, 1,500 psi pressure and 100% flush volume. After the fat removal sample was fractionated and further purified on carbon and activated alumina columns as described earlier (Kiviranta et al., 2004; Parmanne et al., 2006). Internal ^{13}C PCDD/PCDF standards (altogether 16 standards) were used to quantitate the concentrations of PCDDs/PCDFs. ^{12}C PCB 30 and 12 ^{13}C-labelled PCB congeners (PCB 80, 101, 105, 118, 138, 153, 156, 157, 170, 180, 194, and 209), and 4 ^{13}C-labelled co-planar (co-PCB) congeners (PCB 77, 81, 126, and 169) were used as internal standards for PCBs and DL-PCBs.

PCDD/PCDFs and PCBs were analyzed by high resolution gas chromatography-high resolution mass spectrometry (Hewlett-Packard 6890-VG 70-250SE) using selective ion recording (resolution 10,000). Both PCDD/PCDF and PCB congeners were separated on a DB-Dioxin capillary column (J & W Scientific: 60 m, 0.25 mm, 0.15 μm). The method is accreditated (Testing laboratory T077 by FINAS; http://www.finas.fi).

Limits of quantification (LOQ) for PCDD/PCDFs, DL-PCBs, and other PCBs varied between 0.0007 and 0.63, 0.0007 and 0.13, and 0.035 and 13 pg/g fresh weight (fw), respectively, depending on each individual congener and on the individual food stuff (Kiviranta et al., 2004). Results were calculated as upperbound, mediumbound and lowerbound values per gram fresh and lipid weight.

3. Results and Discussion

3.1. FISH

In average Baltic herring (Roots et al., 2004; Roots and Zitko, 2004, 2006; Pandelova et al., 2006; Roots and Simm, 2007), Baltic sprat (Roots and Zitko, 2004, 2006; Roots et al., 2006; Simm et al., 2006; Roots and Simm, 2007) and eel samples had WHO-PCDD/PCDF-TEQ and also combined WHO-PCDD/PCDF-PCB-TEQ upperbound values below the maximum limit values set in Council Regulations (EC) No 2375/2001 and EC No. 199/2006 (4 pg WHO-PCDD/PCDF-TEQ/g fresh weight (fw) and 8 (12 for eel) pg WHO-PCDD/PCDF-PCB-TEQ/g fw). In river lamprey and wild salmon samples EC maximum limit values were exceeded (Table 1).

The summary dioxin content (PCDD/PCDFs and DL-PCBs) in herring and sprat increases along with the age of the analyzed fish (Roots and Zitko,

TABLE 1. Content of dioxins (pg WHO-TEQ/g fresh weight) in fish and river lamprey in years 2002–2006 in Estonia

	No of samples	PCDD/PCDFs	DL-PCBs	Sum of PCDD /PCDFs and DL-PCBs
Wild fish				
Herring	49	2.06 ± 0.16	1.93 ± 0.16	4.18 ± 0.32
Sprat	26	2.50 ± 0.17	2.80 ± 0.20	5.47 ± 0.37
Salmon	3	4.04 ± 0.25	4.96 ± 0.45	9.00 ± 0.65
Eel	3	2.20 ± 0.66	2.24 ± 0.66	4.43 ± 1.15
Perch	13	0.46 ± 0.08	0.60 ± 0.10	1.06 ± 0.17
Pikeperch	6	0.39 ± 0.01	0.43 ± 0.02	0.83 ± 0.03
Flounder	3	0.42 ± 0.08	0.85 ± 0.26	1.28 ± 0.32
River lamprey	6	5.89 ± 0.58	4.83 ± 0.43	10.73 ± 0.98
Aqua culture				
Rainbow trout	2	0.31 ± 0.05	0.46 ± 0.02	0.77 ± 0.07
Eel	2	0.33 ± 0.01	0.85 ± 0.02	1.18 ± 0.03

2004, 2006; Roots and Simm, 2006; Simm et al., 2006). The established EU limit (8.0 pg WHO-TEQ/g fresh weight) is exceeded by the summary content of dioxins in 7.5-year-old herring and 5.3-year-old sprat. Based on PCDD/Fs content, only 5-year-old herring and sprat are hazardous (contain over 4 pg WHO-TEQ/g fresh weight). Apparently, the established limit for the summary content of dioxins does not reduce the age of herring and sprat suitable for human consumption. The content of dioxins in eels from Baltic Sea (Table 1) is significantly lower than the established limits (for eels 12 pg WHO-PCDD/F-PCB-TEQ/g fresh weight). The content of dioxins in eels and rainbow trout from Estonian fish farms is significantly lower than in the wild fish and established limits. Many of the PCDD/PCDF compounds in fish farm eels and rainbow trout are below the decision limit, but it should be noted that in comparison with naturally grown fish the share of the most toxic compound (2,3,7,8-TCDD) is somewhat high. PCDD/PCDF and DL-PCB content in river lamprey exceeded the established limits in both cases. Since there is data that shows that during processing the amount of dioxin content is reduced, then it would be expedient to next year determine the dioxin content of prepared river lamprey products, for instance in preserves.

3.2. MILK AND BUTTER

Milk and milk products are good indicators for the contamination of persistent organic pollutants (POPs) in the food chain. Thus, butter has been used for comparison of the PCDD/PCDF contamination in several countries (Malisch and Dilara, 2004; Weiss et al., 2001; Paradiz and Dilara, 2003). By the literature data, the range of PCDD/PCDF concentrations in 65 butter samples from 39 countries was found to be between 0.02 pg and 2.02 pg WHO-PCDD/PCDF-TEQ/g fat (Paradiz and Dilara, 2003), in butter from 24 countries between 0.06 pg and 4.80 pg WHO-PCDD/PCDF-TEQ/g fat (Weiss et al., 2001).

Compounds analyzed in the two butter samples collected in 2006 are below the PCDD/PCDF quantification limits. Five out of ten PCDF congeners exceeded the quantification limit in sample 000031 but only one in sample 002322 – 2,3,4,7,8-PeCDF. Therefore, the upper and lower bound concentrations differ significantly from each other (Table 2). Based upon concentration, the dominant PCDD/PCDF compounds in sample 000031 are 1,2,3,4,7,8-HxCDF (30%) and 2,3,4,7,8-PeCDF (28%), however, based on toxicity 2,3,4,7,8-PeCDF (69%) is dominant. From analyzed DL-PCB congeners 8 out of 12 in sample 002322 or 9 in sample 000031 are over the quantification limit. In both samples, based on concentration the congener CB-118 (respectively 73% and 80%) was dominant, while based on toxicity the congener CB-126 (respectively 58% and 60%) was dominant. On the basis of concentration there is an excess of mono-ortho PCBs (totalling more than 99%) in the butter, while on the basis of toxicity the relative importance of non-ortho PCB (59% and 61%) is somewhat higher. Based on concentration (Table 2) DL-PCBs are overwhelmingly dominant when compared to PCDD/PCDFs (totalling over 99%) in butter. But based on toxicity, the relative importance of PCDD/PCDFs increases, but the content of DL-PCB compounds is still higher (63% and 56%) than PCDD/PCDFs. In all three of the 2006 milk samples, four PCDF compounds – 2,3,4,7,8-PeCDF and three HxCDF exceed the quantification limit. In addition to these, sample 002563 includes one PCDD, namely OCDD and sample 004889 1,2,3,7,8-PeCDF and two PCDDs (1,2,3,6,7,8-HxCDD and 2,3,7,8-TCDD). The difference between the upper and lowerbound results is nearly threefold (Table 3).

TABLE 2. Occurrence of PCDD/PCDF and PCB compounds in 2006 butter samples

Compound/unit		Samples	
		000031	002322
pg/g fat			
PCDD/PCDF	Lowerbound	0.457	0.095
	Mediumbound	0.945	0.989
	Upperbound	1.433	1.882
DL-PCB	Lowerbound	1,489	1,137
	Mediumbound	1,499	1,155
	Upperbound	1,508	1,174
pg WHO-TEQ/g fat			
PCDD/PCDF	Lowerbound	0.093	0.048
	Mediumbound	0.176	0.176
	Upperbound	0.260	0.304
DL-PCB	Lowerbound	0.445	0.390
	Mediumbound	0.446	0.391
	Upperbound	0.447	0.392
ng/g fat			
other PCB	Lowerbound	1.05	0.68
	Mediumbound	1.48	1.24
	Upperbound	1.92	1.80

The congener 2,3,4,7,8-PeCDF dominated in the analyzed milk samples, based on both toxicity as well as concentration is (respectively 67% and 34%), and based on both concentration and toxicity, PCDFs contribute more in milk than PCDDs (Table 3). Of the 12 DL-PCB compounds analyzed in milk, CB-77, CB167 and CB-189 are below the quantification limit. Based on average data CB-118 (71%) is the dominant one in terms of concentration in milk, while based on toxicity CB-126 (74%) is dominating. Based on concentration, mono-ortho PCBs (totalling over 99%) dominate in milk, while based on toxicity DL-PCB (75%) are dominating congeners. In milk DL-PCBs are overwhelmingly dominating based on concentration (Table 3) but on toxicity basis, the relative importance of PCDD/PCDF increases, however, the content of DL-PCB compounds is still higher (respectively 65% and 35%) than PCDD/PCDFs. In the case of milk, the large difference between samples should be noted: based on toxicity the difference between the highest (sample 004889) and lowest (sample 002563) concentrations in terms of PCDD/PCDF is three times, and in the case of DL-PCBs there is over a tenfold difference (Table 3). In butter and milk samples, the content of dioxins as well as that of dioxins and PCBs combined is significantly lower than the established EU maximum limits (respectively 3.0 and 6.0 pg WHO-TEQ/g for fat). Results in terms of PCDD/PCDF – 0.2–0.3 (in one milk sample 0.8) pg WHO-TEQ/g for fat is comparable to data from other European states (as a rule between 0.3 and 2.1 pg WHO-TEQ/g of fat). The same applies for DL-PCB compounds – our results for butter and the two milk samples was 0.2–0.4, with the European average being 0.2–1.8 WHO-TEQ/g for fat. The DL-PCB content was higher in one milk sample (004889) – 2.17 WHO-TEQ/g for fat.

TABLE 3. Content of PCDD/PCDF and PCB compounds in 2006 milk samples

Compound/unit		Samples		
		000032	002563	004889
pg/g fat				
PCDD/PCDF	Lowerbound	0.573	0.765	1.952
	Mediumbound	1.302	1.348	2.643
	Upperbound	2.030	1.931	3.333
DL-PCB	Lowerbound	954	386	3,877
	Mediumbound	958	392	3,879
	Upperbound	962	399	3,882
pg TEQ/ g fat				
PCDD/PCDF	Lowerbound	0.135	0.084	0.615
	Mediumbound	0.258	0.175	0.706
	Upperbound	0.381	0.267	0.797
DL-PCB	Lowerbound	0.356	0.158	2.167
	Mediumbound	0.356	0.159	2.167
	Upperbound	0.357	0.159	2.167
ng/g fat				
other PCB	Lowerbound	1.60	0.70	5.81
	Mediumbound	2.27	1.53	6.31
	Upperbound	2.94	2.35	6.80

3.3. PORK

In the three analyzed pork samples, most of the PCDD/PCDF compounds are below the quantification limit. In excess of the quantification limit in samples 000605 and 002469 are one PCDD (OCDD) and four PCDF compounds (2,3,7,8-TCDF, 1,2,3,4,7,8-HxCDF, 1,2,3,4,6,7,8-HpCDF, and OCDF), but in sample 0014364 only one PCDD/PCDF compound – OCDD exceeded the limit of quantification. The difference between the upper and lower bound concentrations was nearly double, while based on toxicity the difference is several times higher (Table 4). Based on concentration, the PCDD/PCDF congener dominating in sample 002469 is OCDF (38%), but in the other two samples (000605 and 0014364) OCDD was dominating instead (88% and 100%, respectively). On toxicity basis the one congener exceeding the limit of quantification in sample 0014364, OCDD, counted for all of the toxicity in the sample, while in the other two samples the dominant congeners in TEQ respect were 1,2,3,4,7,8-HxCDF (68% and 49%) and 2,3,7,8-TCDF (23% and 44%) (Table 4).

TABLE 4. Content of PCDD/PCDF and PCB compounds in 2006 pork samples

Compound/unit		Samples		
		000605	002469	0014364
pg/g fat				
PCDD/PCDF	Lowerbound	1.046	0.492	1.810
	Mediumbound	1.504	0.878	2.613
	Upperbound	1.962	1.264	3.415
DL-PCB	Lowerbound	203	299	174
	Mediumbound	214	308	187
	Upperbound	224	317	199
pg TEQ/ g fat				
PCDD/PCDF	Lowerbound	0.008	0.018	0.000
	Mediumbound	0.116	0.097	0.176
	Upperbound	0.225	0.176	0.352
DL-PCB	Lowerbound	0.042	0.063	0.024
	Mediumbound	0.045	0.066	0.031
	Upperbound	0.049	0.068	0.038
ng/g fat				
Other PCB	Lowerbound	0.23	1.70	0.44
	Mediumbound	0.79	1.88	0.90
	Upperbound	1.35	2.07	1.37

Of the 12 analyzed DL-PCB compounds, 5–6 congeners in pork exceeded the quantification limit. Congeners CB-105, CB-118, and CB-156 were present in all samples. Based on concentration there is an excess of

CB-118 (74%) and CB-105 (14%), but based on toxicity congeners CB-118 (39%), CB-126 (28%), and CB-156 (22%) dominated. On the basis of both concentration as well as toxicity there is an excess of mono-ortho PCB (respectively totalling 99% and 71%). Based on concentration (Table 4), DL-PCBs are overwhelmingly dominant ones in pork samples (totalling an average of 99%), however, based on toxicity the relative importance of PCDD/PCDF was higher than that of DL-PCB compounds (respectively 83% and 17%). In all pork samples the content of PCDD/PCDF as well as summary of dioxins and PCBs was lower than the established maximum limits (respectively 1.0 and 1.5 pg WHO-TEQ/g fat). Our results, as far as PCDD/PCDF is concerned (0.15–0.29 pg ITEQ/g of fate), are also comparable with the data from other European countries (as a rule below 0.4 pg ITEQ/g fat). The dioxin content of pork (based on fat) is, as a rule, lower than that of beef or poultry. Therefore, the continued analysis of dioxins in pork, in addition to other meat products, is crucial.

3.4. FISH OIL

Dioxin content was determined in three fish oil samples, and sample No 1 (HA2006SB397, Norway, Peter Möller) was clearly distinguished from samples No 2 and No 3 (HA2006SB399, HA2006SB398, Russia, "BioKontur"). In sample No 1, only two PCDD (1,2,3,4,6,7,8-HpCDD, OCDD) and three PCDF compounds (2,3,7,8-TCDF, 1,2,3,4,6,7,8-HpCDF and OCDF) were over the quantification limit. Based on concentration OCDD (77%) was dominating, while based on toxicity 2,3,7,8-TCDF (59%) was dominant. Dioxin-like PCB compounds CB-77, CB-81, CB-114, and CB-118 were below the quantification limits. Based on concentration, CB-156 (40%) was dominant in fish oil sample No 1, and on TEQ basis the congener CB-126 (65%) was dominant. Differences in the upper and lower limit results are large (nearly 30 times) in the case of PCDD/PCDF toxicity for sample No 1 (Table 5).

In samples No 2 and No 3, one PCDD and two PCDF compounds are below the quantification limit. All DL-PCB compounds were quantifiable in both samples. Therefore, the difference between the upper and lower bound concentrations was practically nonexisting (Table 5). Based on concentration 2,3,7,8-TCDF (50%) and CB-118 (64%) are dominant, while based on toxicity 2,3,4,7,8-PeCDF (43%) and CB-126 (62%) dominated. Of the PCDD/PCDF compounds, PCDD is in excess in both concentration and toxicity in sample No 1, while in samples No 2 and No 3 PCDFs dominate (Table 5).

TABLE 5. Content of PCDD/PCDF and PCB compounds in 2006 fish oil samples

Compound/unit		Samples		
		No 1	No 2	No 3
pg/g fat				
PCDD/PCDF	Lowerbound	2.605	9.548	8.550
	Mediumbound	2.974	9.696	8.736
	Upperbound	3.342	9.843	8.922
DL-PCB	Lowerbound	124	10,753	12,689
	Mediumbound	172	10,753	12,689
	Upperbound	220	10,753	12,689
pg TEQ/ g fat				
PCDD/PCDF	Lowerbound	0.007	2.079	1.798
	Mediumbound	0.119	2.091	1.804
	Upperbound	0.230	2.103	1.810
DL-PCB	Lowerbound	0.121	4.022	4.077
	Mediumbound	0.126	4.022	4.077
	Upperbound	0.131	4.022	4.077
ng/g fat				
Other PCB	Lowerbound	1.16	56.0	56.3
	Mediumbound	1.83	56.0	56.3
	Upperbound	2.50	56.0	56.3

In the case of DL-PCBs, mono-ortho PCB concentration is significantly higher than non-ortho PCB in all samples. In toxicity respect the opposite was true, non-ortho PCBs were in excess. In both concentration as well as toxicity respect PCDD/PCDF content was lower than DL-PCB content. Only in sample No 1, based on the upper bound concentrations and average data, the PCDD/PCDF content was higher than the DL-PCB content (Table 5). The dioxin content in samples No 1 and No 3 in fish oil were lower than the established maximum limits for PCDD/PCDFs and combined content of dioxins and PCBs (2.0 and 10.0 pg WHO-TEQ/g of fat, respectively). In sample No 2, the WHO-PCDD/F-PCB-TEQ content was below the established maximum limit, although PCDD/PCDF content alone (upper bound concentration 2.103 pg WHO-TEQ/g fat) was exceeding the EU maximum limit for fish oil into human consumption.

4. Conclusion

As far as fish are concerned, in 2007 the investigation of industrial fish – herring and sprat – should be continued, fulfilling the requirements of the Commission. The future analyses should give answers to questions at which age of fish the content of dioxins does not exceed the established maximum limits. In other analyzed fish species, excluding salmon, fish oil and river lamprey, the content of dioxins is lower than the established maximum limits and future analysis can be more irregular.

In the case of milk and butter, the content of dioxins is, as a rule, 10 times lower than the maximum limit. An exception was one milk sample (004889, from Harju County, Rae rural municipality), in which the content of dioxins was significantly higher than the average of the other samples – PCDD/PCDF content was nearly double, DL-PCB compounds were eight times higher in concentrations. It can be concluded that the DL-PCB compounds can be a problem in Estonian milk and butter and because of that it is necessary to continue analyzing them in the future.

The content of dioxins in pork is comparable with the results from the other countries in Europe, and on average five times lower than the established maximum limit. The content of dioxins in pork is generally lower than in other meat products, therefore in 2007 attention should be focused on analysis of beef and poultry.

The content of dioxins in fish oil (Omega-3; sample No 1) originating in Norway is lower, by nearly an order of magnitude, than the established maximum limit. Fish oil originating from Russia ("BioKontur," samples No 2 and No 3) on the other hand have a combined content of dioxins two times lower than the established maximum limits, although PCDD/PCDF content in sample No 3 (1.81 pg WHO-TEQ/g fat) is near the corresponding established maximum limit and sample No 2 (2.10 pg WHO-TEQ/g fat) actually exceeds the established limit. It can be concluded that it will be necessary in the future to control the share of dioxin in imported products, too.

In 2007, it would be expedient to analyze 26 samples of food for dioxin. Of the fish, herring and sprat (a total of 10 samples), for which older fish could be intentionally selected to study when the content of dioxins exceeds the established maximum limit with respect of age. In addition, river lamprey preserves attention in 2007 analysis, since during processing of river lamprey the level of dioxins is usually reduced. The large variation in the content of dioxins in milk requires the collection of samples from at least six different regions for analysis. For meat products, the content of dioxins should be determined for two beef and two chicken samples. The final two samples could be eggs. All named samples to be analyzed in 2007 would meet EU requirements for the dioxins food analysis in Estonia, allowing together with these results the assessment of the intake of dioxins by Estonians from diet, and thus determining the health risk posed by dioxins in Estonia. To get an indication on whether exposure to PCDD/PCDFs and DL-PCBs might cause a problem in the new European Union (EU) Countries, a study of their levels in 16 butter samples coming from eight New Europan Union Countries (Cyprus, Czech Republic, Estonia, Lithuania, Poland, Romania, Slovakia, and Slovenia) was commissioned. An important conclusion was obtained, all butter samples were below the EU maximum levels and EU action levels for PCDD/PCDFs and DL-PCBs (Malisch and Dilara, 2004).

Acknowledgments

This project was financially supported by the Ministry of Agriculture of Estonia.

References

EU Commission Regulation (EC) No 199/2006 (OJ L32/34, p1, 04/02/2006)

Kiviranta, H., Ovaskainen, M.-L., and Vartiainen, T., 2004, Market basket study on dietary intake of PCDD/Fs, PCBs, and PBDEs in Finland, *Environment International*, **30**:923 – 932.

Malisch, R. and Dilara, P., 2004, PCDD/Fs and PCBs butter samples from new European Union members states and candidate countries, *Organohalogen Compounds*, **66**:2080–2084.

Pandelova, M., Henkelmann, B., Roots, O., Simm, M., Benfenati, E., and Schramm, K.-W., 2006, Levels of PCDD/F in Baltic fish in different age groups, *Organohalogen Compounds*, **68**:1932–1934.

Paradiz, B. and Dilara, P., 2003, Dioxin Emissions in the Candidate Countries, Emission Inventories, Reduction Policies and Measures. EUR 20779 EN.

Parmanne, R., Hallikainen, A., Isosaari, P., Kiviranta, H., Koistinen, J., Laine, O., Rantakokko, P., Vuorinen, P.J., and Vartianen, T., 2006, The dependence of organohalogen compounds concentrations on herring age and size in the Bothnian Sea, northern Baltic, *Marine Pollution Bulletin*, **52**:149–161.

Roots, O. and Simm, M., 2007, Dioksiini seire toidus 2006. aastal (in Estonian), (http://www.agri.ee/index.php/14298)

Roots, O. and Zitko, V., 2004, Chlorinated dibenzo-p-dioxins, and dibenzofurans in the Baltic herring and sprat of Estonian coastal waters, *Environmental Science and Pollution Research*, **11**:186–193.

Roots, O. and Zitko, V., 2006, The effect of age on the concentration of polychlorinated dibenzo-*p*-dioxins, dibenzofurans and dioxin-like polychlorinated biphenyls in Baltic herring and sprat, *Fresenius Environmental Bulletin*, **15**(3):207–219.

Roots, O., Simm, M., Järv, L., Henkelmann, B., and Schramm, K.-W., 2004, Levels of PCDD/Fs and "dioxin like" PCBs in the perch from the north-eastern part of the Baltic Sea, *Organohalogen Compounds*, **66**:1691–1694.

Roots, O., Schramm, K.-W., Simm, M., Henkelmann, B., and Lankov, A., 2006, Polychlorinated dibenzo-*p*-dioxins and dibenzofurans in Baltic herring and sprat in the northeastern part of the Baltic Sea, *Proceedings of the Estonian Academy of Sciences, Biology, Ecology*, **55:**51–60.

Simm, M., Roots, O., Kotta, J., Lankov, A., Henkelmann, B., Shen, H., and Schramm, K.-W., 2006, PCDD/Fs in sprat (*Sprattus sprattus balticus*) from the Gulf of Finland, the Baltic Sea, *Chemospere*, **65**:1570–1575.

Weiss, J., Päpke, O., and Bergman, A., 2001, PCDDs, PCDFs and related contaminants in butter originating from 39 countries world wide, *Organochlorine Compounds*, **51**:271–274.

Part III

SOURCES, RELEASE PATHWAYS, FATE, AND TRANSPORT

ANALYZING THE GLOBAL FRACTIONATION OF PERSISTENT ORGANIC POLLUTANTS (POPs)

MARTIN SCHERINGER*
Institute for Chemical and Bioengineering, ETH Zurich, Wolfgang-Pauli Street 10, CH-8093 Zurich, Switzerland

Abstract: Persistent organic pollutants (POPs) such as PCBs, but also substances still on the market such as PBDEs are transported over long distances and are present in many environmental media. In principle, differences in the physicochemical properties and environmental half-lives of different POPs would lead to a "fractionation" process, i.e., more volatile substances would be transported more rapidly and over larger distances than less volatile compounds. However, the fractionation pattern to be expected on the basis of chemical properties is often confounded by spatial and temporal variability of the emissions of the chemicals and by the conditions under which samples are taken in the field. Here, modeling results are presented that support the identification of fractionation patterns, and key questions in the area of global fractionation requiring further investigation are identified. Research needs concern the degradation processes of POPs in the gas phase, the interaction of POPs with atmospheric aerosols (aerosol–air partitioning; reactivity of the aerosol-bound fraction; deposition and transport with aerosols), and the characterization of various surface media (soil, vegetation, ice/snow, and water) with respect to their sorptive capacity for POPs (importance of secondary sources).

Keywords: persistent organic pollutants, long-range transport, cold condensation, global fractionation, multimedia models

* To whom correspondence should be addressed. E-mail: scheringer@chem.ethz.ch

E. Mehmetli and B. Koumanova (eds.),
The Fate of Persistent Organic Pollutants in the Environment, 189–203.

1. Cold Trap Effect and Global Fractionation

Persistent Organic Pollutants (POPs) are sufficiently stable and sufficiently volatile to be transported over long distances in the environment. In addition, they partition into many environmental media such as soil, water, air, atmospheric aerosols, vegetation, or ice and snow. This combination of properties has made the group of persistent semivolatile organic chemicals a class of particular concern in the assessment and management of large-scale chemical pollution (Wania and Mackay, 1999).

The problematic combination of persistence and volatility that are sufficient for significant long-range transport has early been recognized as a characteristic of POPs. Goldberg (1975) estimated the amount of DDT that would leave the original area of application and introduced the term "global distillation." In the 1980s, the particularly important aspect of POPs transport to polar regions came into the focus of scientific research. Because of the low temperature in polar regions, vapor pressure, and Henry's law constant (HLC) are lower than in warmer regions, and degradation is slower. This leads to the effect of a "cold trap" in which POPs can remain for long times and which makes them available for uptake by organisms and accumulation in the Arctic food chain (Rahn and Heidam, 1981; Weschler 1981). Ottar (1981) further discussed the reevaporation of persistent semivolatile chemicals and their long-term enrichment in the Arctic.

Because different POPs have different physicochemical properties and environmental half-lives, the efficiency of transport to the Arctic varies among the group of POPs. More volatile compounds are transported over larger distances by relatively fast atmospheric transport, whereas less volatile compounds interact more strongly with surface media such as water, soil, vegetation, ice or snow, which makes their long-range transport slower. Risebrough (1990) proposed the model of a "global gas chromatographic system" to systematically describe the long-range transport of different POP-type chemicals.

Wania and Mackay (1993, 1996) put forward the concept of "global fractionation" and discussed the influence of physicochemical properties on the transport of POPs to the Arctic and their enrichment in this particularly vulnerable region. Since the 1990s it has been an important objective for measurements campaigns to determine whether a fractionation pattern can be observed in groups of different compounds such as a series of PCB homologues, different polybrominated diphenylethers (PBDEs), or different organochlorine pesticides (Simonich and Hites, 1995; Bignert et al., 1998; Meijer et al., 2002, 2003; Kalantzi et al., 2001; ter Schure et al., 2002; Malmquist et al., 2003; Shen et al., 2005, 2006). In some cases, a fractionation pattern could be found (Meijer et al., 2002; Shen et al., 2006;

enrichment of lighter, more mobile PCBs towards the Arctic) whereas in others such a pattern was not discernible (ter Schure et al., 2002; Meijer et al., 2003; no relative enrichment of lighter PCBs towards the Arctic). This is because many confounding factors influence the levels and trends of POPs present in the different latitudinal zones from the tropics to the Arctic, such as different emission patterns (including possible effects of recent and/or local sources), weather conditions during the sampling period, and type of matrix sampled (soil, water, and biota). Therefore, the picture of the long-range transport and Arctic accumulation of POPs is very complex and the factors determining the levels and trends of POPs observed in the environment are still insufficiently understood.

An important complement to measurements in the field are modeling studies (Wania and Mackay 1995; Scheringer et al., 2000, 2004; Beyer et al., 2003; Wania and Su, 2004; Wegmann et al., 2004; Breivik et al., 2006). In a model fewer confounding factors are present than in the field, and in addition to concentrations, also all mass fluxes between different media and regions can be determined and analyzed with respect to the underlying environmental processes.

From a conceptual point of view, it is helpful to distinguish clearly between the cold trap effect[1] and the global fractionation process. The cold trap effect can be observed individually for *single* chemicals. It is caused by the temperature dependence of the Henry's law constant and octanol–air partition coefficient, K_{oa}. In colder regions, the HLC is lower (and the K_{oa} higher) than in warmer regions so that higher concentrations in surface media are observed if the concentrations in air are the same in both regions. The most pronounced cold trap effect can therefore be observed for persistent volatile compounds such as chlorofluorocarbons (CFCs), carbon tetrachloride, and also hexachlorobenzene (HCB). These chemicals exhibit nearly uniform concentrations in the air all over the globe and accordingly their concentration in surface media is highest in cold regions. Figure 1 shows the cold trap effect of carbon tetrachloride in ocean water and of HCB in vegetation, as it has been observed in field measurements.

For semivolatile compounds, this inversion of the concentration profile does not occur because their concentrations in air decrease from the tropics

[1] Also the term "cold condensation" is used to describe this effect. However, because no actual condensation, i.e., transition from gaseous to liquid state, takes place, it is more appropriate to use a term such as "cold trap."

to the Arctic; however, the cold trap effect is effective for SVOCs, too, as is discussed in Section 2.1.[2]

Global fractionation, in contrast to the cold trap effect, always refers to *several* chemicals in comparison. The purpose of this concept is to analyze differences in the transport efficiency among a set of different chemicals. The cold trap effect contributes to global fractionation because it occurs to a different extent for different chemicals but it does not fully explain the patterns of global fractionation. Another key factor contributing to the fractionation process are differences in environmental lifetimes of different POPs.

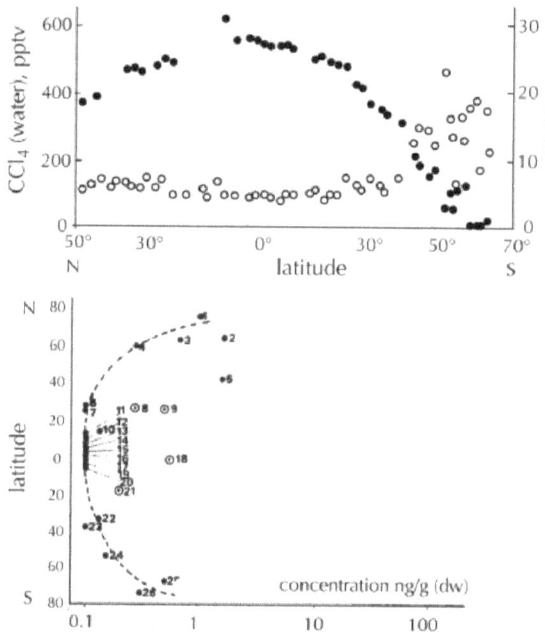

Figure 1. Concentrations increasing towards colder regions, caused by the cold trap effect. Top: CCl_4 in ocean water (open circles: CCl_4 concentration, dots: temperature). (Reproduced from Hunter-Smith et al., 1983. With permission. Copyright: Blackwell Publishing.) Bottom: hexachlorobenzene in vegetation. (Reproduced from Calamari et al., 1991. With permission. Copyright: American Chemical Society, 1991.)

2. Modeling the Global Distribution of POPs

A variety of models have been used to investigate the transport of POPs to the Arctic, ranging from single multimedia box models to highly resolved

[2] Faster degradation in warmer regions can well lead to an inverted concentration profile also for low-volatility chemicals. However, this is different from the purely thermodynamic effect discussed here.

atmospheric transport models (Wania and Mackay, 1995; Scheringer et al., 2000; Leip and Lammel, 2004; Hansen et al., 2004; Müller-Herold et al., 2006). Scheringer and Wania (2003) give an overview of different types of models for the environmental distribution of POPs. Here we use a simple conceptual model to discuss the cold trap effect and the zonally averaged global model CliMoChem (Scheringer et al., 2000) to study the global fractionation of PCBs.

2.1. A CONCEPTUAL MODEL OF THE COLD TRAP EFFECT

To investigate the cold trap effect in more detail, it is useful to consider a simple three-region model (see Figure 2) (Müller-Herold et al., 2006). The model consists of a tropical, temperate, and polar region which all include air, water and soil. The temperature in the three regions is 25°C, 10°C, and –9°C, respectively. All chemicals are continuously released to the air compartment of the tropical region, and on this basis, the model calculates the steady-state concentrations in all nine compartments of the model. Chemical-specific input parameters are the air-water partition coefficient, K_{aw}, and the octanol-water partition coefficient, K_{ow}. The temperature dependence of the K_{aw} is derived from a chemical's enthalpy of vaporization and an assumed enthalpy of dissolution in water of 20 kJ/mol (MacLeod et al., 2007). From these partition coefficients, the relative amounts in the three environmental media are calculated in all three regions. In the model, instant equilibrium between all environmental compartments is assumed.

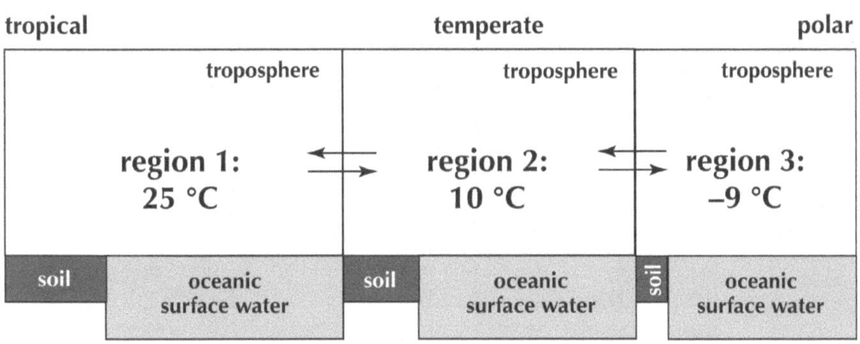

Figure 2. Structure of the conceptual three-region model for investigating the cold trap effect

2.1.1. *Application of the Model to Volatile Chemicals*

For a volatile and persistent chemical such as carbon tetrachloride, the majority of the chemical's amount present in the model is found in the atmosphere. Because the lifetime of the chemical exceeds the time constant for mixing between the three regions, concentrations in air are almost equal in the three regions (see Figure 3). Concentrations in the surface media increase from warmer to colder regions because K_{aw} decreases and K_{oa} increases, favoring the chemical's partitioning to water and soil/vegetation. This corresponds to the concentration increase observed in the field as shown in Figure 1. For volatile compounds, the steady-state assumption is relatively realistic because steady state is achieved within a few years, which is shorter than the historical emission period of compounds such as carbon tetrachloride.

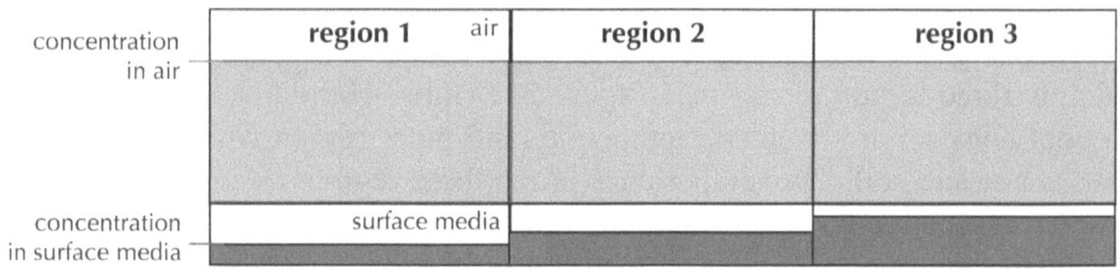

Figure 3. Cold trap effect for a persistent volatile chemical in the three-region model at steady state. Concentration in air uniform, concentration in surface media increasing from south to north

2.1.2. Application of the Model to Semivolatile Chemicals

For semivolatile persistent chemicals, the situation is different because for these chemicals, the majority of the amount present in the model is found in soil. At steady state, the concentration in soil in all three regions are almost identical and the concentrations in the air above the soil are determined by the K_{oa}. In colder regions, the K_{oa} is significantly reduced (by a factor of 5 for PCB 180 in the three-region model used here) and, therefore, the concentrations in air decrease from the tropical to the polar region of the model (see Figure 4).

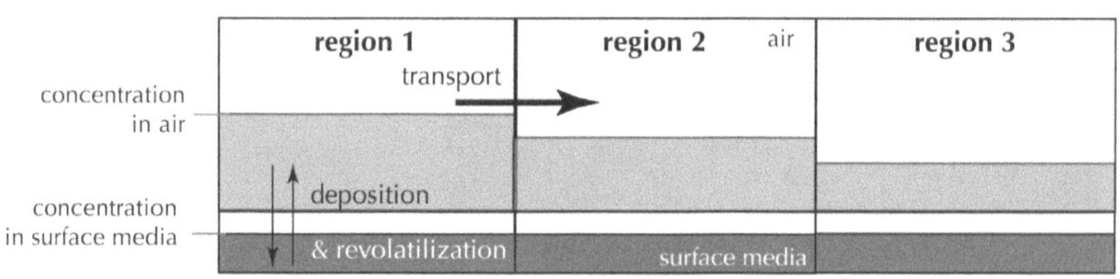

Figure 4. Cold trap effect for a persistent semivolatile chemical in the three-region model at steady state. Concentration in surface media uniform, concentration in air decreasing from south to north

For semivolatile compounds, the model assumption of steady state is not as realistic as for volatile compounds because the transfer from a source region to the Arctic is slower than for volatile compounds and it is not clear whether concentrations in soil in remote regions are already at equilibrium with concentrations in the air. This means that in reality the chemical would not show a uniform distribution in the soil as in Figure 4, but still a decrease in soil concentrations from the source region to the Arctic. In this case, the decrease in atmospheric concentrations would be even stronger than in Figure 4.

In conclusion, the cold trap effect is effective for semivolatile compounds as well as for volatile compounds because for both groups K_{aw} and K_{oa} are strongly temperature dependent. However, for semivolatile compounds, the cold trap effect causes a decrease in the concentrations in air (instead of an increase in the concentrations in the surface media), see Figure 4. Therefore and because concentrations of SVOCs decrease with distance from the source anyway if there is degradation, the cold trap effect is less well discernible in measured concentrations as for volatile compounds (Figures 1 and 3).

2.1.3. Conclusions from the Conceptual Model

In Figure 5, contributions of the cold trap effect to the process of global fractionation are illustrated. Volatile compounds with relatively uniform concentrations in air in the three regions and increasing concentrations in the surface media form a first fraction of chemicals. Semivolatile compounds with a relatively high vapor pressure such as PCB 30 (vapor pressure around 10^{-1} Pa at 25°C) belong to a second fraction. This fraction

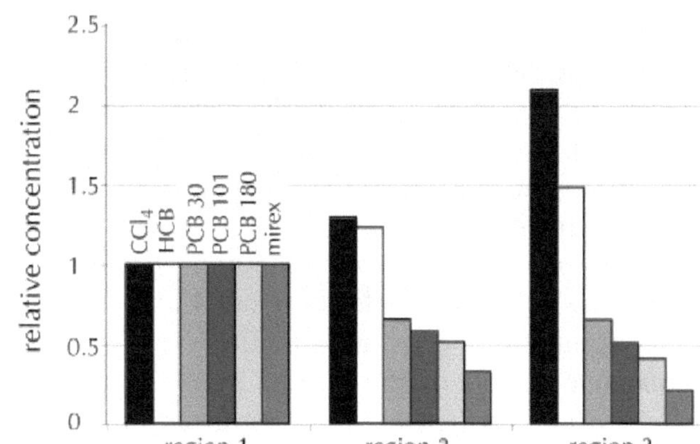

Figure 5. Different chemicals distinguished by the influence of the cold trap effect on their transport to the Arctic. Shown are relative concentrations in the soil from the three-region model at steady state

exhibits decreasing concentrations in surface media but the relative importance of these chemicals among the group of semivolatile compounds increases from the tropical region to the Arctic. Finally, low-volatility chemicals such as highly chlorinated PCBs and mirex (vapor pressure around/below 10^{-4} Pa at 25°C) fall in the third fraction that shows decreasing absolute concentrations in Figure 5 and also decreasing relative importance in the group of semivolatile compounds.

2.2. COLD TRAP EFFECT AND GLOBAL FRACTIONATION IN THE CLIMOCHEM MODEL

The CliMoChem model is a global multimedia fate and transport model that consists of a sequence of latitudinal zones (Scheringer et al., 2000). The zones are individually described in terms of the annual temperature course, land to sea surface ratio, vegetation cover (Wegmann et al., 2004), precipitation, and concentration of hydroxyl radicals. The basic assumption of the model is that transport in east–west direction (occurring on a timescale of weeks) is significantly faster than transport in south–north direction, i.e., between the zones of the model (occurring on a timescale of several months). The purpose of the model is to investigate the influence of temperature and other environmental factors on the transport of POPs to the Arctic.

As chemical-specific input parameters, the model requires the partition coefficients for air/water, K_{aw}, and octanol/air, K_{oa}, the degradation rate constants in soil, water and air, and the temperature dependence of all these properties, given by enthalpies of phase transfer for partition coefficients and activation energies for degradation rate constants.

In a first application of the model, the spatial distribution of carbon tetrachloride was calculated (Scheringer et al., 2000). Because the objective of this calculation was to investigate the basic features of the model performance, no realistic emission scenario was used but the chemical was released in a single pulse event at the equator. One year after the pulse release, carbon tetrachloride has assumed a nearly uniform spatial distribution in the troposphere. The concentration profile in water shows an increase towards the poles by a factor of about 4 as it has been observed in the measured concentrations, comparing Figure 6 to Figure 1.

In a second application of the CliMoChem model, the fractionation pattern of three PCB congeners is investigated: PCB 28 (a trichloro biphenyl), PCB 101 (a pentachloro biphenyl), and PCB 180 (a heptachloro biphenyl). Again, a single pulse release at the equator is used because this emission scenario does not confound the fractionation pattern with the spatial and temporal variability of the actual emissions of PCBs (Breivik et al., 2002). Figure 7 shows the spatial distribution of the three congeners in the five latitudinal zones of the northern hemisphere of the model. Given are the fractions of the three congeners 10 years after the pulse release.

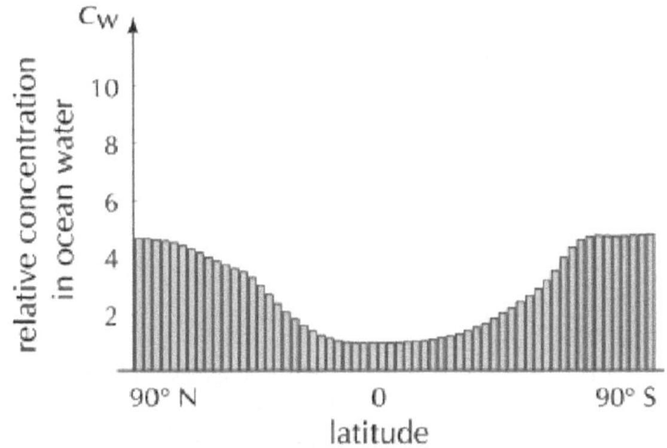

Figure 6. Concentration of carbon tetrachloride in the ocean water of the CliMoChem model, 1 year after a pulse release at the equator. Compare the measured data from Hunter-Smith et al. (1983) in Figure 1

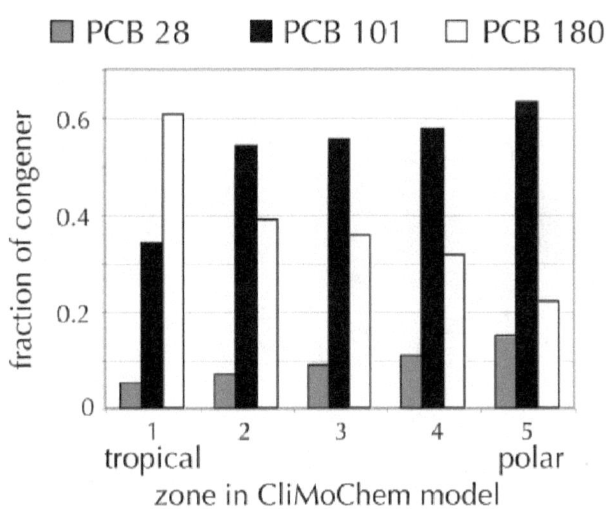

Figure 7. Fractions of PCB 28, 101, and 180 in the five latitudinal zones of the northern hemisphere of the CliMoChem model, 10 years after a pulse release to zone 1

For comparison, in Figure 8 the relative amounts of different PCB homologues are shown as function of latitude as they were measured by Meijer et al. (2002) in European soils. The model results are in good agreement with the field data: there is an increase by almost a factor of 2 for PCB 28/trichloro biphenyls in model and field data, a slight increase in the fraction of PCB 101/pentachloro biphenyls, and a clear decrease in the fraction of PCB 180/heptachloro biphenyls (around a factor of 2). This shows that even with several simplifying assumptions the CliMoChem model is capable of reproducing the fractionation pattern that is observed in the field.

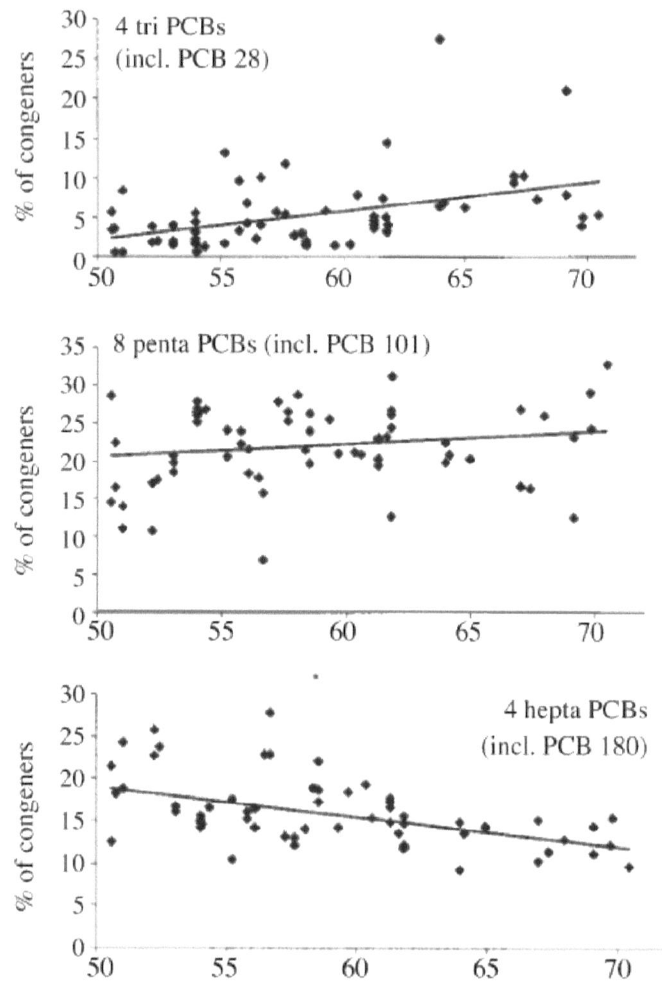

Figure 8. Fractions of different PCB homologue groups as function of latitude (measurements in European soils). (Reproduced from Meijer et al., 2002. With permission. Copyright: American Chemical Society, 2002.)

Currently, we are conducting an extensive evaluation of the CliMoChem model with DDT and its transformation products DDE and DDD as case study (Schenker et al., 2007a). Transformation products such as DDE and heptachlor epoxide contribute to a considerable extent to the environmental and human exposure to POPs (Schenker et al., 2007b). To investigate the influence of transformation products quantitatively, the CliMoChem model has been extended to account for the formation and environmental distribution of transformation products. With this improved version of the model, it is possible to analyze the spatial distribution pattern of parent compound and transformation products in parallel and to evaluate the relative amounts of parent compound and transformation products, e.g., the ratio of DDT to DDE, as an additional dimension of model performance. The agreement of model results and DDT concentrations measured in all regions of the globe is good (Schenker et al., 2007a).

Model results show a slight enrichment of DDE and DDD towards the Arctic, which is mainly caused by the continuous formation of DDE and DDD during the transport of DDT so that the "source region" of these two transformation products is much broader than that of DDT. If a combination of parent compound and transformation products is analyzed for global fractionation, the fractionation process is even more complicated because the chemicals are not released in the same geographical region, but the parent compound is a moving source of the transformation products. This effect may also confound the fractionation pattern of PBDEs because lighter PBDEs can to some extent be formed out of heavier PBDEs by debromination reactions.

3. Conclusions and Research Needs

Several key questions in the context of the environmental distribution of POPs center around the interplay between three competing processes: (i) transport by moving air masses; (ii) net deposition from air to surface media; and (iii) degradation of airborne POPs. The more effective net deposition and/or degradation are, the less is a chemical available for transport out of the region considered. Under conditions of strong degradation (high concentration of hydroxyl radicals) the less persistent compounds such as lighter PCBs are removed predominantly and the more persistent chemicals are more available for atmospheric long-range transport. To accurately describe such situations, the reactivity of POPs in the gas phase, but also in the fraction bound to aerosol particles, needs to be determined. Measurements of reactivity with hydroxyl radicals have been performed for PCBs (Anderson and Hites, 1996), hexachlorocyclohexanes (Brubaker and Hites, 1998; Krüger et al., 2005), DDT, aldrin and mirex (Krüger et al., 2005), and terbuthylazine (Palm et al., 1997; Krüger et al., 2005) but measurements for other organochlorine pesticides, polybrominated diphenylethers and most current-use pesticides are still missing.

The less volatile POPs are predominantly removed by atmospheric deposition. To determine the deposition velocity of POPs, first their interaction with atmospheric aerosols needs to be quantified (Bidleman, 1988; Pankow, 1998; Götz et al., 2007). The fate of the particle-bound fraction of POPs, which can be close to 100% for POPs with high K_{oa}, is strongly linked to the environmental distribution dynamics of aerosols (concentration, composition, and deposition velocity of aerosol particles of different size classes). Finally, the properties of the surface media to which the POPs are deposited need to be known. The content of organic matter in soil, the concentration and settling dynamics of plankton in seawater, the sorptive capacity of vegetation or ice/snow are all influential because they determine the extent of revolatilization of POPs back into the atmosphere (Lee et al., 1998; Jaward et al., 2004; Su and Wania, 2005; Stocker et al., 2007).

In all these areas, there are many open questions. The influence of atmospheric aerosols on the distribution dynamics of POPs is still insufficiently understood. The coupling of POP concentrations in air and in surface media can be strong but to determine the degree of this coupling and to identify the most influential environmental parameters, measurements in both air and surface media with high temporal resolution are needed and the surface media have to be characterized with respect to their sorptive capacity for POPs.

To support research in all these fields (and other areas of importance to the problem of large-scale chemical pollution), an international network of scientists, the International Panel on Chemical Pollution (IPCP) is currently being established (Scheringer et al., 2006). Scientists interested in participating in the work of the IPCP are invited to contact the author at ipcp@chem.ethz.ch. The main objectives of the IPCP are to review and evaluate existing scientific knowledge in the field of environmental chemistry and toxicology, to identify knowledge gaps and research priorities, and to communicate established results, remaining uncertainties, and research needs to decision makers and the public.

Acknowledgments

I thank Konrad Hungerbühler, Fabio Wegmann, Matthew MacLeod, and Urs Schenker for valuable discussions and support.

References

Anderson, P.N., Hites, R.A., 1996, OH radical reactions: the major removal pathway for polychlorinated biphenyls from the atmosphere, *Environ. Sci. Technol.* **30**, 1763–1765.

Beyer, A., Wania, F., Gouin, T., Mackay, D., Matthies, M., 2003, Temperature dependence of the characteristic travel distance, *Environ. Sci. Technol.* **37**, 766–771.

Bidleman, T.F., 1988, Atmospheric processes, *Environ. Sci. Technol.* **22**, 361–367.

Bignert, A., Olsson, M., Persson, W., Jensen, S., Zakrisson, S., Litzén, K., Eriksson, U., Häggberg, L., Alsberg, T., 1998, Temporal trends of organochlorines in Northern Europe, 1967–1995. Relation to global fractionation, leakage from sediments and international measures, *Environ. Pollut.* **99**, 177–198.

Breivik, K., Sweetman, A., Pacyna, J.M., Jones, K.C., 2002, Towards a global historical emission inventory for selected PCB congeners – a mass balance approach. 2. Emissions, *Sci. Total Environ.* **290**, 199–224.

Breivik, K., Wania, F., Muir, D.C.G., Alaee, M., Backus, S., 2006, Empirical and modeling evidence of the long-range atmospheric transport of decabromodiphenyl ether, *Environ. Sci. Technol.* **40**, 4612–4618.

Brubaker, W.W., Hites, R.A., 1998, OH reaction kinetics of gas-phase α- and γ-hexachlorocyclohexane and hexachlorobenzene, *Environ. Sci. Technol.* **32**, 766–769.

Calamari, D., Bacci, E., Focardi, S., Gaggi, C., Morosini, M., Vighi, M., 1991, Role of plant biomass in the global environmental partitioning of chlorinated hydrocarbons, *Environ. Sci. Technol.* **25**, 1489–1495.

Götz, C., Scheringer, M., MacLeod, M., Roth, C., Hungerbühler, K., 2007, Alternative approaches for modeling gas-particle partitioning of semivolatile organic chemicals: model development and comparison, *Environ. Sci. Technol.* **41**, 1272–1278.

Goldberg, E.D., 1975, Synthetic organohalides in the sea, *Proc. R. Soc. Lond. B* **189**, 277–289.

Hansen, K.M., Christensen, J.H., Brandt, J., Frohn, L.M., Geels, C., 2004, Modelling atmospheric transport of α-hexachlorocyclohexane in the Northern Hemisphere with a 3-d dynamical model: DEHM-POP, *Atmos. Chem. Phys.* **4**, 1125–1137.

Hunter-Smith, R.J., Balls, P.W., Liss, P.S., 1983, Henry's law constants and the air–sea exchange of various low molecular weight halocarbon gases, *Tellus* **35B**, 170–176.

Jaward, F.M., Barber, J.L., Booij, K., Dachs, J., Lohmann, R., Jones, K.C., 2004, Evidence for dynamic air-water coupling and cycling of persistent organic pollutants over the open Atlantic Ocean, *Environ. Sci. Technol.* **38**, 2617–2625.

Kalantzi, O.I., Alcock, R.E., Johnston, P.A., Santillo, D., Stringer, R.L., Thomas, G.O., Jones, K.C., 2001, The global distribution of PCBs and organochlorine pesticides in butter, *Environ. Sci. Technol.* **35**, 1013–1019.

Krüger, H.-U., Gavrilov, R., Liu, Q., Zetzsch, C., 2005, Development of a method for determining tropospheric degradation of semivolatile pesticides by OH radicals. German Federal Environment Agency, Berlin, Germany, Report FKZ 201 67 424/02.

Lee, R.G.M., Hung, H., Mackay, D., Jones, K.C., 1998, Measurement and modeling of the diurnal cycling of atmospheric PCBs and PAHs, *Environ. Sci. Technol.* **32**, 2172–2179.

Leip, A., Lammel, G., 2004, Indicators for persistence and long-range transport potential as derived from multicompartment chemistry-transport modelling, *Environ. Pollut.* **128**, 205–221.

MacLeod, M., Scheringer, M., Hungerbühler, K., 2007, Estimating enthalpy of vaporization from vapor pressure using Trouton's rule, *Environ. Sci. Technol.* **41**, 2827–2832.

Malmquist, C., Bindler, R., Renberg, I., van Bavel, B., Karlsson, E., Anderson, N.J., Tysklind, M., 2003, Time trends of selected persistent organic pollutants in lake sediments from Greenland, *Environ. Sci. Technol.* **37**, 4319–4324.

Meijer, S.N., Steinnes, E., Ockenden, W.A., Jones, K.C., 2002, Influence of environmental variables on the spatial distribution of PCBs in Norwegian and U.K. soils: implications for global cycling, *Environ. Sci. Technol.* **36**, 2146–2153.

Meijer, S.N., Ockenden, W.A., Sweetman, A., Breivik, K., Grimalt, J.O., Jones, K.C., 2003, Global distribution and budget of PCBs and HCB in background soils: implications for sources and environmental processes, *Environ. Sci. Technol.* **37**, 667–672.

Müller-Herold, U., Smieszek, T., Peter, P., Scheringer, M., Morosini, M., 2006, A simple measure for precautionary assessment of organic chemicals with respect to global cold condensation, *Environ. Mod.* **194**, 266–273.

Ottar, B., 1981, The transfer of airborne pollutants to the Arctic region, *Atmos. Environ.* **15**, 1439–1445.

Palm, W.-U., Elend, M., Krüger, H.-U., Zetzsch, C., 1997, OH radical reactivity of airborne terbuthylazine adsorbed on inert aerosol, *Environ. Sci. Technol.* **31**, 3389–3396.

Pankow, J.F., 1998, Further discussion of the octanol/air partition coefficient K_{oa}, as a correlating parameter for gas/particle partitioning coefficients, *Atmos. Environ.* **32**, 1493–1497.

Rahn, K.A., Heidam, N.Z., 1981, Progress in Arctic air chemistry, 1977–1981: a comparison of the first and second symposia, *Atmos. Environ.* **15**, 1345–1348.

Risebrough, R.W., 1990, Beyond long-range transport: a model of a global gas chromatographic system, in: *Long-range transport of pesticides*, Kurtz, D.A., ed., Lewis Publishers, Chelsea, MI, pp. 417–426.

Schenker, U., Scheringer, M., Hungerbühler, K., 2007a, Investigating the global fate of DDT: model evaluation and estimation of future trends, *Environ. Sci. Technol.* (under review).

Schenker, U., Scheringer, M., Hungerbühler, K., 2007b, Including degradation products of persistent organic pollutants in a global multi-media box model, *Environ. Sci. Pollut. Res.* **14**, 145–152.

Scheringer, M., Wania, F., 2003, Multimedia models of global transport and fate of persistent organic pollutants, in: *Handbook of Environmental Chemistry 3 O: Persistent Organic Pollutants*, Fiedler, H., ed., Springer, Heidelberg, pp. 237–269.

Scheringer, M., Wegmann, F., Fenner, K., Hungerbühler, K., 2000, Investigation of the cold condensation of persistent organic pollutants with a global multimedia fate model, *Environ. Sci. Technol.* **34**, 1842–1850.

Scheringer, M., Salzmann, M., Stroebe, M., Wegmann, F., Fenner, K., Hungerbühler, K., 2004, Long-range transport and global fractionation of POPs: insights from multimedia modeling studies, *Environ. Pollut.* **128**, 177–188.

Scheringer, M., Fiedler, H., Suzuki, N., Holoubek, I., Zetzsch, C., Bergman, Å., 2006, Initiative for an International Panel on Chemical Pollution (IPCP), *Environ. Sci. Pollut. Res.* **13**, 432–434.

Simonich, S.L., Hites, R.A., 1995, Global distribution of persistent organochlorine compounds, *Science* **269**, 1851–1854.

Shen, L., Wania, F., Lei, Y.D., Teixeira, C., Muir, D.C.G., Bidleman, T.F., 2005, Atmospheric distribution and long-range transport behavior of organochlorine pesticides in North America, *Environ. Sci. Technol.* **39**, 409–420.

Shen, L., Wania, F., Lei, Y.D., Teixeira, C., Muir, D.C.G., Xiao, H., 2006, Polychlorinated biphenyls and polybrominated diphenyl ethers in the North American atmosphere, *Environ. Pollut.* **144**, 434–444.

Stocker, J., Scheringer, M., Wegmann, F., Hungerbühler, K., 2007, Modeling the effect of snow and ice on the global environmental fate and long-range transport potential of semivolatile organic compounds, *Environ. Sci. Technol.* **41**, 6192–6198.

Su, Y., Wania, F., 2005, Does the forest filter effect prevent semivolatile organic compounds from reaching the Arctic? *Environ. Sci. Technol.* **39**, 7185–7193.

ter Schure, A.F.H., Larson, P., Merilä, J., Jönsson, K.I., 2002, Latitudinal fractionation of polybrominated diphenyl ethers and polychlorinated biphenyls in frogs (Rana temporaria), *Environ. Sci. Technol.* **36**, 5057–5061.

Wania, F., Mackay, D., 1993, Global fractionation and cold condensation of low volatility organochlorine compounds in polar regions, *Ambio* **22**, 10–18.

Wania, F., Mackay, D., 1995, A global distribution model for persistent organic pollutants, *Sci. Total Environ.* **160/161**, 211–232.

Wania, F., Mackay, D., 1996, Tracking the distribution of persistent organic pollutants, *Environ. Sci. Technol.* **30**, 390A–396A.

Wania, F., Mackay, D., 1999, The evolution of mass balance models of persistent organic pollutant fate in the environment, *Environ. Pollut.* **100**, 223–240.

Wania, F., Su, Y., 2004, Quantifying the global fractionation of polychlorinated biphenyls, *Ambio* **33**, 161–168.

Wegmann, F., Scheringer, M., Möller, M., Hungerbühler, K., 2004, Influence of vegetation on the environmental partitioning of DDT in two global multimedia models, *Environ. Sci. Technol.* **38**, 1505–1512.

Weschler, C., 1981, Identification of selected organics in the Arctic aerosol, *Atmos. Environ.* **15**, 1365–1369.

EFFECT OF CATASTROPHIC FLOODS ON THE SPATIAL DISTRIBUTION AND VERTICAL MIGRATION OF PAHs IN SOIL OF FLOODED AREAS

IRENA TWARDOWSKA*, KRYSTYNA JANTA-KOSZUTA, SEBASTIAN STEFANIAK
Polish Academy of Sciences, Institute of Environmental Engineering, 34 M.Sklodowska-Curie Street, 41-819 Zabrze, Poland

Abstract: A focus on soil quality issues in the EU has resulted in extensive studies aimed to development of a Soil Framework Directive as well as of other new relevant regulations and standards, in parallel with setting up a harmonized European Soil Monitoring System for routine soil monitoring/ screening with use of reference analytical methods. Polycyclic Aromatic Hydrocarbons (PAHs) belong to the most problematic contaminants from the local sources and diffuse inputs to be monitored and controlled. This study presents the screening survey for 16 PAHs carried out in 2000 in the area severely impacted by the catastrophic flood of 1997 in the Odra River valley in Poland, which affected also Czech Republic and Germany. Within this survey, 16 PAH contents in soils due to river sediments deposition resulting from the flood, as well as the effect of flood on the distribution of PAHs in soils of the affected area were assessed (by the GC-MS reference methods) in view of soil quality standards and need for remediation. The postflood PAH spatial distribution with use of the Geographical Information System (GIS) showed distinct correlation with floodwater flow conditions, while total 16 PAH and specific compounds concentrations in the upper soil layer 0–0.20 m appeared to be mostly within the standard limits. In 17% of composite samples, 16 PAH concentrations were found to be considerably elevated, up to the values >1,000 μg/kg, and exceeded the standards for agricultural soils in particular samples. PAH compounds displayed also different vertical migration potential in soils. The occurrence of the maximum PAH concentrations in the floodwater stagnation areas

* To whom correspondence should be addressed. E-mail: irena@ipis.zabrze.pl

E. Mehmetli and B. Koumanova (eds.),
The Fate of Persistent Organic Pollutants in the Environment, 205–218.

confirmed river sediments to be the major source of these compounds. The qualitative composition of 16 PAHs (ANty < Flth < CHR < BaA < PYR) in humus layer of soils in these areas denoted anthropogenic sources of these compounds.

Keywords: PAHs, soil screening, PAH analytical methods, spatial distribution, 16 PAH quantitative composition

1. Introduction

Catastrophic floods are a new, though recently frequent phenomenon in the map of Europe, affecting vast areas of fertile agricultural lands. Information concerning long-term environmental consequences to soils of the flooded areas is still scarce and incomplete. This brings about the need of assessing the extent of possible impact of flood on the soil status in these areas. Soil quality is a crucial issue for assuring human health and environmental sustainability, thus the requirements with this respect are getting more restrictive. Current activity of the EU Member States is focused on the source control of both point and nonpoint anthropogenic sources of soil contamination, such as dry and wet deposition of air pollutants and application of fertilizers, pesticides or sewage sludge in agriculture. Nevertheless, soil in the EU is still subject to a number of degradation processes, one of which has been identified as local and diffuse contamination from different sources.

In response to concerns about the degradation of soils in the EU, the European Commission issued in 2002 a Communication (COM, 2002), which outlined the first steps that lead to the development of a Thematic Strategy to protect soils in the European Union with an aim of laying down the principles of Community soil protection policy and adopting a Soil Framework Directive. The Technical Working Groups established for the development of the EU soil protection policy, compiled in 2004 recommendations for contamination prevention from local sources and diffuse inputs, as well as for contaminated land management and monitorring (Van-Camp et al., 2004). Among diffuse soil contamination, "water pollution in the case of sediments" was specified, which comprise also such environmental disasters as floods that might contribute to soil contamination by river sediments.

Recommendations for monitoring issued by this body have covered classical soil monitoring, multipurpose monitoring, action-driven monitorring, and regular reporting of these "action-driven" monitoring has been considered the most appropriate approach to assessment of soil contamination. Substances to be measured that considered to pose particular risk to human health and the environment included polycyclic aromatic hydrocarbons (PAHs).

A focus on soil quality issues has resulted in new restrictive national soil quality regulations, as well as in compulsory routine soil monitoring and postemergency screening with use of specified reference analytical methods also in some European countries, among them in Poland. In the national regulations, PAHs are considered to be one of the most problematic groups of the soil contaminants to be monitored and controlled (RMS, 2002).

This study presents the screening survey for 16 PAHs carried out in 2000 in the area severely impacted by the catastrophic flood of 1997 in the Odra River valley in Poland, which affected also Czech Republic and Germany. Within this survey, 16 PAH contents in soils due to river sediments deposition resulting from the flood, as well as the effect of flood on the distribution of PAHs in soils of the affected area were assessed in view of soil quality standards and need for remediation.

2. Surveyed Area and Methods

2.1. AREA CHARACTERISTICS

The screening survey was carried out in 2000, i.e., 3 years after the flood, in the selected representative area of Kuznia Raciborska community at the right bank of the upper Odra River within Silesia land, Poland. The area was typical with respect to soils and agricultural use. Heavy alluvia of pH close to neutral (pH > 6.5) dominated, though weakly acidic (pH 5.6–6.5) brown soils and strongly acidic (pH < 4.5) light sandy soils also occurred there. Predominant part of the flooded area comprised arable lands, pastures and grasslands, while nonflooded lands were mainly covered with forests. The area was drained by the Ruda River, the right-bank tributary of the Odra River. Due to high embankments it was not endangered by frequent floods in the previous decades, but belonged to the areas particularly strongly affected by the catastrophic flood of 1997. The situation, characteristics of the area and the general selection criteria were presented and discussed in detail elsewhere (Twardowska et al., 2004).

The criteria specific for PAH sources were as follows: (i) there are no significant potential sources of PAH contamination in the selected area except local traffic and petrol stations; and (ii) upstream, typical sources of anthropogenic contamination by PAH are sited, among them the Radlin coke plant in Wodzisław Śląski and in particular the carbon electrode plant in Racibrz that might have caused PAH enrichment in the river sediments and soils in the vicinity of these enterprises.

2.2. SAMPLING AND ANALYTICAL METHODS

The sampling procedure and location of sampling cells was also addressed in detail in the referred paper (Twardowska et al., 2004). Composite homogenized samples from the humus soil layer (0–20 cm for arable land and 0–5 cm for meadows and grazing land) for PAHs analysis were prepared in accordance with ISO/DIS 10381 – 1/2/3/4/5 from ≥30 systematically spaced increments poled from every of 24 selected cells 400 m^2 each, uniform with respect to soil and land use, and from the undisturbed soil/subsoil profile O-III in the meadows between the protecting riverbanks.

From the samples, aromatic fraction was eluted, purified and analyzed for 16 PAHs according to ISO 18287 using GC-MS (Varian 3400) technique (ISO CD 18287). GC-MS has been considered also as a European standard with a target date 2007 for adoption (CEN/TC 292, WI 292047). Concentrations of PAHs were presented in micrograms/kilogram (all in dry matter – dm).

For assessing the effect of flood on the PAHs spatial distribution, Geographical Information System (GIS) was applied. Results were presented in the form of isopleth maps of the area for 16 PAHs and benzo[a]pyrene BaP (Figure 1A, B). There was no monitoring data for PAHs in soils of the area in the preflood period, thus the status of soils with respect to PAHs content was evaluated as total.

3. Results and Discussion

3.1. OCCURRENCE OF PAHs IN THE SURFACE SOIL LAYER

The 16 PAHs content in the humus soil layers 0–20 cm (arable land) or 0–5 cm (meadows and grazing land) in the surveyed area appeared to be generally low, though its spatial distribution varied in a wide range, from 52 to 1,918 μg/kg, thus showing diverse contamination level (Table 1, Figure 1). Concentrations of $\sum$7 PAH from the EPA list, of proven carcinogenic or mutagenic properties (benzo[a]pyrene BaP, dibenz[ah]anthracene DBahA, benzo[a]fluoranthene BbF, benz[a]anthracene BaA, benzo[k]fluoranthene BkF, indeno[1,2,3-cd]pyrene INP, chrysene CHR) occurred in the range from <0.1 to 546 μg/kg, and BaP from <0.1 to 21 μg/kg. The concentration range of $\sum$16 PAH in the humus soil layer was found to be in the whole surveyed area below the guide value of 3,000 μg/kg. This value has been provisionally considered as a safe limit, which indicates that the ecological soil multifunctionality is protected independently of land use (Bachmann et al., 1997). While the EU standards concerning soil quality are still under development, several EU Member States, among them Poland, set up national soil quality standards in force that regulate PAH concentrations (RMS, 2002). Polish standards regulate maximum concentration level (MCL) of 10 PAHs in soil and subsoil with regard to its current and planned use, depth of a soil layer below the ground level and hydraulic conductivity. The standardized compounds comprise naphthalene (Nta), phenantrene (Pht), antracene (A), fluoranthene (Flth), chrysene (CHR), benz[a]antrcene (BaA), benzo[a]pyrene (BaP), benzo[a]fluoranthene (BaF) benzo(g,h,i)perylene (BghiP) and their sum (10 PAH). There are MCL differentiation for three soil quality groups (A – protected areas; B – agricultural soils, and C – industrial soil) and within groups B and C in three soil layers below the surface (0–0.3 m; 0.3–15 m and >15 m) and two values of a hydraulic conductivity ($< 10^{-7}$ m/s and $>10^{-7}$ m/s).

TABLE 1. PAH contents (μg/kg dm) in humus soil layer 0–20/0–5 cm of the representative area in the Odra River valley 3 years after the catastrophic flood of 1997 (2000 survey)

Surveyed cells	∑16 PAH	∑7 PAH[a]	BaP	Class[b] (concentration range [μg/kg]) [8,9]	Class (%)[c]
P-6, 9, 16, 21, 25, 27a, 34a	52–193	<0.1–54	<0.1–3.32	I (≤200)	33(33)
P-10,13, 17, 31, 4	224–271	32–108	0.56–4.91	I/II (>200–300)	21
P-8, 34, 39, 46; Profile O-II(0–5 cm)	328–592	70–210	1.43–3.83	II (>300–600)	21
P-15, 19	672–800	172–205	0.32–8.94	III (>600–1,000)	8
P-18, 43; Profiles O-I, -III(0–5 cm)	1,255–1,918	323–546	4.13–21.12	IV(>1,000–3,000)	17(<8)
Poland (216 sites)				> 3,000	(<2)

[a]∑7 PAH = Benzo[a]pyrene BaP, Dibenz[ah]anthracene DBahA, Benzo[b]fluoranthene BbF, Benz[a]anthracene BaA, Benzo[k]fluoranthene BkF, Indeno[1,2,3-cd]pyrene INP, Chrysene CHR

[b]Classification proposed by IUNG – Institute of Cultivation, Fertilizing and Soil Sciences, Poland (Maliszewska-Kordybach et al., 1998; Maliszewska-Kordybach, 1999)

[c]Percentile of classified soils from agricultural lands in All-Poland survey is given in parenthesis (Maliszewska-Kordybach, 1999).

Prior to the enactment of this standard, the IUNG – Institute of Cultivation, Fertilizing and Soil Sciences (Poland) proposed soil classification in I–IV classes according to concentration range that illustrates soil status with regard to PAHs content, and in this respect is a good source of information (Maliszewska-Kordybach et al., 1998; Maliszewska-Kordybach, 1999).

Percentile of sampled soils in the surveyed area containing ∑16 PAH ≤ 1,000 μg/kg accounted for 83%; in one third of soils concentration of ∑16 PAH was < 200 μg/kg, and in 53% was present in the low level of <300 μg/kg that indicated the lack of pollution by PAHs.

These data show a good correlation with the results of screening study carried out for PAHs in the arable lands of Poland by IUNG – Institute of Cultivation, Fertilizing, and Soil Sciences, Poland (Maliszewska-Kordybach et al., 1998; Maliszewska-Kordybach, 1999). The humus soil layer was randomly sampled in the area of 140,000 km^2 of the whole country in 216 sites and analyzed for 13 PAHs. ∑13 PAH content >1,000 μg/kg was detected in less than 8% of sites, and in less than 2% its concentration exceeded precautionary safe level 3,000 μg/kg. In almost 90% of samples taken from the area of 140,000 km^2, ∑13 PAH content was below 1,000 μg/kg, while from 5% to 95% soil samples represented concentration range from 111 to 1,340 μg/kg. Mean concentrations of ∑13 PAHs in Polish arable soils were found to be at the same or lower level than in other European countries (Maliszewska-Kordybach, 1999).

In the studied postflood area the percentile of soils with ∑16 PAH >1,000 μg/kg indicating anthropogenic impact was two times higher, but all the detected concentrations of ∑16 PAH were still within the safe guide limit of 3,000 μg/kg. The concentrations of BaP > 5 μg/kg were found in 16% of samples, in the soils of III–IV class. For the surveyed area, characteristic is lack of strict correlation between ∑16 PAH and BaP contents, and good correlation between ∑7 PAH and ∑16 PAH (Table 1).

In the majority of soil samples the dominating compound was pyrene PYR, less frequently fluorene Fln. In comparable amounts in the soils occurred chrysene CHR and benz[a]anthracene B[a]A, both compounds are proven carcinogens, B[a]A displaying considerably higher carcinogenic potential.

The classification of surveyed surface soil layer for PAHs content according to Polish soil quality standards (RMS, 2002) are presented in Table 2. According to these standards, MCL for all measured individual compounds except benzo[a]pyrene PaP in soils of A and B class accounts for 100 μg/kg dm; for BaP MCL is lower and accounts for 20 μg/kg for A class and 30 μg/kg for B class. For C class MCL values are several orders of magnitude higher.

TABLE 2. Classification of soils (0–20 cm, 0–5 cm) of surveyed postflood area with respect to PAH concentrations according to Polish Soil Standards (RMS, 2002)

PAH	Surveyed cells	Cell number, (% of total)	Concentration range (μg/kg)	Soil group[a] (MCL, μg/kg)
Naphtalene Nta	P-6, 8, 9, 10, 13, 15, 16,17,18,19,21, 25, 27, 27a, 31, 34, 34a, 39, 41, 43, 46, Profiles O-I,O-II, O-III	24 (100%)	0.1–55.23	A,B (<100)
				C (<50,000)
Phenantrene Pht	P-6, 8, 9, 10, 13, 15, 16, 17, 19, 21, 25, 27, 27a, 31, 34, 34a, 39, 41, 46, Profiles O-I, O-III	21(88%)	0.1–89.98	A,B (<100)
	P-18, 43, O-II	3 (12%)	101.25–156.25	C (<50,000)
Anthracene A	P-6, 8, 9, 10, 13, 15, 16, 17, 19, 21, 25, 27, 27a, 31, 34, 34a, 39, 41, 46, Profiles O-I, O-II, O-III	22 (92%)	0.1–80.92	A,B(<100)
	P-18, 43,	2 (8%)	104.7–126,5	C (<50,000)
Fluoranthene Flth	P-6, 8, 9, 10, 13, 15, 16, 17, 19, 21, 25, 27, 27a, 31, 34, 34a, 39, 41, 46, Profiles O-I, O-II, O-III	22 (92%)	0.1–61.55	A,B (<100)
	P-18, 43,	2 (8%)	113.75–151.38	C (<50,000)
Chrysene CHR	P-6, 9, 10, 13, 15, 16, 17, 21, 25, 27, 27°, 31, 34, 34a, 39, 41, O-II	17 (71%)	0.1–84.01	A,B(<100)
	P-8, 18 19, 43, 46 profiles O-I, O-III	7 (29%)	123.87–321.82	C (<50,000)
Benz[a]anthracene BaA	P-6, 8, 9, 10, 13, 15, 16, 17, 19, 21, 25, 27, 27a, 31, 34, 34a, 39, 41, 46,	19 (79%)	0.1–99.64	A,B(<100)
	P-18, 43, profiles O-I, O-II, O-III	5 (21%)	125.32–228.75	C (<50,000)
Benzo[a]piren BaP	P-6, 8, 9, 10, 13, 15, 16, 17, 18, 19, 21, 25, 27, 27a, 31, 34, 34a, 39, 41, 46, profiles O-I, O-II, O-III	23 (96%)	0.1–11.87	A (<20)
	P-43	1 (4%)	21.12	B (<30)
Benzo(g,h,i)perylen e BghiP	P-6, 8, 9, 10, 13, 15, 16, 17, 18, 19, 21, 25, 27, 27a, 31, 34, 34a, 39, 41, 43 46, profiles O-II, O-III	23 (96%)	0.1–4,75	A,B (<100)
	Profile O-I	1 (4%)	485	C (<100,000)
∑10 PAH[b]	P-6, 8, 9, 10, 13, 15, 16, 17, 18, 19, 21, 25, 27, 27a, 31, 34, 34a, 39, 41, 43 P-46, profiles O-I, O-II, O-III	24 (100%)	<0.1–994.18	AB (<1,000)

[a]Soil classes according to Polish Soil Standards (MOS, 2002): A – protected areas; B – agricultural lands; C – industrial lands

[b]Without benzo[a]fluoranthene BaF

In most cells the concentrations of standardized compounds appeared to be below MCL for A and B groups, thus these soils could be considered ecologically safe (Table 2). The values exceeding MCL were detected in floodwater stagnation areas, in particular in cells P-18 and P-43 (Table 2, Figure 1), where MCL values for soils of A and B group were exceeded for 5 of 10 standardized PAHs (Pht, A, Flth, CHR, and BaA). Another group of cells that showed frequent excess of standard MCL values were located in the meadow areas between the protecting riverbanks, where floodwater also stagnated for several days (Profiles O-I, O-II, O-III, 0–5 cm). The highest percentile of soils that exceeded standard MCL values were found for chrysene (29%) and bez[a]anthracene BaA (21%), while both benzo[a]pyrene and 10 PAH concentrations fell within the standards for soils of A and B group (Table 2). The highest excess of MCL was detected for BghiP (fivefold) in one cell (Profile O-I, 0–5 cm), while with respect to other standardized compounds the excess of MCL was rather moderate (25–60%, maximum 2.3-fold for BaA in one cell and 3.2-fold for CHR, also in one cell).

The concentrations of these compounds in the surveyed area were much below RBC – Risk Based Concentrations for human health (Smith, 1994) contamination level of surveyed soils by PAHs may be considered minor.

3.2. SPATIAL DISTRIBUTION OF PAHs IN SOILS OF THE SURVEYED AREA

Spatial distribution of PAHs in the humus soil layer in the form of isopleth maps was assessed with use of GIS program (Figure 1A, B). Due to the lack of data on background concentrations of PAHs in the surveyed area, the effect of flood could not have been evaluated numerically in the form of difference maps and just postflood status was evaluated. The highest PAH concentrations (in the range >600–2,000 μg/kg) were associated with heavy and medium alluvial soils of the inundated terrace along the river bed, in particular in the meadows and grazing lands in the meander areas where due to water stagnation for a longer time, appropriate conditions for river sediments and suspended matter deposition occurred. Spatial distribution of PAH appeared to show a very good compliance with that of $\sum$7 PAH. In general, also BaP concentrations despite weak numerical correlation with $\sum$16 PAH follow the similar spatial distribution pattern (Figure 1A, B).

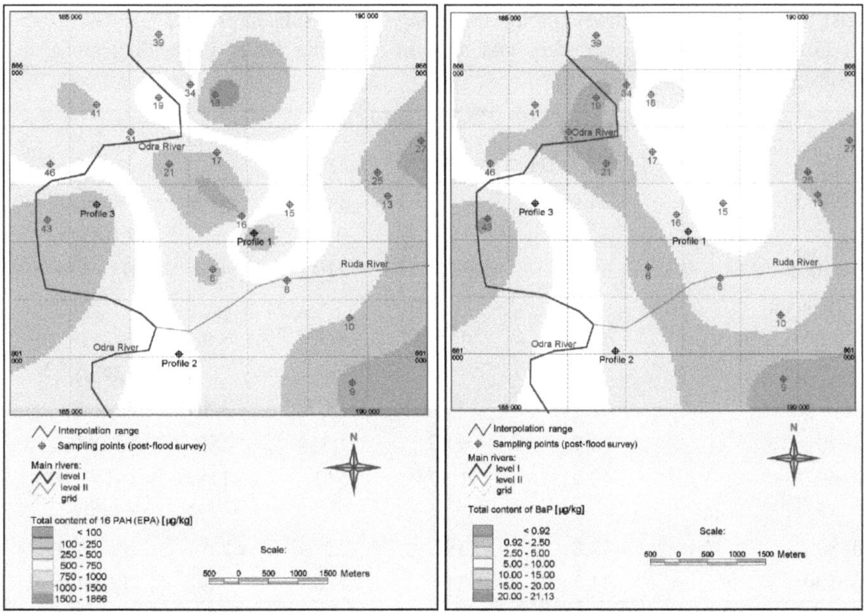

Figure 1. Spatial distribution of (A) ∑16 PAHs and (B) BaP in the humus soil layer 0–20/0–5 cm in the representative postflood area (May 2000 survey)

In this respect, the spatial distribution pattern of PAHs in soils after the flood shows high similarity to the spatial distribution of trace metals presented in another paper (Twardowska et al., 2004), which indicates a strong association of both organic and inorganic compounds with suspended matter in floodwater and the conditions of its sedimentation or washout in the flooded area.

3.3. VERTICAL DISTRIBUTION OF PAHs ALONG THE SOIL PROFILE

While concentrations of 16 PAHs in humus soil layer 0–5 cm were found to be well below the ecologically safe guide value (Bachmann et al., 1997), and contents of standardized PAH compounds were mostly within Polish quality standards of soils (0–30 cm) of Group A and B (RMS, 2002), vertical distribution of PAHs in the undisturbed soil profile (O-III, 0–100 cm) showed very high contamination level of the soil layer 5–40 cm that can be attributed to flood (Table 3, Figures 2 and 3).The contamination of the layer 5–20 cm exceeded many times both ecologically safe guide value 3,000 µg/kg, dm, and standard MCL for soils of A and B group. PAHs concentrations displayed different intensity of migration along the soil profile.

TABLE 3. PAH concentrations (μg/kg dm) along the soil/subsoil profiles O-I, O-II (0–5 cm) and O-III (0–100 cm) in the surveyed area in the Odra River valley 3 years after the catastrophic flood of 1997

Soil profile, layer (cm)	% ∑7 PAH[a, b]	∑16 PAH	∑7 PAH[a]	BaP	Class[c]
O-I, 0–5	24.4	1,416	346	4.13	IV(>1,000–3,000)
O-II, 0–5	35.5	592	210	4.01	II (>300–600)
O-III, layer					
0–5	43.5	1,255	546	8.98	IV(>1,000–3,000)
5–20	47.7	29,207	13,920	3,473	RBC 8800 (BaP)
20–40	13.7	9,871	1,354	189.7	>3,000
40–60	40.9	1,571	643	118.6	IV(>1,000–3,000)
60–80	38.8	587	228	Nd	II (>300–600)
80–100	34.3	344	118	Nd	II (>300–600)

[a]∑7 PAH = Benzo[a]pyrene BaP, Dibenz[ah]anthracene DBahA, Benzo[a]fluoranthene BbF, Benz[a]anthracene BaA, Benzo[k]fluoranthene BkF, Indeno[1,2,3-cd]pyrene INP, Chrysene CHR;

[b]Percentage of carcinogenic 7 PAH share in total 16 PAH content in soil

[c]Classification proposed by IUNG – Institute of Cultivation, Fertilizing and Soil Sciences, Poland

RBC – Risk-based concentration (BaP = 8,800 μg/kg dm) [10]; Nd – not detected

Most compounds occurred in the highest concentrations in the layer 5–20 cm (fluorene Fln, phenantrene Pht, anthracene A, fluoranthene Flth, pyrene PYR, chrysene CHR, benz[a]anthracene BaA, benzo[b + k]fluoranthene B(b + k)F, benzo[a]pyrene BaP, benzo(ghi)perylene BghiP), of these concentrations Flth < B(b + k)F < CHR ≈ BaP < BaA < PYR exceeded 1,000 μg/kg dm and ranged from 2,848 to 10,306 μg/kg. The weakest vertical migration showed acetynaphtylene ANty with maximum concentration in 0–5 cm layer, while the best migrants were naphtalene Nta and acenaphtene ANte with maximum in 20–40 layer (Figure 3).

Therefore, quantitative pattern indicates slow gradual redistribution of PAHs downward the profile, diverse for different compounds. While in the surface layer 0–5 cm of all three profiles Σ16 PAH contents were at the toxicologically safe, though elevated level <1,500 μg/kg, ∑16 PAH concentrations in the layer 5–20 cm was as high as 30,000 μg/kg that was tenfold higher than the recommended guide value 3,000 μg/kg. Very high concentrations of carcinogenic compounds draw special attention (∑7 PAH ≈ 14,000 μg/kg., BaP ≈ 3,500 μg/kg., close to the RBS value) In the deeper layers the reduction of PAH concentrations was observed, up to $\sum 16$ PAH = 350 μg/kg (Figure 2).

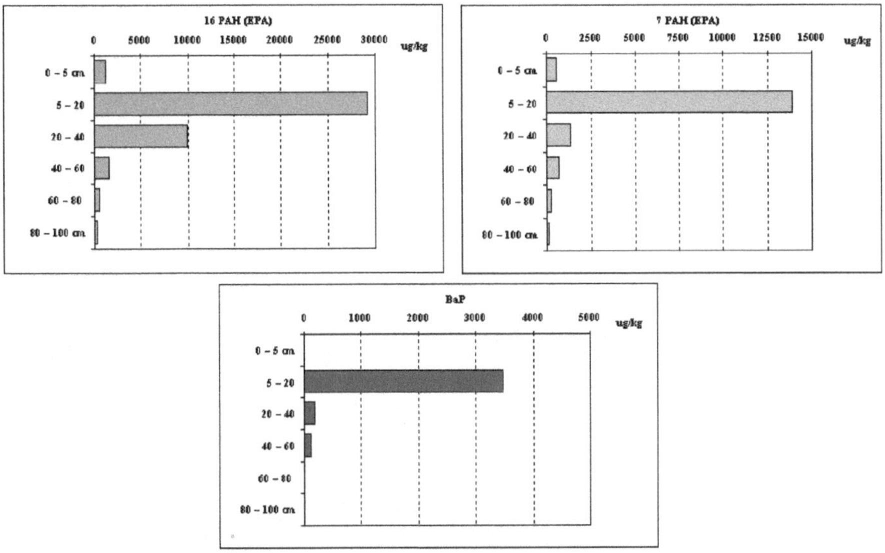

Figure 2. Contents of $\sum 16$ PAH, $\sum 7$ PAH (US EPA), and BaP along the soil profile O-III in the postflood period (May 2000 survey)

The qualitative pattern of PAHs denotes strong upstream source of river sediment contamination, presumably Carbon Electrode Plant in Racibórz, and vertical redistribution of contaminants in time, different for particular compounds. In the surface 0–5 cm layer acenaphtylene Any > CHR > BaA > PYR dominated, while in the next 5–20 cm layer pyrene was the prevailing compound: PYR > BaA ≈ BaP ≈ CHR ≈ B(b + k)F > Flth (Figure 3).

Qualitative composition of PAHs differed considerably from that commonly occurring in the soils, which usually shows domination of fluoranthene Flth. Changing qualitative pattern of PAH composition along the vadose zone profile suggests different dynamics of their downward migration. Considering the low infiltration rate in heavy and medium alluvia, the migration of PAH loads along the vadose zone towards the Quaternary ground water table may last several years. Therefore, though detected PAH concentrations in the uppermost undisturbed 0–5 cm layer are ecologically safe, contents and pattern of PAHs in the layer 5–40 cm and downward migration trends pose a hazard to the soil organisms, plants and potentially to ground water.

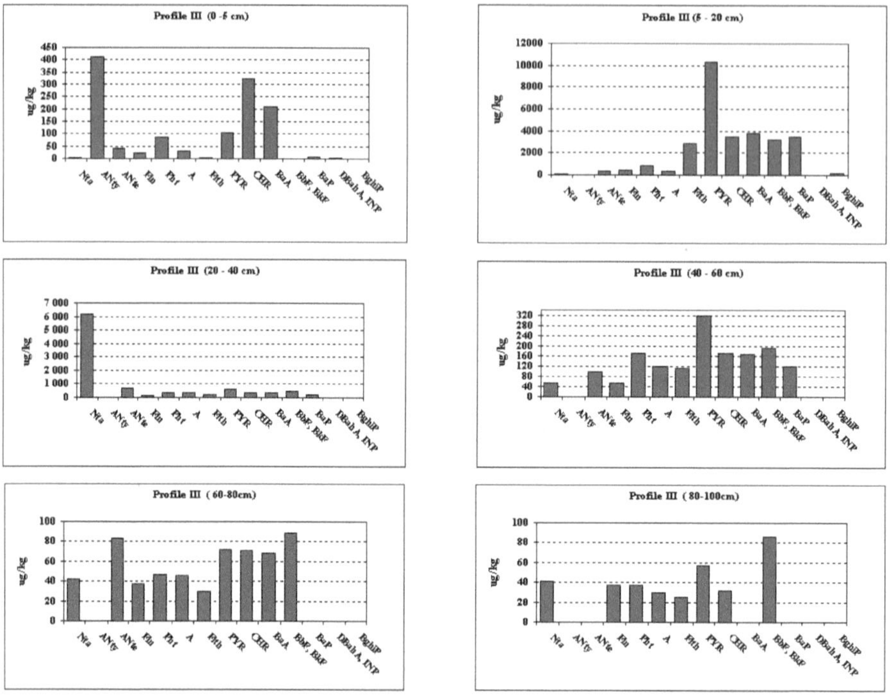

Figure 3. Distribution pattern of 16 PAHs along the vadose zone profile O-III (May 2000 survey)

For better evaluation of flood impact on soil contamination by PAHs originated from river sediments and suspended matter, the further more detailed monitoring studies and close observations of spatial PAH quantitative and qualitative distribution trends along the soil/subsoil profile of the vadose zone in postflood area are required. For these purposes, the fast prescreening with use field rod techniques with Raman and SERS monitors for PAH analysis in pore solution, e.g., those proposed by Vo-Dinh (2004), would be of great use.

4. Conclusions

Despite of location of anthropogenic PAHs sources upstream of the surveyed postflood area, the concentration range of ∑16 PAH in the uppermost humus soil layer (0–20 cm in arable land, 0–5 cm in meadow and pasture land) of this area was found to be much below the guide value 3,000 μg/kg, which is considered to assure ecological soil function, and show good correlation with the general PAH concentration pattern in Poland.

The postflood PAH spatial distribution with use of the Geographical Information System (GIS) showed distinct correlation with floodwater flow conditions and was the highest in the long-term water stagnation areas.

Most of PAH concentrations detected in the humus layer did not exceed Polish soil standards for soils of A and B group, though in 21% of surveyed cells benz[a]anthrscene BaA that is proven carcinogen, and in 29% of cells chrysene CHR of a weaker carcinogenic potential, exceeded Polish standards for agricultural soils. Concentrations of the strongest carcinogens benzo[a]pyrene BaP and dibenz[ah]anthracene DBahA did not occur in concentrations exceeding the standards in the whole surveyed area.

Observed occurrence of PAH compounds within the inundation terrace downward the alluvial soil profile (with maximum in 5–20 cm layer) in concentrations much above the guide value and close to the Risk-Based Concentrations poses a hazard to soil organisms and potentially to ground waters. Qualitative pattern of PAHs suggests a strong anthropogenic source of these compounds upstream of the surveyed area.

Spatial distribution of PAHs in the soils of postflood area proves river sediments and suspended matter to be a major mean of PAH transport and soil enrichment. For these reasons, the further more detailed monitoring studies and close observations of PAH quantitative and qualitative distribution trends in pore solutions along the soil/subsoil profile of the vadose zone in postflood areas with use of field rod techniques with Raman or SERS monitors would be of great use as a source of prescreening information and detecting possible "hot spots."

Acknowledgments

The study was funded by the Polish Committee of Scientific Research, Grant 3 TO9C02415. The PAHs analysis was performed by M. Czaplicka.

The results of the study were for the first time presented by the authors in a paper: Twardowska, I., Janta-Koszuta, K., Stefaniak, S., Kyziol, J., 2005, Results of PAH screening in floodplain soils vs soil quality standards, in *Advanced Environmental, Chemical and Biological Sensing Technologies III*, edited by Tuan Vo-Dinh, R.A. Lieberman, and G. Gauglitz. *Proceedings of SPIE,* Vol. 5993, SPIE, Bellingham, WA, 2005, 59930H.

References

Bachmann, G., Terytze, K.F., Rück, F., and Bannick, C.G., 1997, Precautionary values for the protection of soil; derivation and recommendation for some heavy metals, PCB and PAH. In: *Technological and Economic Aspects of Soil Bio/Phytoremediation.* Materials of Training Course. Plovdiv, Bulgaria, 5–7 October.

CEN/TC 292, Target Date 2007-03, Work Item WI 292049 *Characterization of waste – Determination of Polycyclic Aromatic Hydrocarbons(PAH)s in Soil, Sludge and Waste using High-Performance Liquid Chromatography (HPLC) or Gas Chromatography – Mass Spectrometry (GC-MS)*, WG 5.

COM, 2002, 179 final, Commission of the European Communities, Communication from the Commission to the Council, the European parliament, the Economic and Social Committee and the Committee of the Regions, *Towards a Thematic Strategy for Soil Protection,* Brussels, 16 April 2002.

ISO CD 18287, Determination of polycyclic aromatic hydrocarbons (PAH): gas chromatographic method with mass spectrometric detection. In *International Standards: Soil Quality ISO TC 190.*

ISO-DIS 10381-1/2/3/4. Part 1: guidance on the design and sampling programmes. Part 2: guidance on sampling techniques. Part 3: guidance on safety. Part 4: guidance on procedures for the investigation of soil contamination of urban and industrial sites.

Maliszewska-Kordybach, B., 1999, Persistent organic contaminants in the environment. PAHs as a case study, in *Bioavailability of Organic Xenobiotics in the Environment and Practical Consequences for Bioremediation*, Block, J.C. and Goncharuk, V.V., eds., Kluwer Academic, Dordrecht/Boston/London, 1999.

Maliszewska-Kordybach, B., Terelak, H., and Smreczak, B., 1998, Wpływ przemysłu, komunikacji i urbanizacji na zawartość WWA w glebach z terenu gruntów ornych w Polsce (Impact of industry, transport and urbanization on PAH content in soils from arable lands.in Poland)., in *Zwiazki Organiczne w Srodowidku i Metody ich Oznaczania. Mater. Symp., Jachranka 1998,* Biblioteka Monitoringu Srodowiska, Warszawa, pp. 233–234 (in Polish).

RMS – Directive of Minister of Environment of 9 September 2002 on the standards of soil quality and the standards of earth quality. *Dz.U.* 02.165.1359 of 4 October 2002.

Smith, R.L., 1994, *Risk-Based Concentrations: a Method to Prioritize Environmental Problems Using Limited Data.* US EPA Region III, Philadelphia.

Twardowska, I., Janta-Koszuta, K., Stefaniak, S., and Kyzioł, J., 2004, Screening and monitoring of metal contamination in soils of environmental disaster areas: available techniques and needs, in *Advanced Environmental, Chemical and Biological Sensing Technologies II*, Tuan Vo-Dinh, Gauglitz, G. and Lieberman, R.A., eds. *Proceedings of SPIE*, Vol. 5586, SPIE, Bellingham, WA, pp. 116–125.

Twardowska, I., Janta-Koszuta, K., Stefaniak, S., Kyziol, J., 2005, Results of PAH screening in floodplain soils vs. soil quality standards, in *Advanced Environmental, Chemical and Biological Sensing Technologies III*, Tuan Vo-Dinh, Lieberman, R.A., and Gauglitz, G., eds., *Proceedings of SPIE,* Vol. 5993, SPIE, Bellingham, WA, 59930H.

Van-Camp, L., Bujarrabal, B., Gentile, A.-R., Jones, R.J.A., Montanarella, L., Olazabal, C., Selvaradjou, S.-K., 2004, *Reports of the Technical Working Groups Established under the Thematic Strategy for Soil Protection.* EUR 21319 EN/4, Office for Official Publications of the European Communities, Luxembourg, p. 872.

Vo-Dinh, T., 1994, Remote monitors for in situ characterization of hazardous wastes, in *Solid Waste: Assessment, Monitoring and Remediation*, Twardowska, I., Allen, H.E., Kettrup, A.A.F., and Lacy, W.J., eds., Elsevier, Amsterdam/Boston/Heidelberg/ London/New York/Oxford/Paris/San Diego/San Francisco/Singapore/Sydney/Tokyo, pp. 485–504.

SPATIAL AND TEMPORAL TRENDS OF GLOBAL, REGIONAL, AND LOCAL POPs DISTRIBUTION

IVAN HOLOUBEK,* JANA KLÁNOVÁ
RECETOX, Faculty of Science, Masaryk University, Regional Central and Eastern European POPs Centre, National POPs Centre of the Czech Republic, Kamenice 126/3, 625 00 Brno, Czech Republic

Abstract: The Stockholm Convention on Persistent Organic Pollutants was adopted with the objective of protecting human health and the environment from persistent organic pollutants. The effectiveness evaluation of Convention measures needs the comparable monitoring data on the presence of the POPs including the spatial and temporal trends evaluation and their regional and global environmental transport. Air monitoring for POPs has conventionally been conducted at a very limited number of sites using "active" or high volume air samplers or broadly with application of passive air sampling. Persistent organic pollutants have been monitored using the integrated monitoring approach in all environmental matrices in the Kosetice observatory, Czech Republic since 1988. This facility serves as a Central European background monitoring station for the purpose of various national and international monitoring program and national environmental research projects. Using the integrated monitoring approach, a set of monitoring parameters and ecosystem indicators was applied to determine the state of the environment, to detect anthropogenic impacts and influences, to determine the time-related trends of POPs levels, and to predict the future changes of terrestrial and freshwater ecosystems in a long-term perspective.

Keywords: POPs, temporal and spatial trends, monitoring

* To whom correspondence should be addressed. E-mail: holoubek@recetox.muni.cz

E. Mehmetli and B. Koumanova (eds.),
The Fate of Persistent Organic Pollutants in the Environment, 219–228.

1. Persistent Organic Pollutants

1.1. BASIC CHARACTERISTICS

POPs are persistent, bioaccumulative, and toxic substances, likely to cause the adverse human health or environmental effects.[1] In this context, "a substance" means a single chemical species or a number of chemical species which form a specific group by virtue of (i) having similar properties and being emitted together into the environment or (ii) forming a mixture marketed as a single product. Depending on their mobility in the environment, PBTs can be of a local, regional or global concern.[2] POPs are a group of compounds prone to the long-range atmospheric transport and deposition in the distant regions. A global impact of the POP pollution became apparent when POPs were detected in the polar regions where they have never been used or produced, at the levels posing risks to both wildlife and humans. In 1992, a newly established initiative of United Nations Economic Commission for Europe in the UN-ECE region (Eastern and Western Europe, Canada, and USA) had prepared the Protocol on POPs (includes 16 POPs)[3] with the goal to control, reduce or eliminate their discharges, emissions and losses. Beside that, a similar program of United Nations Environment Program (covering 12 POPs) was introduced in cooperation with the International Forum for Chemical Safety (UNEP/ IFCS).[1] Based on this, the Stockholm Convention was developed, signed and ratified 141 countries.

1.2. GLOBAL POPs MONITORING

The objective of the Stockholm Convention on POPs[1] can be stated as to protect human health and the environment from persistent organic pollutants by reducing or eliminating releases to the environment. Parties have agreed that they need a mechanism to measure whether this objective is reached. According to Article 16 of the Convention, its effectiveness shall be evaluated starting 4 years after the date of entry into force of the Convention and periodically thereafter at intervals to be decided by the Conference of the Parties (COP). For the global monitoring as a one part of effectiveness evaluation procedure, the Guidance on global monitoring Document was developed. The Guidance is focused on the development and implementation of arrangements to provide comparable monitoring information on the presence of the chemicals listed in Annexes A, B, and C of the Convention, as well as their regional and global environmental transport.

The focus of the global program to support the effectiveness evaluation of the Stockholm Convention is on environmental background concentrations in media with a high potential for comparability. The COP has decided that the air monitoring and human exposure through breast milk or blood serum will be used as core media for the first evaluation. For future evaluations, the COP has also decided to endeavor to supplement the core data with data from other media such as biota, water, soil, and sediments (SC-2/13). The present guidance is aimed at the core media for the first assessment and the document will be revised for future evaluations.

1.3. ENVIRONMENTAL FATE OF POPs

1.3.1. Environmental Fate of Persistent Organic Pollutants

For persistent organic pollutants the reversibility of deposition processes and the consequent reemissions make the soils and vegetation relevant sinks as well as the sources.[2,4,5] The inventory of current POP concentrations in various environmental compartments and assessment of their trends is, therefore, an important step in development of effective control measures based on the improved understanding of a current status, relevant pathways, and potential effects of chemical substances in the environment. A global nature of these processes requires an international effort to generate a network of comparable measurements on the base of which the evaluation of burdens and fluxes, transformation, and transport can be performed, and the fate of the chemicals predicted. The simultaneous and comparable measurements of the concentrations in the atmosphere, wet, bulk and dry deposition, soils, sediments, water and vegetation (lichens, mosses, needles, leaves, and bark) are desirable for a range of compounds in various regions and climatic conditions. Such information would improve the estimates of local pools and fluxes as well as of substance and site specific parameters for the exposure assessments, and also the precision of using the vegetation samples as indicators of environmental pollution. Generation of additional data for validation and improvement of regional models of the long range atmospheric transport is another deliverable of a great value. However, there are still only a few sites where the set of POPs is continuously measured in several matrices over large time periods.

A risk of irreversible changes in the terrestric and aquatic ecosystems as well as a danger of the global climate change caused by the environmental pollution was first recognized in the early 1960s. However, detection of such changes in the natural environment at regional and global levels requires a coordinated monitoring effort based on the broad international cooperation. First international monitoring programs were introduced in the

1960s and 1970s by the international institutions, such as WMO, ECE, and UNEP, and they focused on various environmental aspects including effects of human activities on the climate change, a transboundary transport of pollutants, or an exchange of substances between the environmental compartments.

European Monitoring and Evaluation Programme (EMEP) was established with a main goal to provide the governments and subsidiary bodies under the Convention on Long Range Trans-boundary Air Pollution (CLRTAP, signed in 1979)[2] with qualified scientific information supporting the development and evaluation of the international protocols on emission reductions negotiated within the Convention. 15 EMEP stations monitoring POPs exist, only six sites report POPs in both air and deposition. Kosetice observatory in the Czech Republic is the only site where POPs are also determined in other environmental matrices, such as surface waters, sediments, soils, mosses, and needles and realized POPs part of integrated monitoring programs.[4,5]

2. RECETOX POPs Monitoring and Study Strategy

2.1. MONITORING OF POPs

A risk of irreversible changes in the terrestric and aquatic ecosystems as well as a danger of the global climate change caused by the environmental pollution was first recognized in the early 1960s. However, detection of such changes in the natural environment at regional and global levels requires a coordinated monitoring effort based on the broad international cooperation. First international monitoring programs were introduced in the 1960s and 1970s by the international institutions, such as WMO, ECE, and UNEP, and they focused on various environmental aspects including effects of human activities on the climate change, a transboundary transport of pollutants, or an exchange of substances between the environmental compartments.

European Monitoring and Evaluation Programme (EMEP)[3–5] was established with a main goal to provide the governments and subsidiary bodies under the Convention on Long Range Trans-boundary Air Pollution (CLRTAP, signed in 1979) with qualified scientific information supporting the development and evaluation of the international protocols on emission reductions negotiated within the Convention. 15 EMEP stations monitoring POPs exist, only six sites repor POPs in both, air and deposition. Kosetice observatory in the Czech Republic is the only site where POPs are also determined in other environmental matrices, such as surface waters, sediments, soils, mosses, and needles and realized POPs part of integrated monitoring programs.[4,5]

2.2. MONITORING AS A TOOL FOR THE STUDY OF POPs ENVIRONMENTAL FATE

Persistent organic pollutants remain in the center of scientific attention due to their slow rates of degradation, their toxicity, and potential to both, long-range transport, and bioaccumulation in living organisms. POPs can be transported in ecosystems as dissolved, or bound to different-sized particles. But due to their overall low solubility, POPs tend to appear mainly in association with particulate matter. Thus, in terrestric ecosystems the major fraction of POPs is bound to the soil organic matter rather than dissolved in water. In the aquatic environment, significant fraction of POPs is connected to suspended material. As many of those particles settle down to the sediment layer, sediment also serves as a long-term sink for various environmental pollutants, especially hydrophobic organic contaminants, including POPs.

The accumulation of persistent organic pollutants in the soils and sediments may present a significant risk for the future. In many regions, freshwater sediments were found to be a major reservoir of POP compounds, and under specific conditions they may turn into potential "chemical time bombs" – a significant source of the pollutants, posing risk to the downstream sites.[4,5] Polluted river sediments could become a contamination source because they can be subject to remobilization, transport, and redistribution during certain environmental events such as bioturbation, flood events, or human activities such as dredging and relocation of sediments. The remobilization processes of sediments also significantly affect the availability of the contaminants and their distribution throughout the aquatic ecosystem. The effects of major floods were examined from the continuous set of data for floodplain soils and sediments from an industrial area in the Czech Republic where 100-year flood occurred in 1997 and 2002.[6] The changes in the distribution of pollutants have been assessed by multiple sampling of sediments and alluvial soils. Pollutants concentrations and relative distribution in sediments and floodplain soils were significantly affected by the flooding resulting in a decrease of the studied contaminants in sediments and a significant rise of pollution in the flooded soils.[6] Therefore, any information on the levels of pollutants as well as on their distribution, mobility, and accumulation is crucial for evaluation of future fate and effects of those compounds in the terrestric and aquatic ecosystems as well as risks connected with this contamination.

A long-term project of integrated monitoring including both aquatic and terrestric environments has been carried out by the Centre RECETOX, Masaryk University, Brno, CR in the area of a background observatory of the Czech Hydrometeorological Institute in Kosetice, Czech Republic, since 1988.[4,5,7] A small catchment of Anensky brook offers a diverse terrestrial ecosystem linked to the aquatic ecosystem of adjacent water bodies, and it includes all interacting components: atmosphere and depositions, plants and soils, brooks and ponds. While soils have more stable long-term environmental memory related to contamination level, pattern, and distribution, sediments are more dynamic as they provide information on immediate situation within the watercourse. This article presents a second part of the work focused on the interpretation of the results from the Kosetice monitoring project.[5] Spatial and temporal trends in the background levels of persistent organic compounds in various matrices of terrestric and aquatic environments are assessed here.

3. Polyurethane Foam Passive Samplers for the Sampling of POPs in Ambient Air

3.1. PASSIVE SAMPLERS AS A TOOL FOR THE STUDY OF POPs GLOBAL, REGIONAL, AND LOCAL TRENDS

As the air pollution is an issue of great public health concern, new methods for air quality monitoring have been developed in the last years.[8,9] POPs are due to their wide distribution, ability to bioaccumulate in the fatty tissues, and carcinogen, mutagen and endocrine disruption potentials, remain in the center of our attention. They are emitted from various primary and secondary sources, and the atmosphere often plays a key role in their transport within the immediate vicinity of POPs sources as well as over great distances.[10]

Although a number of regional and global monitoring programs have been established to report on the presence of POPs in the environment, there is very little previous experience of POPs monitoring designed to help evaluate the effectiveness of a legally binding international agreement. Moreover, the establishment of an appropriate monitoring capacity in areas where it does not exist yet will take several years to become operational.

Air monitoring for POPs has conventionally been conducted at a very limited number of sites using "active" or high volume air samplers. These are expensive, require electricity, and a trained operator. Regulatory and other developments mean there will be a pressing need to obtain more POPs data for air, in a much more routine and cost-effective way, to ensure compliance. This provides the incentive to develop new and cheaper passive air sampler (PAS) options.

Passive air sampling as a cheap and versatile alternative to the conventional high volume air sampling is one of the methods currently considered as suitable for the purpose of such monitoring programs.

3.2. RECETOX CONCEPTUAL APPROACH

It was demonstrated previously[8,9] that passive air samplers using PUF filters are suitable to study vapor-phase air concentrations of some types of POPs, particularly more volatile compounds from the group of polycyclic aromatic hydrocarbons, polychlorinated biphenyls and organochlorinated pesticides, and they were successfully applied as a tool for POPs monitoring on the global and regional levels. RECETOX studies confirmed that they are sensitive enough to mirror even small-scale differences, which makes them capable of monitoring of spatial, seasonal, and temporal variations.[9,11] Passive samplers can be used for point sources evaluation in the scale of several square kilometers or even less – from the local plants to diffusive emissions from transportations or household incinerators – as well as for evaluation of diffusive emissions from secondary sources. While not being sensitive to short time accidental releases passive air samplers are suitable for measurements of long-term average concentrations at various levels.

On the other hand, this sensitivity to local effects can lead to some limitations toward its application in large-scale monitoring. Sampling site selection seems to be crucial for the success of such projects since small-scale variability in each region can exceed the continental variability. To develop a monitoring network, the local conditions must be evaluated very carefully since only detail characterization of potential local effects for every sampling site can assure the successful selection of sites for larger (regional or global) scale monitoring.

Very good capability of passive air samplers to reflect temporal and spatial fluctuation in concentrations of persistent organic pollutants in the ambient air was confirmed by our studies. While this sensitivity makes them suitable for the monitoring of local sources, it also needs to be considered when designing large-scale monitoring networks.

4. Conclusions

Data from 10 years of integrated monitoring in the Kosetice observatory were used in this project to assess the long-term trends in the European background levels of persistent organic pollutants in the ambient air and wet deposition. As can be seen from our results, most of the selected compounds exhibit decreasing trends in the last decade. This is consistent with data reported from other European sites.[4,5]

Results of our project proved that the long-term background monitoring is not only an excellent way to study the regional levels and trends, but also a powerful tool for evaluation of an impact of various local, regional or global events – from industrial accidents to natural disasters. As such, it has a crucial role in the effectiveness evaluation of various global measures and international conventions focused on persistent toxic substances, reduction of their emissions and environmental impacts.

Besides all the others, a monitoring data from the Kosetice observatory are currently being used for the assessments of the sources, distribution and transport processes, for the calibrating of various passive air samplers, and for the validation of various distribution and transport models. This study was carried out as a contribution to the ongoing national POPs inventory in the Czech Republic.

The POP concentrations in sediment, surface water, soil, moss, and needle samples have been monitored in 16 sampling sites near the background observatory Kosetice over the period of 18 years. There was a significant variability in occurrence and distribution of selected groups of persistent pollutants in the environmental matrices. PAHs are emitted from a variety of primary sources, and the influence of the local heating systems, for instance, can be easily detected even in the background station – both in the typical seasonal fluctuations of the ambient air concentrations, and in the long-term trends. The atmospheric concentrations of PCBs and OCPs are more homogenous in time and space since they are more influenced by the diffusive sources such as evaporation from contaminated soils. However even here we can detect the events with the high environmental impact (floods, industrial disasters, and constructions) transporting significant amounts of persistent compounds from their primary sources, storage places or contaminated sites to the surface waters and soils where they are a subject of evaporation in the following years. The moss or needle samples precisely reflect a current contamination of the atmosphere, while the sediment and soil samples (especially the organic carbon-rich forest soils which can act as an efficient sink of the POP pollution) provide us with the long-term record of the regional pollution. At the same time we need to be aware that the matrices with high accumulation potential can turn from the sinks to the local sources of POPs. Such complex background monitoring data can be used not only to evaluate the long-term trends in the environmental pollution or the impact of various short-term events. They are very valuable for the assessments of the fate of persistent compounds in the environment – their cycling between compartments, long-range transport or accumulation in various matrices – as well as for the validation of various distribution and transport models. Results of our project presented in this article strongly support the argument that the long-term monitoring programs are of the high importance for the decision making and legislation as well as for the effectiveness evaluation of various measures.

It was demonstrated previously[8,9] that passive air samplers using PUF filters are suitable to study vapor-phase air concentrations of some types of POPs, particularly more volatile compounds from the group of polycyclic aromatic hydrocarbons, polychlorinated biphenyls and organochlorinated pesticides, and they were successfully applied as a tool for POPs monitoring on the global and regional levels. This study suggests they are sensitive enough to mirror even small-scale differences, which makes them capable of monitoring spatial, seasonal, and temporal variations.

Passive samplers can be used for point source evaluation in the scale of several square kilometers or even less from the local plants to diffusive emissions from transportations or household incinerators, as well as for evaluation of diffusive emissions from secondary sources. While not being sensitive to short time accidental releases passive air samplers are suitable for measurements of long-term average concentrations at various levels.

On the other hand, this sensitivity to local effects can lead to some limitations toward its application in large-scale monitoring. Sampling site selection seems to be crucial for the success of such projects since small-scale variability in each region can exceed the continental variability. To develop a monitoring network, the local conditions must be evaluated very carefully since only detailed characterization of potential local effects for every sampling site can assure the successful selection of sites for larger (regional or global) scale monitoring.

Very good capability of passive air samplers to reflect temporal and spatial fluctuation in concentrations of persistent organic pollutants in the ambient air was confirmed in this study. While this sensitivity makes them suitable for the monitoring of local sources, it also needs to be considered when designing large-scale monitoring networks.

Acknowledgments

This project was/is supported by Ministry of Education of the Czech Republic MSM 0021622412 INCHEMBIOL, and EUDG Centre of Excellence EVK1-CT-2002-80012.

References

1. UNEP. Stockholm Convention on POPs, Text and Annexes, Interim Secretariat for the Stockholm Convention on Persistent Organic Pollutants, UNEP Chemicals, Geneva, Switzerland, 2001.
2. H. W. Wallack, D. J. Bakker, I. Brandt, E. Brostrom-Lundén, A. Brouwer, K. R. Bull, C. Gough, R. Guardans, I. Holoubek, B. Jansson, R. Koch, J. Kuylenstirna, A. Lecloux, D. Mackay, P. McCutcheon, P. Mocarelli, and R. D. F. Taalman. *Environ. Toxicol. Pharmacol.* 6, 143–175 (1998).
3. UN-ECE, The Convention on Long-range Transboundary Air Pollution, 1998.
4. I. Holoubek, J. Klánová, J. Jarkovský, and J. Kohoutek. Assessment of the Central European trends in the background levels of persistent organic pollutants based on the results of the long-term integrated monitoring in Kosetice observatory, Czech Republic. Part I. Ambient air and wet deposition 1988–2005. *J. Environ. Monitoring* 9, 557–563 (2007).
5. I. Holoubek, J. Klánová, J. Jarkovský, V. Kubík, and J. Helešic. Assessment of the Central European trends in the background levels of persistent organic pollutants based on the results of the long-term integrated monitoring in Kosetice observatory, Czech Republic. Part II. Aquatic and terrestric environments 1988–2005. *J. Environ. Monitoring* 9, 564–571 (2007).
6. K. Hilscherova, L. Dusek, V. Kubik, J. Hofman, P. Cupr J. Klanova, and I. Holoubek. Redistribution of organic pollutants in river sediments and alluvial soils related to major floods. *J. Soils Sediments* 7(3), 167–177 (2007).
7. M. Váňa, I. Holoubek, A. Pacl, J. Pekárek, V. Smrčková, P. Machálek, J. Helešic, Z. Šeda, V. Adamec, M. Janouch, J. Honzák, A. Ansorgová, J. Kohoutek, I. Holoubková, V. Šatalov, S. Dutčak, D. Fottová, J. Hruška, J. Hofman, and P. Anděl. Quality of the Natural Environment in the Czech Republic at the Regional Level. Results of the Košetice Observatory. ČHMÚ Praha, 2001, 189 s., ISBN 80-85813-88-2.
8. T. Harner, M. Bartkow, I. Holoubek, J. Klanova, A. Sweetman, F. Wania, and K. C. Jones: Passive air sampling for persistent organic pollutants: introductory remarks to the Special Issue. *Environ. Pollut.* 144/2, 361–364 (2006).
9. J. Klánová, J. Kohoutek, L. Hamplová, P. Urbanová, and I. Holoubek: Passive air sampler as a tool for long-term air pollution monitoring: Part 1. Performance assessment for seasonal and spatial variations. *Environ. Pollut.* 144/2, 393–405 (2006).
10. Wania, F. Assessing the potential of persistent organic chemicals for long-range transport and accumulation in Polar regions. *Environ. Sci. Technol* 37, 1344–1351 (2003).
11. P. Čupr, P., Klánova, J., Bartoš, T., Flegrová, Z., Kohoutek, J., and Holoubek, I. Passive air sampler as a tool for long-term air pollution monitoring: Part 2. Air genotoxic potency screening assessment. *Environ. Pollut.* 144, 406–413 (2006).

HALOGENATED FLAME RETARDANTS AS SOURCES FOR POPs IN THE ENVIRONMENT

MÜFIT BAHADIR*
Institute of Ecological Chemistry and Waste Analysis, Braunschweig University of Technology, Hagenring 30, D-38106 Braunschweig, Germany

Abstract: Brominated flame retardants (BFR) are commonly used for reducing the risk of fire. Bromine radicals formed at elevated temperatures terminate the radical-chain reaction in fire. Remaining aromatic BFR fragments can partially recombine to halogenated POPs including PBDD/F. This paper deals with the formation of POPs from BFR in laboratory-scale combustion experiments and discusses the reaction mechanism involved. It is shown that the mechanism of POPs formation is mostly independent of the type of BFR incorporation into the polymer matrix as monomeric additive or through chemical bonds to the polymer backbone. Furthermore, an improved clean-up procedure is shown which enables an almost complete removal of the remaining BFR from the analyte solution in order to avoid their interferences with the traces of PBDD/F and/or formation of artifacts in the ionization chamber of GC/MS equipment.

Keywords: brominated flame retardants, BFR, POPs, PBDD/F, fire accident, cleanup

1. Introduction

Flame retardants are applied to plastic compounds and technical products in order to reduce the risk of fire accidents. Brominated flame retardants (BFR) are most commonly used for these purposes. During thermal stress like pyrolysis,[1] extrusion,[2] or recycling[3] plastics protected by BFR often produce polybrominated dibenzo-*p*-dioxins and dibenzofurans (PBDD/F). Some of the most commonly used brominated flame retardants are:

* To whom correspondence should be addressed. E-mail: m.bahadir@tu-bs.de

E. Mehmetli and B. Koumanova (eds.),
The Fate of Persistent Organic Pollutants in the Environment, 229–237.

- Tetrabrombisphenol-A (TBBP-A) (e.g., in epoxy resins EP and polycarbonates PC)
- Polybrominated biphenyls (PBB) (e.g., in polystyrene PS and acrylonitrile–butadiene–styrene copolymers – ABS)
- Polybrominated diphenyl ethers (PBDPE) (e.g., in PS and ABS)
- Hexabromo cyclododecane (HBCD) (e.g., in PS foams)
- Brominated phenyl ethers and Bisphenol-A ethers (e.g., in poly-α-olefins like LDPE)

Under thermal stress, brominated flame retardants release HBr. HBr inhibits the spreading of fire by substituting the reactive H·and HO·radicals in the radical-chain reaction for less reactive Br·radicals.[4] Antimony(III)oxide increases the effect of fire inhibition as a synergist (Figure 1).

R – Br	→	R· + Br·	Thermal decomposition of BFR
Br· + Polymer – H	→	HBr + Polymer·	Formation of HBr
H· + O_2	→	HO· + O·	Radical-chain reaction
O· + H_2	→	HO· + H·	Radical-chain reaction
HBr + H·	→	Br· + H_2	Chain termination
HBr + HO·	→	Br· + H_2O	Chain termination
Br· + Polymer – H	→	HBr + Polymer·	Re-formation of HBr

Formation of HBr is additionally supported by the synergist Sb_2O_3

Sb_2O_3 + nHBr → → $SbBr_3$ → → + H· → HBr + $SbBr_2$ etc.

Figure 1. Mode of action of brominated flame retardants

During this radical-chain reaction the formation of PBDD/F by de novo synthesis is one possible mechanism.[5] The second and in case of PBDPE and PBB the likely more important route via condensation or recombination of fragments without de novo synthesis is shown in Figure 2 for decabromo diphenyl ether as an example.

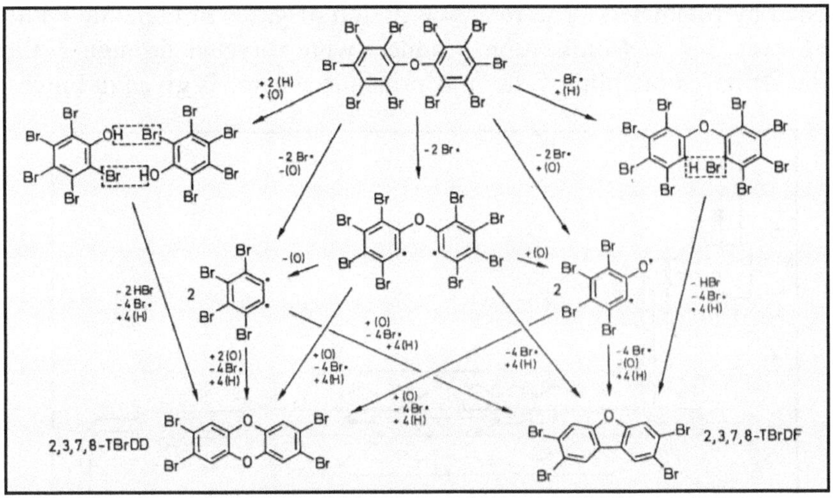

Figure 2. Formation of PBDD/F from decabromo diphenyl ether

This paper deals with different aspects of POPs generation from BFR. PBDD/F formation was investigated based on thermolysis experiments in a laboratory-scale incinerator. A comparison of polymer linked and of free TBBP-A was conducted and discussed whether there are any differences in their likely PBDD/F formation. While aiming at to analyze the formation of trace amounts of PBDD/F in the presence of high percentages of BFR one is regularly confronted with artifact formation in GC/MS. Another aim of this paper is, therefore, to show an improved method of cleanup for the removal of the BFR in order to reduce the analytical interferences with PBDD/F as analytes.

2. Materials and Methods

TBBP-A (Merck), TBBP-S, TBBP-A-bis-2,3-dibromopropylether, TBBP-S- bis-2,3-dibromopropylether (Great Lakes or Marubishi Oil), polybrominated benzenes (PBBz) and polybrominated phenols (PBP) (Aldrich), PBDD/F (Campro Scientific, Promochem), and PAH (Hewlett Packard) were used as standards and for the experiments. LDPE with incorporated brominated or nonbrominated epoxy resins and with nonbrominated EP plus TBBP-A were produced in an extrusion press (200°C, 300 kPa). These materials as well as the plastic matrices EP and LDPE (Techno-Polymer) were pulverized before use in an ultra centrifuge mill (Retsch) while cooling with liquid nitrogen.

The combustion apparatus consisting of a quartz tube of 1 m length was inserted into a furnace with temperature control. A constant gas flow was adjusted by rotameters. The retention period of gases in the zone with T ≥ 500°C was 3–4 s. Combustion products were directed through a Liebig condenser into a sampling system. A principle scheme is given in Figure 3.

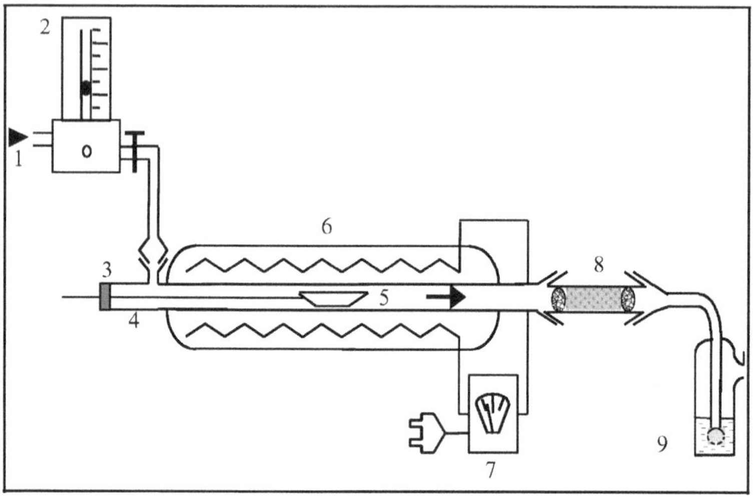

Figure 3. Laboratory combustion device for testing POPs formation from BFR (principle scheme). 1 – synthetic air inlet, 2 – rotameter, 3 – quartz gasket for push rod, 4 – quartz combustion tube, 5 – quartz combustion boat, 6 – quartz tubular furnace, 7 – oven heat control, 8 – absorption unit, 9 – impinger with toluene

Polyurethane foam as absorbent for PBDD/F was spiked with 2,3,7,8-$^{13}C_{12}$-TeBDD/F standards. The flow rate of synthetic air was adjusted to 1 L/min and the furnace was heated to 600°C. 0.5 g of the sample was placed in a quartz boat and first inserted into the front part of the combustion tube. Additional HBr flow rate was adjusted to 80 mL/min, when required. The quartz boat containing the sample was pushed into the hot zone after 10 min of forerun time. Combustion was restricted to 10 min. When cooled down, the whole system was rinsed or refluxed with toluene for 16 h and with acetone for another 4 h. Toluene from the impinger was added to the extracts.

The sample preparation, modules of chromatographic column cleanup, and analyses were conducted after Hagenmair[6] and its modifications.[7,8] In order to reduce interferences of BFR with PDD/F formed a new clean-up procedure was also developed and applied.[9]

After cleanup, the fractions were evaporated to 1 mL by slight vacuum and then carefully to dryness in nitrogen stream, in order to minimize losses of low brominated compounds. The residue was resolved in 100 µL 1,2,3,4-$^{13}C_{12}$-TeCDD 25 pg/µL in iso-octane and determined by using HRGC/LRMS (Agilent) on a DB-5 column (30 m, 0.25 mm i. d., 0.25 or 0.1 µm film). For quality assurance all experiments were performed at least twice.

3. Results and Discussion

In order to form flame-retarded epoxy resins, TBBP-A as brominated diglycidyl ether was mixed with nonbrominated diglycidyl ether, and the mixture was then cured. It was the first time, when an investigation of PBDD/F formation from polymer linked versus as additive incorporated TBBP-A was performed. In the past, it was generally assumed, that TBBP-A bound into the polymer backbone should not play a significant role in PBDD/F formation due to sterical hindrances. Any formation of the latter under thermal stress was expected due to the nonreacted part of monomeric TBBP-A moiety. Experiments reported here were the first one exemplarily showing the impact of sterical fixation of TBBP-A on POPs formation at high temperatures.

It is difficult to directly compare the polymer linked with the additive incorporated TBBP-A in epoxy resin because of the reactivity of the monomers with each other. We, therefore, produced the starting materials for the combustion experiments by incorporation of already cured nonbrominated epoxy resin and monomeric TBBP-A into polyethylene via extrusion, on the one hand, and fire retarded brominated epoxy resin into LDPE, on the other. In the following, they are called as additive incurporated and polymer linked TBBP-A. Both starting materials had an identical bromine content of 3.6 wt. %. Additionally, a mixture of nonbrominated epoxy resin and polyethylene was combusted in an atmosphere consisting of synthetic air and HBr, in order to elucidate the influence of the polymer matrix on PBDD/F formation in the presence of HBr only.

Before starting the combustion experiments all starting materials had been analyzed for precontamination with PBDD/F which was far below their contents in the combustion products (by factor > 1,000), and there was no correlation with concentrations of any relevant precursor, as well. That means, neither already occurring PBDD/F nor any monoaromatic precursor contents of the fire loads were of any significance.

In all cases an ignition of the polymers tested occurred. That means, the reaction conditions were almost identical. Related to the applied fraction of TBBP-A, the PBDD/F concentrations formed in the mixture with the additive incorporated TBBP-A were at the same range as in the case of the polymer linked BFR (PBDD/F: 17.5 mg/kg vs 19.6 mg/kg. The combustion of the epoxy resin in LDPE matrix in the HBr containing atmosphere resulted in 12.1 mg/kg PBDD/F. It is obvious, that more PBDD/F were formed during combustion of the polymer linked and the additive incorporated TBBP-A than during the combustion of the plastic mixture without BFR but in an HBr enriched atmosphere. In the condensate of the latter, the

PBDD to PBDF ratio was (11:1) and, hence, completely different from that when TBBP-A was applied (1:7) (Table 1).

TABLE 1. PBDD/F formation from BFR – additive vs reactive

Sample	**Precontamination ∑PBDD/F (µg/kg)**	**After incineration ∑PBDD/F (mg/kg)**
TBBP-A	9.2	215
TBBP-A dibromopropylether	23.6	91
TV housing	25.4	146*
brominated EP resin + PE	17.4	19.6*
EP resin + PE + TBBP-A	26.7	17.5*
EP resin + PE + HBr	–	12.1

*Based on the BFR content

It is assumed that in case of both experiments when TBBP-A was applied as BFR a precursor reaction dominated. In contrast, the completely different PBDD/PDDF ratio of HBr experiment suggested a de novo reaction mechanism.[5]

While analyzing trace amounts of PBDD/F in the presence of percent-tage amounts of BFR like PBDPE and PBB there are several problems to be solved:

- High detection limits of PBBD/F compared with the PCDD/F
  (Te- to OCDD/F 1–5 pg/µL)
  Mo- to TrBDD/F 1–5 pg/µL
  Te- to PeBDD/F 5–50 pg/µL
  HxBDD/F 100–250 pg/µL
  HpBDD/F 250–500 pg/µL
  OcBDD/F >1 ng/µL
- Low-resolution efficiency for isomer-specific analysis of single congeners.
- Interferences with PBDF, which are formed during the GC/MS measurements of PBDF as nondistinctive fragment ions.

A significant amount of artifact formation in the ionization chamber of GC/MS may occur. It is, unfortunately, not possible to distinguish between these artifacts and the PBDD/F formed during the combustion process. In order to avoid misinterpretation it is suggested to remove the remaining BFR from the sample by an improved clean-up method before GC/MS measurement.

By the application of mixed acid–alkaline silica gel column it was possible to substitute the toxic *n*-hexane through *n*-heptane without any restrictions. Recovery rates of PBDD/F and PCDD/F were above 70%. But no separation of PBDPE was observed (Figure 4).

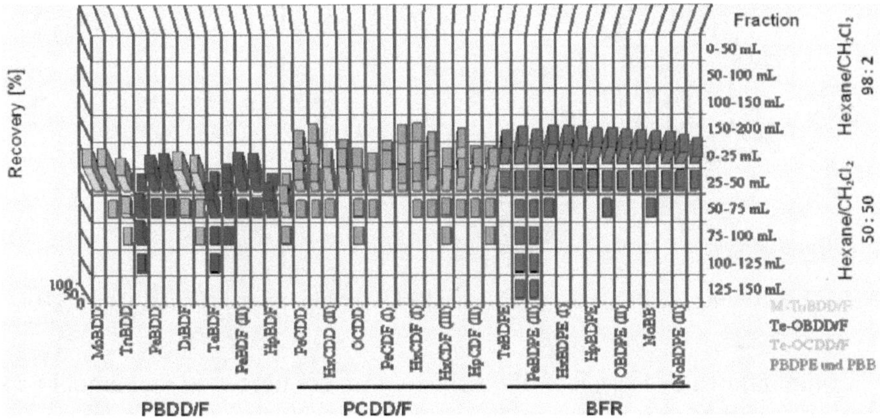

Figure 4. Conventional dioxin cleanup for PBDD/F from BFR

Using 25 g aluminum oxide as adsorbent, it was possible, as well, to substitute the solvent mixture *n*-hexane/dichloromethane similar to the analysis of PCDD/F. All substances examined eluted in the PBDD/F containing fraction of *n*-heptane/ethyl acetate 9:1. The distribution behavior of PBDD/F, PCDD/F and selected flame retardants like PBDPE and some PBB in these fractions were insufficient as well. No separation of PBDPE from PBDD/F was observed. The determination of HxBDF was disturbed due to the appearance of PBDPE.

By elution with 15 mL ethyl acetate a large amount of PBDPE was removed by a charcoal/silica gel column. The PBDPE amount remaining in the toluene fraction was less than 20%. The recovery rates of PBDD/F in the toluene fraction were above 60% and no selective losses of certain congeners was observed. A moderate separation of PBDPE from PBDD/F was found.

An almost complete separation of PBDPE from PBDD/F was achieved 5 g florisil with 1% water content was applied as adsorption phase (Figure 5). Fifty milliliters of *n*-heptane is necessary to remove PBDPE. PBDD/F were eluted with further 50 mL toluene/diethylether 9:1. When activated florisil without water is applied, a complete separation could not be observed. With a water content of 2% or more Mo- to TriBDD/F already eluted in the *n*-heptane fraction.

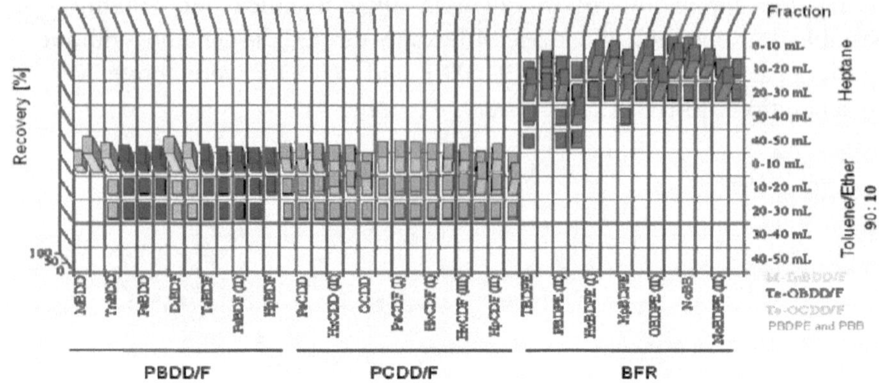

Figure 5. Separation of PBDD/F, PCDD/F, and BFR on mini florisil column with 1% H_2O

Combinations of florisil as the first column connected to a charcoal or a second florisil column gave similar good or even better results of separation (Figure 6).

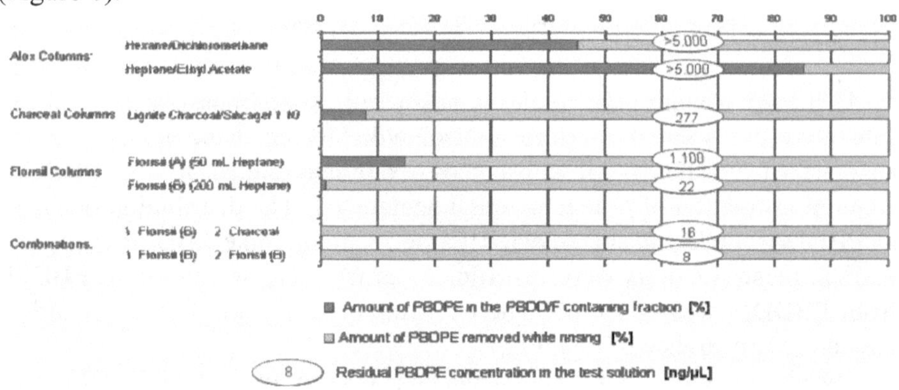

Figure 6. Removal of PBDPE during PBDD/F cleanup with different columns

4. Conclusions

When searching the literature it is obvious that flame retardants are essential in order to protect life and the everyday's goods. But brominated and also chlorinated flame retardants are proved to form polyhalogenated POPs like PBDD/F and PCDD/F.[10,11] It is, thereby, not decisive whether these substances are cross-linked to the matrix by means of chemical bonds or just even incorporated as monomeric additives. Significant problems occur regularly while analyzing trace amounts of PBDD/F in the presence of brominated flame retardants at percentage levels. Due to the formation of artifacts in the ionization chamber of the GC/MS the BFR have to be removed in an improved clean-up procedure.

In order to avoid environmental and health problems by the POPs emitted in case of fire or using elevated temperatures, e.g., during the recycling processes, halogenated flame retardants should not longer be applied for material protection purposes. They should better be replaced by halogen-free flame retardants.

References

1. H. Thoma et al., Polybrominated dibenzofurans (PBDF) and dibenzodioxins (PBDD) from the pyrolysis of neat brominated diphenylethers, biphenyls and plastic mixtures of these compounds, *Chemosphere*, **16(1)**, 277–285 (1987)
2. H. Meyer et al., Flame retardants determine recyclability (German), *Kunststoffe*, **83(4)**, 253–257 (1993)
3. R. Luijk et al., Formation of polybrominated dibenzofurans during extrusion of high-impact poly–styrene/decabromodiphenyl ether/antimony(III)oxide, *Environmental Science and Technology*, **26(11)**, 2191–2198 (1992)
4. J. Troitzsch, Combustion behavior of plastics. Carl Hanser, Munich, Germany, pp. 47–67 (1982)
5. F. Korte, M. Bahadir, et al., Lehrbuch der Ökologischen Chemie (Textbook of Ecological Chemistry), 3rd edn., Georg Thieme Verlag, Stuttgart, New York, p.139 (1992)
6. H. Hagenmaier et al., Status of dioxin-analysis (German), *VDI-Berlin*, **634**, 61–89 (1987)
7. J. Höckel et al., Modified clean-up for PCDD/PCDF under toxicological, ecological and economical points of view, *Organohalogen Compounds*, **22**, 433–436 (1995)
8. F. T. Dettmer et al., Optimization of dioxin clean-up by using lignite coke-silica gel columns and solvent substitutions, *Fresenius Environmental Bulletin*, **7**, 216–223 (1998)
9. J. Ebert et al., Optimization of the analytical performance of polybrominated dibenzo-*p*-dioxins and dibenzofurans (PBDD/F). *Chemosphere*, **39**, 977–986 (1999)
10. J. Ebert and M. Bahadir, Formation of PBDD/F from flame-retarded plastic materials under thermal stress. *Environment International*, **29**, 711–716 (2003)
11. M. Bahadir et al., Isomer specific synthesis using the Suzuki coupling – polychlorinated terphenyls (PCT) as standards for environmental analysis. *Chemosphere*, **50**, 1151–1156 (2003)

DISTRIBUTION OF POPs IN AQUATIC ECOSYSTEMS AND PROCESSES FOR THEIR REMOVAL

BOGDANA KOUMANOVA*
University of Chemical Technology and Metallurgy, Department of Chemical Engineering, 8 Kliment Ohridski blvd., Sofia 1756, Bulgaria

Abstract: Persistent Organic Pollutants (POPs) are chemicals that are produced and released into the environment by human activity. Some of them are specially synthesized chemicals for agricultural application like pesticides, other are unwanted by-products of chemical and mainly combustion processes. POPs characteristics, their metabolites and their distribution into water bodies are described. The measures for the decreasing of POPs emissions in the environment are also discussed. POPs in the environment may undergo transformations or degradation due to reactions caused by light, chemical reactions, biological reactions or a combination of these phenomena. The effective removal of POPs and their metabolites from wastewaters are very important for preserving of the aquatic environment.

Keywords: POPs, metabolites, MSW combustion, water treatment, aquatic environment

1. Introduction

Persistent Organic Pollutants (POPs) are chemicals that are produced and released into the environment by human activity. Some of them are specially synthesized chemicals for agricultural application like pesticides, other are unwanted by-products of chemical and mainly combustion processes. The Stockholm Convention on POPs entered into force 17 May

* To whom correspondence should be addressed: E-mail: bkk@uctm.edu

E. Mehmetli and B. Koumanova (eds.),
The Fate of Persistent Organic Pollutants in the Environment, 239–249.

2004. Its objective is to protect human health and the environment from the persistent organic pollutants. The effectiveness of the Stockholm Convention on POPs will be evaluated within 4 years after the entry into force, e.g., before 17 May 2008. The aim of the specialists from the countries all over the world for the meaningful evaluation of the POPs distribution is based on the consideration of their characteristics and the realization of the risk for the human health. Very important result will be the evaluation of their regional and global environmental transport. Twelve POPs are on the focus, usually named “Dirty Dozen” POPs:

- Aldrin, Dieldrin, Endrin, Chlordane, DDT, Heptachlor, Mirex, Toxaphene, and Hexachlorobenzene (HCB)
- Polychlorinated Biphenyls (PCBs)
- Chlorinated dioxins and furans

These chemicals are characterized by their long lifetimes (persistence) in the environment (air, water, soil, and biota), their capacity to build up to dangerous levels in some larger predatory species through bioaccumulation, and their potential for long-range transport through air, water, or migratory species. As a result POPs can be found all over the world thus disrupting ecosystems and affecting human health in regions far distant from where they first entered the environment. Main local sources are usually hazardous waste sites, incinerators and pesticide-intensive agriculture.

2. Fate of the POPs

The determination of the fate of these chemicals includes their physical and chemical properties, transformation processes, and mobility.

2.1. POPs CHARACTERISTICS

POPs physical and chemical properties are essential for the prediction of their environmental behavior. These properties include water solubility, which is useful in predicting their mobility and deposition into sediments in an aquatic environment. Vapor pressure is an important indicator of the potential of a compound to volatilize. The octanol/water partitioning coefficient (Kow) indicates the possible transfer from soil or water to organisms, as well as the potential for bioaccumulation.

Organochlorine pesticides are hydrocarbon compounds in which various numbers of hydrogen atoms have been replaced by chlorine atoms. Commonly used in the 1960s their application has been restricted because of their toxicities, environmental persistence, and accumulation in food chains. They show some general similarities in properties. They have low water solubility, high lipophilicity and are comparatively persistent in the natural environment. In addition, they bioaccumulate in individual organisms and may biomagnify in food chains.

DDT (dichlorodiphenyltrichloroethane) was used extensively following World War II for controlling mosquito-borne malaria and as an agricultural insecticide. Breakdown products of DDT include DDE (dichlorodiphenyldichloroethylene) and DDD (dichlorodiphenyldichloroethane), which are also highly persistent and have similar chemical and physical properties (Table 1). Water solubility is 0.003 mg dm^{-3} and octanol/water partition coefficient Kow 1.6×10^6.

TABLE 1. Chemical structures of DDT, DDE, and DDD (US Department of health and human services, Public health service agency for toxic substances and disease registry, September 2002)

Characteristic	*p,p'*-DDT	*p,p'*-DDE	*p,p'*-DDD
Synonym(s)	4,4'-DDT; 1,1,1-trichloro-2,2-bis (*p*-chlorophenyl)ethane; dichlorodiphenyl trichloroethane; DDT; 1,1'-(2,2,2-trichloroethylidene)bis(4-chlorobenzene); α-α-bis(*p*-chlorophenyl)-β, β, β -trichloroethane	4,4'-DDE; dichlorodiphenyl-dichloroethane; 1,1-dichloro-2,2-bis(*p*-chlorophenyl) ethylene; 1,1'-(2,2-dichloroethylidene)bis(4-chlorobenzene); DDE	4,4'-DDD; DDD; 1,1-dichloro-2,2-bis(*p*-chlorophenyl)ethane; 1,1-bis (4-chlorophenyl)-2,2-dichloroethane; TDE; tetrachlorodiphenylethane
Registered trade name(s)	Genitox, Anofex, Detoxan, Neocid, Gesarol, Pentachlorin, Dicophane, Chlorophenothane[b]	No data	DDD; Rothane; Dilene, TDE
Chemical formula	$C_{14}H_9Cl_5$	$C_{14}H_8Cl_4$	$C_{14}H_{10}Cl_4$
Chemical structure	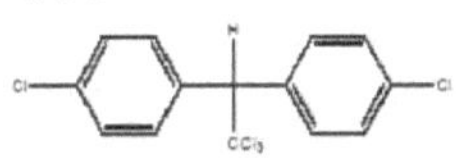	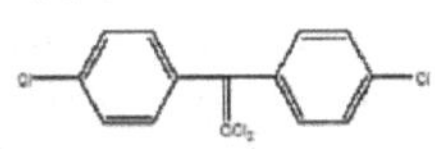	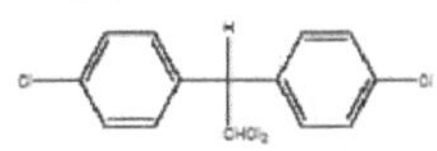

DDT is very persistent in the environment with a reported half-life between 2 and 25 years (US EPA, 1989; Augustijn-Beckers et al., 1994) and is immobile in most soils. Routine loss and degradation include runoff, volatilization, photolysis, and biodegradation (both aerobic and anaerobic) (ATSDR, 1994a). Degradation processes are generally very slow (WHO, 1989; Augustijn-Beckers et al., 1994). Due to its extremely low solubility in water, DDT is retained to a greater degree by soils and soil fractions with higher proportions of soil organic matter (WHO, 1989). Generally, DDT is tightly sorbed by soil organic matter, but it (along with its metabolites) has been detected in many locations in soil and groundwater where it may be available to organisms (US EPA, 1989; WHO, 1989). Over very long periods of time DDT may be able to eventually leach into groundwater, especially through soils with little soil organic matter. Volatilization losses may be significant and rapid in soils with very low organic matter content (desert soils) and high irradiation of sunlight (Jorgensen et al., 1991). DDT may reach surface waters primarily by runoff, atmospheric transport, drift, or by direct application (ATSDR, 1994a). The reported half-life for DDT in the water environment is 56 days in lake water and approximately 28 days in river water (US EPA, 1989). The main pathways for loss are volatilization, photodegradation, adsorption to waterborne particulates and sedimentation. Aquatic organisms also readily take up and store DDT and its metabolites.

Chlordane (1,2,4,5,6,7,8,8a-octachlor-2,3,3a,4,7,7a-hexahydro-4,7-methanoindane) was discovered in 1945 as a mixture of over 140 different compounds (Kirk-Othmer, 1995). Its primary uses included soil and household insecticide (ATSDR, 1994b), preservative for wood and underground cables, and fire ant control in power transformers (ASTDR, 1989). It is viscous amber liquid with chlorine like odor. Reported water solubility is 0.1 mg dm^{-3} and octanol/water partition coefficient 1×10^6. Chlordane enters the environment through the direct application to crops, lawns, and to control termites. It is very persistent in soil, sorbing strongly to organic particles. The half-life in soil is 4–20 years (ATSDR, 1994). Volatilization soil may be the only major route of chlordane entering air. Because it is insoluble in water and very rapid bind to soil, potential for surface and groundwater contamination is very low.

Hexachlorobenzene (HCB) *is* used as a pesticide, but also arises from incomplete combustion at dumpsites. Studies in USA have identified two main industrial sources: the manufacture of chlorinated solvents (e.g., trichloroethylene, tetrachloroethylene, and carbon tetrachloride) and the manufacture of specific pesticides where HCB remains as an impurity. HCB emissions can also result from the use of hexachloroethane tablets as a degassing agent in secondary aluminum melting. HCB emissions may also arise from combustion sources, but other than waste incineration. Practically it is insoluble in water (50 μg dm^{-3} at 20°C). Its metabolites are shown on Figure 1.

Mirex is one of the most persistent pesticides (half-life 10 years). Practically it is not soluble in water (0.07 μg dm^{-3} at 25°C). In the 1950s, it was used as insecticide but also for producing of fireproof plastics, rubber, paper, etc.

Toxaphene is a mixture of about 300 congeners (polychlorinated bornanes and camphenes) containing more than 60% chlorine. Its solubility in water is 550 μg dm^{-3} at 20°C.

Heptachlor (1,4,5,6,7,8,8-heptachlor-3a,4,7,7a-tetrahydro-4,7-methanoindene) is soluble in water – 180 μg dm^{-3} at 25°C and its half-life in soil is 0.75–2 years. Easily metabolites into more persistent heptachlore epoxide.

Aldrin was produced in the 1950s and was used in many countries until the early 1970s. It easily metabolizes into *Dieldrin*. It was classified as moderately persistent and its half-life in soils and surface waters is 20 days to 1.6 years. The solubility of Aldrine in water is 27 μg dm^{-3} at 25°C and that of dieldrine is 140 μg dm^{-3} at 20°C.

Endrin has been used in the 1950s and has high persistency in soil. In some cases its half-life is near 12 years. The solubility in water is 220–260 μg dm^{-3} at 25°C.

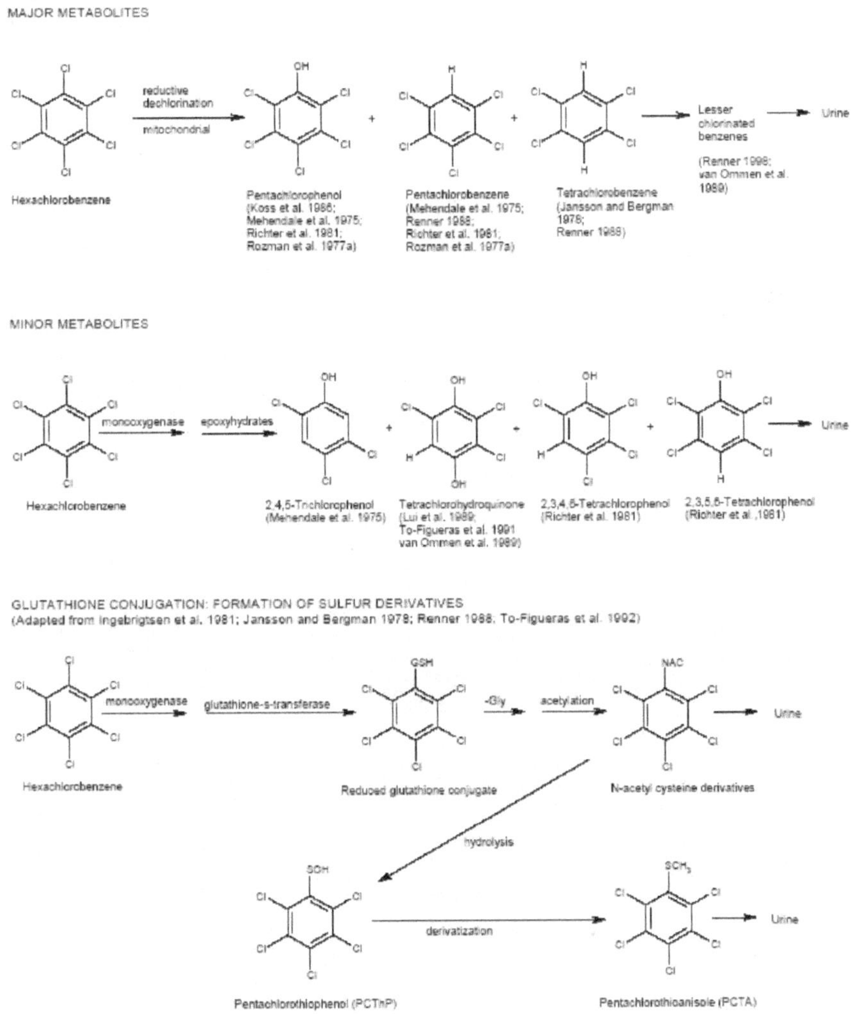

Figure 1. Hexachlorobenzene metabolites (US Department of health and human services, Public health service agency for toxic substances and disease registry, September 2002)

Polychlorinated biphenils (PCBs) are synthetic organic compounds that have mainly been used in electrical equipment as dielectric insulating media. They are classified as probably carcinogenic to humans. PCBs are extremely persistent in the environment and possess the ability to accumulate in the food chain. These compounds are highly insoluble in water but accumulate in body fat. The commercial synthesis of PCBs involves the progressive chlorination of biphenyl to give a variety of products with differing chlorine content. The commercial products are graded according to chlorine content. In Figure 2 are shown metabolic pathways for PCBs.

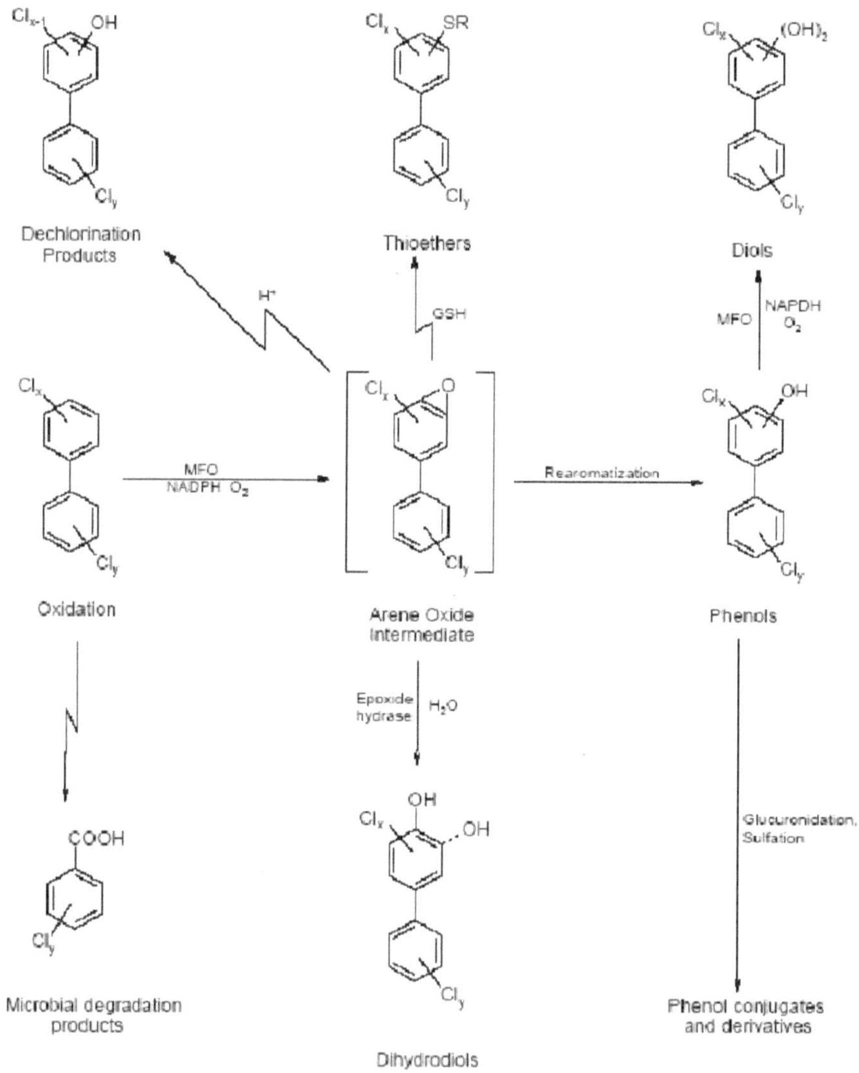

Figure 2. Metabolic pathways for PCBs (US Department of health and human services, Public health service agency for toxic substances and disease registry, November 2000)

Polychlorinated dibenzo-p-dioxins (PCDD) and *polychlorinated dibenzofurans* (PCDF) are formed unintentionally as by-product of many combustion and industrial thermal processes. Today PCDD/PCDF is known as one of the most toxic group of anthropogenic organic substances. Although there are 210 PCDD/PCDF compounds in total, the emissions of importance are those of the 17 congeners (7 PCDDs and 10 PCDFs), among them 2,3,7,8-TCDD is the most toxic. In Figure 3, the basic chemical structures are shown.

Figure 3. PCDD and PCDF structures

PCDD/PCDF can arise from any thermal process where chlorine is present in any form. For example, coal and other solid fuels contain trace amounts of chlorine compounds which can under certain combustion conditions result in PCDD/PCDF formation. Primary sources of environmental contamination with PCDD/PCDF in the past was due to production and use of chloroorganic chemicals, including the pulp and paper industry. In wet-chemical processes the propensity to generate PCDD/PCDF during synthesis of chemical compounds decreases in the following order:

Chlorophenols < Chlorobenzenes < Aliphatic chlorinated compounds <
< Inorganic chlorinated compounds

Factors favorable for the formation of PCDD/PCDF are high temperatures, alkaline media, presence of UV light, and presence of radicals in the reaction mixture/chemical process (US Department of health and human services, Public health service agency for toxic substances and disease registry, December 1998).

Previously the largest source of PCDD/PCDF emission was waste incineration. Three possibilities have been proposed to explain the presence of dioxins and furans in incinerator emissions:

- PCDD/PCDF are already present in the incoming waste and are incompletely destroyed or transformed during combustion.
- PCDD/PCDF are produced from related chlorinated precursors (predioxins) such as PCB, chlorinated phenols, and chlorinated benzenes.
- PCDD/PCDF are formed via new synthesis from the pyrolysis of chemically unrelated compounds such as polyvinyl chloride (PVC) or other chlorocarbons, and/or the burning of nonchlorinated organic matter such as polystyrene, cellulose, lignin, coal, and particulate carbon in the presence of chlorine donors.

From the knowledge gained from MSW incinerators it can be concluded that PCDD/PCDF can be formed in other thermal processes in which chlorine-containing substances are burnt together with carbon and a suitable catalyst (preferably copper) at temperatures above 300°C in the presence of excess air or oxygen. Preferentially, dioxin formation takes place in the zone when combustion gases cool down from about 450–250°C. Possible sources of the chlorine input are PVC residues as well as chloroparaffines in waste oils and inorganic chlorine.

However, emissions have fallen after introduction of significant control measures. New designs of MSW incinerators result in significantly lower levels of PCDD/PCDF emissions. The use of rotary kilns and the incorporation of a secondary combustion chamber in the process to destroy organic contaminants contribute to the relatively low emissions. Emissions from power stations are fairly low because the combustion is efficient and the postcombustion fly ash temperatures are rapidly reduced. The accidental fires and open agricultural burning are also emission sources.

2.2. TRANSFORMATION PROCESSES

POPs in the environment may undergo transformations or degradation due to reactions caused by light, chemical reactions, biological reactions, or a combination of these phenomena. Therefore, it is important to study the phototransformation (reactions caused by light), hydrolysis (chemical reactions), and biotransformation (biological reactions) in order to determine their significance in different environmental media.

2.3. MOBILITY

Investigation of mobility and leaching potential is important to understand the ability of terrestrial-use pesticides and their major transformation products to move through soils and their potential to contaminate aquatic environments by leaching into groundwater or moving in surface runoff or eroding soil.

3. Instrumental Analysis of POPs

Very important condition to obtain comparable data from all regions for investigations of POPs is to use reliable analytical procedures. Usually analytical procedures consist of the following four steps:

- Sample collection and extraction
- Cleanup using partition and chromatographic fractionation

- Separation on gas chromatography
- Detection with selective and sensitive detectors

In the early 1960s, POPs have been determined using gas chromatography (GC) techniques with electron capture detection (ECD), initially using packed columns. More advanced methods, such as capillary GC-ECD and GC coupled with mass spectrometry (GC-MS) have been used in more resent studies to identify the individual congeners. In Table 2 are given the requirements for the instrumental analysis.

TABLE 2. Instrumental analysis of POPs

Equipment	Chemicals
Basic sample extraction and cleanup equipment, capillary GC/ECD[a]	Most PCB and all OCPs except toxaphene
Sample extraction and cleanup equipment, capillary GC/LRMS[b]	Most PCB and all OCPs; toxaphene if negative chemical ionization is available
Sample extraction and cleanup equipment, capillary GC/HRMS[c]	PCDD/PCDF, all PCB, all OCPs except toxaphene

[a]GC/ECD – gas chromatography/electron capture detection
[b]GC/LRMS – gas chromatography/low-resolution mass spectrometry
[c]GC/HRMS – gas chromatography/high-resolution mass spectrometry

The requirements for the strict control on the toxic compounds into environment need well equipped laboratories. Now it is possible to determine traces of compounds (sensitivity in pg). According to Guidance for a Global Monitoring Programme for POPs, UNEP, 2004, the most correct data for POPs can be obtained using the following instruments:

- Capillary GC with Electron Capture Detector
- Quadrupole mass spectrometry in Electron Ionization (EI) mode
- Quadrupole mass spectrometry in Electron Capture Negative Ionization (ECNIMS) mode
- Ion trap mass spectrometry using MS/MS mode
- High resolution magnetic sector mass spectrometry in Electron Ionization (EI) mode

4. POPs Distribution in Aquatic Environment

The hydrosphere is a major reservoir for POPs and their metabolites. Pesticides enter into the hydrosphere via many pathways including:

- Direct application for pest and disease control
- Urban and industrial wastewater discharges

- Runoff from nonpoint sources including agricultural soils
- Leaching through soil
- Aerosol and particulate deposition, rainfall
- Absorption from the vapor phase at the air–water interface

It is difficult to assess the relative inputs from these sources. Generally, water bodies associated with urban regions receive substantial pesticide inputs from industrial and domestic effluents. The major input probably originates from agricultural and forestry practices.

Inputs of organic contaminations to the urban wastewater system occur from three general sources: domestic, commercial and urban runoff. Inputs of the main persistent organic pollutants of concern, including PCBs and PCDD/PCDFs to urban wastewater are principally from atmospheric deposition onto paved surfaces and runoff. PCDD/PCDFs are released during waste incineration and also by coal combustion. Soil acts as a long-term repository for these contaminant types and remobilization by volatilization from soil is an important mechanism responsible for recycling and redistributing them in the environment. PCDD/PCDFs or PCBs enter urban wastewater principally from diffuse atmospheric deposition and environmental cycling. Being strongly hydrophobic these pollutants are efficiently removed during urban wastewater treatment and bind to the sludge solids.

Special attention has to be kept on the quality of ground waters. It is quite possible leachates from agriculture activities as well as from toxic deposits to reach ground water aquifers. Common practice is in many countries to use ground waters as sources for drinking water. Because of application of pesticides in agriculture for long periods many ground water sources are polluted with them. Suitable processes for the removal of the persistent compounds have to be applied. These processes have to destroy these strong compounds. They are ozonation, wet oxidation, catalytic photooxidation usually followed by adsorption of the residual impurities onto activated carbon.

Also it is important to remove the metabolites of POPs from wastewaters. Such metabolites, as shown before, are chlorinated phenols. Many investigations have been done for the application of low-cost adsorbents. Among them are the widely spread deposits of natural materials like bentonite, perlite, etc., in the Balkan area (Koumanova and Peeva-Antova, 2001; Koumanova and Yaneva, 2005; Koumanova et al., 2005).

References

Agency for Toxic Substances and Diseases Registry (ATSDR), 1989. Toxicological Profile for Chlordane. US Department of Human and Health Services, Public Health Service, Atlanta, GA.

Agency for Toxic Substances and Diseases Registry (ATSDR), 1994a. Toxicological Profile for 4,4′-DDT, 4,4′-DDE, 4,4′-DDD (update), US Department of Human and Health Services, Public Health Service, Atlanta, GA.

Agency for Toxic Substances and Diseases Registry (ATSDR), 1994b. Toxicological Profile for Chlordane (update), US Department of Human and Health Services, Public Health Service, Atlanta, GA.

Augustijn-Beckers, P.W.M., Hornsby, A.G., and Wauchope, R.D., 1994. SCS/ARS/CES Pesticide Properties Database for Environmental Decision making II, Additional Properties Reviews of Environmental Contamination and Toxicology, 137, 1–82.

Jorgensen, S.E., Jorgensen, L.A., and Nielsen, S.N., 1991. Handbook of Ecological Parameters and Ecotoxicology, Elsevier, Amsterdam, The Netherlands.

Kirk-Othmer, 1995. Insect Control Technology. In: Encyclopedia of Chemical technology, Ed.: kroschwitz, J.I. Wiley, New York.

Koumanova, B. and Yaneva, Z., 2005a. Sorption modeling of chloro- and nitrophenols by natural materials, 1st South East European Congress of Chemical Engineering, 25–28 September 2005, Belgrade, Serbia/Montenegro.

Koumanova, B. and Yaneva, Z., 2005b. The application of natural materials in the biotreatment of phenolic wastewaters, Journal of the University of Chemical Technology and Metallurgy (Sofia), 40(2), 123–128.

Koumanova, B., P. Peeva-Antova, Z. Yaneva, 2005. Adsorption of 4-chlorophenol from aqueous solutions on activated carbon – kinetic study, Journal of the University of Chemical Technology and Metallurgy(Sofia), 40(3), 213–218.

Toxicological profile for chlorinated dibenzo-*p*-dioxins, 1998. US Department of health and human services, Public health service agency for toxic substances and disease registry, December 1998.

Toxicological profile for DDT, DDE, and DDD, 2002. US Department of health and human services, Public health service agency for toxic substances and disease registry, September 2002.

Toxicological profile for polychlorinated biphenyls (PCBs), 2000. US Department of health and human services, Public health service agency for toxic substances and disease registry, November 2000.

Toxicological profile for hexachlorobenzene, 2002. US Department of health and human services, Public health service agency for toxic substances and disease registry, September 2002.

US Environmental Protection Agency, 1989. Environmental Fate and Effects Division, Pesticide Environmental Fate One Line Summary: DDT (p,p′). Washington, DC.

World Health Organization (WHO), 1989. Environmental Health Criteria 83, DDT and its derivatives, Environmental Effects. World Health Organization, Geneva.

VARIATIONS OF ORGANOCHLORINE CONTAMINANTS IN ANTARCTICA

VLADIMIR BOGILLO,* MARIYA BAZYLEVSKA
Department of Antarctic Geology and Geoecology, Institute of Geological Sciences, National Academy of Sciences of Ukraine, 55B Olesya Gonchara St., 01054 Kiev, Ukraine

Abstract: The transport of organochlorine contaminants (OCs) to Antarctica, the latitudinal, seasonal, and spatial variations of their content in air, the temporal variations in air, seawater, snow, and sea ice are discussed. The temporal declines in OCs level in the mobile compartments of Antarctic environment reflect the global declines in use and emission of the contaminants in Southern Hemisphere.

Keywords: Antarctica, Southern ocean, temporal, seasonal, spatial, latitudinal variations of organochlorine contaminants, "cold condensation," "cold fractionation," half-lives of contaminants in air, seawater, snow, sea ice

1. Introduction

Antarctica has traditionally played a pivotal role in discussions of global environmental change. The key atmospheric issues in the Antarctica are the depletion of the stratospheric ozone layer, the long-range transport of air pollutants and warming associated with global climate change. These problems are mainly due to anthropogenic activities in other parts of the world.

The persistent semivolatile or low-volatile organochlorine contaminants (OCs), such as pesticides (aldrin, dieldrin, endrin, chlordane (CHL), heptachlor, dichlorodiphenyltrichloroethane (DDT) and its metabolites (DDE and DDD), toxaphenes, mirex, hexachlorobenzene (HCB) and hexachlorocyclohexanes (HCHs)), polychlorinated biphenyls (PCBs),

* To whom correspondence should be addressed. E-mail: vbog@carrier.kiev.ua

E. Mehmetli and B. Koumanova (eds.),
The Fate of Persistent Organic Pollutants in the Environment, 251–267.

polychlorinated dibenzo-*p*-dioxins (PCDDs), and polychlorinated dibenzofurans (PCDFs) were found in various abiotic compartments of Antarctica and its biota. These compounds are neither produced nor applied in the Antarctica.

With the exception of low-level and usually highly localized emissions of PCDDs, PCDFs and PCBs within the region, introduction of the compounds to Antarctica and the Southern Ocean come entirely from extraregional sources.

Both the Arctic and Antarctic regions act as a form of chemical sink for these contaminants through a process known as "cold condensation."[1] The contaminants emitted in warmer, midlatitude climates make their way to the poles via atmospheric streams. Being much colder at the poles, they condense out of the atmosphere and settle there in large quantities, in the water, snow, ice, soil, sediment, and biota. The calculations for emitted in tropics or temperate latitudes PCB, p,p′-DDT, HCB, γ-HCH,[2] CCl_4, α-HCH, and mirex,[3] γ-HCH and p,p′-DDT,[4] and the specific distribution patterns of these chemicals in global environment were carried out using the global non-steady-state, multicompartmental mass balance models. The model results confirm that condensation at low temperatures can result in elevated contaminant concentrations in the polar regions.

Two scenarios are proposed for "global fractionation" of persistent organic contaminants in the environment.[5] In first of them, "primary source," air concentrations would be dominated by emission from primary sources. After release from a primary source, the contaminant is deposited and subsequently prevented from volatilizing through permanent retention in the soil, snow cover-firn-ice cover, while shelf sediments,[6] or burial in deep oceanic waters.[7] The environmental reservoirs would as sinks in this scenario. The contaminants have different long-range atmospheric potential. This would result in fractionation of a contaminants mixture away from the primary source. In this primary source scenario, absolute amounts would therefore be expected to decrease with latitude/distance from the primary source.

In second scenario, "secondary source," emission of the contaminants from the environmental reservoirs would control levels in the atmosphere. Repeated air–surface exchange would see the contaminant move in a series of "hops." This periodic movement between mobile and nonmobile media has been termed the "grasshopper effect."[4] This would also result in fractionation, becoming more pronounced over time. In this scenario, absolute concentrations of some contaminants may become higher at higher latitudes ("cold condensation"). Temporal trends in atmospheric concentrations would be expected to show a shift in relative concentrations over time at one site, e.g., more volatile contaminants becoming more abundant over time in higher latitudes.

In the Southern Ocean the apparent final fate of the OCs is to be collected in the fauna and the sediments of the abyss. Although the models predict the soil in Antarctica as main final sink for the contaminants on ground of the continent, the mass fraction of the soil is negligible in comparison with that for the snowpack and firn and the last snow-firn-ice cover would be main final sink for the OCs on the continent.

The DDT was first recognized in Antarctic snow in the 1967.[8] After that tens analytical studies were performed during last 40 years on levels of several OCs in such abiotic compartments of the environment as air, sea and lake water, snow, sea ice, soil, sea, and lake sediments in different locations of the coast, Southern Ocean, and sub-Antarctic islands. In the chapter we discussed transport pathways for the OCs in Antarctica, the latitudinal, seasonal, and spatial variations of their content in air, their temporal variations in air, seawater, snow and sea ice, and their agreement and contradiction with the "global condensation" hypothesis. The apparent half-lives of the OCs in the compartments were derived from these variations and they are compared with emission half-lives for OCs in air.

2. Variations of OCs in Mobile Compartments of Antarctic Environments

2.1. LATITUDINAL VARIATIONS

The Southern Ocean is bounded by the Antarctic Convergence to the north. The Convergence is the strongest of a series of eastward-flowing jets of the Antarctic Circumpolar Current and it is a strong barrier to free north–south exchange of water. Comparable levels of HCHs contamination in seawater[9] and for p,p′-DDE and PCBs in penguin eggs[10] from north and south of the Antarctic Convergence, which separates sharply defined and distinct water masses, indicated that the atmosphere, not the water, was the dominant pathway for the transport of the contaminants to the Antarctica.

At first the latitudinal variations of PCBs, DDTs, and HCHs in air were found in the studies.[11,12] The marked decrease along north to south was established for HCHs family and this was not observed for DDTs and PCBs species.

Similar north–south gradients were found for PCBs, HCHs, and DDTs in air over open ocean between Syowa Station (69° 00′ S, 39° 35′ E) and Mauritius (23° S) during later summer of 1982.[13] The concentrations of the atmospheric OCs tended to decline toward the Antarctica. In contrast to their concentrations, little variation was found in the composition of HCHs isomers, DDT, and its metabolites, PCBs congeners and their chlorine content. Such constancy in distribution of PCB congeners can be an argument against the popular hypothesis of "global fractionation" for persistent organic pollutants in polar environment.[1]

More low levels in air and seawater concentrations for HCHs, CHLs, DDTs, and PCBs were obtained in Southern Ocean in comparison with Eastern Indian Ocean and other northern seas during survey cruises in 1989–1990.[14]

Concentrations of α- and γ-HCH, HCB, heptachlor epoxide, *trans*- and *cis*-chlordanes, *trans*- and *cis*-nonachlors, p,p′-DDT and p,p′-DDE in air over Pacific Ocean were measured from New Zealand to Ross Island in Antarctica during austral summer 1990.[15] The levels of the compounds appeared lowest at Ross Island, highest near Australia, and New Zealand. The significance of negative correlations between concentration and latitude was high for α-HCH, chlordanes + nonachlors, and DDE + DDT, and weaker for γ-HCH.

Concentrations of 11 PCBs congeners, HCB, HCHs, and DDTs were determined in air during cruise routes of research vessels in Atlantic Ocean from South America (23° S, 43° W) to West Antarctic Peninsula (62° S, 55° W) in austral summers 1987 and 1995.[16,17] For most contaminants there was a decline in their level with increasing of southern latitude. The HCHs levels were 10 times lower in 1995 than their previous data of 1987. Also, DDT/DDE ratio was 0.3–0.6 against 20–29 obtained in 1987. In contrast to other OCs, concentration of HCB increased slightly towards the Antarctica. Although, this can be an evidence that HCB in contrast to other organochlorines where levels in air decrease with increasing the latitude or decreasing the temperature – is undergoing "cold condensation" and actually accumulating in seawater of Southern Ocean, this was explained[17] by better retention of HCB in the adsorption column of polyurethane foam (PUF) for the colder sampling stations. Also, the breakthrough of HCB from front to back PUF plugs was evident in the summer season samples, 2001, with HCB levels on the back plug exceeding 33% of quantities found on the front PUF in most samples.[18]

We calculated the changes of relative contributions for PCB congeners to total PCBs level in percents versus southern latitude. The dependences of absolute concentrations of PCBs congeners and the relative contribution on southern latitude are presented in Figure 1a and b. It is shown that absolute concentrations for all PCB congeners decrease from tropics to Antarctica, whereas relative contribution of light congeners increases in this direction and that for heavy congeners decreases towards Antarctica. These patterns are agreed with "cold condensation" hypothesis.

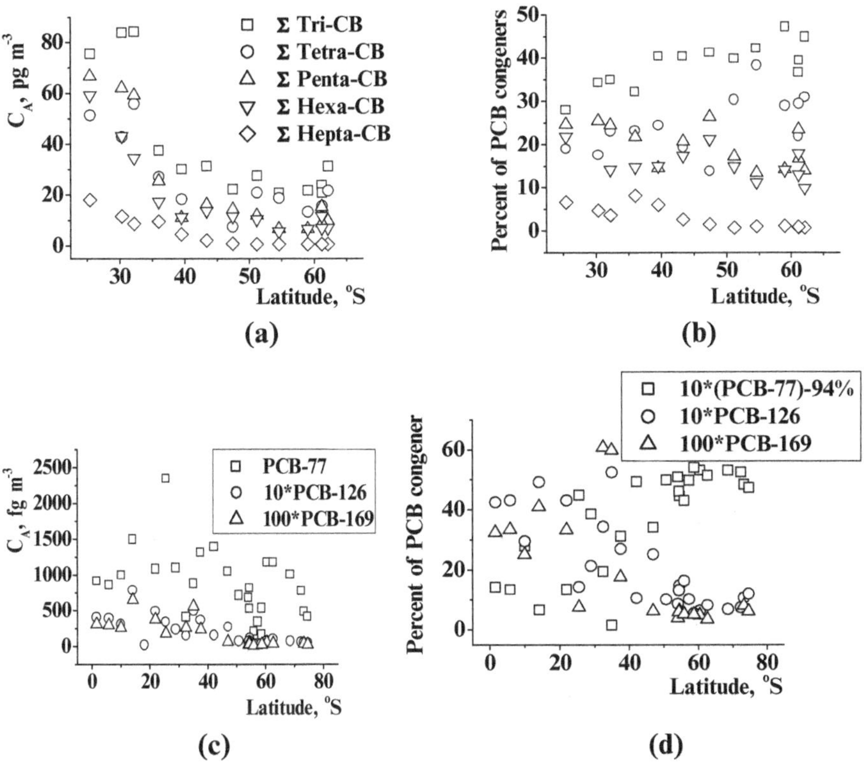

Figure 1. (a) Latitudinal variations of absolute concentrations for PCBs congeners in air[17] ; (b) the relative contributions for the congeners in air[17]; (c) absolute concentrations for non-ortho PCBs congeners in air[19]; (d) the relative contributions for the non-ortho congeners in air[19]

Concentrations in air for PCDD/Fs and coplanar dioxin-like PCBs (77, 126, and 169) during an Atlantic cruise from the UK to Antarctica (Halley Station) were measured in October – December, 1998.[19] Lowest concentrations, measured in the remote Southern Hemisphere, were generally two orders of magnitude lower than their highest concentrations over the North Atlantic.

The dependences of absolute concentrations for coplanar PCBs congeners and the relative contributions to total PCBs level on southern latitude are presented in Figure 1c and d. It is clear that absolute concentrations of the congeners decrease towards Antarctica, while the relative contribution for light PCB 77 increases in this direction, but those for heavy PCB 126 and 169 decrease towards Antarctica.

Concentrations of 6 PCBs congeners were determined in air over Halley Research Station on the Brunt Ice Shelf, Coats Land (75° 35′ S, 26° 30′ W) and at Moody Brook, Falkland Islands (51° 25′ S, 57° 56′ W), and over water near these sites during austral summer, 1999.[20] The PCBs concentrations were consistently lower at Halley in comparison with Falkland Islands.

Levels of α- and γ-HCH in air over Atlantic Ocean during north–south transect from Arctic to Antarctica in 1999 are reported[9] and demonstrate strong latitudinal decline between Cape Town and Neumayer Station. It was noted that the station is influenced to a minor extent to several long-range transport episodes of air masses from South America, because it is located in the inner part of the polar vortex.

The strong latitudinal gradients for the organochlorines in air towards the Antarctic coast are comparable with similar gradients for such components of atmospheric aerosols as particulate organic matter, elemental carbon and sea salts, measured in November–December 1999 during a cruise over the Atlantic Ocean between the English Channel and the Antarctic coast, along with a corresponding data measured in January 2000 at the Finnish Antarctic research station Aboa (73° 23′ S, 14° 09′ W).[21]

2.2. SEASONAL VARIATIONS

They were discussed for PCBs, DDTs, and HCHs,[13] for atmospheric PCBs, DDTs and γ-HCH,[22] and for HCHs, heptachlor, and heptachlor epoxide.[18] High OCs concentrations were generally obtained during the austral summer.[13] This seasonal difference was explained by removing of OCs by snowfall in winter with intensive snowfalls, more intensive application of the pesticides on lands in summer and active evaporation of the OCs into atmosphere from their various sources on lands due to high temperature in summer season.

Similar results were obtained for γ-HCH[22] with significant correlation between concentration of γ-HCH and mean temperature of air ($r = 0.548$). No significant relationship was found between levels of DDTs and PCBs in the air and the temperature. These findings were explained by lower vapor pressures of PCBs and DDTs compounds compared with that of lindane which leads to more easily remobilization of last compound from the ground to the atmosphere and to more great potential of lindane to long-range transport over the Indian Ocean.

We estimated dependence of ΣDDT level in air on monthly air temperature using the air concentration[13] and monthly air temperatures on Syowa Station.[23] The dependencies of air concentrations for γ-HCH[22] and ΣDDT[13] on air temperature are displayed in Figure 2a and b. They demonstrate increasing the OCs level as the temperature growths.

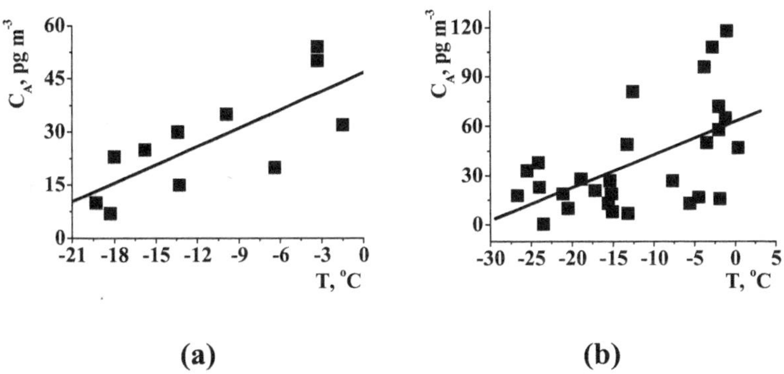

Figure 2. Seasonal dependencies for level of ΣDDT (a) and γ-HCH (b) in Antarctic air on the air temperature

In contrast to above studies, levels of HCHs, heptachlor, and heptachlor epoxide in air over Western Antarctic Peninsula were higher in winter versus summer samples.[18] These differences were explained by different seasonal sources for these compounds or their removal rates/processes.

Above distinguished summer concentration maximum for the OCs and winter minimum for those coincide with a summer maximum and winter minimum for concentrations of black carbon in air measured at Halley station and at South Pole.[24] Then, additional reason to the seasonal variations may be break down of the circumpolar vortex and weakness of surface inversion during summer months in Antarctica.

We checked the temperature trends of OCs concentrations in air from their latitudinal and seasonal variations in accordance with following relationships:

$$C_{\mathrm{abs}} = a_1 + b_1 T_{\mathrm{air}} \quad (1a)$$

$$C_{\mathrm{rel}} = a_2 + b_2 T_{\mathrm{air}} \quad (1b)$$

where C_{abs} is the atmospheric concentration of OC, pg m^{-3}, C_{rel} is the contribution of the congeners in total amount of OC, in percents, and T is the air temperature, in °C. The coefficients of the relationships and the correlation coefficients are presented in Table 1.

It is obvious that slopes for plot of absolute OCs concentrations versus temperature drop at transition from light to heavy congeners of PCBs, PCDDs, and PCDFs, whereas the slopes for plot of congener contribution versus temperature increase at such transition. The slope for plot of absolute OCs concentrations versus temperature is always positive for all OCs, except for HCB. The b_1 coefficient decreases and b_2 coefficient increases as the volatility of OCs falls. These patterns agree with "cold condensation" hypothesis.

TABLE 1. Coefficients of the relationships 1a (a_1, b_1) and 1b (a_2, b_2), the correlation coefficients (r_1, r_2)

OC	a_1	b_1	r_1	a_2	b_2	r_2	Reference
PCB 77[a]	635 ± 112	22.9 ± 7.4	0.536	99.2 ± 0.3	−0.12 ± 0.02	0.800	19
PCB126[a]	5.1 ± 3.7	1.3 ± 0.2	0.746	0.89 ± 0.26	0.109 ± 0.017	0.805	19
PCB169[a]	0.2 ± 0.4	0.13 ± 0.02	0.807	0.056 ± 0.048	0.011 ± 0.003	0.682	19
Cl_2DD[a]	13.0 ± 3.1	1.8 ± 0.2	0.897	25.4 ± 2.6	0.16 ± 0.17	0.204	19
Cl_3DD[a]	25.4 ± 8.2	0.4 ± 0.8	0.135	6.2 ± 2.0	0.35 ± 0.12	0.543	19
Cl_4DD[a]	18.3 ± 4.9	1.3 ± 0.5	0.546	9.8 ± 1.3	0.3 ± 0.1	0.535	19
Cl_5DD[a]	6.8 ± 1.7	0.5 ± 0.2	0.616	5.0 ± 0.8	0.06 ± 0.05	0.246	19
Cl_6DD[a]	6.4 ± 1.3	0.11 ± 0.15	0.193	4.2 ± 0.8	0.03 ± 0.05	0.127	19
Cl_7DD[a]	15.6 ± 3.7	0.8 ± 0.4	0.442	11.8 ± 1.8	0.01 ± 0.11	0.011	19
Cl_8DD[a]	24.9 ± 4.2	0.3 ± 0.3	0.214	36.2 ± 4.6	−0.5 ± 0.3	0.325	19
Cl_2DF[a]	289 ± 204	52 ± 13	0.664	60 ± 42	−0.99 ± 0.27	0.625	19
Cl_3DF[a]	−85 ± 268	94 ± 17	0.740	34.1 ± 5.8	0.7 ± 0.4	0.355	19
Cl_4DF[a]	5.2 ± 8.9	3.3 ± 0.6	0.754	3.5 ± 0.8	0.005 ± 0.053	0.017	19
Cl_5DF[a]	4.4 ± 3.3	1.2 ± 0.2	0.777	1.9 ± 0.6	0.006 ± 0.038	0.036	19
Cl_6DF[a]	2.9 ± 1.4	0.6 ± 0.1	0.810	1.1 ± 0.3	−0.02 ± 0.02	0.198	19
Cl_7DF[a]	6.3 ± 2.8	0.4 ± 0.2	0.427	2.7 ± 0.9	−0.06 ± 0.06	0.194	19
Cl_8DF[a]	3.8 ± 1.1	0.26 ± 0.08	0.725	1.9 ± 0.8	0.04 ± 0.06	0.190	19
Cl_3B	17.2 ± 5.7	2.2 ± 0.4	0.842	43.7 ± 1.6	−0.57 ± 0.11	0.833	17
Cl_4B	11.5 ± 3.9	1.3 ± 0.3	0.807	28.4 ± 2.3	−0.37 ± 0.15	0.581	17
Cl_5B	3.5 ± 5.1	2.1 ± 0.4	0.856	15.8 ± 1.9	0.46 ± 0.12	0.728	17
Cl_6B	3.4 ± 4.0	1.5 ± 0.3	0.833	13.6 ± 1,3	0.22 ± 0.09	0.585	17
Cl_7B	−8.1 ± 23	1.0 ± 0.2	0.927	1.0 ± 0.8	0.17 ± 0.05	0.701	17
HCB	17.5 ± 2.5	−0.58 ± 0.16	0.715				17
α-HCH	4.2 ± 1.5	0.26 ± 0.10	0.611				17
γ-HCH	3.4 ± 0.9	0.49 ± 0.07	0.919				17
DDE	2.1 ± 0.7	0.24 ± 0.05	0.803				17
ΣDDT	46.6 ± 6.3	1.7 ± 0.5	0.761				13
γ-HCH	62.8 ± 8.7	2.0 ± 0.6	0.560				22
α-HCH	3.5 ± 0.3	0.1 4± 0.04	0.783				15
γ-HCH	4 ± 23	0.31 ± 0.29	0.370				15
HCB	57.6 ± 2.3	−5.1 ± 0.7	0.991				15
p,p'DDE	0.27 ± 0.06	0.026 ± 0.009	0.810				15
p,p'DDT	0.38 ± 0.10	0.049 ± 0.015	0.827				15

[a]Air concentrations in fg m^{-3}, and in pg m^{-3} in other cases

2.3. SPATIAL VARIATIONS

Spatial variations for concentrations of PCBs, DDTs, and HCHs in Antarctic air were discussed by Tanabe et al.[13] More high concentrations were obtained for samples off Sabrina Coast (61–65° S, 121–125° E) and near Balleny Islands (60–67° S, 155–164° E) in comparison with off Syowa Station (68–69° S, 39° E), and near Amundsen Bay (66° S, 47–50° E). These findings were explained by prevailing northwest winds in coastal Antarctica and preferential transport of the pesticides through the Indian Ocean with pesticide-rich air to off Sabrina Coast.

Maximum concentrations for the pesticides and PCBs in air were observed near Antarctica Peninsula, King George Island, 61.16° S, 55.7° W[17,28], and Signy Island, 60° 72′ S, 45° 60′ W[29] in comparison with more eastern locations (Halley Station, 75° 35′ S, 26° 30′ W[20], Neumayer Station, 70° 38′ S, 8° 16′ W,[25] Terra Nova Buy, 75° S, 164° 06′ E[30]), and Ross Island, 77° 38.1′ S, 166° 24.6′ E.[15,22]

By analogy with atmospheric tracers, this west–east gradient may be explained by prevailing northwest winds and west–east direction of cyclones in coastal Antarctica,[25] by more short time for transport of air masses from South America in comparison with those from South Africa or Australia/New Zealand,[26] and gradual decrease for level of atmospheric tracers from the west coastal area to east of inland area.[27]

2.4. TEMPORAL VARIATIONS

As a rule, the seasonal and spatial OCs concentration variations in air are from two to four times. Most changes for OCs level were obtained in time of the observations. We developed database for variations of concentrations of DDTs, HCHs, HCB, PCBs, and CHLs in air,[9,11–18,20,22,28–33] seawater,[9,11–14,18,34–37] snow[8,10,13,18,37–41], and sea ice[13,18,35] of Antarctica and Southern Ocean. The temporal trends are presented in Figures 3a–f, 4a–d, and 5a–d.

It is shown that distinct declines are observed during last decades in Antarctica for ΣDDT and ΣHCH in air, seawater and snow, HCB, ΣPCB and ΣCHL in air, ΣHCH in sea ice, DDT/DDE ratio in air and snow, and for α-HCH to γ-HCH ratio in seawater. We treated these trends using a relationship for decay of the compounds in the compartments assuming first-order kinetics:

$$\ln C_i = A - k_{\mathrm{ap}} \times (\mathrm{year}) \qquad (2a)$$

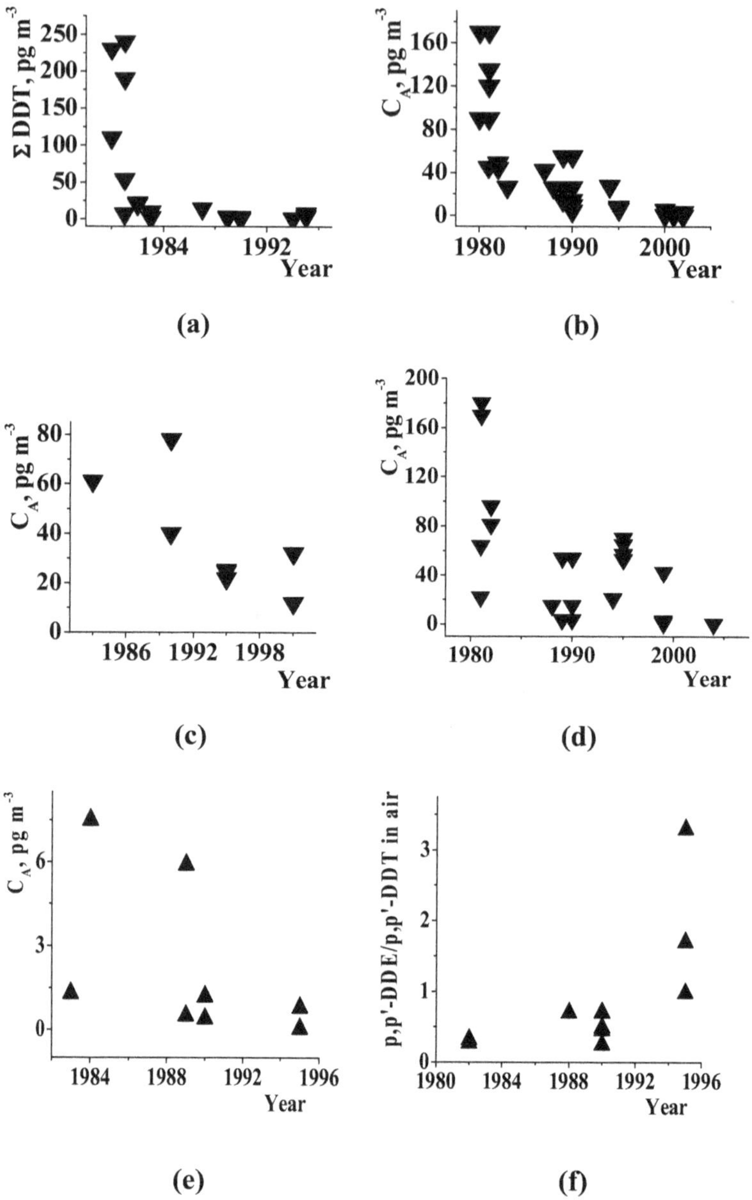

Figure 3. Temporal variations of OCs in Antarctic air: ΣDDT (a), ΣHCH (b), HCB (c), ΣPCB (d), ΣChlordane (e), and p,p′-DDE/p,p′-DDT ratio (f)

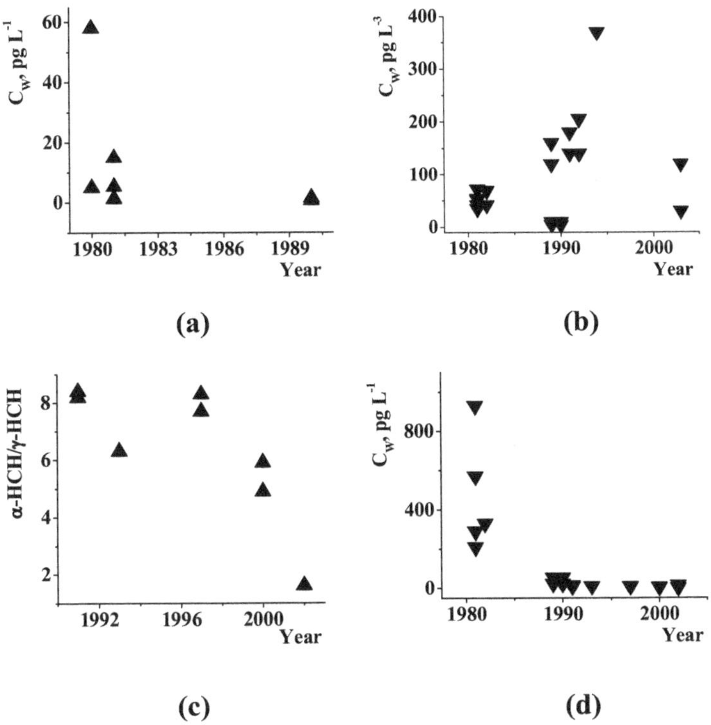

Figure 4. Temporal variations of OCs in seawater of Southern Ocean: ΣDDT (a), ΣPCB (b), α-HCH/γ-HCH ratio (c), and ΣHCH (d)

where C_i is the current concentration of a chemical in *i*-compartment at current year, A is the constant for given chemical and compartment, and k_{ap} is the apparent rate coefficient for fate the chemical in the compartment.

Also, we estimated the first-order half-lives for the chemicals in the compartments, $\tau_{ap1/2}$, using the k_{ap} values:

$$\tau_{ap1/2} = (\ln 2)/k_{ap} \quad (2b)$$

The regression results for of Eq. 2a and b, and correlation coefficients for relationship 2a are presented in Table 2. It is shown that the half-lives for ΣDDT and ΣHCH families are close with each other in such different compartments as air, seawater and snow. The ΣCHL family demonstrates the short half-live in air, whereas HCB is most long-lived contaminant in the air. The half-live for sum of PCBs congeners in the Antarctic air (4.6 years) is comparable with half-lives for the PCBs congeners in Norwegian, UK, and Great Lakes background air ($\tau_{ap1/2}$ = 1.7–6 years).[5]

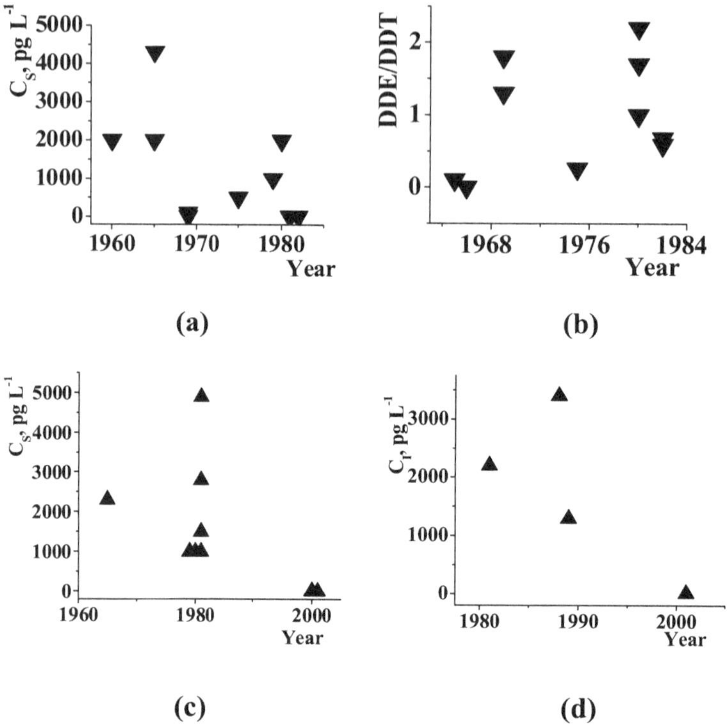

Figure 5. Temporal variations of OCs in Antarctic snow and sea ice: ΣDDTs in snow (a), DDE/DDT ratio in snow (b), ΣHCHs in snow (c), and ΣHCHs in sea ice (d)

TABLE 2. Parameters of Eq. 2a and b, and correlation coefficients for Eq. 2a (*r*) for temporal trends of organochlorine contaminants in abiotic compartments of Antarctic environment

Compounds	Media	Units	*A*	k_{ap} (year^{-1})	$\tau_{ap1/2}$ (years)	*R*
ΣDDTs	Air	pg m^{-1}	480 ± 116	0.240 ± 0.058	2.9 ± 0.7	0.697
ΣDDTs	Seawater	pg L^{-1}	454 ± 217	0.228 ± 0.109	3.0 ± 1.4	0.619
ΣDDTs	Snow	pg L^{-1}	261 ± 180	0.129 ± 0.091	5.4 ± 3.8	0.449
ΣHCHs	Air	pg m^{-3}	437 ± 39	0.218 ± 0.020	3.2 ± 0.3	0.897
ΣHCHs	Seawater	pg L^{-1}	442 ± 42	0.220 ± 0.021	3.2 ± 0.3	0.925
ΣHCHs	Snow	pg L^{-1}	427 ± 64	0.212 ± 0.032	3.3 ± 0.5	0.911
ΣHCHs	Sea ice	pg L^{-1}	689 ± 148	0.343 ± 0.074	2.0 ± 0.4	0.937
HCB	Air	pg m^{-3}	156 ± 52	0.076 ± 0.026	9.1 ± 3.1	0.768
ΣPCBs	Air	pg m^{-3}	307 ± 83	0.152 ± 0.042	4.6 ± 1.3	0.635
ΣCHLs	Air	pg m^{-3}	486 ± 232	0.245 ± 0.117	2.8 ± 1.3	0.650
DDT/DDE	Air		255 ± 105	0.129 ± 0.053	5.4 ± 2.2	0.653
DDT/DDE	Snow		117 ± 108	0.059 ± 0.054	11.7 ± 10.7	0.378
α-HCH/γ-HCH	Seawater		181 ± 76	0.09 ± 0.04	7.7 ± 3.4	0.694

Also, the DDT to DDE ratio in air declines with half-live ~5 years. The degradation products of p,p′-DDT are mainly the dechlorination products p,p′-DDE, and p,p′-DDD. The water solubility of p,p′-DDT is approximately 9.6 times higher in comparison to p,p′-DDE.[14] Then increase of DDE to DDT ratio on time is explained by the preferential transport of the DDE from emission sources in temperate latitudes of southern hemisphere and global continuing drop in atmospheric DDT between 1980 and 1995.

The half-lives for the organochlorines in different compartments of Antarctic environment ($\tau_{ap1/2}$) are not related to their decomposition half-lives ($\tau_{1/2}$) in air, water, and sediments as a result of their OH radical degradation in air, hydrolysis, and microbial decomposition in water and sediments.[42–45] For example, the $\tau_{1/2}$ values at 25°C for α-HCH are 43 days in air, 1.9 years in water (and, presumably, in snow), and 6.3 years in sediment, whereas they are almost coincide ($\tau_{ap1/2}$ = 3.2 years) in air, seawater and snow in Antarctica. Also, p,p′-DDT quickly decomposes in air ($\tau_{1/2}$ = 2.3 days), not decomposes in water and slowly decomposes in sediments ($\tau_{1/2}$ = 4.4 years), but it has close $\tau_{ap1/2}$ values (3–5 years) in air, seawater and snow in Antarctica. The chlordane has much more short half-live in air ($\tau_{1/2}$ = 0.12 days) in comparison with HCB ($\tau_{1/2}$ = 1.9 years), but their $\tau_{ap1/2}$ values in air are comparable (2.8 and 9 years, respectively). Then, the temporal declines in organochlorines level in the mobile compartments of Antarctic environment reflect the global declines in use and emission of the contaminants in Southern Hemisphere.

The $\tau_{ap1/2}$ values for HCHs in Antarctic air and seawater from Table 2 (3.2 ± 0.3 years) almost coincide with the half-live for HCHs in the Antarctic air estimated in[18] ($t_{ap1/2}$ = 3.0 ± 0.7 years). It was noted[18] with reference to[46] that "biodegradation is responsible for the shorter half-life for HCHs in Antarctic seawater (2 years[46]) compared to air (3 years)." However, foregoing half-life for HCHs in water[46] in fact was derived from temporal trend of HCHs concentrations (in ng/g_{lipid}) in sea plankton collected in the Antarctic Peninsula region during 1983–1984,[47] 1986, 1987,[48] and 2002[46] years. Our estimate of the half-life on basis of the trend from 1983 to 1987 years gives $\tau_{ap1/2}$ = 6.7 years. Moreover, there were different concentration units for level of HCHs in sea plankton in these studies: ng/g fresh mass,[47] ng/g dry mass,[48] and ng/g lipid[46]. As was mentioned,[47] "highest levels of HCHs were in the Admiralty Bay samples, collected closer to land than other samples and they were explained by release of HCHs stored in snow by glacial runoff, acting as a 'delay loop' between the time of local deposition in snowfall and subsequent release to the marine environment. In this case, the compounds released at the time of sampling may have originated at the time of maximum pesticide use." Then the half-life for HCHs derived from those levels is not related to actual $\tau_{ap1/2}$ value in seawater, neither to its biodegradation half-live in seawater.

We estimated the emission half-lives ($\tau_{em1/2}$) for p,p′-DDT and HCB using temporal variations of their total usage (DDT) and emission (HCB) in European countries during 1970–1995.[49] Also, the emission half-live for γ-HCH can be estimated from difference of its usage in Europe: 25,000 t in 1970 year and 671 t in 1990 year.[50] It turned out, that these half-lives (3.3 ± 0.4 years for DDT, 7.0 ± 1.1 years for HCB and 3.8 years for γ-HCH) are close to the $\tau_{ap1/2}$ values for these chemicals in Antarctic air. The emission half-lives for PCB congeners are varied from 5 to 10 years,[51,52] and they in general exceed the $\tau_{ap1/2}$ value for PCBs in Antarctic air.

In general, the differences in the decline rates of OCs in air are related to the removal of the OCs from air by different processes, or to the differences in the decline rates of OCs emissions from primary sources. The model simulations[53] suggest that the rate of decline of PCBs air concentrations is entirely driven by the rate of decline in PCBs emissions. This is consistent with the observation that the congeneric PCBs composition in the air is largely a result of the congeneric composition of the emissions to the atmosphere.[54]

3. Conclusions

The atmosphere, not the water, is the dominant pathway for the transport of the organochlorine contaminants (OCs) to the Antarctica.

The slopes for plot of absolute OCs concentrations versus temperature drop at transition from light to heavy congeners of PCBs, PCDDs, and PCDFs, whereas the slopes for plot of congener contribution versus temperature increase at such transition. The slope for plot of absolute OCs concentrations versus temperature is always positive for all OCs, except for HCB. These patterns agree with "cold condensation" hypothesis.

The apparent half-lives of the OCs in air, seawater, snow, and sea ice were derived from temporal variations in the compartments. The half-lives for ΣDDT and ΣHCH families are close with each other in air, seawater, and snow. The ΣCHL family demonstrates the short half-live in air, whereas HCB is most long-lived contaminant in the air. These half-lives in air are comparable with their European emission half-lives and temporal declines in OCs level in the mobile compartments of Antarctic environment reflect the global declines in use and emission of the contaminants in Southern Hemisphere.

References

1. F. Wania and D. Mackay, Global fractionation and cold condensation of low volatility organochlorine compounds in polar regions, *Ambio* 22, 10–18 (1993).
2. F. Wania and D. Mackay, A global distribution model for persistent organic chemicals, *Sci. Total Environ.* 160/161, 211–232 (1995).
3. M. Scheringer, F. Wegmann, K. Fenner, and K. Hungerbühler, Investigation of the cold condensation of persistent organic pollutants with a global multimedia fate model, *Environ. Sci. Technol.* 34, 1842–1850 (2000).
4. V. S. Semeena and G. Lammel, The significance of the grasshopper effect on the atmospheric distribution of persistent organic substances, *Geophys. Res. Let.* 32, L07804, doi:10.1029/2004GL022229 (2005).
5. S. N. Meijer, W. A. Ockenden, E. Steinnes, B. P. Corrigan, and K. C. Jones, Spatial and temporal trends of POPs in Norwegian and UK background air: Implications for global cycling, *Environ Sci. Technol.* 37, 454–461 (2003).
6. A. Jonsson, O. Gustafsson, J. Axelman, and H. Sundberg, Global accounting of PCBs in the continental shelf sediments, *Environ. Sci. Technol.* 37, 245–255 (2003).
7. R. Lohmann, E. Jurado, M. E. Q. Pilson, and J. Dachs, Oceanic deep water formation as a sink of persistent organic pollutants, *Geophys. Res. Lett.* 33, doi: 10.1029/2006GL025953 (2006).
8. T. J. Peterle, DDT in Antarctic snow, *Nature* 224, 620 (1969).
9. S. Lakaschus, K. Weber, F. Wania, R. Brun, and O. Schrems, The air-sea equilibrium and time trend of hexachlorocyclohexanes in the Atlantic Ocean between the Arctic and Antarctica, *Environ. Sci. Technol.* 36(2), 138–145 (2002).
10. R. W. Risebrough, W. I. Walker, T. T. Shmidt, B. W. Delappe, and C. W. Connors, Transfer of chlorinated biphenyls to Antarctica, *Nature* 364, 738–739 (1976).
11. S. Tanabe, M. Kawano, and R. Tatsukawa, Chlorinated hydrocarbons in the Antarctic, Western Pacific and Eastern Indian Oceans, *Trans. Tokyo Univ. Fish.* 5, 97–109 (1982).
12. S. Tanabe, R. Tatsukawa, M. Kawano, and H. Hidaka, Global distribution and atmospheric transport of chlorinated hydrocarbons: HCH (BHC) isomers and DDT compounds in the Western Pacific, Eastern Indian and Antarctic Oceans, *J. Oceanograph. Soc. Japan* 38, 137–148 (1982).
13. S. Tanabe, H. Hidaka, and R. Tatsukawa, PCBs and chlorinated hydrocarbon pesticides in Antarctic atmosphere and hydrosphere, *Chemosphere* 12(2), 277–288 (1983).
14. H. Iwata, S. Tanabe, N. Sakal, and R. Tatsukawa, Distribution of persistent organochlorines in the oceanic air and surface seawater and the role of ocean on their global transport and fate, *Environ. Sci. Technol.* 27, 1080–1098 (1993).
15. T. F. Bidleman, M. D. Walla, R. Roura, E. Carr, and S. Schmidt, Organochlorine pesticides in the atmosphere of the Southern Ocean and Antarctica, January–March, 1990, *Marine Pollut. Bull.* 26(5), 258–262 (1993).
16. R. R. Weber and R. C. Montone, in: *Longe-range transport of pesticides*, edited by D. A. Kurtz (Lewis Publishers, Boca Raton, FL, 1990), pp. 185–197.
17. R. C. Montone, S. Taniguchi, C. Boian, and R. R. Weber, PCBs and chlorinated pesticides (DDTs, HCHs and HCB) in the atmosphere of the southwest Atlantic and Antarctic oceans, *Marine Pollut. Bull.* 50, 778–786 (2005).
18. R. M. Dickhut, A. Cincinelli, M. Cochran, and H. W. Ducklow, Atmospheric concentrations and air-water flux of organochlorine pesticides along the Western Antarctic Peninsula, *Environ. Sci. Technol.* 39, 465–470 (2005).
19. R. Lohmann, W. A. Ockenden, J. Shears, and K. C. Jones, Atmospheric distribution of polychlorinated dibenzo-*p*-dioxins, dibenzofurans (PCDD/Fs), and non-*ortho* biphenyls (PCBs) along a North – South Atlantic transect, *Environ. Sci. Technol.* 35(20), 4046–4053 (2001).

20. W. A. Ockenden, R. Lohmann, J. R. Shears, and K. C. Jones, The significance of PCBs in the atmosphere of the Southern Hemisphere, *Environ. Sci. Pollut. Res.* 8(3), 189–194 (2001).
21. A. Virkkula, K. Teinil, R. Hillamo, V.-M. Kerminen, S. Saarikosk, M. Aurela, J. Viidanoja, J. Paatero, I. K. Koponen, and M. Kulmala, Chemical composition of boundary layer aerosol over the Atlantic Ocean and at an Antarctic site, *Atmos. Chem. Phys. Discuss.* 6, 455–491 (2006).
22. P. Larsson, C. Jarnmark, and A. Sodergren, PCBs and chlorinated pesticides in the atmosphere and aquatic organisms of Ross Island, Antarctica, *Marine Pollut. Bull.* 25(12), 281–287 (1992).
23. W. Schwerdtfeger, *Weather and climate of the Antarctic* (Elsevier, Amsterdam, 1984).
24. E. W. Wolff and H. Cachier, Concentrations and seasonal cycle of black carbon in aerosol at a coastal Antarctic station, *J. Geophys. Res.* 103(D9), 11033–11041 (1998).
25. *Atlas of the Oceans, v. 6. Antarctic* (Ministry of Defense, Federation of Russian Navy, St. Petersburg, 2005).
26. G. Krinner and C. Genthon, Tropospheric transport of continental tracers towards Antarctica under varying climatic conditions, *Tellus* 55B, 54–70 (2003).
27. D. H. Qin, P. A. Mayevski, J. W. Ren, C. D. Xiao, and J. Y.Sun, The Weddell Sea region: an important precipitation channel in the interior of the Antarctic ice sheet as revealed by glaciochemical investigation of surface snow along the longest trans-Antarctic route, *Ann. Glaciol.* 29, 55–60 (1999).
28. R. C. Montone, S. Taniguchi, and R. R. Weber, PCBs in the atmosphere of King George Island, Antarctica, *Sci. Total Environ.* 308(1–3), 167–173 (2003).
29. R. Kallenborn, M. Oehme, D. D. Wynn-Williams, M. Schlabach, and J. Harris, Ambient air levels and atmospheric long-range transport of persistent organochlorines to Signy Island, Antarctica, *Sci. Total Environ.* 220, 167–180 (1998).
30. A. Gambaro, L. Manodori, R. Zangrando, A. Cincinelli, G. Capodaglio, and P. Cescon, Atmospheric PCB concentrations at Terra Nova Bay, Antarctica, *Environ. Sci. Technol.* 39, 9406–9411 (2005).
31. E. L. Atlas and C. S. Giam, in: *Chemical oceanography*, edited by R. A. Duce, J. P. Riley, and R. Chester (Academic Press, New York, 1989), pp. 339–378.
32. M. Kawano, S. Tanabe, T. Inoue, and R. Tatsukawa, Chlordane compounds found in the marine atmosphere from the southern hemisphere, *Trans. Tokyo Univ. Fish.* 59–66 (1985).
33. R. Lohmann, F. M. Jaward, L. Durham, J. L. Barber, W. Ockenden, K. C. Jones, R.Bruhn, S. Lakaschus, J. Dachs, and K. Booij, Potential contamination of shipboard air samples by diffusive emissions of PCBs and other organic pollutants: implications and solutions, *Environ. Sci. Technol.* 38, 3965–3970 (2004).
34. R. Fuoco, M. P. Colombini, A. Ceccarini, and C. Abete, Polychlorbiphenyls in Antarctica, *Microchem. J.* 54, 384–390 (1996).
35. P. Desideri, R. Lepri, D. Santianni, and L. Checcini, Chlorinated pesticides in sea water and pack ice in Terra Nova Bay (Antarctica), *Ann. Chim.*, 81, 533–540 (1991).
36. R. C. Montone, S. Taniguchi, and R. R. Weber, Polychlorinated biphenyls in marine sediments of Admiralty Bay, King George Island, Antarctica, *Marine Pollut. Bull.* 42(8), 611–614 (2001).
37. R. Fuoco, S. Giannarelli, Y. Wei, C. Abete, S. Francesconi, and M. Termine, Polychlorobiphenyls and polycyclic aromatic hydrocarbons in the sea-surface micro-layer and the water column at Gerlache Inlet, Antarctica, *J. Environ. Monit.* 7, 1313–1319 (2005).
38. D. A. Peel, Organochlorine residues in Antarctic snow, *Nature* 254, 324–325 (1975).
39. F. Ja. Rovinsky, L. D. Voronova, M. I. Afanasyev, A. V. Denisova, and I. G. Pushkar, in: *Background monitoring of ground ecosystems contamination of organochlorine compounds* (Gidrometeoizdat, Leningrad, No 1, 1990), pp. 90–98.

40. R Bargagli, *Antarctic ecosystems* (Springer, Berlin, Heidelberg, 2005).
41. S. Lakaschus and O. Schrems, Air – snow/firn transfer of HCHs at Neumayer station/ Antarctica, Abstracts of SETAC-2002.
42. A. Leip and G. Lammel, Indicators for persistence and long-range transport potential as derived from multicompartment chemistry – transport modelling, *Environ. Pollut.* 128, 205–221 (2004).
43. M. Scheringer, M. Salzmann, M. Stroebe, F. Wegmann, K. Fenner, and K. Hungerbühler, Long-range transport and global fractionation of POPs: insights from multimedia modelling studies, *Environ. Pollut.* 128, 177–188 (2004).
44. S. Sinkkonen and J. Paasivirta, Degradation half-life times of PCDDs, PCDFs and PCBs for environmental fate modelling, *Chemosphere* 40, 943–949 (2000).
45. M. Stroebe, M. Scheringer, H. Held, and K. Hungerbühler, Inter-comparison of multimedia modelling approaches: modes of transport, measures of long range transport potential and the spatial remote state, *Sci. Total Environ.* 321, 1–20 (2004).
46. A. L. Chiuchiolo, R. M. Dickhut, M. A. Cochran, and H. W. Ducklow, Persistent organic pollutants at the base of the Antarctic marine food web, *Environ. Sci. Technol.* 38, 3551–3557 (2004).
47. A. B. Lukowski and R. Ligowski, Accumulation of chloroorganic insecticides by Antarctic marine diatoms, *Polish Polar Res.* 8, 167–177 (1987).
48. C. Joiris and W. Overloop, PCBs and organochlorine pesticides in phytoplankton and zooplankton in the Indian sector of the Southern Ocean, *Antarctic Sci.* 3, 371–377 (1991).
49. J. M. Pacyna, K. Breivik, J. Münch, and J. Fudala, European atmospheric emissions of selected persistent organic pollutants, 1970–1995, *Atmos. Environ.* 37(1), Suppl., S119–S131 (2003).
50. *Environmental cycling of selected persistent organic pollutants (POPs) in the Baltic region (POPCYCLIN-Baltic)*, edited by J. M. Pacyna. Final report, 1999.
51. K. Breivik, A. Sweetman, J. M. Pacyna, and K. C. Jones, Towards a global historical emission inventory for selected PCB congeners – a mass balance approach. 1. Global production and consumption, *Sci. Total Environ.* 290, 181–198 (2002).
52. A. Sweetman, J. M. Pacyna, and K. C. Jones, Towards a global historical emission inventory for selected PCB congeners – a mass balance approach. 2. Emissions, *Sci. Total Environ.* 290, 199–224 (2002).
53. H. Hung, S. C. Lee, F. Wania, P. Blanchard, and K. Brice, Measuring and simulating atmospheric concentration trends of polychlorinated biphenyls in the Northern Hemisphere, *Atmos. Environ.* 39, 6502–6512 (2005).
54. F. Wania and Y. Su, Quantifying the global fractionation of polychlorinated biphenyls, *Ambio* 33, 161–168 (2004).

DIOXIN CHARACTERISATION, FORMATION, AND MINIMISATION

MU'TAZ AL-ALAWI*
Mu'tah University, P.O. Box 3, Karak 61710, Jordan

Abstract: The present paper initially discussed chemical structures and properties of the dioxins (polychlorinated dibenzo-*para*-dioxins or PCDDs) and furans (polychlorinated dibenzo-*para*-furans or PCDFs). Also, presents information concerning the sources and origins of PCDDs/PCDFs. A detailed discussion on the emissions of dioxins from incinerators and in particular, their formation in MSW incinerators. The final section of the paper mention approaches to dioxin minimisation during incineration.

Keywords: Municipal solid waste incineration, polychlorinated dibenzodioxins, polychlorinated dibenzofurans

1. Introduction

The demands of our society generate wastes that tend to increase in quantity as the standard of living increases. The most effective means of dealing with the problem is to reduce the amount of wastes generated. The trend with continually improving living standards has been contrary to this. The opportunities for landfilling as a disposal method for municipal solid waste (MSW) are rapidly declining with depleting available cheap land resources and the wasteful nature of disposing useful resources in the landfill operation. The options for public sector waste reduction translate into separation, recycling and resource recovery. Due to the limited economic benefits of separation and recycling, resource recovery in the form of heat and power production has gained favour in the last 20 years [1]. During this

*To whom correspondence should be addressed. Tel.: +962-795340079; fax: +962-3-2375540; E-mail: alawi1979@yahoo.com

E. Mehmetli and B. Koumanova (eds.),
The Fate of Persistent Organic Pollutants in the Environment, 269–282.

period incineration of MSW has seen turbulent times in terms of popularity, but it is an attractive alternative for disposal and has significant benefits [2].

1. The volume and mass of MSW is reduced to a fraction of its original size (by 85–90% by volume).
2. The waste reduction is immediate and not dependent on long biological break-down reaction times.
3. Incineration facilities can be constructed closer to the MSW sources or collection points, reducing transportation costs.
4. Using heat recovery technology, the cost of the operation can be offset by energy sales.
5. Air discharges can be controlled to meet environmental legislative limit values.

However, incineration does have its problems and consequently its critics due to the following reasons.

1. Some materials should not be incinerated because they are more valuable for recycling, they are non-combustible or their by-products may give rise to harmful emissions.
2. Poor operating practices and the presence of chlorine in the MSW may lead to emissions containing highly toxic dioxins and furans.
3. The control of metal emissions may be difficult for inorganic wastes containing heavy metals, such as arsenic, cadmium, chromium, copper, lead, mercury, nickel, etc.
4. Incinerators require a high capital costs and trained operators leading to moderately high operating costs.
5. Supplementary fuels are required to achieve the necessary high combustion temperatures.

2. Dioxins and Benzofurans

2.1. CHEMICAL STRUCTURES AND PROPERTIES

Dioxin is one of the most toxic chemicals known. A report released for public comment in September 1994 by the US Environmental Protection Agency [3] clearly describes dioxin as a serious public health threat in the 1960s. Dioxins, as they are commonly called, are PCDDs and PCDFs are compounds with similar chemical properties. Each compound comprises two benzene rings interconnected by oxygen atoms. In the case of PCDDs, the benzene rings are joined by two oxygen bridges, and in the case of the PCDFs, the benzene rings are connected by a carbon bond and an oxygen bridge. Figure 1a and b show the generic structures of PCDDs and PCDFs, respectively.

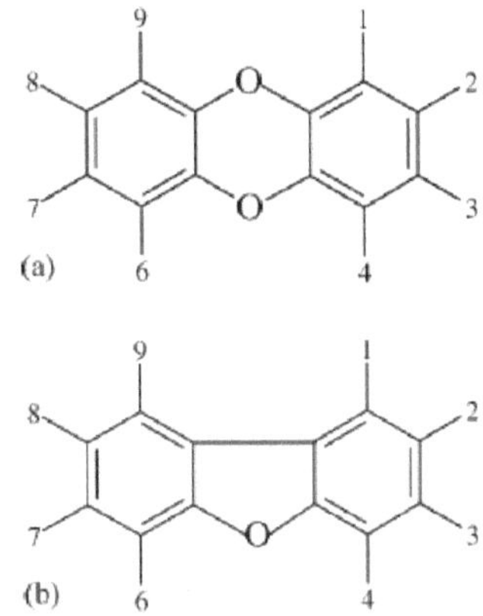

Figure 1. (a) Dibenzo-*para*-dioxin; (b) dibenzofuran

PCDD and PCDF congeners with chlorine atoms in the positions 2, 3, 7, and 8 are of particular environmental concern, especially the tetrachloro-CDD congener 2,3,7,8-TCDD, which achieved notoriety following its release from the ICSEMA plant at Seveso, Italy in 1997. The structure of this target PCDD is shown in Figure 2.

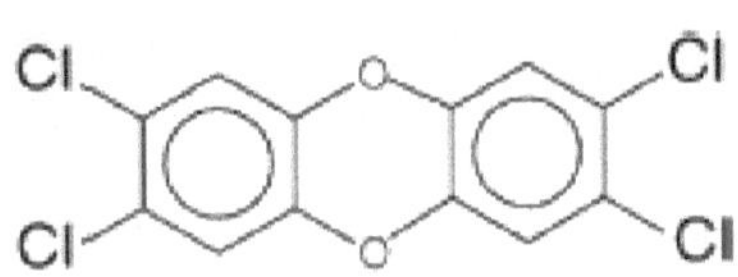

Figure 2. Structural formula of 2,3,7,8-tetra CDD

All PCDDs and PCDFs are organic solids with high melting points and low vapour pressures. They are characterised by extremely low water solubilities, and have a tendency for being strongly adsorbed on surfaces of particulate matter. The water solubility of dioxin and furans decreases and the solubility in organic solvents and fats increase with increasing chlorine content. There are 75 PCDDs and 135 PCDFs, each differing in the number and position of the chlorine atoms. Each individual PCDD or PCDF is termed a congener (giving 210 in total), while groups of congeners with the same number of chlorine atoms are called homologues. The number of congeners in each homologue group is shown in Table 1. The homologue groups are often abbreviated for convenience; for example, tetrachloro

CDDs and CDFs (PCDD/Fs with four substituted chlorine atoms) are abbreviated to TCDDs (tetrachlorinated dibenzodioxin) and TCDFs (tetrachlorinated dibenzofurans), respectively, while the fully chlorinated octachloro congeners (eight substituted chlorine atoms) are abbreviated to OCDD (octachlorinated dibenzodioxin) and OCDF (octachlorinated dibenzofuran), respectively.

TABLE 1. Homologues and congeners of PCDDs and PCDFs

Homologue (abbreviation)	Number of congeners	
	PCDDs	PCDFs
Monochloro (M)	2	4
Dichloro (D)	10	16
Trichloro (Tr)	14	28
Tetrachloro (T)	22	38
Pentachloro (Pe)	14	28
Hexachloro (Hx)	10	16
Heptachloro (Hp)	2	4
Octachloro (O)	1	1
Total	**75**	**135**

2.2. DIOXIN EMISSION SOURCES

Many natural sources can release dioxins. For instance, before the large-scale manufacturing and use of chlorinated chemicals, the presence of these compounds was confirmed [4] in historical times. Biological formation of PCDDs and PCDFs in sediments and soils, especially forest soils and sediments [5] has been confirmed. Examination of PCDD/F congener profiles in uncontaminated forest soils and sediments and comparison with those of known sources has suggested that the presence of PCDD/Fs in these samples cannot be readily explained by inputs from man-made sources [6]. Oberg et al. [7] and Schramm et al. [8] have also noted the potential for the natural formation of PCDD/Fs in sewage sludge and compost under normal environmental conditions; these are synthesised by peroxidates from chlorinated organic precursors such as chlorophenols.

Since the 1930s, there has been a steady increase in environmental levels of dioxins coinciding with the large-scale production and use of chlorinated chemicals [9]. Man-made sources of PCDD/Fs can be divided into a number of main categories and countries are now beginning to compile data and establish dioxin inventories. The USEPA [10] have derived these emission estimates and identified the following major sources of dioxin releases in the USA.

2.2.1. Combustion Sources

PCDD/PCDFs are formed in most combustion systems. These can include waste incineration (e.g., MSW, sewage sludge, medical waste, and hazardous wastes), burning of various fuels, such as coal, wood, and petroleum products [11] other high temperature sources such as cement kilns, and poorly controlled combustion sources such as building fires, burning any chlorine compounds [12].

2.2.2. Metal Smelting and Refining Sources and Processing Sources

PCDD/PCDFs can be formed during various types of primary and secondary metals operations including iron ore sintering, steel production, and scrap metal recovery [13].

2.2.3. Chemical Manufacturing

PCDD/PCDFs can be formed as by-products from the manufacture of chlorine bleached wood pulp [14] and chlorinated phenols, e.g., pentachlorophenol (PCP), PCBs (polychlorinated biphenyls), phenoxy herbicides (e.g., 2,4,5-trichloro-phenoxyacetic acid), and chlorinated aliphatic compounds (e.g., ethylene dichloride).

2.2.4. Biological and Photochemical Processes

Recent studies have suggested that PCDD/PCDFs can be formed under certain environmental conditions, e.g., composting from the action of microorganisms on chlorinated phenolic compounds. Similarly, PCDD/PCDFs have been reported to be formed during photolysis of highly chlorinated phenol [15].

2.2.5. Reservoir Sources

Reservoirs are material or pieces which contain previously formed PCDD/PCDFs or dioxin-like PCBs and have been the potential for redistribution and circulation of these compounds into the environment. Potential reservoirs include soils, sediments, vegetation, and PCP-treated wood [16]. Recently, PCDD/PCDFs have been discovered in ball clay deposits. Although the origin of the PCDD/PCDFs in these clays has not been confirmed, natural occurrence is a possibility.

3. Formation of PCDD/PCDF in MSW Incinerators

3.1. GENERAL

- Incomplete combustion of organic wastes in the combustion chamber leads to the formation of organic fragments, commonly termed products of incomplete combustion, and these can act as organic precursors to the dioxin/dibenzofuran molecule.
- The MSW waste components provide a source of chlorine, and of metals. The latter are incorporated into fly ash [17] which carries over to the cooler (250–400°C) post-combustion zone of the incineration system.
- The organic precursors (potential dioxin forming compounds) adsorb onto the surface of the fly ash in the post-combustion zone, and following a complex series of reactions which are catalysed by metals (mainly copper) in the fly ash, leading to the formation of PCDD/PCDFs along with other chlorinated trace organics.

3.2. PCDD/F FORMATION AND FACTORS AFFECTING EMISSIONS FROM COMBUSTION SOURCES

These factors will be critical and provide the basis for the development of PCDD/F emission control strategies. The factors affecting emissions from combustion sources are the following:

1. PCDD in feed
2. Precursors in feed
3. Chlorine in feed
4. Combustion temperature
5. Residence time
6. Oxygen availability
7. Feed processing
8. Supplemental fuel

3.2.1. PCDD in Feed

PCDD/Fs enter the environment as contaminants of commercial products, such as wood preservatives and pesticides. The widespread use of these products increases the possibility of finding PCDDs in the feed of a combustion process. For example, PCP-treated wood may be used to fire boilers. Run-off may carry pesticide to water treatment facilities where the organics are incorporated into sludge. The sludge may then be incinerated. Likewise, contaminated waste streams from manufacturing processes may be incinerated as an energy recovery procedure. One example is PCP sludge incinerators used at wood preserving facilities. If PCDDs are found in the feed of an inefficient or poorly controlled combustion process, it is very likely that they will be released to the atmosphere. The burning of any organic material containing organochlorine compounds or even inorganic chlorides, such as sodium chloride can generate dioxins.

3.2.2. Precursors in Feed

Many hundreds of studies have focused on the formation of PCDDs and PCDFs from precursors. The majority of the experimental work to date has centred on three classes of precursors: chlorinated phenols, chlorinated benzenes, and PCBs.

PCDD formation from the combustion of chlorinated phenols has been tested extensively. Chlorinated phenols are used as wood preserves, herbicides, and sap stain control products. Wood or vegetation sprayed with chlorophenol may be disposed of by incineration or used as a supplemental fuel in boilers. In addition, chlorophenol, i.e., wastes, have the potential to be disposed of in sludge incinerators and industrial boilers.

Chlorobenzenes are used in solvents, dyes, pharmaceuticals, and rubber production. These products make up much of the organic chlorine found in the feed of municipal waste incinerators. The associated waste products may also be disposed of in an incinerator or boiler.

Until 1975, PCBs were used as dielectric fluids in transformers and capacitors. PCBs have also been used in hydraulic fluids, plasticizers, and dyes. The low quality incineration of PCBs at waste disposal facilities or in boilers may result in PCDD and PCDF emissions.

3.2.3. Chlorine in Feed

The chlorine content of fuel is obviously an important parameter affecting the formation of PCDDs or PCDFs.

Moche and Thanner [18] have demonstrated the effect of chlorine on PCDD emissions and it has been shown that PCDDs were emitted from coal combustion only when chlorine was added and PCDD formation occurred during the combustion of pine in the presence of HCl, but no PCDDs were detected during the combustion of pine alone. Liberti et al. [19] studied the combustion of vegetables and found that when inorganic chlorine or PVC is added, PCDDs and PCDFs were detected in the ash. However, it can be generally stated that chlorine must be present for the formation of PCDD and general trends indicate that increased chlorine concentrations in the feed improve the possibilities of PCDD emissions. Although laboratory studies have indicated a correlation between chlorine input and dioxin emissions, no significant statistical correlation has been shown in field tests. Whatever effect chlorine has on stack gas PCDD/F concentrations in commercial scale systems, it is masked by other parameters such as particulates carry-over and the particulate matter filtration device temperature [20]. Therefore the role of chlorine is still uncertain.

Further studies [21] have shown the presence of sulphur dioxide reduces the level of PCDD/F formation during incineration processes. A number of mechanisms have been proposed but the basic concept is that sulphur [22] can scavenge the chlorine molecule in the presence of moisture producing SO_3 and HCl. An alternative mechanism [23] in that sulphur blocks the activity of metallic catalysts in the ash thus reducing their PCDD/F activity formation.

3.2.4. *Combustion Conditions*

The remaining factors identified in the literature that affect PCDD emissions are combustion conditions. These include combustion temperature, residence time, supplemental fuel, fuel processing, and oxygen availability. Combustion efficiency is a function all of these factors. In order to destroy PCDDs or prevent their formation, the combustion efficiency must be high. This requires a combination of high temperatures, available oxygen, high heating value fuel, and long residence times [24]. Even with these optimised combustion conditions end-of-pipe flue gas treatment is still required to meet PCDD/F emission limits of 0.1 ng/Nm^3.

3.2.5. *Combustion Temperatures*

Experimental evidence suggests that temperatures of 500–800°C promote PCDD formation, while temperatures greater than 900°C destroy PCDDs. However, pyrolysis at temperatures greater than 700°C causes 99% destruction of PCBs and no PCDF formation. Combustion temperature is a function of the heating value of the fuel or supplemental fuel, the available air, and the degree of fuel or supplemental fuel, the available air, and the degree of fuel processing.

3.2.6. *Residence Time*

The residence time necessary to destroy PCDDs and the combustion temperature are inversely related. The higher the combustion temperature, the shorter the required residence time for PCDD destruction. Likewise, a low temperature source will require a long residence time for destruction of PCDDs. Sakurai et al. [25] showed that an increase in both temperature and residence time decreased the formation of PCDDs from chlorophenol combustion. Similar results have been found at hazardous waste incinerators that run with 1.5–2 s residence times. Combustion sources with longer residence times and high temperatures are less likely to form products of incomplete combustion, such as PCDDs.

3.2.7. Oxygen Availability

Oxygen availability is a function of both the air/fuel ratio and air/fuel mixing efficiency, both of which are of concern when burning solid fuels. Solid fuels and high viscosity liquid fuels such as waste tars burn as particulates or large droplets; therefore, portions of the fuel are burned in low oxygen or pyrolysis conditions. An insufficient supply of oxygen or poor air/fuel mixing will promote poor combustion conditions and PCDD formation. Insufficient air supply increases PCDD emissions from chlorophenol combustion. Municipal waste incinerators are usually fired with excess air. However, large mass burn units may have poor air/fuel mixing due to the lack of fuel processing or poorly designed air distribution systems.

3.2.8. Feed Processing

The feed material for a combustion source may be a liquid, a solid, or a gas. Both liquid and gas fuels can be easily mixed with air and resulting in high combustion efficiency; solid feeds usually require some processing to improve combustion. Often solid feeds require drying, shredding, or separation to improve combustion. Similarly, high viscosity fuels (i.e., waste tars) require preparations such as preheating and atomisation prior to combustion.

Feed processing will determine in part both oxygen availability and residence time. Fine, homogenous feed particles will improve air/fuel mixing and combustion. Larger particles will require longer residence times and may result in local oxygen deficiencies due to poor mixing. High moisture will also decrease combustion efficiency. Therefore, highly processed homogenous feeds are less likely to emit products of incomplete combustion, such as PCDDs.

3.2.9. Supplemental Fuel

When burning a low calorific value (CV) fuel, the addition of supplemental fuel will increase the combustion temperature and improve combustion. Dow Chemical Corporation [26] tested an industrial incinerator burning waste tars without supplemental fuel and found ppb levels of TCDDs in the fly ash. After the addition of a supplemental fuel, no TCDDs were detected.

4. Dioxin Minimisation during Incineration

It is now possible to specify a set of operating condition guidelines which will lead to minimum formation of dioxins. These conditions should be considered only as guidelines for best operating practice and are discussed in the following sections.

4.1. PRECURSOR CONCENTRATIONS

The presence of precursors in the feed in most instances is now not considered as a significant source of dioxin formation in a high temperature, highly effective combustion system. Although the presence of precursors such as the chlorophenols is considered insignificant as a source of dioxins, they should be monitored periodically as sources of HCl and chlorine.

4.2. COMBUSTION TEMPERATURE

The major source for initiation of dioxin/furan formation is due to residues present from incomplete combustion. It is generally considered that at temperatures above 850°C any dioxins/furans present in the feed will be destroyed, but for effective and complete combustion of particulate carbonaceous material a temperature greater than 1,000°C should be achieved.

4.3. RESIDENCE TIME

In addition to a high combustion temperature, it is recommended that a sufficient residence time should be allowed to ensure complete destruction of combustible material. At 1,000°C, a gas residence time in the combustion zone greater than 1 s is specified and at 850°C a residence time of 2 s is recommended.

4.4. EXCESS OXYGEN

A further stipulation to ensure the complete destruction and combustion of any fuel particulates in the hot flue gas is the recommendation that a 3–6% (v/v) excess oxygen level is utilised.

4.5. TURBULENCE IN THE COMBUSTION CHAMBER

Further efforts to ensure complete combustion can be made by creating turbulence in the combustion chamber. A Reynolds number greater than 10.000 is recommended and the most effective incineration furnaces have Reynolds number of 50.000 or greater.

4.6. PRECURSOR FORMATION

The early views on dioxin/furan formation consider the "precursor theory" as the main source of dioxins and furans. In more recent years, it was considered that precursors in the feed would now be destroyed in modern incinerators and the de novo synthesis became popular. However, it was apparent that many researchers now favour the theory of PCDD/PCDF formation from precursors, themselves formed after the incineration process, such as chlorophenols, chlorobenzenes, and several other molecules. Consequently, any route leading to the formation of hydrogen chloride or unburnt or unsaturated ring systems could lead to dioxin/furan formation. Therefore, there are a number of steps to ensure minimisation:

- The chloride level in the MSW feed should be limited and controlled, this can be done both by periodic analysis of the feed or by monitoring chloride compounds in the flue gas outlet.
- Increasing the lime through put in order to remove more chlorinated compounds, particularly hydrogen chloride.
- Increasing the combustion temperature will lead to more effective destruction of hydrocarbons, thus minimising the quantity and type of unburnt hydrocarbons available for formation of dioxins/furans in the downstream processes.
- Activated carbon adsorption is used for final polishing to remove chlorinated compounds including furans and dioxins. Various types of carbon adsorbers can be used but the best performance is by using direct spray injection of powdered activated carbon. New processes based on photocatalytic destruction are currently under investigation.

4.7. FLUE GAS MONITORING

In order to ensure effective process control is being achieved and compliance is being attained monitoring is essential. Due to a combination of economics and legislative requirements a number of process sampling systems should be adopted. Continuous analysis of oxygen, SO_X, NO_X, hydrogen chloride, carbon monoxide, and carbon dioxide should be implemented. In addition, monthly dioxin/furan samples and PCB levels should be taken for compliance analysis and weekly samples of Cl_2, HF, and H_2S. Dioxin equivalence concentrations should be less than 0.1 ng I-TEQ/m^3. The dioxin/furan concentration should be less than the widely accepted environmental limit of 1 mg I-TEQ/g ash.

4.8. TURN-DOWN AND START-UP RESTRICTIONS

For best operating practice, MSW incinerators should not be fed with MSW during the start-up or shutdown operating cycles. A conventional coal, oil, or gas co-combustion system should be utilised and the supplementary fuel should be used until the furnace structure has achieved its thermal capacity.

4.9. FEED INPUT PREPARATION

It is important try to supply MSW feed, which is homogenous in texture, size, composition and which has been blended by shredding into a regular small sized feedstock. It is known that lumpy feed causes erratic combustion and generates unburnt material in the hot flue gases. Current thinking, that restricting chlorinated compounds in the feedstock is important, is now considered to be insignificant and high chlorine content feeds can be combusted effectively. However, the method of presentation of the feed to the incinerator is regarded as important.

4.10. OPTIMUM SIZE OF MSW INCINERATION PLANT

Small MSW incineration plants have difficulty in achieving combustion and are not very economic to operate. Very large individual incinerators can suffer from uneven temperature gradients, areas of poor feed distribution and localised cold spots, which support poor combustion and therefore are conducive to dioxin/furan formation.

4.11. AIR POLLUTION CONTROL DEVICES FOR MSW

MSW incinerators are commonly equipped with one or more post-combustion air pollution control devices (APCDs) to remove various pollutants to release from the tank (e.g., particulate matter, heavy metals, acid gases, and/or organic contaminants) [27,28].

These APCDs include:

- Electrostatic precipitator (ESP)
- Fabric filter (FF)
- Dry scrubber (DS)
- Dry sorbent injection (DSI)
- Wet scrubber (WS)
- Entrained flow adsorber
- Circulating fluidised bed
- Moving-bed adsorber
- Selective catalytic reactor

5. Conclusions

The paper has presented an introduction to dioxins, their characteristic properties. The key conclusions were that effective dioxin minimisation can be achieved by applying the following:

1. Combustion temperature >1,000°C
2. Chamber turbulence Re >50.000
3. Combustion residence time >2 s
4. Air pollution control devices:
   (a) Semidry scrubber
   (b) Bag-filter
   (c) Activated carbon injection

Following these process conditions will minimise PCDD/PCDF formation and lead to emission concentrations well below 0.1 ng I-TEQ/m^3.

References

1. USEPA, Report to the US Congress-Resource Recovery and Source Reduction, USEPA Report SW-118, February 1978.
2. C.R. Brunner, Hazardous Waste Incineration, 2nd Edition, McGraw-Hill, New York, 1994.
3. USEPA, Combustion Emissions Technical Resource Document (CETRED), Report No. EPA 530-R-94-014, Washington, DC, 1994.
4. R.E. Alcock, R. Gemmill, and K.C. Jones, Improvements to the UK PCDD/F and PCB atmospheric emission inventory following an emissions measurement programme, *Chemosphere* 38, 759–770 (1998).
5. E.J. Hoekstra, H. de Weerd, E.W.B. de Leer, and U.A.T.h. Brinkman, Natural formation of chlorinated phenols, dibenzo-*p*-dioxins and dibenzo furans in soil of Douglas fir forest, *Environ. Sci. Technol.* 33m, 2543–2549 (1999).
6. C. Rappe, R. Andersson, M. Bonner, K. Cooper, H. Fiedler, C. Lau, and M. Bonner, PCDDs and PCDFs in lake sediments from a rural area in the USA, *Organohalogen Comp.* 32, 88–93 (1997).
7. L.G. Oberg, R. Anderson, and C. Rappe, De novo formation of hepta- and octachlorodibenzo-*p*-dioxins from pentachlorophenol in municipal sewage sludge, *Organohalogen Comp.* 9, 351–354 (1992).
8. K.-W. Schramm, C. Klimm, B. Henkelmann, A. Kettrup, Formation of octa- and heptachlorodibenzo-*p*-dioxins during semi-anaerobic digestion of sewage sludge, *Organohalogen Comp.* 27, 84–87 (1996).
9. C. Fortin, D. and Caldbick, Are dioxins and furans predominantly anthropogenic? *Organohalogen Comp.* 32, 417–429 (1997).
10. USEPA, The Inventory of Sources of Dioxin in the United States, Review Draft, EPA/600/P, 98002Aa (NTIS PB98-137037), National Centre for Environmental Assessment, Washington, DC, April 1998.
11. B. Schatzwitz, G. Brandt, F. Gafner, E. Schlump, R. Buhler, P. Hasler, and T. Nussbaumer, Dioxin emissions from wood combustion, *Chemosphere* 29, 2005–2013 (1994).

12. P. Ruokojarvi, M. Aatamila, and J. Ruushanen, Toxic chlorinated and polyaromatic hydrocarbons in simulated house fires, *Chemosphere* 41, 825–828 (2000).
13. A. Buekens, H. Huang, and L. Stieglitz, Dioxins from the sintering process. I. Particle characteristaion and SEM/wet analysis of samples, *Organohalogen Comp.* 41, 109–112 (1999).
14. C. Rappe, R. Andersson, P.A. Bergqvist, C. Brohede, M. Hansson, L.O. Kjeller, G. Lindstrom, S. Marklund, M. Nygen, S.E. Swanson, M. Tysklind, and K. Wiberg, Overview on environmental fate of chlorinated dioxins and dibenzofurans-sources, levels and isomeric pattern in various matrices, *Chemosphere* 16, 1603–1618 (1987).
15. M. Tysklind, I. Fangmark, S. Marklund, A. Lindskog, L. Thaning, and C. Rappe, Atmospheric transport and transformation of polychlorinated dibenzo-*p*-dioxins and dibenzofurans, *Environ. Sci. Technol.* 27, 2190–2197 (1993).
16. W. Rotard, W. Christman, and W. Knoth, Background levels of PCDD/F in soils of Germany, *Chemosphere* 29, 2193–2200 (1994).
17. I. Fangmark, B. Stromberg, N. Berge, and C. Rappe, The influence of fly ash load and particle size on the formation of PCDD, PCDF, PCBz and PCB in a pilot incinerator, *Waste Manage. Res.* 13, 259–272 (1995).
18. W. Moche and G. Thanner, PCFF/F-emissions from coal combustion in small residential plants, *Organohalogen Comp.* 36, 329–332 (1998).
19. L. Liberti, A. Tursi, N. Costantino, L. Ferrara, and G. Nuzzo, Optimisation of infectious hospital waste management in Italy. 1. Wastes production and characterization study, *Waste Manage. Res.* 12(5), 373–385 (1994).
20. G.H. Rigo, A.J. Chandler, and W.S. Lanier, The relationship between chlorine in waste streams and dioxin emissions from waste combuster stacks, ASME Research Report CRTD 36, 1.1–1.7, 2.1–2.63 (1995).
21. K. Raghunathan and B.K. Gullett, Role of sulfur in reducing PCDD and PCDF formation, *Environ. Sci. Techno*l. 30, 1827–1834 (1996).
22. D. Verhust, A. Buekens, P.J. Spancer, and G. Erikson, Thermodynamic behavior of metal chlorides and sulfates under the conditions of incineration sources, *Environ. Sci. Technol.* 30, 50–56 (1996).
23. J.-C. Chen, M.-Y. Wey, and M.-H. Yan, Theoretical and experimental study of metal capture during incineration process, *J. Environ. Eng.* 123, 1100–1106 (1997).
24. I. Fangmark, B. van Bavel, S. Marklund, B. Stromberg, N. Berge, and C. Rappe, Influence of combustion parameters on the formation of polychlorinated dibenzo-*p*-dioxins, dibenzofurans, benzenes, biphenyls and polyaromatic hydrocarbons in a pilot incinerator, *Environ. Sci. Technol.* 27, 1602–1610 (1993).
25. T. Sakurai, T. Kobayashi, T. Watanabe, T. Kondo, Formation of PCFF/Fs from chlorophenols (CPs) on fly ash produced by municipal solid waste incinerators, *Organohalogen Comp.* 27, 183–186 (1996).
26. Dow Chemicals, The Trace Chemistries of Fire-A Source of and Routes for the Entry of Chlorinated Dioxins into the Environment, The Chlorinated Dioxin Task Force, Michigan Division, Dow Chemicals, USA, 1978.
27. S. Marklund, I. Fangmark, and C. Rappe, Formation and degradation of chlorinated aromatic compounds in air pollution control devices for MSW combustion, *Chemosphere* 25, 139–142 (1992).
28. A. Buekens and H. Huang, Comparative evaluation of techniques for controlling the formation and emission of chlorinated dioxins/furans in municipal waste incineration, *J. Hazard. Mater.* 62, 1–33 (1999).

OCCURRENCE, FATE, AND BEHAVIOR OF BROMINATED FLAME RETARDANTS IN EBRO RIVER BASIN (SPAIN)

ETHEL ELJARRAT,* DAMIÀ BARCELÓ
Environmental Chemistry Department, IIQAB-CSIC, Jordi Girona 18-26, 08034 Barcelona, Spain

Abstract: This work summarizes brominated flame retardant research in the Ebro River basin. Within a monitoring program at different risk zones in this river, two high-contaminated areas were detected. The first one is located along the Cinca River, a tributary of Ebro River, downstream a heavily industrialized town (Monzón), and data showed a high HBCD contamination in this area. The second one is located along the Vero River, a tributary of Cinca River, downstream an industrial park in Barbastro. In this case, high contamination of deca-BDE-209 was found. Our work included the analysis of sediments and biota, with special attention on aspects such as temporal trends, bioavailability and bioaccumulation of these contaminants. Moreover, an attempt of identification of source contamination was carried out, with the analysis of industrial effluents. In both cases, the industry responsible of the contamination was identified.

Keywords: brominated flame retardant, hexabromocyclododecane, polybrominated diphenyl ether

1. Introduction

Brominated flame retardants (BFRs) are comprised of diverse classes or chemical compounds used in a variety of commercial applications. They are used in plastics, textiles, electronic circuitry and other materials to prevent fires. The estimated global consumption of BFRs shows that their usage is on the rise: during the period from 1990 to 2000, there was more than a doubling of usage from approximately 145–310 kt.[1]

*To whom correspondence should be addressed: E-mail: eeegam@cid.csic.es

E. Mehmetli and B. Koumanova (eds.),
The Fate of Persistent Organic Pollutants in the Environment, 283–299.

Currently, the contamination in aquatic environment includes, together with other persistent organic pollutants (POPs), a variety of BFRs. The studies performed in the last 15 years have mainly concerned polybrominated diphenyl ethers (PBDEs), and in particular those congeners which derive from the Penta-mix formulation.[2,3] PBDEs are typically produced at three different degrees of bromination, i.e., Penta-BDE, Octa-BDE, and Deca-BDE. Of the three main technical mixtures in use, the penta-BDE and octa-BDE mixtures are currently being phased out in Europe. Consequently, a shift in production of deca-BDE mixtures for these lower brominated PBDE mixtures has taken place. A recent report of The Bromine Science Environmental Forum (BSEF) estimated the total market demand for the major commercial BFRs in 2001.[4] This shows the dominance of tetrabromobisphenol-A (TBBPA) (59% of total world usage) and the deca-mix PBDE formulation (27% of total world usage) in volume terms. Recent reports also suggest that usage of hexabromocyclododecane (HBCD) is increasing[5] and that because attention is now switching to this compound which is widely used.

Similarly to other POPs, BFRs (except TBBPA) appear to be lipophilic and bioaccumulate in biota and humans.[5] A considerable number of studies of PBDEs in wildlife have been undertaken since the mid-1980s, when Jansson et al.[6] first indicated that these compounds were present in samples collected remote from local sources and so may have become ubiquitous environmental contaminants.[7,8]

2. Area of Study

The study area is located in the northeast of Spain, along the Cinca and Vero Rivers, in the Ebro River basin (Figure 1). Four different sampling stations were selected at the Cinca River: site C1 (Puente de las Pilas) and C2 (La Boquera), 20 and 12 km, respectively upstream from Monzón; site C3, just downstream from Monzón, a heavily industrialized town with a very important chemical industry; and site C4 (Chalamera), 30 km downstream of site C3. Moreover, three sampling stations were selected at the Vero River: site V1 (Castillazuelo), 11 km upstream from an industrial park; site V2 (Barbastro), just 1 km downstream from the industrial park, and site V3 (La Boquera), 4 km downstream of site V2.

Different sampling campaigns were carried out from 2002 to 2005 (Table 1). Surficial sediments (0–2 cm) were collected at each different selected site. Attempts were made to collect several fishes from the same locations as the sediment samples. Fish specimens were collected by DC electric pulse. Fish were killed, weighted and the fork length of each fish was measured. Different fish species were sampled: barbels (*Barbus graellsii*), bleaks (*Alburnus alburnus*), southwestern nases (*Chondrostoma toxostoma*), and carps

(*Cyprinius carpio*). Whole fish was analyzed for bleaks and southwestern nases, whereas muscle and liver tissues were analyzed for barbels and carps. Muscle samples were collected from below the dorsal fin. The body cavity of each fish was opened and the liver was removed and preserved frozen at –20°C with the muscle sample until analysis. In order to determine the sources of contamination, some water samples, including wastewaters and effluents from different industries were also sampled. Along the Cinca River and in the industrial area of Monzón, industrial effluents from two different industries were selected: the first one produced EPS (expandable polystyrene) treated with flame retardants and ABS (acrylonitrile–butadiene–styrene), and the second one produced PVC (polyvinyl chloride). As regards the Vero River, three industries were sampled: the first one, a textile industry which produced polyester fibers treated with flame retardants; the second one produced epoxy resins; and the third one is focused on the polyamide polymerization. Moreover, the effluent of the industrial park at the discharge site to the Vero River was also collected.

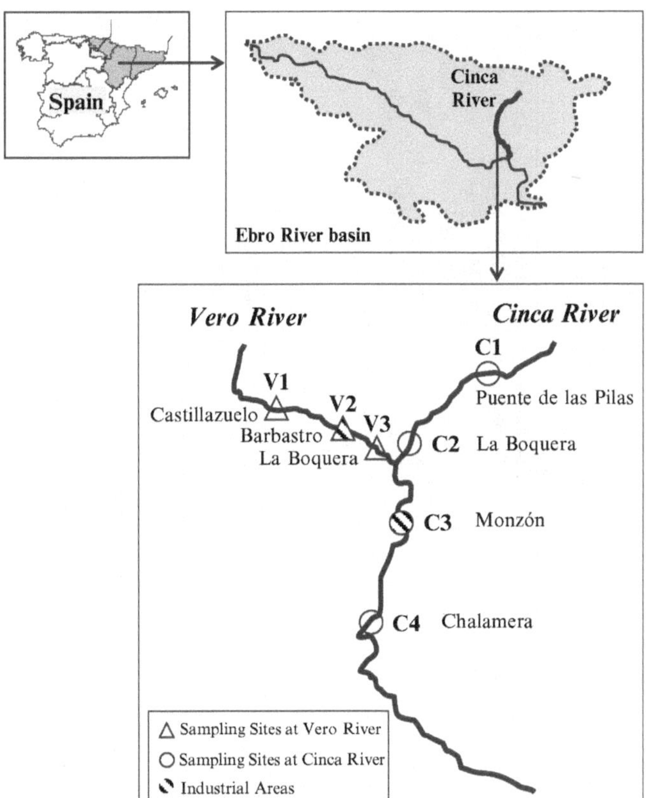

Figure 1. Geographical location of the area of study and sampling stations along the Cinca and Vero Rivers

TABLE 1. Sampling campaigns carried out in the area of study between 2002 and 2005

	Cinca River				Vero River		
	Site 1	Site 2	Site 3	Site 4	Site 1	Site 2	Site 3
October 2002	1 Sediment 4 Barbel[a] (M + L) –	1 Sediment 8 Barbel (M + L) 12 Bleak[b] (WF)	1 Sediment 6 Barbel (M + L) 12 Bleak (WF)	1 Sediment 5 Barbel (M + L) 8 Bleak (WF)		-	
November 2004	1 Sediment 6 Southwestern nase[c] (WF) –	1 Sediment 6 Southwestern nase (WF) –	1 Sediment – –	1 Sediment 6 Southwestern nase (WF) –	1 Sediment 6 Barbel (M) –	1 Sediment – –	1 Sediment 8 Barbel (M) 2 Carp[d] (M)
November 2005		–			1 Sediment 8 Barbel (M)	1 Sediment –	1 Sediment 5 Barbel (M)

[a]Barbel = *Barbus graellsii*
[b]Bleak = *Alburnus alburnus*
[c]Southwestern nase = *Chondrostoma toxostoma*
[d]Carp = *Cyprinus carpio*
M = muscle tissue; L = liver tissue; WF = whole fish

3. Analytical Methodology

Forty-seven PBDE congeners were included in the analytical work: 3 monoBDEs (1,2,3), 7 diBDEs (7,8,10,11,12,13,15), 8 triBDEs (17,25,28,30, 32,33,35,37), 6 tetraBDEs (47,49,66,71,75,77), 7 pentaBDEs (85,99,100, 116,118,119,126), 5 hexaBDEs (138,153,154,155,166), 3 heptaBDEs (181,183,190), 4 octaBDEs (194,196,197,203), 3 nonaBDEs (206,207,208), and the decaBDE (209).

A rapid and simple method for PBDE and HBCD determinations in sediment and fish samples was used. The analytical method was based in selective pressurized liquid extraction (SPLE)[9] without further cleanup step and analysis by gas chromatography coupled to mass spectrometry (GC-MS), working with negative ion chemical ionization (NICI).[10,11]

3.1. SAMPLE PREPARATION

Sediments and fishes were freeze dried. Lyophilized sediment samples were ground and homogenized by sieving through a stainless steel 0.2 mm sieve, and stored in sealed containers at –20°C until analysis. One gram dry weight (dw) for sediment samples and 0.5 g dw for fish samples were spiked with 3–30 ng of ^{13}C-BDE-209 (150 pg/μL, Wellington Labs., Guelph, Ontario, Canada). Spiked samples were kept overnight to equilibrate. SPLE[9] was carried out using a fully automated ASE 200 system (Dionex, Sunnyvale, California, USA). A 22 mL extraction cell was loaded by inserting two cellulose filters into the cell outlet, followed by 6 g of alumina (0.063–0.200 mm, from Merck, Darmstadt, Germany). Spiked sediment samples were ground with alumina and cooper (<63 μm, from Merck, Darmstadt, Germany) (1:2:2), whereas fish samples only with alumina (1:2). Mixtures were loaded into the extraction cell on top of alumina. The dead volume was filled with Hydromatrix (Varian Inc., Palo Alto, USA), and the cell was sealed with the top cell cap. Extraction cell was heated to 100°C and filled with hexane/CH_2Cl_2 (1:1) mixture until the pressure reached 1,500 psi. After an oven heat-up time of 5 min under these conditions, two static extractions of 10 min at constant pressure and temperature were developed. After this static period, fresh solvent was introduced to flush the lines and cell, and the extract was collected in the vial. The flush volume amounted to 80–100% of the extraction cell. The extraction was cycled twice. The volume of the resulting extract was about 35 mL.

Water samples (2 L) were filtered through 1 μm glass microfiber filters followed by 0.45 μm nylon membrane filters (Whatman). Filtrate water was then mixed with methanol (30%). Extraction of the filtrate water was performed using an automated sample processor Aspec XL (Automated Sample Preparation with Extraction Columns) fitted with a 817 switching valve and an external 306 LC pump, for selection and dispensing of samples, respectively, through the solid phase extraction (SPE) cartridges, from Gilson (Villiers-le-Bel, France). Cartridges (C_{18} (200 mg), from Isolute) were conditioned by passing 6 mL of hexane (1 mL/min) followed by 6 mL of CH_2Cl_2 (1 mL/min), 6 mL of methanol (1 mL/min), and 6 mL of LC-grade water (1 mL/min). 500 mL of the water sample were percolated through the cartridge at 5 mL/min. After sample loading, cartridges were dried by applying vacuum and subsequently eluted with 3 mL of methanol (1 mL/min) followed by 3 mL of CH_2Cl_2 (1 mL/min) and 3 mL of hexane (1 mL/min). The particulate matter retained on filters was extracted using PLE without alumina. An additional cleanup step was performed by SPE with alumina cartridges (5 g, from Isolute). Cartridges were conditioned with 20 mL hexane. The sample volume loaded was 1 mL, and the elution step was performed with 30 mL hexane:CH_2Cl_2 (1:2).

All the extracts (sediment, fish, and water) were finally concentrated to incipient dryness and redissolved with 15 μL of PCB-209 at 1 ng/μL (Lab. Dr. Ehrenstorfer, Augsburg, Germany), 15 μL of Cl-BDE-208 at 2 ng/μL,[12] and 20 μL of CH_2Cl_2 prior to the analysis by GC-NICI-MS.

3.2. INSTRUMENTAL ANALYSIS

GC-NICI-MS analyses were performed on a gas chromatograph Agilent 6890 connected to a mass spectrometer Agilent 5973 Network (Agilent Technologies España, Madrid, Spain). A HP-5ms (30 × 0.25 mm i.d., 0.25 μm film thickness) containing 5% phenyl methyl siloxane (model HP 19091S-433) capillary column was used for the determination of congeners from mono- to hepta-BDEs. The temperature program was from 110°C (held for 1 min) to 180°C (held for 1 min) at 8°C/min, then from 180°C to 240°C (held for 5 min) at 2°C/min, and then from 240°C to 265°C (held for 6 min) at 2°C/min, using the splitless injection mode during 1 min and injecting 2 μL. The operating conditions were as follow: ion source temperature = 250°C, ammonia as chemical ionization moderating gas at an ion source pressure of 1.9×10^{-4} torr.[10] For the determination of congeners from octa- to deca-BDE, a DB-5ms (15 × 0.25 mm i.d., 0.1 μm film thickness) containing 5% phenyl methyl siloxane capillary column was used with helium as the carrier gas at 10 psi. The temperature program was from 140°C (held for 1 min) to 325°C (held for 1 min) at 10°C/min (held for 10 min), using the splitless injection mode during 1 min.[11] Both experiments were carried out monitoring the two most abundant isotope peaks from the mass spectra corresponding to m/z = 79 and 81 ($[Br]^-$). For octa- to deca-BDE experiments, m/z = 487 and 489 for nona- and deca-BDE, and m/z = 497 and 499 for ^{13}C-deca-BDE, were also included. Confirmation criteria for the detection and quantification of PBDEs should include the following: (i) all m/z monitored for a given analyte should maximize simultaneously ±1 s, with signal to noise ratio ≥3 for each; (ii) the ratio between the two monitored ions should be within 15% of the theoretical. Quantification of mono- to hepta-BDEs was carried out by internal standard procedure using PCB-209 as internal standard, whereas octa- to deca-BDEs were quantified using ^{13}C-BDE-209 and Cl-BDE-208 as internal standards.

HBCD determinations were carried out simultaneously with mono- to hepta-BDEs. Technical HBCD consists of three diastereomers: α, β, and γ-HBCD. HBCD can be determined by GC-MS, but until now, the three diastereomers have not been separated by this technique. However, as apparently the response factors of the three diastereomers do not differ very much,[13] HBCD can be quantified as total HBCD by GC-MS. Quantification was carried out by external standard.

Using the described methodology, recoveries ranged from 53% to 84% for sediment, and between 52% and 103% for fish samples. Detection limits were in the range of 6–50 pg/g dw for sediments, and in the range of 2–19 pg/g ww for fish samples. Relative standard deviations of the method were in the range of 1–13%.

4. Results and Discussion

4.1. CINCA RIVER

4.1.1. Sediment

PBDEs were detected in all the sediment samples at concentrations ranging from 2 to 131 ng/g dw (Table 2). HBCD was detected only in samples collected downstream of Monzón, with levels ranging between 90 and 1,613 ng/g dw. In these samples, HBCD contamination was greater than that observed for PBDEs. Site C3 was found to be the most contaminated zone followed by site C4 > site C1 = site C2. As expected, PBDE and HBCD levels were greater near the site of industrial impact.[14]

Our PBDE results were consistent with reported data for river sediments. PBDEs were determined in Swedish river sediments at 8–50 ng/g dw.[15] Similar values were found in Japanese river sediments, with concentration levels between 21 and 59 ng/g dw.[16] Higher levels up to 1,400 ng/g dw were found in a downstream area of a manufacturing plant in UK[17] and at 120 ng/g dw downstream of an area with textile industries.[15] As regards data for HBCD,

Sellström et al.[15] reported concentration levels between nd and 1,600 ng/g dw in river sediments from a Swedish river with numerous textile industries.

Of 47 congeners included in the analytical work, eight different PBDEs were detected, ranging from tetra- to deca-brominated compounds: tetra-BDE-47, penta-BDE-99, penta-BDE-100, penta-BDE-118, hexa-BDE-153, hexa-BDE-154, hepta-BDE-183, and deca-BDE-209. However, BDE-99 could not be quantified due to coelutions with breakdown products of HBCD. This coelution has been previously described by Covaci et al.[18]

TABLE 2. Results (expressed in ng/g dw and ng/g ww for sediment and biota samples, respectively) obtained for samples collected along the Cinca River

				Site C1	Site C2	Site C3	Site C4
October 2002	Sediment	Total PBDEs		2.4	2.6	42	40
		HBCD		nd	nd	514	90
	Barbel[a]	Total PBDEs	Muscle	1.3	4.5	281	96
			Liver	0.2	0.4	237	75
		HBCD	Muscle	nd	nd	530	90
			Liver	nd	nd	554	432
	Bleak[a]	Total PBDEs		–	4.7	555	232
		HBCD		–	nd	1,501	760
November 2004	Sediment	Total PBDEs		14	38	131	87
		HBCD		nd	nd	1,613	866
	Southwestern nase[a]	Total PBDEs		2.0	38	–	520
		HBCD		nd	nd	–	4,863

[a]Mean values

4.1.2. Biota

Table 2 shows the mean concentrations of BDE congeners, as well as of HBCD, in the selected sites. PBDEs were detected in all the biota samples at concentrations ranging from 0.2 to 555 ng/g wet weight (ww). HBCD was detected only in samples corresponding to sites downstream Monzón. In these samples, HBCD contamination was similar or greater than that observed for PBDEs. HBCD levels ranged from 90 to 4,863 ng/g ww. Similar to our findings in sediment samples, site C3 was found to be the most contaminated zone followed by site C4 > site C1 = site C2.[16, 19]

The majority of published data in fish samples expressed PBDE levels in ng/g lipid weight (lw), making difficult the comparison with our results expressed in ng/g ww. However, some works also used the same unities of this study. Zennegg et al.[20] analyzed whitefish from different Swiss lakes, and reported PBDE levels between 1.6 and 7.4 ng/g ww. Carps from Detroit River and Des Plaines River[21] presented PBDE concentration levels around 5 and 12 ng/g ww, respectively. All these values were in accordance with our results in sites out of industrial impact (sites C1 and C2).

Of 47 congeners included in the analytical work, 17 different PBDEs were detected, ranging from di- to hepta-brominated compounds. However, BDE-209, the dominant congener of the total PBDE contamination in sediments, was not detected in biota samples. Luross et al.[22] found a BDE congener distribution in aquatic biological matrices dominated by BDE-47 and followed by BDE-99, BDE-100, BDE-153, BDE-154, BDE-28, and BDE-183. However, the distribution found in our study showed a clear predominance of hexa-BDE congeners 153 and 154, as well as of hepta-BDE-183. These data correlate well with distribution observed in sediment samples. The same findings were reported by Rice et al.,[21] who analyzed carps from Des Plaines River (USA). The dominant isomers were two hepta-BDE congeners (BDE-181 and BDE-183) as well as two hexabromo congeners (BDE-153 and BDE-154).

Comparison between levels found in muscle tissues and livers showed that similar contamination was detected in both matrices. However, slightly higher BFR levels were found in muscle tissues. Ratios between total PBDE levels in muscle and in liver corresponding to the same individual fish ($R_{M/L}$) were calculated for samples collected at the most contaminated site (site C3). The mean value of these $R_{M/Ls}$ was 1.1. The same $R_{M/Ls}$ were calculated for HBCD contamination, with a similar mean value ($R_{M/L}$ =1.3). In previous study on cod and whiting fishes, PBDE concentrations have been found to be lower in muscle tissue than in liver, whereas higher values were detected in muscle of herring.[23]

4.1.3. Bioavailability and Bioaccumulation

When sediment concentrations were compared to those in biota collected at the same sites along Cinca River, high fish to sediment ratios were seen. Specially, levels of BDE-47, BDE-153 and BDE-154 in fish were high compared with levels in sediment (13–21 fold) indicating the high bioavailability of these congeners. Moreover, our calculated fish to sediment ratios correlated well with those reported by Sellström et al.[15] These large fish to sediment ratios indicate that these contaminants are bioavailable and are taken up readily by the fish. This is also supported by the fact that for some BDE congeners detectable levels were found in fish but not in corresponding sediments: 17 different PBDEs were detected in fish samples, whereas only eight PBDE congeners were found in sediments. Solely BDE-209, the main BDE congener found in sediments from the area, was not found in biota samples. The main reason for their absence in biota seems to be its relatively low bioaccumulation potential. This may be due either to a low uptake rate for this very large molecule or a relatively rapid excretion after biotransformation. The low bioconcentration hazard of BDE-209 was also reported by other authors.[24]

The fact that HBCD is found in fish indicates that it is also bioavailable. However, the low fish to sediment ratio (below 1) does not agree with the high ratios reported by Sellström et al.[15] for pikes from a Swedish River. Pike is a predator at the top of the food chain and the higher fish to sediment ratio in this specie could indicate the biomagnification of this compound through the aquatic trophic chain.

Previous studies showed that PBDE levels increased with the age of the fish, indicating bioaccumulation.[25] Fish length is directly related to fish age. Figure 2 shows the measured concentrations of HBCD as a function of fish length at sites C3 and C4. The correlation coefficients indicate acceptable correlations ($R^2 = 0.71$).

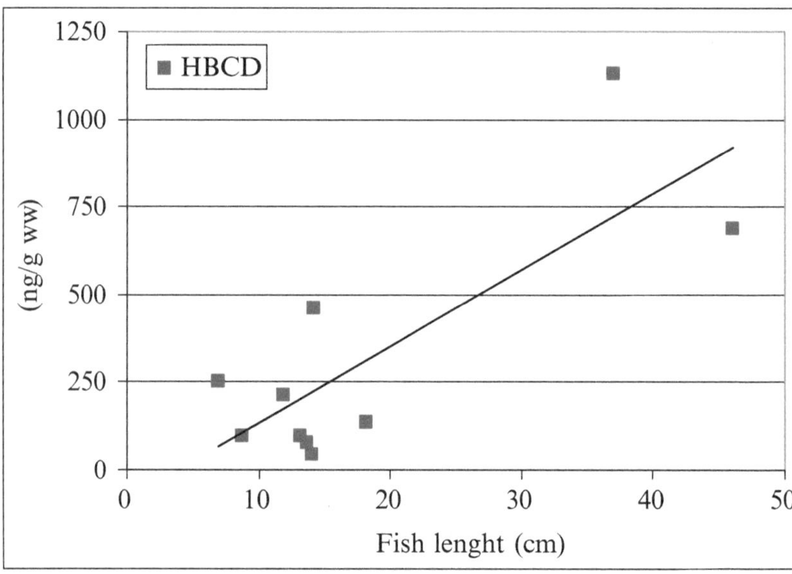

Figure 2. Length of Cinca River barbel (sites C3 and C4) versus concentrations of HBCD in the muscle. Line is linear regression on the data

4.1.4. Source of Contamination

Two industries were suspected to cause the HBCD contamination along the Cinca River: the first one produced EPS treated with flame retardants and ABS, and the second one produced PVC. Analysis of industrial effluents of each industry revealed that industry which produced EPS and ABS is the responsible of the HBCD contamination, with concentration levels around 5 μg/L (Table 3).

TABLE 3. HBCD and BDE-209 contamination in industrial effluents

	HBCD (ng/L)	BDE-209 (ng/L)
EPS and ABS production	4,980	nd
PVC production	nd	nd
Polyester fibers production	nd	5
Epoxy resins production	nd	45
Polyamide polymerization	nd	2,600
Industrial park effluent	nd	105

4.2. VERO RIVER

4.2.1. Sediment

Table 4 shows the concentrations of BDE congeners from the different selected sites during the two sampling campaigns. PBDEs were detected in all the sediment samples at concentrations ranging from 11 to 14,400 ng/g dw. The site V2 was found to be the most contaminated zone followed by site V3 > site V1. As expected, PBDE levels were greater near the site of industrial impact.[26]

Sediments affected by the source contamination (sites V2 and V3) showed a congener pattern clearly dominated by the BDE-209, which was present at concentrations varying from 1,910 to 12,500 ng/g dw. The highest BDE-209 level was found at the sampling site closest to the source of contamination (V2-November 2005, located 5 m downstream from the effluent discharge) and contamination decreased with distance from the industrial area. High levels of BDE-154 (844 ng/g dw) were also detected in this sediment sample. Moreover, some octa- and nona-BDEs were also present at relatively high levels: 37–39 ng/g dw for octa-BDEs, and 169–375 ng/g dw for nona-BDEs. The distribution of octa- and nona-BDE congeners in this sediment sample correlates well with the pattern observed in the technical decaBDE formulation containing 97–98% of BDE-209 and up to 3% nona-BDEs.[27]

Some studies have also reported PBDE levels in sediments collected near industry facilities. Sellström et al.[15] reported total PBDE concentration levels between nd and 364 ng/g dw (nd-360 ng/g dw for BDE-209) in river sediments from a Swedish river with numerous textile industries. Higher PBDE levels up to 1,400 ng/g dw (nd-399 ng/g dw for BDE-209) were found downstream from a foam manufacturing plant using PBDEs in the United Kingdom.[17] BDE-209 burdens up to 4,600 ng/g dw have also been reported on suspended particulates from Dutch surface waters, decreasing with distance from a textile facility.[28] Our PBDE results were considerably higher than those reported for sediments collected near industrial areas, probably because of the small dilution factor of the Vero River at this area, which has an average flow of 2.1 m^3/s.

TABLE 4. Results (expressed in ng/g dw and ng/g lw for sediment and biota samples, respectively) obtained for samples collected along the Vero River

			Site 1	Site 2	Site 3
November 2004	Sediment	BDE-209	7.5	5,395	1,911
		Total PBDEs	11	5,531	1,930
	Barbel[a]	BDE-209	nd	–	67
		Total PBDEs	54	–	791
	Carp[a]	BDE-209	–	–	80
		Total PBDEs	–	–	1,560
November 2005	Sediment	BDE-209	27	12,459	7,454
		Total PBDEs	30	14,395	7,767
	Barbel[a]	BDE-209	nd	–	195
		Total PBDEs	83	–	1,007

[a]Mean values

4.2.2. Biota

PBDEs were detected in all the fish muscle samples at concentrations ranging from 28 to 2,092 ng/g lw (Table 4). Similar to our findings in sediment samples, samples from the site downstream of the textile industry (V3) were found to be much more contaminated than those collected at the site upstream of the factory (V1). The comparison between the two sampling campaigns showed that levels during 2005 were slightly higher than those found in 2004, probably reflecting the same situation observed for sediment samples with an increase of contamination with time.

Of 23 congeners included in the analytical work, 12 different PBDEs were detected, ranging from tri- to deca-brominated compounds. The most relevant finding is the presence of significant concentrations of BDE-209 in fish samples collected downstream of the factory. In published works related to PBDEs in aquatic biota samples, BDE-209 was not detected or detected only in a minority of the samples and always in concentrations around the limit of detection. The study of BDE-209 levels in aquatic biota samples is of special interest because its bioavailability has been questioned due to its large molecular size resulting in decreased uptake rates.[17] Another possible explanation of low BDE-209 concentrations in aquatic biota may be due to relatively rapid biotransformation. Therefore, and despite high concentrations in abiotic matrices, some authors argue that BDE-209 would not pose an environmental threat. Our work demonstrated that BDE-209 is bioavailable: 14 out of 15 biota samples (barbels and carps) collected downstream the source of contamination, showed BDE-209 levels ranging between 20 and 707 ng/g lw. BDE-209 values in fish collected during the second sampling campaign, with a mean value of 195 ng/g lw and a median value of 86 ng/g lw, were higher than those found in barbels collected during November 2004 (mean value of 67 ng/g lw and median value of 32 ng/g lw). These results are in accordance with the higher values found also in sediment samples collected at the same site (V3).

During the last 5 years, a few studies have been published demonstrating that aquatic biota is able to accumulate BDE-209. Lepom et al.[29] first reported the occurrence of BDE-209 in freshwater fish in Europe, with concentrations up to 37 ng/g lw. Similar results were obtained by Burreau et al.,[30] who detected BDE-209 in three different fish species from rural waters of the Lumparn estuary in the Baltic Sea with concentrations up to 116 ng/g lw. The BDE-209 levels in fish reported here are on average higher than those reported elsewhere.

It is interesting to note the different congener distribution found in fish collected up- and down-stream the source of contamination (Figure 3). For samples at the control site (V1), the distribution showed a clear predominance of the tetra-BDE-47 (69% and 77% for the first and the second sampling, respectively). The predominance of BDE-47 is consistent with other studies on freshwater fish species.[5,31] However, in fish from the contaminated area (V3), the BDE-47 contribution to the total PBDE burden decreased to 29%. In contrast, samples from V3 contained a high contribution of hexa-BDE-154, with values between 16% (sampling of 2004) and 20% (sampling of 2005). The contribution of this hexa-BDE congener was only 5% and 7% for the first and the second sampling, respectively, at the site V1. The higher predominance of BDE-154 in the contaminated area could be related to high contamination of this congener in the sediment of this area. But, it could be also attributed to biotransformation of deca-BDE-209. Stapleton et al.[32] studied the debromination of BDE-209 by caged carp following dietary exposure. They concluded that debromination of BDE-209 occurs in carp. Although they did not detect any accumulation of BDE-47 and BDE-99, accumulation of other PBDEs, such as BDE-154, occurs. Finally, it should be pointed out that the deca-BDE-209 contribution in samples collected downstream of the industrial park ranged from 6–7% during the first sampling campaign, to 19% in the sampling carried out during 2005. For fish collected during 2005, BDE-209 was the third highest contributing PBDE congener, indicating the relevance of the deca-congener in these samples.

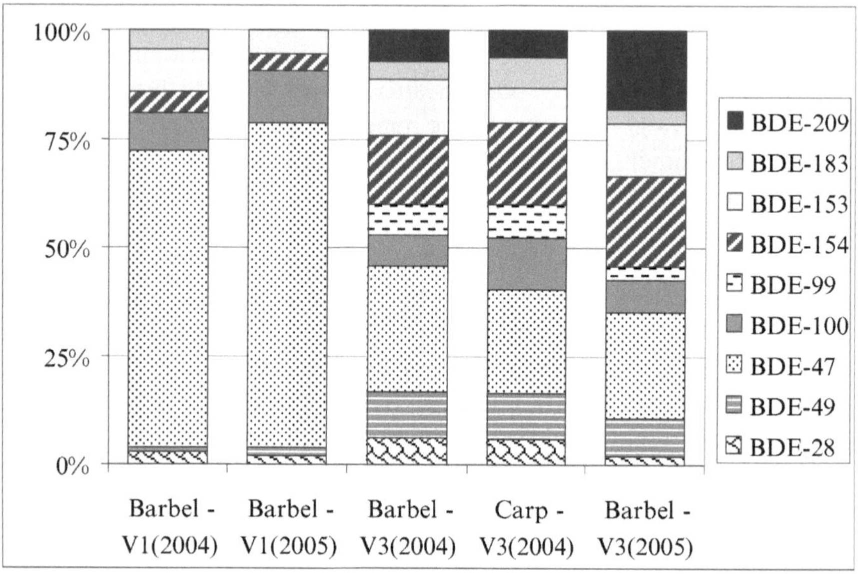

Figure 3. Percentage contribution of various congeners to the total PBDEs detected in fish samples from the control site (V1) and contaminated area (V3) during the two sampling campaigns (2004 and 2005)

4.2.3. Bioavailability and Bioaccumulation

When sediment concentrations were compared to those in barbel or carp collected at the same site along the Vero River, fish to sediment ratios were obtained. Levels of BDE-47 in fish were high compared with levels in sediment indicating the high bioavailability of this congener, and in accordance with previous ratios obtained in the Cinca River study. The bioavailability decreases with increase in bromination degree, with a fish-to-sediment ratio around 0.1 for the hepta-congener BDE-183. BDE-209 is bioavailable but the bioavailability potential is lower than that observed for other BDE congeners. Much lower are the fish to sediment values: 0.0013 and 0.0011 for the 2004 and 2005 sampling campaigns, respectively. It should be pointed out that an important factor affecting the ratios may be that the system is not in equilibrium because PBDEs are released directly to the river in textile industry effluents, and new inputs are ongoing. The low fish to sediment ratios obtained could be an indication of a recent release of BDE-209 from the industrial park that has contaminated the sediment but not yet been taken up by barbels or carps.

4.2.4. Source of Contamination

Three industries were suspected to cause the BDE-209 contamination along the Vero River: the first one, a textile industry which produced polyester fibers treated with flame retardants, the second one produced epoxy resins, and the third one is focused on the polyamide polymerization. Moreover, the effluent of the industrial park at the discharge site to the Vero River was also analyzed. Analysis of industrial effluents of each industry revealed that industry focused on the polyamide polymerization is the main responsible of the BDE-209 contamination, with concentration levels around 2,600 ng/L (Table 3). Nevertheless, the two other industries also contribute in some way to the total contamination.

Acknowledgments

This research study was funded by the European Union under the Global Change and Ecosystems (FP6) Water Cycle and Soil Related Aspects: Integrated modeling of the river–sediment–soil–groundwater system; advanced tools for the management of catchment areas and river basins in the context of global change (AQUATERRA, Project number 505428), and by the Spanish Ministry of Education and Science (Project numbers CTM2005-25168-E, and CTM2005-07402-C02-02/TECNO). This work reflects only the author's views and the European Community is not liable for any use that may be made of the information contained therein. Confederación Hidrográfica del Ebro is thanked for its assistance in the industrial effluent sampling.

References

1. P.A. Arias, Proceedings of 2nd International Workshop on Brominated Flame Retardants, Stockholm, Sweden, 2001, 17–19.
2. B. Strandberg, N.G. Dodder, I. Basu, and R.A. Hites, Concentrations and Spatial Variations of Polybrominated Diphenyl Ethers and Other Organohalogen Compounds in Great Lakes Air, *Environ. Sci. Technol.* 35, 1078–1083 (2001).
3. R.C. Hale, M. Alaee, J.B. Manchester-Neesvig, H.M. Stapleton, and M.G. Ikonomou, Polybrominated diphenyl ether flame retardants in the North American environment, *Environ. Int.* 29, 771–779 (2003).
4. R.J. Law, C.R. Allchin, J. de Boer, A. Covaci, D. Herzke, P. Lepom, S. Morris, J. Tronczynski, and C.A. de Wit, Levels and trends of brominated flame retardants in the European environment, *Chemosphere* 64, 187–208 (2006).
5. C. de Wit, An overview of brominated flame retardants in the environment, *Chemosphere* 46, 583–624 (2002).
6. B. Jansson, L. Asplund, and M. Olsson, Brominated flame retardants – Ubiquitous environmental pollutants? *Chemosphere* 16, 2343–2349 (1987).

7. J.B. Manchester-Neesvig, K. Valters, and W.C. Sonzogni, Comparison of Polybrominated Diphenyl Ethers (PBDEs) and Polychlorinated Biphenyls (PCBs) in Lake Michigan Salmonids, *Environ. Sci. Technol.* 35, 1072–1077 (2001).
8. S. Voorspoels, A. Covaci, and P. Schepens, Polybrominated diphenyl ethers in marine species from the Belgian North Sea and the Western Scheldt Estuary: levels, profiles, and distribution, *Environ. Sci. Technol.* 37, 4348–4357 (2003).
9. A. de la Cal, E. Eljarrat, and D. Barceló, Determination of 39 PBDE congeners in sediment samples using fast selective pressurized liquid extraction and purification, *J. Chromatogr. A* 1021, 165–173 (2003).
10. E. Eljarrat, S. Lacorte, and D. Barceló, Optimization of congener-specific analysis of 40 polybrominated diphenyl ethers by gas chromatography/mass spectrometry, *J. Mass Spectrom.* 37, 76–84 (2002).
11. E. Eljarrat, A. de la Cal, and D. Barceló, Determination of decabromodiphenyl ether in sediments using selective pressurized liquid extraction followed by GC-NCI-MS, *Anal. Bioanal. Chem.* 378, 610–614 (2004).
12. A. Christiansson, D. Teclechiel, J. Eriksson, A. Bergman, and G. Marsh, Methods for synthesis of nonabromodiphenyl ethers and a chloro-nonabromodiphenyl ether, *Chemosphere* 63, 562–569 (2006).
13. QUASIMEME. Report Workshop Analysis Brominated Flame Retardants, Barcelona, Spain, 21–23 October. Aberdeen, UK: FRS Marine Laboratory, 2002.
14. E. Eljarrat, A. de la Cal, D. Raldúa, C. Duran, and D. Barceló, Occurrence and bioavailability of polybrominated diphenylethers and hexabromocyclododecane in sediment and fish from Cinca River, a tributary of Ebro River (Spain), *Environ. Sci. Technol.* 38, 2603–2608 (2004).
15. U. Sellström, A. Kierkegaard, C. de Wit, and B. Jansson, Polybrominated diphenyl ethers and hexabromocyclododecane in sediment and fish from a Swedish river, *Environ. Toxicol. Chem.* 17, 1065–1072 (1998).
16. I. Watanabe, M. Kawano, R. Tatsukawa, Polybrominated and mixed polybromo/chlorinated dibenzo-*p*-dioxins and –dibenzofurans in the Japanese environment, *Organohalogen Compd.* 24, 337–340 (1995).
17. C.R. Allchin, R.J. Law, and S. Morris, Polybrominated diphenylethers in sediments and biota downstream of potential sources in the UK, *Environ. Pollut.* 105, 197–207 (1999).
18. A. Covaci, S. Voorspoels, and J. de Boer, Determination of brominated flame retardants, with emphasis on polybrominated diphenyl ethers (PBDEs) in environmental and human samples – a review, *Environ. Int.* 29, 735–756 (2003).
19. E. Eljarrat, A. de la Cal, D. Raldúa, C. Duran, and D. Barceló, Brominated flame retardants in *Alburnus alburnus* from Cinca River basin (Spain), *Environ. Pollut.* 133, 501–508 (2005).
20. M. Zennegg, M. Kohler, A.C. Gerecke, and P. Schmid, Polybrominated diphenyl ethers in whitefish from Swiss lakes and farmed rainbow trout, *Chemosphere* 51, 545–553 (2003).
21. C.P. Rice, S.M. Chernyak, L. Begnoche, R. Quintal, and J. Hickey, Comparisons of PBDE composition and concentration in fish collected from the Detroit River, MI and Des Plaines River, IL, *Chemosphere* 49, 731–737 (2002).
22. J.M. Luross, M. Alaee, D.B. Sergeant, C.M. Cannon, D.M. Whittle, and K.R. Solomon, Spatial distribution of polybrominated diphenyl ethers and polybrominated biphenyls in lake trout from the Laurentian Great Lakes, *Chemosphere* 46, 665–672 (2002).
23. J. Boon, W. Lewis, M. Tjoen-A-Choy, C. Allchin, R. Law., J. de Boer, C. Ten Hallers-Tjabbes, and B. Zegers, Levels of polybrominated diphenyl ether (PBDE) flame retardants

in animals representing different trophic levels of the North Sea food web, *Environ. Sci. Technol.* 36, 4025–4032 (2002).
24. O. Andersson and G. Blomkvist, Polybrominated aromatic pollutants found in fish in Sweden, *Chemosphere* 10, 1051–1060 (1981).
25. P.S. Haglund, D.R. Soc, H.R. Buser, and J. Hu, Identification and quantification of polybrominated diphenyl ethers and methoxy-polybrominated diphenyl ethers in Baltic biota, *Environ. Sci. Technol.* 31, 3281–3287 (1997).
26. E. Eljarrat, A. Labandeira, G. Marsh, D. Raldua, and D. Barceló, Decabrominated diphenyl ether in river fish and sediment samples collected downstream a textile industry, *Chemosphere* (under revision).
27. M. Alaee, P. Arias, A. Sjödin, and A. Bergman, An overview of commercially used brominated flame retardants, their applications, their use patterns in different countries/regions and possible modes of release, *Environ. Int.* 29, 683–689 (2003).
28. J. de Boer, P.G. Wester, A. van der Horst, and P.E.G. Leonards, Polybrominated diphenyl ethers in influents, suspended particulate matter, sediments, sewage treatment plant and effluents and biota from the Netherlands, *Environ. Pollut.* 122, 63–74 (2003).
29. P. Lepom, T. Karasyova, and G. Sawal, Occurrence of polybrominated diphenyl ethers in freshwater fish from Germany, *Organohalogen Compd.* 58, 209–212 (2002).
30. S. Burreau, Y. Zebühr, D. Broman, and R. Isaac, Biomagnification of polychlorinated biphenyls (PCBs) and polybrominated diphenyl ethers (PBDEs) studied in pike (*Esox lucius*), perch (*Perca fluviatilis*) and roach (*Rutilus rutilus*) from the Baltic Sea, *Chemosphere* 55, 1043–1052 (2004).
31. A. Labandeira, E. Eljarrat, and D. Barceló, Congener distribution of polybrominated diphenyl ethers in feral carp (*Cyprinus carpio*) from the Llobregat River, Spain, *Environ. Pollut.* 146, 188–195 (2007).
32. H.M. Stapleton, M. Alaee, R.J. Letcher, and J.E. Baker, Debromination of the flame retardant decabromodiphenyl ether by juvenile carp (*Cyprinus carpio*) following dietary exposure, *Environ. Sci. Technol.* 38, 112–119 (2004).

MILK CONTAMINATION FROM DIOXINS IN ITALY: SOURCE IDENTIFICATION AND INTERVENTION STRATEGIES

GIANFRANCO DILETTI, ROBERTA CECI, ANNAMARIA CONTE, ALFONSO DE BENEDICTIS, GIACOMO MIGLIORATI, GIAMPIERO SCORTICHINI*
Istituto Zooprofilattico Sperimentale dell'Abruzzo e del Molise "G. Caporale," Campo Boario 64100 Teramo, Italy

Abstract: PCDDs and PCDFs levels exceeding the European Union tolerance limits were detected in milk from different animal species (cows, sheep, goats, and buffaloes) and in animal feed samples from Campania region in southern Italy. A monitoring plan was performed in the period 2002–2004 to investigate the extension of the contamination. A total of 595 milk samples were analyzed and the percentage of noncompliant samples (PCDD/Fs > 3 pg WHO-TEQ/g fat) was 27% (maximum value 30.43 pg WHO-TEQ/g fat for a sheep's milk). The number of tested feedingstuffs was 434 but only feed likely grown in the area of sampling such as grass, silage and hay (267 samples) showed a significant contamination with 13% of samples containing PCDD/F > 0.75 ng WHO-TEQ/kg product (tolerance limit referred to 12% moisture content). The spatial autocorrelation analysis based on the calculation of the I-Moran index gave significant values for both milk and animal feed and highlighted an area characterized by a widespread practice of illegal waste dumping and burning. The problem of waste management in the area of study is particularly worrying from the sanitary point of view with a significant excess of cancer mortality and occurrence of malformations revealed by a recent research.

Keywords: polychlorinated dibenzo-*p*-dioxins, polychlorinated dibenzofurans, gas chromatography, high-resolution mass spectrometry, milk, feedingstuffs, statistical analysis

E. Mehmetli and B. Koumanova (eds.),
The Fate of Persistent Organic Pollutants in the Environment, 301–314.

1. Introduction

Polychlorinated dibenzo-*p*-dioxins (PCDDs) and polychlorinated dibenzofurans (PCDFs) collectively referred to as "dioxins" are trace by-products from many industrial and nonindustrial processes, such as combustion (e.g., incineration of wastes, combustion of solid and liquid fuels) and other thermal processes (metals sinter plants, electric arc furnaces) and the production of chlorinated pesticides. Secondary sources such as landfills, contaminated soils and sediments can also release dioxins in the environment (DEFRA, 2002).

These substances are resistant to environmental degradation through chemical, biological, and photolytic processes. Due to these properties, they have been observed to persist in the environment and to be capable of long-range transport (Lohmann and Jones, 1998). PCDDs and PCDFs tend toward bioaccumulation in human and animal tissues and biomagnification in the food chain (Karl et al., 2002; Macdonald et al., 2005; Cok et al., 2007), with potential significant impacts on human health and the environment (Fiedler, 1996; Kogevinas, 2000).

At present the primary source of human dioxins exposure is food that contributes for more than 90% to the total daily intake. Of this, about 90% derives from foods of animal origin (European Commission, November 2000).

The availability of data on the levels of dioxins measured in various foods is fundamental for the estimation of the dietary exposure of the general population. The European Union (EU) has established that the production process of animals and primary products of animal origin must be monitored for the purpose of detecting the presence of certain substances and residues including environmental contaminants as dioxins (Council of the EU Directive, 1996).

In the framework of the National Residues Surveillance Plan activities, PCDD/Fs levels exceeding the EU tolerance limit were detected in two sheep's milk samples (6.15 and 4.95 pg WHO-TEQ/g fat against a limit of 3 pg WHO-TEQ/g fat) from Campania region in southern Italy collected in November 2001 (Diletti et al., 2003). Repetition of the analysis on samples from the same flocks confirmed unacceptable dioxins concentrations.

To investigate this phenomenon a monitoring plan was conducted from April 2002 to September 2004. The sampling area covered a large part of the Provinces of Naples and Caserta, broadening the control to the other milk-producing animal species existent in the territory (cows, goats, and buffaloes). Animal feed contamination levels were also determined taking samples at the same farms where milk samples were collected. The aim of this paper is to conclusively define the geographical extension of the contamination and to assess the levels of dioxins and furans in milk and feedingstuffs. Statistical analysis of the PCDD/Fs congener profiles was used in a tentative identification of the dioxins sources in a complex context of environmental contamination linked to waste management. This study also illustrates the adopted intervention strategies for reducing the PCDD/Fs concentrations in milk and dairy products and, consequently, the human exposure to these contaminants.

2. Materials and Methods

2.1. SAMPLING

Milk and feedingstuff samples were taken at farms sited in rural areas and near urban and industrial zones in Campania region. Milk samples were stored in glass bottles and subsequently frozen. Animal feed samples were packaged in plastic bags. A total of 595 milk samples (191 from cows, 223 from buffaloes, 131 from sheep, 22 from goats, and 28 mixed sheep/goats milk) and 434 feedingstuffs samples (including grass, silage, hay, cereals and other farinaceous feed, premixes, and mixed feeds) were analyzed for PCDDs and PCFDs.

2.2. SAMPLE PREPARATION, EXTRACTION, AND CLEANUP

Feedingstuff samples containing a high water amount were left to dry at room temperature before the analysis and then grinded. Their moisture content was determined in drying oven at 103°C allowing the calculation of PCDD/Fs concentrations considering 12% moisture content, as requested by the EU legislation (Council of the EU Directive, 2001). Milk samples were homogenized by handshaking.

A representative aliquot of sample was spiked with the specific PCDD/Fs standard solution, a mixture of $^{13}C_{12}$-labeled 2,3,7,8-congeners (Wellington Laboratories, Ontario, Canada).

Feedingstuffs were extracted by accelerated solvent extraction (ASE) using an ASE 300 Dionex (Sunnyvale, California) instrument with a mixture of n-hexane and acetone 80:20 (v/v). The collected extract was evaporated to dryness in a rotary evaporator and the fat residue weighed. Milk samples were first added of ethyl alcohol in order to precipitate proteins and then fat was extracted by means of a mixture of diethyl ether and petroleum ether 1:1 (v/v) (AOAC, 1973).

After a double liquid–liquid partitioning process (the first with sulphuric acid, the second with potassium hydroxide) the extract was purified by means of an automated clean-up process with Power-Prep system (Fluid Management System, Waltham, Massachusetts)) using disposable columns (multilayer silica, alumina, and carbon) (Focant et al., 2001).

Eluates containing PCDD/Fs congeners were taken to dryness under nitrogen and the remainder was dissolved in nonane (10 μl) and spiked with 10 μl of the internal standard solution (containing $^{13}C_{12}$-1,2,3,4 TCDD and $^{13}C_{12}$-1,2,3,7,8,9-HxCDD).

2.3. INSTRUMENTAL ANALYSIS

PCDD/Fs were separated by high-resolution gas chromatography (HRGC) on a DB-5 MS capillary column (60 × 0.25 mm, 0.10 μm film thickness, J&W Scientific, California) and determined by high-resolution mass spectrometry (HRMS), at a resolution of 10,000 operating with electron ionization (EI) at 55 eV in the selected ion monitoring (SIM) mode.

The HRGC/HRMS system consisted of a GC Trace Series 2000 coupled with a MAT 95 XL (Thermo Finnigan, Bremen, Germany).

Operation conditions were optimized in order to obtain the complete separation of the seventeen 2,3,7,8-chlorosubstituted PCDD/F congeners. The injector temperature was 280°C and the interface temperature 290°C. The oven temperature program was as follows: 120°C (2 min), 10°C/min to 220°C, 3°C/min to 235°C, 7 min at 235°C, 4°C/min to 315°C. The solutions were injected in the splitless mode (split flow 70 ml/min). The ion source temperature was 270°C.

The concentrations of the different congeners were calculated on the basis of a five points calibration curve obtained by diluting five times calibration solutions recommended by the EPA method 1613 (US EPA, 1994).

Toxic equivalent (TEQ) values were calculated using the World Health Organization Toxic Equivalency Factors (WHO-TEFs) (Van den Berg et al., 1998). In agreement with the European legislation, WHO-TEQs were calculated as upper bound concentrations assuming that all values of specific dioxins congeners below the limit of determination (LOD) are equal to the respective LOD (Council of the EU Regulation, 2001; Council of the EU Directive, 2001).

2.4. QUALITY CONTROL/QUALITY ASSURANCE

The quality control/quality assurance included:

- Blank and spiked samples analysis in each test series
- Check of the instrumental detection limit before the analysis of each test series (signal to noise ratio better than 60:1 injecting 100 fg 2,3,7,8-TCDD, ion m/z 319.8965)

- Check of gas chromatographic separation of isomers (overlapping <25% between 1,2,3,4,7,8-HxCDD and 1,2,3,6,7,8-HxCDD and between 2,3,7,8-TCDD and the most closely eluted TCDD isomer)
- The recovery estimate for the labeled standards (range of tolerance 20–150%)
- Duplicate analysis of milk samples exceeding 80% of maximum permitted limit (2.40 pg WHO/TEQ/g fat)
- Duplicate analysis of feed samples exceeding 80% of maximum permitted limit (0.60 ng WHO/TEQ/kg product, moisture 12%, for feedingstuffs)
- The analysis of a certified reference material consisting in powdered milk (CRM 607, Institute for Reference Materials and Measurements, Geel, Belgium)
- The participation obtaining proficient results in internationally recognized interlaboratory testing schemes, such as "Food Analysis Performance Assessment Scheme" (FAPAS, Central Science Laboratory, York, UK) and "Dioxins in Food" (Norwegian Institute of Public Health, Oslo, Norway)

3. Results and Discussion

3.1. ANALYTICAL LEVELS

The analytical results are reported as pg WHO-TEQ/g fat for milk and ng WHO-TEQ/kg product for feedingstuffs referring to a moisture content of 12%, in conformity with Council Regulation (EC) 2375/2001 and Council Directive 2001/102/EC (now substituted by Regulation (EC) 1881/2006 and Directive 2003/57/EC, respectively).

The results for milk are presented in Tables 1–3. Out of 595 samples, 162 (accounting for 27%) exceeded the EU maximum tolerance level of 3 pg WHO-TEQ/g fat. Taking into account the different animal species the highest percentage of not compliant samples was for sheep's milk (45%), followed by goat's and mixed sheep/goat's milk (40%), buffalo's (29%) and cow's milk (9%).

For sheep's milk were also recorded the highest PCDD/Fs mean concentration (3.77 pg WHO-TEQ/g fat) and the most contaminated sample containing 30.43 pg WHO-TEQ/g fat.

Table 2–3 illustrate the results for animal feed samples, divided into two groups: the first group embodies feed likely grown in the area of sampling such as silage, hay and grass (267 samples), in the second group are included all the other kinds of feed (167 samples), comprising cereals and other farinaceous feed, premixes, and mixed feeds. This distinction was made on the basis of a marked difference in the contamination levels between the two groups: the former showed a relatively high PCDD/Fs content (mean 0.43 ng WHO-TEQ/kg), the latter exhibited a PCDD/Fs presence at background level (mean 0.04 ng WHO-TEQ/kg). A percentage of 13% (34 out of 267 samples) of the first group exceeded the EU tolerance limit of 0.75 ng WHO-TEQ/kg, while none of the second group was above this level.

TABLE 1. Number of milk samples of different animal origin grouped for PCDD/Fs concentration (years 2001–2004)

Parameter	Milk-producing animal species			
	Cow	Sheep	Goat sheep/goat	Buffalo
N. of samples >3 pg WHO-TEQ/g fat	18	59	20	65
N. of samples 2–3 pg WHO-TEQ/g fat	9	16	4	40
N. of samples 1–2 pg WHO-TEQ/g fat	32	15	11	54
N. of samples <1 pg WHO-TEQ/g fat	132	41	15	64

TABLE 2. Number of feed samples grouped for PCDD/Fs concentration intervals (years 2002–2004)

Parameter[a]	Silage, hay ,and grass	Cereals, premixes, and mixed feeds
N. of samples >0.75 ng WHO-TEQ/kg	34	0
N. of samples 0.50–0.75 ng WHO-TEQ/kg	16	2
N. of samples 0.30–0.50 ng WHO-TEQ/kg	43	1
N. of samples <0.30 ng WHO-TEQ/kg	174	164

[a]ng WHO-TEQ/kg referred to a 12% moisture content

TABLE 3. PCDD/Fs levels in milk (pg WHO-TEQ/g fat) and animal feed (ng WHO-TEQ/kg[a]) samples (years 2001–2004)

Parameter	Milk-producing animal species				Animal feed	
	Cow	Sheep	Sheep/ goat	Buffalo	Silage, hay, and grass	Cereals, premixes, and mixed feeds
Mean	1.39	3.77	3.24	2.85	0.43	0.04
Median	0.61	2.66	1.90	1.84	0.16	0.02
Minimum value	0.11	0.12	0.10	0.17	0.005	0.003
90th Percentile	2.92	6.88	6.27	6.42	0.94	0.06
95th Percentile	6.15	11.42	11.29	8.47	1.68	0.10
Maximum value	18.03	30.43	16.25	20.50	9.80	0.57

[a]ng WHO-TEQ/kg referred to a 12% moisture content

3.2. STATISTICAL ANALYSIS

In a previous paper (Diletti et al., 2004) a statistical evaluation of PCDD/Fs levels and profiles in milk and feedingstuffs was performed. The main conclusions from this study were as follows:

(a) A positive correlation between PCDD/Fs WHO-TEQ values in 197 paired milk and feedingstuffs samples collected in the same farms was obtained (Spearman coefficient 0.670, $p < 0.01$).

(b) A statistical significance ($p < 0.05$) in the difference of the normalized concentration for 12 out of the 17 PCDD/Fs congeners was observed by applying the Mann Whitney (M–W) test (Siegel and Castellan, 1988) on two groups of contaminated (PCDD/Fs > 3 pg WHO-TEQ g/fat) and background level (PCDD/Fs < 0.5 pg WHO-TEQ g/fat) milk samples. The M–W values for the highest toxic congeners (2,3,7,8-TCDD and 1,2,3,7,8-PeCDD, WHO-TEF = 1) were among those more statistically significant.

(c) A hierarchical cluster analysis (Morrison, 1990) was applied on 49 paired milk (PCDD/Fs > 2 pg WHO-TEQ/g fat) and feedingstuffs (PCDD/Fs > 0.30 ng WHO-TEQ/kg) samples. Two main clusters were identified for both milk and animal feed samples, as shown in Figures 1 and 2, where PCDD and PCDF homologues are normalized to the most abundant congener. PCDFs were predominant in the majority of milk samples (44 out of 49). In feedingstuffs 21 samples showed a mean profile characterized by a prevalent abundance of PCDF congeners (in particular 1,2,3,4,6,7,8-HpCDF) while OCDD prevailed in the other 28 samples.

The spatial autocorrelation analysis based on the calculation of the I-Moran index (Cliff and Ord, 1981) was also performed in our previous study (Diletti et al., 2004). This analysis was repeated on the basis of the larger data set now available. The maximum value of contamination in the sample collection (milk and animal feed samples) was selected as representative of each municipality: a total of 153 and 64 municipalities were considered for milk and animal feed (locally grown only), respectively. The I–Moran index values were significant for both milk (I-Moran = 0.48 $p < 0.01$) and feedingstuffs (I-Moran = 0.50 $p < 0.01$). The PCDD/F levels in milk and feedingstuffs are presented in Figures 3 and 4.

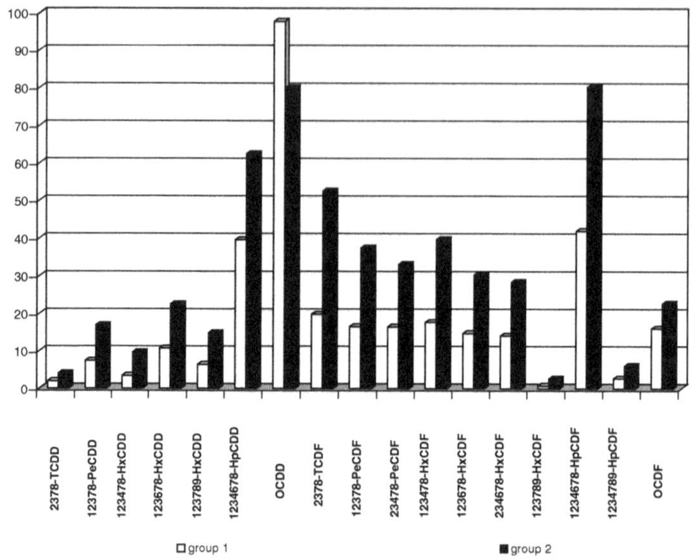

Figure 1. PCDD/Fs normalized congener profiles in the two clusters identified – MILK

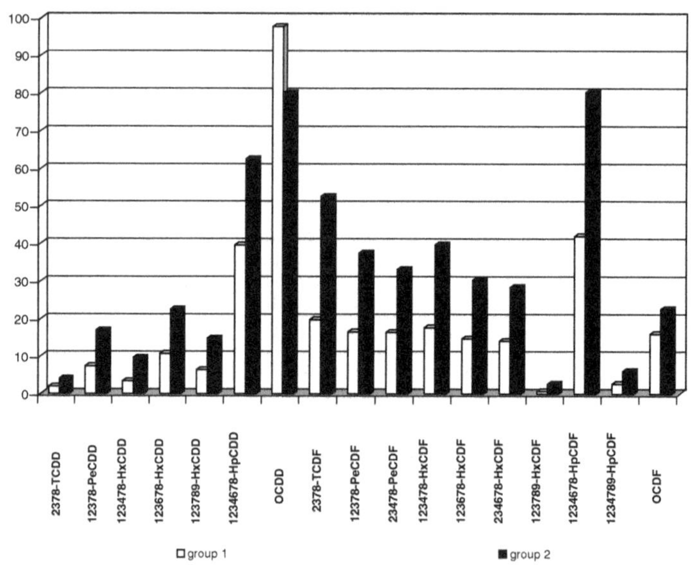

Figure 2. PCDD/Fs normalized congener profiles in the two clusters identified – FEED

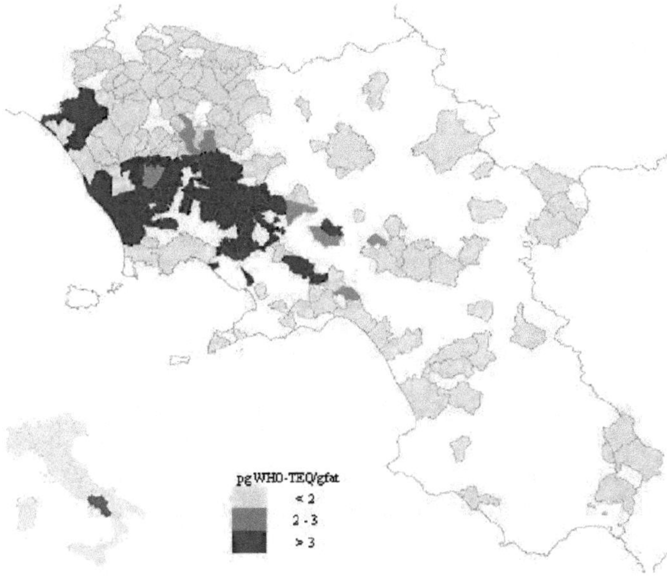

Figure 3. PCDD/Fs levels in milk from Campania region

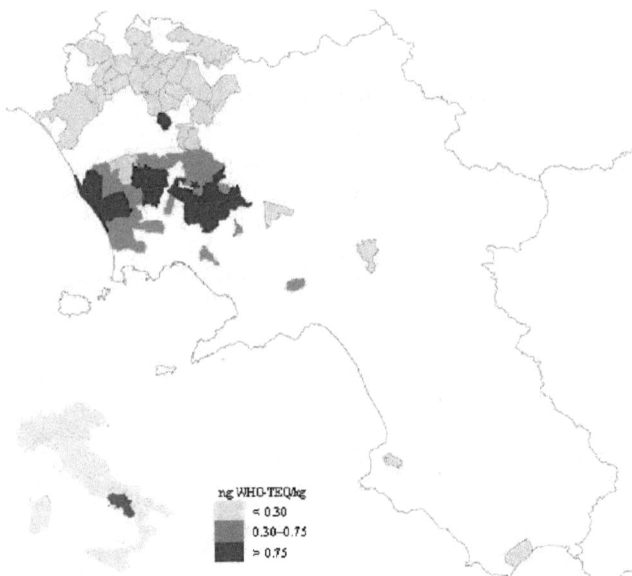

Figure 4. PCDD/Fs levels in feed from Campania region

3.3. CONGENER PROFILES EVALUATION

As mentioned above, the PCDD/F congeners in the majority of highly contaminated milk samples were characterized by a predominant PCDFs relative abundance (65%) compared to PCDDs (35%) independent of milk-producing animal species that is typical in the vicinity of contamination sources of thermal origin, while PCDDs usually prevail in milk from rural areas where only background contamination is present (Focant et al., 2002). As an example, a representative PCDD/Fs profile for cow's milk was compared with that of a grass sample taken in the same farm (Figure 5). These profiles are very similar to those obtained for cow's milk and grass from a site near a municipal solid waste incinerator in a dioxin bioavailability study to support the hypothesis of a contamination deriving from thermal processes (Slob et al., 1995). The dissimilarity between congener profiles in milk and grass is due to the different carryover rates (intended as the fraction of the daily intake of a contaminant that is excreted via the milk) among PCDD/Fs depending on the degree of chlorination. The absorption in the digestive tract decreases when an increased number of chlorine atoms is present thus justifying the low relative abundance of highly chlorinated congeners (e.g., OCDD and 1,2,3,4,6,7,8-HpCDF) in milk while they predominate in grass (McLachlan and Richter, 1998). The major abundance of these congeners in vegetation is explained by their higher thermodynamic stability in comparison to PCDD/Fs with a lower number of chlorine atoms more subjected to undergo chemical transformation reactions.

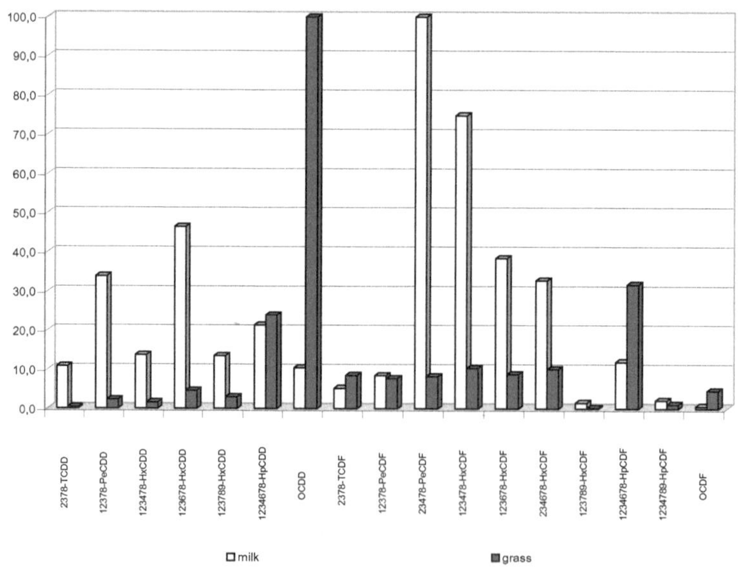

Figure 5. PCDD/Fs profile comparison between cow's milk and grass sample

3.4. CONTAMINATION OF THE AGRICULTURAL FOOD CHAIN AND INTERVENTION MEASURES

Due to the atmospheric transport and deposition processes, PCDD/Fs generated from combustion fallout are dispersed onto soils and plant surface. Grazing animals can take up vegetable material together with considerable amounts of soil. Tracing PCDD/Fs back through the agricultural food chain, it has been found that a major contribution to the dairy cow's uptake arises from the consumption of grass and grass products such as hay and silage (European Commission, September 2000). Permanent pasture should be the predominant source of nutrition for the sheep flocks (Umberger, 1996) and it could explain the contamination levels found in sheep's milk, the highest recorded among the animal species included in this study. Therefore, the accumulation of PCDD/Fs in grassland is a key factor for human exposure. The regional health authority adopted several measures to limit human exposure, first of all to seize the animal holding in the farms where contaminated milk was produced even in those farms lacking of an official certificate of analysis assessing the compliance of milk/dairy products to the aforesaid legal limits. Further actions included the delimitation of the areas at risk identified by means of georeferentiation of the seized farms, the prohibition of pasture in the areas at risk, the withdrawal and destruction of contaminated milk and dairy products. Contaminated feed products were also eliminated and substituted with feed considered "dioxin-free." Having regard to the depletion rate of PCDD/Fs in milk after switching from contaminated feed to a "dioxin-free" diet, data are available from a study on cows: the time needed to decrease PCDD/Fs concentration from a level about four times higher than the legal limit to an acceptable level was 4 months (Duhem and Fraisse, 1999).

3.5. WASTE TREATMENT AND HEALTH ASPECTS

The tentative to correlate the identified PCDD/Fs profiles clusters to specific contamination sources was not feasible for the presence of multiple complex sources in the area of interest, characterized by a long-term waste disposal crisis with a consequent widespread practice of illegal waste dumping and burning (Italian Civil Protection, 2005). Uncontrolled urban and industrial waste incineration has taken place for many years in the Provinces of Naples and Caserta, becoming a real national emergency. The area of PCDD/Fs contamination highlighted by the spatial autocorrelation analysis is overlapping with the area were wastes burning is more extensive. Consequently there is a strong possibility that the dioxins contamination could be linked to wastes incineration.

From the sanitary point of view, recent data seem to confirm the alarms claimed by people living in the area. Data from the Cancer Registry of the Sanitary Local Unite (ASL) Naples 4 (included in the area of study) show a very high level of liver cancer, leukemia, and lymphoma mortality compared with rates seen in Campania and in Italy as a whole (Senior and Mazza, 2004). The national authority responsible for the management of waste crisis in Campania requested a study to investigate the impact of waste treatment on human health (WHO, 2006). A geographic study on cancer mortality and occurrence of malformations was conducted in 196 municipalities constituting the Provinces of Naples and Caserta (Comba et al., 2006). This research detected consistent and significant excesses of the outcomes considered in an area where most illegal waste treatment activities take place.

4. Conclusions

The high levels of PCDDs and PCDFs in milk and feedingstuffs found in Campania region may result in a significant human exposure. The high probability of a linking between dioxins contamination and illegal waste dumping and burning widespread in the area of study should determine the adoption of radical measures to manage wastes of urban and industrial origin in a safe and cost-effective way. Recent studies confirmed the environmental factors influence on human health even though further researches are needed to confirm the hypothesis of correlation between human pathologies and the presence of pollutants due to waste disposal.

References

Association of Official Analytical Chemists. 1973. AOAC Official Method 905.02: Fat in milk.

Cliff A., Ord J. 1981. Spatial processes Models and Applications. Pion, London.

Cok I., Donmez M.K., Hakan Satiroglu M., Aydinuraz B., Henkelmann B., Kotalik J., Schramn K.-W. 2007. Concentration of polychlorinated dibenzo-*p*-dioxins, polychlorinated dibenzofurans and dioxin-like PCBs in human adipose tissue from Turkish men. Chemosphere 66, 1955–1961.

Comba P., Bianchi F., Fazzo L., Martina L., Menegozzo M., Minichilli F., Mitis F., Musmeci L., Pizzuti R., Santoro M., Trinca S., Martuzzi M. 2006. Cancer mortality in an area of Campania (Italy) characterized by multiple toxic dumping sites. Annals of the New York Academy of Sciences 1076 (1), 449–461.

Council of the European Union. 1996. Council Directive 96/23/EC of 29 April 1996 on measures to monitor certain substances and residues thereof in live animals and animal products and repealing Directives 85/358/EEC and 86/469/EEC and Decisions 89/187/EEC and 91/664/EEC. Official Journal L 125, 10–32.

Council of the European Union. 2001. Council Directive 2001/102/EC of 27 November 2001 amending Directive 1999/29/EC on the undesirable substances and products in animal nutrition. Official Journal L 6, 45–49.

Council of the European Union. 2001. Council Regulation (EC) No 2375/2001 of 29 November 2001 amending Commission Regulation (EC) No 466/2001 setting maximum levels for certain contaminants in foodstuffs. Official Journal L 321, 1–5.

DEFRA (Department for Environment, Food and Rural Affairs). 2002. Dioxins and dioxin-like PCBs in the UK environment. DEFRA Publications. Admail, London.

Diletti G., Torreti L., De Massis M.R., Migliorati G., Scortichini G. 2003. A case of milk contamination by PCDD/Fs in Italy: analytical levels and contamination source identification. Organohalogen Compounds 64, 1–4.

Diletti G., Scortichini G., Conte A., Migliorati G., Caporale V. 2004. Statistical evaluation of PCDD/Fs levels and profiles in milk and feedingstuffs from Campania region – Italy. Organohalogen Compounds 66, 2122–2127.

Duhem K., Fraisse D. 1999. Decrease of PCDD-PCDF content in contaminated cow's milk. Organohalogen Compounds 43, 317–320.

European Commission. September 2000. Evaluation of the occurrence of PCDD/PCDF and POPs in wastes and their potential enter the foodchain. Available from <http://ec.europa.eu/environment/dioxin/ pdf/001_ubt_final. pdf>

European Commission. November 2000. Opinion of the SCF on the risk assessment of dioxins and dioxin-like PCBs in food. Available from <http://europa.eu.int/ comm./food/ fs/sc/scf/index_en.html>

Fiedler H. 1996. Sources of PCDD/PCDF and impact on the environment. Chemosphere 32(1), 55–64.

Focant J.-F., Eppe G., Pirard C., De Pauw E. 2001. Fast clean-up for polychlorinated dibenzo-*p*-dioxins, dibenzofurans and coplanar polychlorinated biphenyls analysis of high-fat-content biological samples. Journal of Chromatography A 925, 207–221.

Focant J.-F., Eppe G., Pirard C., Massart A.-C, Andrè J.-E., De Pauw E. 2002. Levels and congeners distributions of PCDDs, PCDFs and non-*ortho* PCBs in Belgian foodstuffs. Assessment of dietary intake. Chemosphere 48, 167–179.

Karl H., Ruoff U., Blutghen A. 2002. Levels of dioxins in fish and fishery products on the German market. Chemosphere 49, 765–773.

Kogevinas M. 2000. Studied of cancer in humans. Food Additives and Contaminants 17(4), 317–324.

Italian Civil Protection. 2005. Waste treatment in Campania region (Italy) and human health impact. Available from <http://www. protezionecivile.it/cms/attach/editor/rischi-nucleare/la.pdfla.pdf> (in Italian).

Lohmann R. Jones K.C. 1998. Dioxins and furans in air and deposition: a review of levels, behaviour and processes. The Science of the Total Environment 219, 53–81.

Macdonald R.W., Harner T., Fyfe J. 2005. Recent climate change in the arctic and its impact on contaminant pathways and interpretation of temporal trend data. The Science of the Total Environment 342, 5–86.

McLachlan M., Richter W. 1998. Uptake and transfer of PCDD/Fs by cattle fed naturally contaminated feedstuffs and feed contaminated as a result of sewage sludge application. Journal of Agricultural and Food Chemistry 46, 1166–1172.

Morrison D.F. 1990. Multivariate statistical methods. 3rd edition. McGraw-Hill, New York.

Senior K., Mazza A. 2004. Italian "triangle of death" linked to waste crisis. The LANCET Oncology 5(9) 525–527.

Siegel S., Castellan S.J. Jr. 1988. Non-parametric statistics for the behavioural sciences. 2nd edition. McGraw-Hill, New York.

Slob W., Olling M., Derks H.J.G.M., de Jong A.P.J.M. 1995. Toxicokinetics of eight ^{13}C-labelled PCDDs and PCDFs in lactating cows. Chemosphere 31(8), 3827–3838.

Van den Berg M., Birnbaum L., Bosveld A.T.C., Brunstrom B., Cook P., Feely M., Giesy J.P., Hanberg A., Hasegawa R., Kennedy S.W., Kubiak T. Larsen J.C., van Leeuwen F.X.R., Liem A.K.D, Nolt C., Peterson R.E., Poellinger L., Safe S., Schrenk D., Tillit D., Tysklind M., Younes M., Waern F., Zacharewski T. 1998. Toxic equivalency factors (TEFs) for PCBs, PCDDs, PCDFs for humans and wildlife. Environmental Health Perspective 106(12), 775–792.

Umberger S.H. 1996. Feeding sheep. Available from <http://www.ext.vt. edu/pubs/sheep/ 410-853/410-853.html>

US EPA (United States Environmental Protection Agency). 1994. Method 1613 Rev. B: tetra through octa-chlorinated dioxins and furans by isotope dilution HRGC/HRMS.

World Health Organization (WHO) Regional Office for Europe. 2006. Investigating the impact of waste treatment on human health: a pilot study in Campania region (Italy). Available from <http://www.euro.who.int/ healthimpact/ MainActs/20050207_1

PERSISTENT ORGANIC POLLUTANTS (POPs) AND AIR–SOIL EXCHANGE: CASE STUDIES FOR DDTs

PERIHAN B. KURT-KARAKUS,
TERRY F. BIDLEMAN*
Centre for Atmospheric Research Experiments (CARE), Environment Canada, 6248 Eighth Line, Egbert, L0L 1N0, ON, Canada

KEVIN C. JONES, SANDRA N. MEIJER
Department of Environmental Science and Centre for Chemical Management, Lancaster Environment Centre, Lancaster University, Lancaster, LA1 4YQ, UK

RALF M. STAEBLER
Environment Canada, 4905 Dufferin Street, Downsview, ON, M3H 5T4, Canada

Abstract: Background soils (~230 samples) were analyzed for DDT and its derivatives to determine the distribution of these compounds on a global scale. Total concentrations of DDT and its metabolites (ΣDDTs) ranged over four orders of magnitude, from <0.01 to 426 ng g^{-1} dry weight and correlated poorly with soil organic matter content (SOM%) and latitude. The average composition of soil residues was similar in the Northern and Southern hemispheres and followed the order: DDT > DDE > DDD. The fraction F_{DDTe} = *p,p′*-DDT/(*p,p′*-DDE + *p,p′*-DDE) averaged 0.60 ± 0.30, or an average *p,p′*-DDT/*p,p′*-DDE ratio of 1.5. Soil-to-air fluxes of DDT compounds were measured at an experimental farm in Ontario, Canada, where DDT was last applied more than three decades ago. Results showed that agricultural soils are still a significant source for emission of DDTs to the atmosphere. DDT compound ratios in air can be used to distinguish fresh applications from soil emissions, but fractionation of the DDTs occurs during volatilization. The ratio of *p,p′*-DDT/*p,p′*-DDE in air over soil can be

* To whom correspondence should be addressed. E-mail: Terry.Bidleman@ec.gc.ca

E. Mehmetli and B. Koumanova (eds.),
The Fate of Persistent Organic Pollutants in the Environment, 315–331.

predicted from soil residues, assuming that their volatilities are in a direct relation to liquid-phase vapor pressure (P_L). Results at the Ontario farm and at other farms in Canada and the USA where soil–air exchange studies have been done showed that P_L was a better predictor for F_{DDTe} than K_{OA}.

Keywords: DDT, background soils, agricultural soils, soil–air exchange

1. Introduction

It is estimated that 4.5 Mt of DDT has been produced from the 1940s to 1990s (Li and Macdonald, 2005), of which 2.6 Mt was used in agriculture and 1.9 Mt in forestry and public health (Li and Macdonald, 2005; Voldner and Li, 1995). Due to its adverse environmental effects and bioaccumulation properties, DDT use in agriculture was stopped in most countries in the 1970s (Curtis and Lines, 2000). By 1995, DDT had been banned in 59 countries and severely restricted in 20 others (Li and Macdonald, 2005). Although DDT is one of 12 Persistent Organic Pollutants (POPs) to be eliminated under the Stockholm Convention (UNEP, 2001), its use is still permitted for control of vector-borne diseases under the guidance of World Health Organization (WHO, 2005). WHO recently announced that continued use of DDT is needed to fight malaria and recommended that indoor spraying of DDT, stopped 30 years ago, be resumed (WHO, 2006). Thus, usage of DDT in tropical countries may increase in the near future.

Signatories to the Stockholm Convention have the responsibility to conduct environmental monitoring of POPs. The atmosphere responds quickly to changes in pesticide usage and emissions, as shown by the tenfold decrease in airborne DDT concentrations in the Mississippi Delta 2 years after its 1972 deregistration in the USA (Arthur et al., 1976). A similar decline rate was found for α-hexachlorocyclohexane (α-HCH) in arctic air following bans or controls of technical HCH by the major producing and using countries (Li and Bidleman, 2003). For this reason, air monitoring has been proposed as way of tracking the effectiveness of the Convention (Pozo et al., 2006).

Soils are important in the global cycling of POPs since they have a role in supplying and receiving chemicals to/from the atmosphere. It is well documented that agricultural soils are an important reservoir (see Table 3 for references) and emission source (Bidleman and Leone, 2004; Bidleman et al.,

2006; Kurt-Karakus et al., 2006; Leone et al., 2001; Scholtz and Bidleman, 2006, 2007; Spencer et al., 1996) of "legacy" organochlorine pesticides, including DDT. Among monitoring sites worldwide, concentrations of the *p,p′*-DDT metabolite *p,p′*-DDE in air followed the order: agricultural > urban > background > polar (Pozo et al., 2006).

POPs are retained in soils for several years due to their high affinity for soil organic matter, and are subject to losses through biological degradation, erosion, leaching and volatilization. Globally, soils have a huge storage capacity for POPs (Dalle Valle et al., 2005), and therefore soil–air exchange may act to buffer atmospheric concentrations. The interaction between soil and air is one of the key issues in the global transport of POPs, and there are gaps in knowledge about the soil processes and the influence of physical and chemical properties on exchanges.

Today, DDT stands at the crossroads. Mexico, which relied on DDT for malaria control only a few years ago, has recently eliminated its usage (NACEC, 2003). Yet, as pointed out above, applications of DDT may be increasing in other tropical countries. In the spirit of the Stockholm Convention, there is a need to determine the distribution of DDT in the global environment to provide a baseline for assessing future trends. This paper considers the global distribution of DDT and its metabolites in soils.

2. DDTs in Global Background Soils

2.1. GLOBAL SOILS SURVEY

In 1998, the Environmental Science Department of Lancaster University conducted a soil sampling campaign for background soils. Sampling sites were aimed to be remote from urbanized, industrialized and agrochemical application areas. Samples were collected from 191 sites worldwide and 41 sites from a UK–Norway transect. Samples have been analyzed for DDT and its metabolites (DDE, DDD) and other POPs: PCBs, hexachlorobenzene (HCB), and HCHs. Results for PCBs and HCB are reported by Meijer et al. (2002, 2003a). Sampling and analytical methodolgy are reported in these papers. Here, results for 202 soil samples in which one or more DDTs were detected are discussed and placed in context with similar studies in agricultural soils. It is important to note that some of the countries where DDT is/might be still in use are poorly represented with no/few samples in the current data set.

TABLE 1. ΣDDT[a] concentrations in background soils

	ΣDDTs (ng g^{-1} dw)					
	AM[b]	GM[c]	Median	Min.	Max.	*n*[d]
			North America			
Canada	50.0	12.0	10.0	0.20	210	13
USA	18.0	3.0	4.0	0.03	76.0	8
Mexico	33.0	5.0	5.0	0.3	93.0	3
			South America			
Bolivia	5.0	5.0	5.0	5.0	5.0	1
Brazil	12.0	2.0	1.0	0.2	93.0	15
Chile	3.0	0.2	0.05	0.02	14.0	5
Peru	3.0	3.0	3.0	3.0	3.0	1
			Europe			
France	2.0	1.0	1.0	0.2	5.0	4
Germany	0.5	0.3	0.5	0.08	0.8	2
Ireland	50.0	3.0	50.0	0.1	100.0	2
Norway	5.0	2.0	3.0	0.04	43.0	49
Poland	160	86.0	52.0	32.0	393.0	3
Scotland	0.07	0.07	0.07	0.07	0.07	1
Spain	93.0	23.0	45.0	0.9	310.0	6
Switzerland	69.0	6.0	4.0	0.2	426.0	11
UK	10.0	2.0	3.0	0.1	159.0	37
			Asia			
Bhutan	1.0	0.6	0.3	0.3	0.3	3
Brunei	0.8	0.8	0.8	0.8	0.8	1
Japan	3.0	3.0	3.0	2.0	5.0	2
Sri Lanka	3.0	2.0	4.0	0.5	6.0	3
Thailand	5.0	4.0	6.0	1.0	7.0	4
Vietnam	11.0	3.0	6.0	0.2	24.0	5
			Africa			
Ethiopia	0.8	0.8	0.8	0.8	0.8	1
Ghana	3.0	3.0	3.0	3.0	3.0	1
South Africa	2.0	1.0	1.0	0.5	4.0	10
Zambia	2.0	2.0	2.0	0.7	3.0	2
			Australia			
Australia	0.4	0.2	0.4	0.04	0.8	4
			Antarctica[e]			
Antarctica	6.0	0.7	7.0	0.01	14.0	5

[a]Sum of *o,p′*- and *p,p′*- congeners of DDT, DDD, and DDE
[b]Arithmetic mean
[c]Geometric mean
[d]Number of samples
[e]Including samples from South Georgia (UK Island in South Atlantic Ocean)

2.2. RESULTS AND DISCUSSION

The concentrations of ΣDDTs (one or more of *o,p′*-DDE, *p,p′*-DDE, *o,p′*-DDD, *p,p′*-DDD, *o,p*-DDT, *p,p′*-DDT) in background soils ranged from 0.013 to 430 ng g^{-1} dry weight (dw), with the lowest and highest residues found in Antarctica and Switzerland, respectively (Table 1).

The ΣDDTs (log C) was weakly ($r^2 = 0.03$), but significantly ($p < 0.01$) correlated with percent soil organic matter (SOM%) and with latitude ($r^2 = 0.03$, $p = 0.01$) Correlations between ΣDDTs and HCB, ΣPCBs, and homolog groups of PCBs (data from Meijer et al., 2003) were also investigated in the same soil samples where available. Statistical analysis of the data on a log concentration basis showed that ΣDDTs had statistically significant correlations with HCB, ΣPCBs and the homolog groups of PCBs except trichlorobiphenyls. Values of r^2 and p for these correlations are given in Table 2.

TABLE 2. Correlations between ΣDDTs and other POPs in background soils

	Σ DDTs		
	r^2	p	n
HCB	0.26	<0.001	86
Σ PCBs	0.22	<0.001	83
Σ tri-	0.29	0.23	19
Σ tetra-	0.16	0.003	57
Σ penta-	0.34	<0.001	82
Σ hexa-	0.48	<0.001	79
Σ hepta-	0.24	<0.001	82
Σ octa-	0.28	<0.001	79

Concentrations of ΣDDTs in the global background soils are compared in Table 3 with levels in agricultural and nonagricultural soils reported in other studies.

Profiles of individual DDT compounds as percentage of ΣDDTs in the northern and southern hemispheres are shown in Figure 1. Maxima of 100% indicate that only one compound type was found in some cases; e.g., only DDD was found in two soils and only DDT in 32 samples. The compound distributions are similar in both hemispheres, and similar profiles are observed when the analysis is done on a continent basis.

Converting ratios to fractions is preferred for statistical purposes (Ulrich et al., 2003), so a quantity F_{DDTe} is defined (Kurt-Karakus et al., 2006):

TABLE 3. Reported concentrations of ΣDDTs in soils of various locations

			Σ DDTs (ng g^{-1} dw)		
Land use	Country	*n*	Mean[a] or median	Min.	Max.
Background	Global[c]	202	2.4	0.013	430
Agricultural	USA[d]	1	4,100	8.0	30,000
Agricultural	USA[d]	95	506	nd‡	11,800
Agricultural	USA[e]	40	9.6	nd	706
Agricultural	USA[f]	100	13	2.2	26
Cemetery	USA[g]	3	17	1,800	3,400
Forest[b]	USA[h]	6	850	270	1,900
Agricultural	Canada[i]	11	780	5.0	8,000
Agricultural	Canada[j]	23	7.0	1.0	53
Agricultural	Canada[k]	13	690	11	70,000
Agricultural	New Zealand[l]	61	610	nd	34,500
Grazing	New Zealand[l]	15	35	nd	13,001
Mixed[b]	New Zealand[m]	23	810	250	1,660
Agricultural	China[n]	30	5.1	nd	95
Agricultural	China[o]	188	11.7	0.7	972
Mixed	China[p]	47	38	0.77	2,180
Agric.(crop)	China[q]	37	16	0.27	707
	China[q]	14	22	2.0	257
Agric.(paddy)	China[q]	23	2.7	0.19	9.0
	China[r]	66	0.52	0.13	1.85
Nonagricultural					
Mixed					
Industrial/urban	Poland[s]	24	200	4.3	2,400
Rural	Romania[t]	20	123	3.5	1,331
Urban	Romania[t]	26	194	9	1,542
Agricultural	Germany[u]	11	60	24	173
Forest	Germany[v]	9	16.3	1.4	62
National parks	Costa Rica[w]	10	0.005	nd	1.8
Agricultural	Congo[x]	4	135	29	266
Urban	Congo[x]	15	1.1	nd	166
Antarctica	Antarctica[y]	4	0.28	0.06	0.88

ΣDDTs indicates 1–6 of the DDT compounds and metabolites, depending on the study
[a]Mean is arithmetic unless stated as geometric
[b]DDT applied deliberately; nd: not detected; na: not reported in the reference
[c]Geom. mean, this study
[d]California, geom. mean, Mischke et al. (1985)
[e]Midwest states, geom. mean, Aigner et al. (1998)
[f]Southern states, geom. mean, Bidleman and Leone (2004); Kannan et al. (2003)
[g]Harner et al. (1999)
[h]Geom. mean, sampled in 1993, Dimond and Owen (1996)
[i]Geom. mean, British Columbia, farm and orchard, Bidleman et al. (2006)
[j]Geom. mean, Saskatchewan, Bailey et al. (2005)
[k]Geom. mean, Ontario, farm and orchard, Kurt-Karakus et al. (2006); Meijer et al. (2003b); Webber and Wang (1995)
[l]"Agricultural" includes farm, orchard and vineland; geom. mean, Gaw et al. (2006)

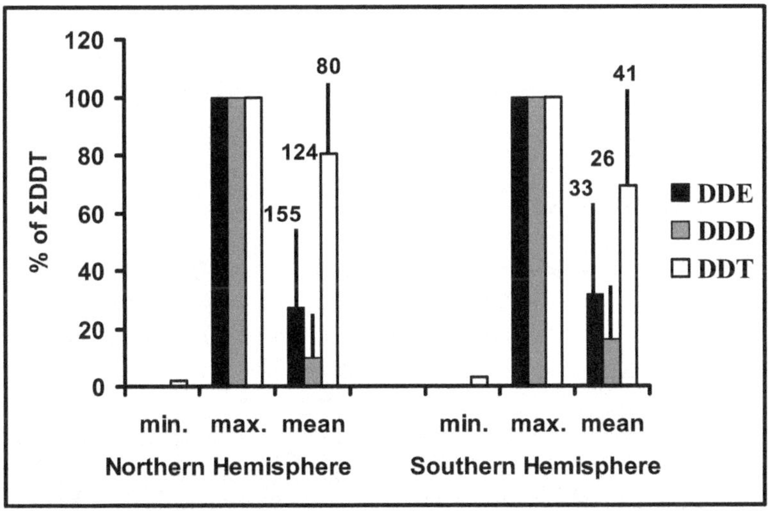

Figure 1. Compound profiles of DDTs in background soils of the northern and southern hemispheres (sum of *o,p'*- and *p,p'*-congeners for each homolog group), numbers indicate the number of samples

$$F_{DDTe} = p,p'\text{-DDT}/(p,p'\text{-DDT} + p,p'\text{-DDE}) \quad (1)$$

The average F_{DDTe} for the background soils = 0.60 ± 0.30, which corresponds to a *p,p'*-DDT/*p,p'*-DDE ratio of 1.5. A literature survey of DDT residues in agricultural soils in different regions of the USA and Canada shows F_{DDTe} ranging from 0.35 to 0.80 with a mean of 0.58 (Section 3.3).

3. Emissions of DDTs from a Historically Treated Farm in Ontario, Canada

Several studies have shown that DDTs which have been in agricultural soils for decades are continually emitted into the atmosphere. These studies used direct measurements of air concentrations above agricultural soils (Bidleman and Leone, 2004; Bidleman et al., 2006; Leone et al., 2001; Spencer et al.,

[m]Geom. mean, Boul et al. (1994)
[n]Median, Zhang et al. (2005)
[o]Geom. mean, Gong et al. (2004)
[p]Median, Zhu et al. (2005)
[q]Medians, Li et al. (2006)
[r]Mean for five land-use classes, min. and max. reflect the averages for each class, Zhang et al. (2006)
[s]Falandysz et al. (2001)
[t]Covaci et al. (2001)
[u]Geom. mean, Manz et al. (2001)
[v]Wenzel et al. (2002)
[w]Daly et al. (2007)
[x]Geom. mean, Ngabe and Bidleman (2006)
[y]Borghini et al. (2005)

1996) and modeling of soil–air fluxes (Scholtz and Bidleman, 2006, 2007) as evidence. To put soil emissions on a quantitative basis, we carried out measurements of DDT fluxes from a historically treated agricultural soil in southern Ontario.

3.1. EXPERIMENTAL APPROACH

In fall 2004 and spring 2005, a study was conducted to measure the soil–air fluxes of aged DDT residues from a DDT-treated agricultural soil in the Holland Marsh area of Ontario, Canada (Figure 2) (Kurt-Karakus et al., 2006). Holland Marsh is a 2,830 ha vegetable-growing area with high

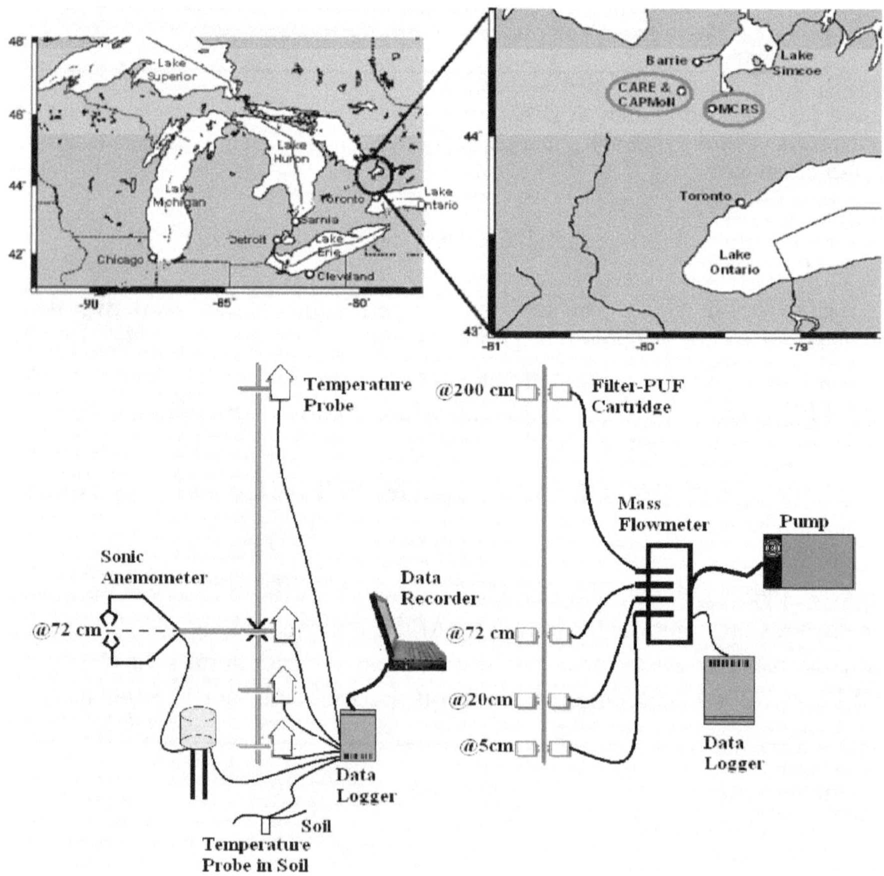

Figure 2. Sampling design at Muck Crops Research Station. Top: location of Muck Crop Station. Bottom: right – gradient air sampling above the soil, left – temperature gradient and sonic anemometer measurements

organic matter soils (~70%) and heavy usage of DDT in the past. The last application date of DDT is not known, but it is assumed that DDT was not used in the area after mid-1970s when Canada phased out agricultural DDT usage.

DDT fluxes from soil to air were measured at Muck Crops Research Station (MCRS), a research farm operated by University of Guelph, using a gradient approach. Air samples were taken at 5, 20, 72 and 200 cm above the soil and concurrent micrometeorological data were taken with a sonic anemometer to obtain estimates of heat and momentum fluxes (Figure 2). Rural air concentrations were also measured at the CARE-CAPMoN station, about 30 km away from MCRS, in a forested area surrounded by farms. Sampling, analytical and data treatment details are given in (Kurt-Karakus et al., 2006).

3.2. FLUX RESULTS

Soil concentrations of ΣDDTs at MCRS averaged 19,000 ng g^{-1} dw, which is similar to previous measurements at the same location (24,000 ng g^{-1}) (Meijer et al., 2003b). Concentrations of ΣDDT in air ranged from 200 to 17,000 pg m^{-3} and decreased with height above the soil (Figure 3). Fall concentrations were lower compared to those in spring.

Figure 4 shows the average fluxes of ΣDDTs from soil to air at MCRS in spring (n = 8) and fall (n = 6). Average fluxes in fall and spring ranged from 50 to 1,800 ng m^{-2} h^{-1} with mean values of 90 and 660 ng m^{-2} h^{-1}, respectively for events respectively for sampling periods with acceptably consistent micrometeorological conditions. Higher fluxes in spring were associated with higher soil temperatures, which increased the volatilization of the DDTs.

If it is assumed that the volatilization is the only loss process from soil to air, consideration of the dissipation rate and the mass of DDT in the upper 5 cm of soil leads to an estimated half-life of 220 years (Kurt-Karakus et al., 2006). In a fortunate manner, dissipation rates of DDT compounds from soil are usually faster, on the order of decades, due to microbial degradation (Boul et al., 1994; Martijn et al., 1993; Meijer et al., 2001; Spencer et al., 1996) as well as lower soil organic matter content (Scholtz and Bidleman, 2007). If the ΣDDTs concentration measured in MCRS soil is representative of the Holland Marsh agricultural area, the total volatilized amount of ΣDDTs from the area can be estimated assuming the average fluxes in fall (90 ng m^{-2} h^{-1}) and spring (640 ng m^{-2} h^{-1}), 90 days in each season, and volatilization during a 12 h daytime period. Total emissions estimated for fall and spring were 2.3 and 16 kg.

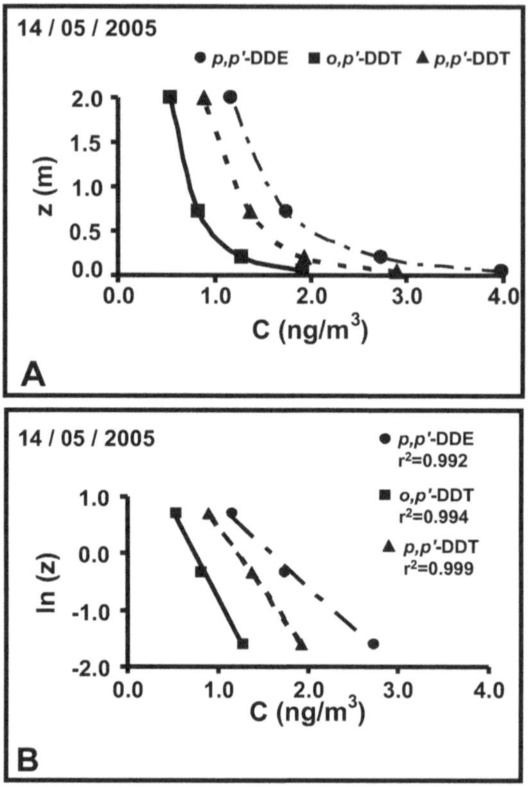

Figure 3. A. Height above soil (Z) vs concentrations of DDTs (C) in air. B. linearized ln Z vs C relationships

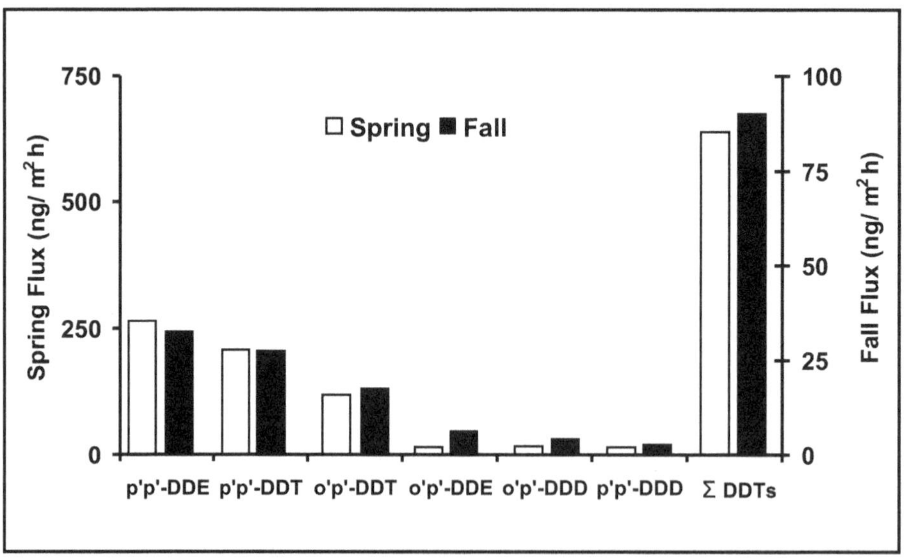

Figure 4. DDT compound fluxes at MCRS during spring and fall (ng m^{-2} h^{-1}) (see Kurt-Karakus et al., 2006 for details)

3.3. FRACTIONATION OF DDT COMPOUNDS DURING VOLATILIZATION

DDT isomers and metabolites have different volatilities and evaporate from soil at different rates. Studies with agricultural soils show that ratios of DDT compounds in soil can be used to predict the ratios in vapor-phase emissions (this study and Kurt-Karakus et al., 2006). In order to predict the ratios of emitted compounds from soil, a direct relation to liquid-phase vapor pressure (P_L) or an inverse relation to octanol–air partition coefficient (K_{OA}) (Bidleman et al., 2006; Harner et al., 2001; Shoeib and Harner, 2002) can be used:

$$(C_1/C_2)_{air} = (C_1/C_2)_{soil} \times P_{L,1}/P_{L,2} \tag{2}$$

$$(C_1/C_2)_{air} = (C_1/C_2)_{soil} \times K_{OA,2}/K_{OA,1} \tag{3}$$

where 1 is *p,p′*-DDT and 2 is *p,p′*-DDE. The predicted ratios in air are then converted to F_{DDTe} according to Eq. 1.

Results from the MCRS study showed that P_L was a better predictor for F_{DDTe} (predicted/measured = 0.82 ± 0.15) than K_{OA} (predicted/measured = 1.86 ± 0.33). Figure 5 shows the predicted (Eq. 2) and measured F_{DDTe} in air from the MCRS study as well as in other locations (data from Bidleman and Leone, 2004; Bidleman et al., 2006), with an average predicted/measured agreement of 99%.

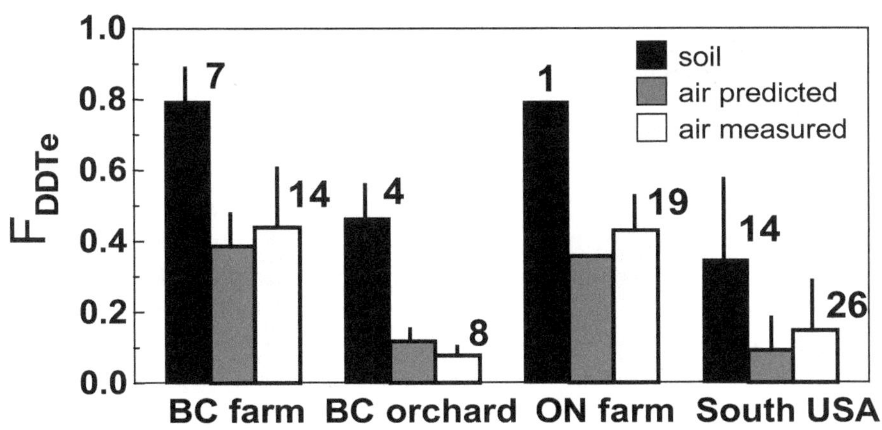

Figure 5. F_{DDTe} in air above agricultural soils, measured and predicted from soil residues using Eq. 2. ON (Ontario) farm = MCRS (Kurt-Karakus et al., 2006). BC farm and BC orchard are agricultural sites in British Columbia, Canada (data from Bidleman et al., 2006); South USA are farms in Alabama, Louisiana and Texas (data from Bidleman and Leone, 2004). Numbers indicate soil sites and air measurements; e.g., four BC orchards and two air measurements at each orchard

F_{DDTe} can be used to estimate whether soil residues are fresh or have been subjected to metabolism. Typical compositions of technical DDT are given as 7.1% *p,p'*-DDT, 14.9% *o,p'*-DDT and 4% *p,p'*-DDE by WHO (1975), and 77% *p,p'*-DDT, 22% *o,p'*-DDT and 0.1% *p,p'*-DDE by Spencer and Cliath (1972). Thus, fresh DDT may be characterized by F_{DDTe} of 0.95–0.99. It can be expected that freshly contaminated soils would contain residues with F_{DDTe} close to this range (Ngabe and Bidleman, 2006) whereas older DDT contamination should display lower values.

F_{DDTe} in background soils ranged from 0.03 to 1, and average regional values ranged from 0.45 to 0.70 (Figure 6A). F_{DDTe} tended to be lower in Asia and Africa compared to the Americas, but such differences may be reflective of the countries where the samples were taken rather than an overall eastern vs western hemispheric difference. For example, all African and most Asian soils were from subtropical and tropical areas. Even within a country, proportions of the DDTs in background soils can differ according to geography. A study of soils at remote forest sites in Austria found that ratios of DDT/DDE increased with altitude (Weiss et al., 2000). In general, values of F_{DDE} in background soils were similar to those in agricultural soils (Figure 6B).

Dissipation of DDT is faster from tropical soils than temperate soils (Khan, 1994), which may be due to enhanced volatilization, degradation, or both. DDT is degraded in soil by chemical and microbial processes to the persistent and toxic metabolites DDE and DDD. The proportions of DDT compounds and their breakdown products vary greatly in agricultural soils and are affected by soil management practices such as tilling, fertilizer additions, the presence of aerobic or anaerobic conditions and reactions with clay minerals (Boul et al., 1994; Edwards, 1966; Lalah et al., 2001; Spencer et al., 1996). Some soils in the southwestern USA retain a high proportion of parent DDT for decades after the end of DDT applications because they lack the microbial capability for DDT degradation (Hitch and Day, 1992). There was no correlation between soil organic matter and F_{DDTe} values in the global background soils ($p > 0.05$). Li et al. (2006) found that the proportion of parent DDT to its metabolites, DDT/(DDE + DDD), increased with the ΣDDTs concentration in background soils of China.

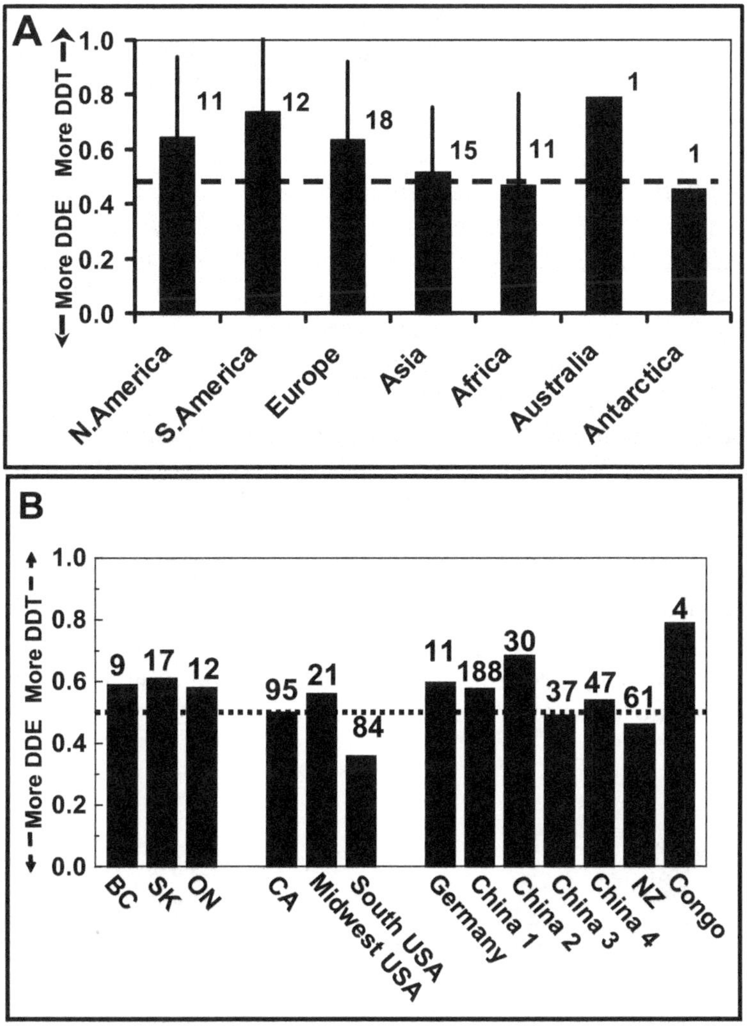

Figure 6. F_{DDTe} in background (this study, A) and agricultural soils (B). Numbers indicate soil samples. Germany (Manz et al., 2001), China 1 (Gong et al., 2004), China 2 (Zhang et al., 2005), China 3 (Li et al., 2006), China 4 (Zhu et al., 2005), BC = British Columbia (Bidleman et al., 2006), SK = Saskatchewan (Bailey et al., 2005), and ON = Ontario (Kurt-Karakus et al., 2006; Meijer et al., 2003b; Webber and Wang, 1995), CA = California (Mischke et al., 1985), Midwest USA (Aigner et al., 1998), south USA (Bidleman and Leone, 2004; Kannan et al., 2003), NZ = New Zealand (Gaw et al., 2006), Congo (Ngabe and Bidlema, 2006).

4. Conclusions

Levels of DDTs in background soils in global background soils vary over four orders of magnitude, from ~0.01 to ~400 ng g^{-1} dw, and the range overlaps reported measurements in agricultural soils. Factors responsible for such variation are difficult to ascertain, but SOM% appears to exert only a minor influence. It is suggested that a country's DDT usage on a per-area basis may have some influence, and improvement in DDT usage data would aid the interpretation.

Agricultural soils which last received DDT applications decades ago continue to emit DDTs to the atmosphere. This is shown in several studies of elevated air concentrations directly above such soils. Flux measurements put these observations on a quantitative basis and enable emissions estimates to be made, but they are very labor intensive. Coupling flux measurements with soil–air exchange modeling (Harner et al., 2001; Scholtz and Bidleman, 2006, 2007) can extend emissions estimates over a broad scale and allow the impact of agricultural soils on regional air concentrations to be assessed. In the absence of other dissipation processes such as microbial degradation, treated soils will continue to be an important source of atmospheric contamination for a long time.

The fractionation of *p,p′*-DDT and *p,p′*-DDE which occurs on evaporation of residues from soils can be adequately predicted from the relative liquid-phase vapor pressures of the two compounds. Using this information, proportions of the two chemicals found in ambient air can be compared to those due to soil emissions.

References

Aigner, E.J., Leone, A.D., and Falconer, R.L., 1998, Concentrations and enantiomeric ratios of organochlorine pesticides in soils from the US corn belt, *Environ. Sci. Technol.* **32**:1162–1168.

Arthur, R., Cain, J., and Barrentine, B., 1976, Atmospheric levels of pesticides in the Mississippi Delta, *Bull. Environ. Contam. Toxicol.* **15**:129–134.

Bailey, P., Waite, D., Quinnette-Abbott, L., and Ripley, B., 2005, Residues of DDT and other selected organochlorine pesticides in soils from Saskatchewan, Canada (1999), *Can. J. Soil Sci.* **85**:265–272.

Bidleman, T.F. and Leone, A., 2004, Soil-air exchange of organochlorine pesticides in the southern United States, *Environ. Pollut.* **128**:49–57.

Bidleman, T.F., Leone, A.D., Wong, F., van Vliet, L., Szeto, S., and Ripley, B.D., 2006, Emission of "legacy" chlorinated pesticides from agricultural and orchard soils in British Columbia, Canada, *Environ. Toxicol. Chem.* **25**(6):1448–1457.

Borghini, F., Grimalt, J.O., Sanchez-Hernandez, J.C., and Bargagli, R., 2005, Organochlorine pollutants in soils and mosses from Victoria Land (Antarctica), *Chemosphere* **58**:271–278.

Boul, H.L., Garnham, M.L., Hucker, D., Baird, D., and Aislable, J., 1994, Influence of agricultural practices on the levels of DDT and its residues in soils, *Environ. Sci. Technol.* **28**(8):1397–1402.

Covaci, A., Hura, C., and Schepens, P., 2001, Selected persistent organochlorine pollutants in Romania, *Sci. Total Environ.* **280**:143–152.

Curtis, C.F. and Lines, J.D., 2000, Should DDT be banned by international treaty? *Parasitol. Today* **16**(3):119–121.

Dalle Valle, M., Jurado, E., Dachs, J., Sweetman, A.J., and Jones, K.C., 2005, The maximum reservoir capacity of soils for persistent organic pollutants: implications for global cycling, *Environ. Pollut.* **134**(1):153–164.

Daly, G.L., Lei, Y.D., Teixeira, C., Muir, D.C.G., Castillo, L.E., Jantunen, L.M.M., and Wania, F., 2007, Organochlorine pesticides in the soils and atmosphere of Costa Rica, *Environ. Sci. Technol.* **41**(4):1124–1130.

Dimond, J.B. and Owen, R.B., 1996, Long-term residue of DDT compounds in forest soils of Maine, *Environ. Pollut.* **92**:227–230.

Edwards, C.A., 1996, Insecticide residues in soils. *Residue Rev.* **13**:82–132.

Falandysz, J., Brudnowska, B., Kawano, M., and Wakimoto, T., 2001, Polychlorinated biphenyls and organochlorine pesticides in soils from the southern part of Poland, *Arch. Environ. Contam. Toxicol.* **40**:173–178.

Gaw, S.K., Wilkins, A.L., Kim, N.D., Palmer, G.T., and Robinson, P., 2006, Trace element and ΣDDT concentrations in horticultural soils from the Tasman, Waikato and Auckland regions of New Zealand, *Sci. Total Environ.* **355**:31–47.

Gong, Z.M., Tao, S., Xu, F.L., Dawson, R., Liu, W.X., Cui, Y.H., Cao, J., Wang, X.J., Shen, W.R., Zhang, W.J., Qing, B.P., and Sun, R., 2004, Level and distribution of DDT in surface soils from Tianjin, China, *Chemosphere* **54**:1247–1253.

Harner, T., Wideman, J.L., Jantunen, L.M.M., Bidleman, T.F., and Parkhurst, W.J., 1999, Residues of organochlorine pesticides in Alabama soils, *Environ. Pollut.* **106**:323–332.

Harner, T., Bidleman, T.F., Jantunen, L.M.M., and Mackay, D., 2001, Soil-air exchange model of persistent pesticides in the United States Cotton Belt, *Environ. Toxicol. Chem.* **20**:1612–1621.

Hitch, R.K. and Day, H.R., 1992, Unusual persistence of DDT in some western USA soils, *Bull. Environ. Contam. Toxicol.* **48**:259–264.

Kannan, K., Battula, S., Loganathan, B.G., Hong, C.S., Lam, W.H., Villeneuve, D.L., Sajwan, K., Giesy, J.P., and Aldous, K.M., 2003, Trace organic contaminants, including toxaphene and trifluralin, in cotton field soils from Georgia and South Carolina, USA, *Arch. Environ. Contam. Toxicol.* **45**:30–36.

Khan, S.U., 1994, FAO/IAEA appraisal of overall programme accomplishments, *J. Environ. Sci. Health* **B29**:205–226.

Kurt-Karakus, P.B., Bidleman, T.F., Staebler, R.M., and Jones, K.C., 2006, Measurement of DDT fluxes from a historically treated agricultural soil in Canada, *Environ. Sci. Technol.* **40**(15):4578–4585.

Kurt-Karakus, P.B., Lay, J., Meijer, S., Bidleman, T., and Jones, K.C., 2007, DDT and HCH residues in global background soils (in preparation).

Lalah, J.O., Kaigwara, P.N., Getenga, Z., Mgenyi, J.M., and Wandiga, S.O., 2001, The major environmental factors that influence rapid disappearance of pesticides from tropical soils in Kenya, *Toxicol. Environ. Chem.* **81**:161–197.

Leone, A., Amato, S., and Falconer, R., 2001, Emission of chiral pesticides from agricultural soils in the Cornbelt Region of the US, *Environ. Sci. Technol.* **35**:4592–4596.

Li, J., Zhang, G., Qi, S., Li, X., and Peng, X., 2006, Concentrations, enantiomeric compositions, and sources of HCH, DDT and chlordane in soils from the Pearl River Delta, South China, *Sci. Total Environ.* **372**:215–224.

Li, Y.-F. and Bidleman, T.F., 2003, Correlation between global emissions of alpha-HCH and its concentrations in arctic air, *J. Environ. Informatics* **1**:52–57.

Li, Y.F. and Macdonald, R.W., 2005, Sources and pathways of selected organochlorine pesticides to the Arctic and the effect of pathway divergence on HCH trends in biota: a review, *Sci. Total Environ.* **342**:87–106.

Manz, M., Wenzel, K.D., Dietze, U., and Schüürmann, G., 2001, Persistent organic pollutants in agricultural soils of central Germany, *Sci. Total Environ.* **277**:187–198.

Martijn, A., Bakker, H., and Schreuder, R.H., 1993, Soil persistence of DDT, dieldrin and lindane over a long period, *Bull. Environ. Contam. Toxicol.* **51**:178–184.

Meijer, S.N., Halsall, C.J., Harner, T., Peters, A.J., Ockenden, W.A., Johnston, A.E., and Jones, K.C., 2001, Organochlorine pesticide residues in archived UK soil, *Environ. Sci. Technol.* **35**(10):1989–1995.

Meijer, S.N., Steinnes, E., Ockenden, W.A., and Jones, K.C., 2002, Influence of environmental variables on the spatial distribution of PCBs in Norwegian and UK soils: implications for global cycling, *Environ. Sci. Technol.* **36**(10):2146–2153.

Meijer, S.N., Ockenden, W.A., Sweetman, A., Breivik, K., Grimalt, J.O., and Jones, K.C., 2003a, Global distribution and budget of PCBs and HCB in background surface soils: Implications for sources and environmental processes, *Environ. Sci. Technol.* **37**(4):667–672.

Meijer, S.N., Shoeib, M., Jantunen, L.M.M., Jones, K.C., and Harner, T., 2003b, Air-soil exchange of organochlorine pesticides in agricultural soils. 1. Field measurements using a novel in situ sampling device, *Environ. Sci. Technol.* **37**(7):1292–1299.

Mischke, T., Brunetti, K., Acosta, V., Weaver, D., and Brown, M., 1985, Agricultural sources of DDT residues in California's environment, California Department of Food and Agriculture, Sacramento, CA.

NACEC, 2003, DDT no longer used in North America. Fact sheet DDT-04-2003, North American Commission for Environmental Cooperation, Montreal, Canada, http://www.cec.org/files/PDF/POLLUTANTS/DDT_en.pdf

Ngabe, B. and Bidleman, T.F., 2006, DDT concentrations in soils of Brazzaville, Congo, *Bull. Environ. Contam. Toxicol.* **76**:697–704.

Pozo, K., Harner, T., Wania, F., Muir, D.C.G., Jones, K.C., and Barrie, L.A., 2006, Toward a global network for persistent organic pollutants in air: Results from the GAPS Study, *Environ. Sci. Technol.* **40**(16):4867–4873.

Scholtz, M.T. and Bidleman, T.F., 2006, Modelling of the long term fate of pesticide residues in agricultural soils and their surface exchange with the atmosphere: Part I. Model description and evaluation, *Sci. Tot. Environ.* **368**:823–838.

Scholtz, M.T. and Bidleman, T.F., 2007, Modelling of the long term fate of pesticide residues in agricultural soils and their surface exchange with the atmosphere: Part II. Projected long-term fate of pesticide residues, *Sci. Tot. Environ.* **377**:61–80.

Shoeib, M. and Harner, T., 2002, Using measured octanol-air partition coefficients to explain environmental partitioning of organochlorine pesticides, *Environ. Toxicol. Chem.* **21**(5):984–990.

Spencer, W.F. and Cliath, M.M., 1972, Volatility of DDT and related compounds, *J. Agric Food Chem.* **20**:645–649.

Spencer, W.F., Singh, G., Taylor, C.D., LeMert, R.A., Cliath, M.M., and Farmer, W.J., 1996, DDT persistence and volatility as affected by management practices after 23 years, *J. Environ. Qual.* **25**:815–821.

Ulrich, E.M., Helsel, D.R., Foreman, W.T., 2003, Complications with using ratios for environmental data: comparing enantiomeric ratios (ERs) and enantiomer fractions (EFs), *Chemosphere* **53**:531–538.

UNEP, 2001. Text of the Stockholm Convention on Persistent Organic Pollutants for adoption by the Conference of Plenipotentiaries. United Nations Environment Program, Geneva, Switzerland.

Voldner, E.C. and Li, Y.F., 1995, Global usage of selected persistent organochlorines, *Sci. Total Environ.* **160/161**:201–210.

Webber, M.D. and Wang, C., 1995, Industrial organic compounds in selected Canadian soils, *Can. J. Soil Sci.* **75**:513–524.

Weiss, P., Lorbeer, G., and Scharf, S., 2000, Regional aspects and statistical characterisation of the load with semivolatile organic compounds at remote Austrian forest sites, *Chemosphere* **40**:1159–1171.

Wenzel, K.D., Manz, M., Hubert, A., and Schüürmann, G., 2002, Fate of POPs (DDX, HCHs and PCBs) in upper soil layers of pine forests, *Sci. Total. Environ.* **286**:143–154.

World Health Organization (WHO), 2005. WHO position on DDT use in disease vector control under the Stockholm Convention on persistent organic pollutants. World Health Organization, Roll Back Malaria Department, Geneva, Switzerland, Who/HTM/RMB/2004.53.

World Health Organization (WHO), 2006, WHO gives indoor DDT use a clean bill of health for controlling malaria. news release, media centre, http://www.who.int/mediacentre/news/releases/2006/pr50/en/

Zhang, H., Lu, Y., Dawson, R.W., Shi, Y., and Wang, T., 2005. Classification and ordination of DDT and HCH in soil samples from the Guantning Reservoir, China, *Chemosphere* **60**:762–769.

Zhang, H.B., Luo, Y.M., Zhao, Q.G., Wong, M.H., and Zhang, G.L., 2006, Residues of organochlorine pesticides in Hong Kong soils, *Chemosphere* **63**:633–641.

Zhu, Y., Liu, H., Xi, Z., Cheng, H., and Xu, X., 2005, Organochlorine pesticides (DDTs and HCHs) in soils from the outskirts of Beijing, China, *Chemosphere* **60**:770–778.

PARTITIONING AND EXCHANGE OF ORGANOCHLORINE CONTAMINANTS BETWEEN ABIOTIC COMPARTMENTS IN ANTARCTICA

VLADIMIR BOGILLO,* MARIYA BAZYLEVSKA
Department of Antarctic Geology and Geoecology, Institute of Geological Sciences, National Academy of Sciences of Ukraine, 55B Olesya Gonchara Street, 01054 Kiev, Ukraine

Abstract: The partition coefficients for OCs at typical temperatures in Antarctica and Southern Ocean are calculated and they are used to estimate the partitioning of the OCs between atmospheric particulates and gas phase, scavenging the OCs from gas phase by snow, atmospheric persistence of the OCs, the direction and strength of air/seawater, air/snow, air/soil, and seawater/sediment fluxes for the OCs in Antarctica. The air/soil and seawater/sediment fluxes of OCs are agreed with "cold condensation" hypothesis. The deep water formation in Weddell and Ross Seas, the shelf sediments and the Antarctic snowpack would be important sinks for the OCs.

Keywords: Antarctica, Southern ocean, persistent organic pollutants, "cold condensation," "cold fractionation," air/seawater, air/snow, air/soil and seawater/sediment exchange, atmospheric persistence of organochlorine contaminants, global warming, POPs fate

1. Introduction

Antarctica is a harsh continent by any frame of reference. It is the highest (mean elevation = 2.300 m), coldest (minimum temperature = –89.6°C), geographically most isolated land mass on Earth. The ice cap covers 13.8 ×

* To whom correspondence should be addressed. E-mail: vbog@carrier.kiev.ua

E. Mehmetli and B. Koumanova (eds.),
The Fate of Persistent Organic Pollutants in the Environment, 333–351.

10^6 km^2 of Antarctica's 14 × 10^6 km^2 surface area.[1] By virtue of its geographical isolation and unique meteorological conditions, this most southerly of continents provides unparalleled opportunities for monitoring the long-range transport of air pollutants.

Since 1960s, persistent semivolatile or low-volatile organochlorine contaminants (OCs), such as chlordane (CHL), DDT and its metabolites (DDE and DDD), hexachlorobenzene (HCB), α- and γ-hexachlorocyclohexanes (HCHs), polychlorinated biphenyls (PCBs), polychlorinated dibenzo-*p*-dioxins (PCDDs) and polychlorinated dibenzofurans (PCDFs) were found in abiotic compartments of Antarctica and its biota. The introduction of the compounds to Antarctica and the Southern Ocean comes entirely from extraregional sources.

The Antarctic region acts as a sink for the OCs through their "cold condensation."[2] The OCs emitted in warmer midlatitudes transferred to the Antarctica via atmospheric streams. They condense out of the atmosphere at the cold region and settle there in large quantities, in the water, snow, ice, soil, sediment, and biota of Antarctica and Southern Ocean.

All from the OCs show a strong tendency to partition between different abiotic compartments. This partitioning is controlled by their physicochemical properties, the characteristics of the compartments as well as temperature. Then, the partition coefficients for the OCs at typical temperatures in Antarctica and Southern Ocean were calculated and they used to estimate the atmospheric particulate – air partition, scavenging from air by snow and atmospheric persistence, while the direction and strength of air/seawater, air/snow, air/soil, and seawater/sediment fluxes for the OCs in Antarctica. The impact of current climate changes on reemission of the contaminants in Antarctica is discussed also.

2. Behavior of OCs at Cold Antarctic Temperatures

2.1. PARTITIONING BETWEEN COMPARTMENTS

Generally, air temperature is varied in Antarctica from –70°C to –20°C (Central Plateau) and from –30°C to 0°C (Coastal areas), whereas seawater temperature is near 0°C.[3] To estimate the partitioning and exchange of the OCs between abiotic compartments in Antarctica, the *n*-octanol – air (K_{OA}), air – water (K_{AW}), *n*-octanol – water (K_{OW}), and water/air interface – air (K_{IA}) partition coefficients for the OCs and their dependencies on the temperature are required. The log K_{OA}, log K_{AW} and log K_{OW} quantities for the OCs at 25°C were taken from references.[4–9]

To estimate the dependence of these partition coefficients on temperature, we suggested the maintenance of "isoequilibrium" relationship or "compensation" effect between partition enthalpy and entropy for compounds of same type (e.g., OCs)[10]:

$$\Delta H_{ij} = \Delta G_{ij}(T_{is}) + T_{is}\Delta S_{ij} \quad (1)$$

where T_{is} is the "isoequilibrium" or "compensation" temperature and ΔG_A (T_{is}) is the molar free partition energy at this temperature.

Then:

$$-2.3RT\log K_{ij} = a\Delta H_{ij} + b \quad (2)$$

where R is the universal gas constant, $a = 1 - T/T_{is}$ and $b = \Delta G_A(T_{is})\ T/T_{is}$.

The partition coefficients for the OCs at low temperatures were estimated using known partition enthalpies (ΔH_{ij}) for 6 PCBs and following relationships between the log K_{ij} and ΔH_{ij}:

$$\Delta H_{OA} = -38.9 \pm 8.4 - (4.89 \pm 0.96)\log K_{OA}\ (N = 6, r = -0.931) \quad (3a)$$

$$\Delta H_{OW} = -36.9 \pm 15.7 + (3.4 \pm 2.6)\log K_{OW}\ (N = 6, r = 0.540) \quad (3b)$$

The ΔH_{AW} values were calculated from difference between ΔH_{OW} and ΔH_{OA} quantities. The plot in Figure 1 for experimental ΔH_{AW} values versus those estimated by this way demonstrates satisfactory agreement between these quantities.

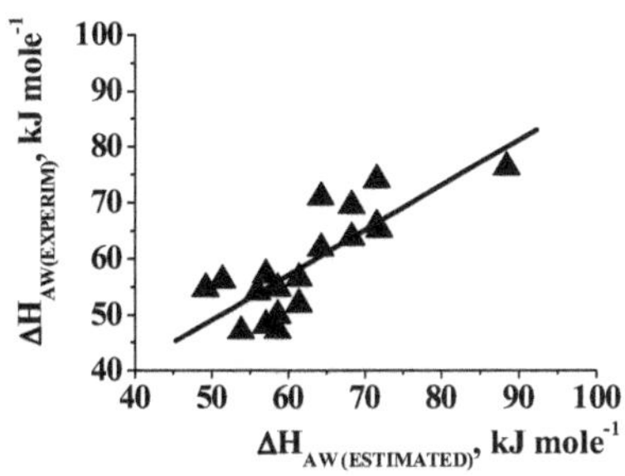

Figure 1. Plot of experimental ΔH_{AW} values for OCs versus their estimated ΔH_{AW} values

Main modes for transport of OCs in range from 25°C to 0°C can be distinguished from dependences of their log K_{AW} on log K_{OA} quantities in accordance with Wania.[11] Such dependences are displayed in Figure 2. It is shown that DDTs, heavy PCBs, PCDDs and PBDE congeners are "single hoppers" and they have potential to easily irreversible deposition to Antarctic surface, whereas HCB, HCHs, and light PCBs congeners are "multiple hoppers" and they would be readily exchanged between air and underlying surfaces in response to the temperature changes.

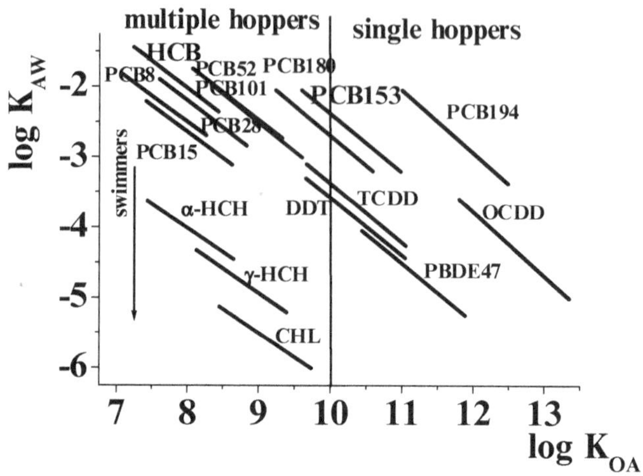

Figure 2. Dependencies between log K_{AW} and log K_{OA} quantities for the OCs. Major modes of the OCs transport in range 25–0°C are quantified in accordance with Wania[11]

The log K_{IA}, m, at 15°C and corresponding partition enthalpies (ΔH_{IA}) were estimated using the relationships derived in Roth et al.[12]:

$$\log K_{IA}(15°C) = 0.635 \times \log L^{16} + 5.11 \times \Sigma\beta_2^{H} + 3.60 \times \Sigma\alpha_2^{H} - 8.4 \quad (4)$$

$$\Delta H_{IA} = -5.07 \times \ln K_{IA}\,(15°C) - 108 \quad (5)$$

where log L^{16} is the compound' *n*-hexadecane–air partition coefficient, $\Sigma\beta_2^{H}$ is the overall compound hydrogen bond basicity and $\Sigma\alpha_2^{H}$ is the overall compound hydrogen bond acidity.

The log L^{16}, $\Sigma\beta_2^{H}$, and $\Sigma\alpha_2^{H}$ quantities for the OCs were taken from studies[13,14] and were estimated for DDTs, PCDDs, PCDFs, and HCB using a group contribution approach.[15]

The partition coefficients at low temperatures were calculated as:

$$\ln K_{ij}(T) = \ln K_{ij}(T_{\text{ref}}) - \frac{\Delta H_{ij}}{R}\left(\frac{1}{T} - \frac{1}{T_{\text{ref}}}\right) \quad (6)$$

where T_{ref} is the reference temperature, in K.

The log K_{OA}, log K_{AW}, log K_{OW}, and log K_{IA} values at reference temperatures, and the partition enthalpies are presented in Table 1.

Key primary processes affecting on the behavior of the OCs at cold Antarctic temperatures are partitioning the OCs between atmospheric particulates and gas phase, scavenging the OCs from gas phase by snow flakes and high atmospheric persistence of the OCs.

TABLE 1. Logarithms of *n*-octanol – air (K_{OA}), air – water (K_{AW}), *n*-octanol – water (K_{OW}) and water/air interface – air (K_{IA}) partition coefficients for OCs at reference temperatures, and corresponding partition enthalpies (ΔH_{ij})

OCs	log K_{OA}	ΔH_{OA}	Log K_{AW}	ΔH_{AW}	log K_{OW}	ΔH_{OW}	log K_{IA}	ΔH_{IA}
Units		kJ/mol		kJ/mol		kJ/mol	M	kJ/mol
T, °C	25		25		25		15	
p,p′DDT	9.7	−86.1	−3.3	70.8	6.4	−15.3	−0.9	−97.8
p,p′DDE	8.4	−79.9	−2.7	62.4	5.7	−17.5	−1.5	−91.1
p,p′DDD	9.0	−82.9	−3.5	64.7	5.5	−18.2	−1.2	−93.9
α-HCH	7.4	−75.3	−3.6	51.4	3.8	−23.9	−0.2	−105.7
γ-HCH	8.1	−78.7	−4.3	54.7	3.8	−24.0	−0.3	−105.1
HCB	7.2	−74.3	−1.4	57.1	5.8	−17.2	−3.3	−70.0
PCB15	7.4	−75.2	−2.2	56	5.2	−19.2	−2.6	−77.8
PCB28	7.6	−76.1	−1.9	58.6	5.7	−17.5	−2.5	−79.3
PCB 31	8.0	−78.1	−2.0	60.3	5.6	−17.8	−1.8	−86.5
PCB 52	8.1	−78.4	−1.8	61.4	5.9	−17.0	−2.2	−81.9
PCB101	8.3	−79.6	−2.0	64.3	6.4	−15.3	−2.0	−84.5
PCB 105	10.0	−87.8	−2.4	74	6.8	−13.8	−1.5	−90.3
PCB138	8.9	−82.5	−2.2	68.3	6.7	−14.2	−1.3	−93.1
PCB 153	9.6	−85.8	−2.1	71.9	6.8	−13.9	−1.6	−89.5
PCB180	9.2	−84.1	−2.1	71.6	7.2	−12.5	−1.1	−95.7
TCDD	9.7	−86.2	−3.1	72.4	6.8	−13.8	−0.6	−101.4
PeCDD	11.4	−94.5	−4.0	82.8	7.4	−11.7	0.04	−108.4
OCDD	11.8	−96.6	−3.6	87.6	8.2	−9.0	0.1	−109.0
PCDF	10.2	−88.8	−3.9	74	6.5	−14.8	−1.0	−96.5
HxCDF	11.2	−93.9	−3.7	82.5	7.5	−11.4	−0.2	−105.8
OCDF	12.1	−98.1	−4.1	88.4	8.0	−9.7	−0.5	−102.1

2.2. SORPTION TO ATMOSPHERIC PARTICULATES

The atmospheric particle-bound fractions of the OCs, Φ_P, were calculated as follows[16]:

$$\Phi_P = \frac{K'_P}{1+K'_P} \quad (7)$$

$$K'_P = f_{OM} v_Q (\rho_{part} / \rho_{oct}) K_{OA} \quad (8)$$

where v_Q is the volume fraction of particles in the atmosphere, f_{OM} is the volume fraction of organic matter in the particle, ρ_{part} and ρ_{oct} are the densities of the particles and *n*-octanol, respectively.

We used here the same parameters for atmospheric particles as in Wania et al.[16]. The Φ_P quantities at 25°C and at –30°C are presented in Table 2 and plots of Φ_P quantity versus temperature for the OCs are given in Figure 3a. It is obvious, that the heavy PCBs congeners (starting with penta-CB), TCDDs, TCDFs and their more heavy congeners, and DDT would be entirely particle-bounded at typical Antarctic temperature. These OCs, called "single hoppers,"[11] are usually deposited irreversibly to the Earth surface. The rest of the OCs from Table 1 are referred to "multiple hoppers" and they readily undergo air – underlying surface exchange.

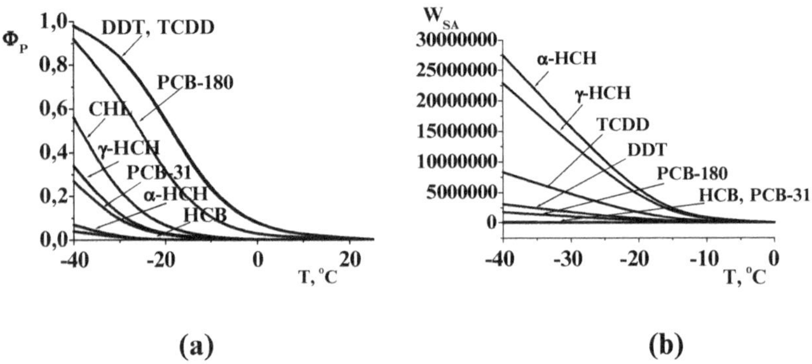

Figure 3. Temperature dependencies for atmospheric particle-bound fractions of the OCs (a) and snow – gas-phase scavenging ratios for the OCs (b)

Concentrations in air for PCDD/Fs and coplanar PCBs (77, 126, and 169) in gas phase (C_G) and in atmospheric particulates (C_P) during an Atlantic cruise from the UK to Antarctica are reported by Lohmann et al.[17] Using these data, we calculated the slopes for plots of $\ln(C_P/C_G)$ versus $1/T$ (T, in K) and obtained the apparent particle/air partition enthalpies, ΔH_{PA}. The ΔH_{PA} values for the OCs are all negative and they ranged from –20 to –270 kJ mole^{-1}, but in case of Cl_5–, Cl_6–, Cl_8DFs ($-\Delta H_{PA}$ = 81–83 kJ mole^{-1}), Cl_4–, Cl_6DDs ($-\Delta H_{PA}$ = 82–87 kJ mole^{-1}), and PCB-126 ($-\Delta H_{PA}$ = 71 kJ mole^{-1}) they are close to range of corresponding octanol/air partition enthalpies ($-\Delta H_{OA}$ = 80–98 kJ mole^{-1}). This confirms the validity for choice of *n*-octanol as a surrogate for organic matter in the atmospheric particulates.

TABLE 2. Atmospheric particle bound fractions (Φ_P), snow – gas scavenging ratios ($W_{gas,snow}$), mean atmospheric half-lives ($\tau_{1/2(A)}$) and those, estimated taking into account for sorption on atmospheric particulates ($\tau_{1/2(A,P)}$) for OCs at reference temperatures and at −30°C

OC	Φ_P	Φ_P	$W_{gas,snow}$	$W_{gas,snow}$	$\tau_{1/2(A)}$	$\tau_{1/2(A,P)}$	$\tau_{1/2(A)}$	$\tau_{1/2(A,P)}$
Units					Years	Years	Years	Years
T, °C	25	−30	0	−30	25	25	−30	−30
PCB15	0.00	0.01	1.9×10^2	1.3×10^4	0.04	0.04	0.4	0.5
PCB28	0.00	0.02	2.6×10^2	1.9×10^4	0.07	0.07	1.9	2.0
PCB 31	0.00	0.07	1.3×10^3	1.4×10^5	0.06	0.06	0.8	0.9
PCB 52	0.00	0.08	4.6×10^2	4.0×10^4	0.1	0.1	2.1	2.2
PCB101	0.00	0.14	8.1×10^2	8.1×10^4	0.2	0.2	5.7	6.6
PCB 105	0.01	0.94	2.9×10^3	4.0×10^5	0.2	0.2	5.7	101
PCB138	0.00	0.46	5.4×10^3	8.3×10^5	0.4	0.4	14.5	27
PCB 153	0.00	0.85	2.5×10^3	3.2×10^5	0.4	0.4	14.5	95
PCB180	0.00	0.67	9.5×10^3	1.7×10^6	0.7	0.7	28	87
TCDD	0.00	0.87	3.4×10^4	8.3×10^6	0.07	0.07	0.4	2.9
PeCDD	0.11	1.00	3.7×10^4	1.0×10^7	0.1	0.1	0.9	658
OCDD	0.26	1.00	1.8×10^5	6.6×10^7	1.5	2.0	25	6×10^4
PCDF	0.01	0.97	1.1×10^4	2.2×10^6	0.2	0.2	2.2	65
HxCDF	0.09	1.00	8.7×10^4	2.7×10^7	0.5	0.6	6.2	3×10^3
OCDF	0.41	1.00			2.4	4.1	50	3×10^5
p,p'-DDT	0.00	0.87	1.5×10^4	3.1×10^6	0.5	0.5	6.0	46
p,p'-DDE	0.00	0.17	3.5×10^3	5.0×10^5	0.8	0.8	11.2	13.5
α-HCH	0.00	0.01	8.7×10^4	2.7×10^7	0.5	0.5	10.6	10.8
γ-HCH	0.00	0.09	7.6×10^4	2.3×10^7	0.4	0.4	10.9	12.0
HCB	0.00	0.01	33	1.5×10^3	2.7	2.7	199	201

2.3. SCAVENGING BY SNOW FROM AIR

The snow – gas-phase scavenging ratio ($W_{gas,snow}$) being given by Hansen et al.[18]:

$$W_{gas\text{-}snow} = K_{IA} SA_s \rho_w \tag{9}$$

where SA_s is the snow specific surface area and ρ_w is the density of water (1,000 kg m^{-3}).

An SA_s value of 120 m^2 kg^{-1} was chosen for freshly fallen snow.[18] The $W_{gas\text{-}snow}$ values at 0 and −30°C are presented in Table 2 and plots of $W_{gas\text{-}snow}$ values versus temperature for the OCs are given in Figure 3b. It is evident that snow scavenging at low Antarctic temperatures would be very effective way for removal of the OCs from atmosphere, with the exception of HCB and light PCBs congeners. By virtue of the fact that usually snow precipitation rate decreases with the temperature drop,[3] the total wet deposition flux of OCs not strongly changes when the temperature descends.

2.4. ATMOSPHERIC PERSISTENCE

The drop of temperature strongly influences on atmospheric half-live of the OCs. Reaction of gaseous OCs with OH radical in atmosphere is the major loss process. Mean atmospheric half-live due to atmospheric degradation, $\tau_{1/2(A)}$, is $\ln 2/k_A$, where $k_A = k_{OH}[OH]$. The k_{OH} is the bimolecular rate coefficient for reaction of a chemical with OH radical and [OH] is the concentration of OH radical in air. Lowering the temperature leads to deceleration of the reaction due to reduction both of the rate coefficient and the atmospheric concentration of OH radical. Second effect of the temperature lowering is the increase of nonreactive particle-bound fraction for the OCs. Sorption of the OCs on atmospheric particulates can lead to growth of their gas phase half-lives. The half-live of OCs in presence of the atmospheric particulates, $\tau_{1/2(A,P)}$, is determined as follows[16]:

$$\tau_{1/2(A,P)} = \frac{\ln 2\{1 + f_{OM}(\rho_{part} / \rho_{oct})K_{OA}v_Q\}}{k_{OH}[OH]} \quad (10)$$

The $\tau_{1/2(A)}$ and $\tau_{1/2(A,P)}$ quantities were estimated using the same parameters for atmospheric particles as in Wania et al., study.[16] The rate coefficients at 25°C, $k_{OH(298)}$, for the OCs were taken from studies[4,19–24]. The k_{OH} values at low temperatures, $k_{OH(T)}$, were calculated as:

$$k_{OH(T)} = k_{OH(298)} \exp\left(\frac{E_{OH}}{R}\right) \times \left(\frac{1}{T} - \frac{1}{298}\right) \quad (11)$$

where E_{OH} is the activation energy for reaction of a chemical with OH radical.

The E_{OH} values are known for 24 OCs, including PCBs, HCHs, and HCB.[23] We checked the realization of "isokinetic" relationship or "compensation" effect[10] between the activation energy and logarithm of the preexponential factor, A_{OH}, for the OCs and obtained following relationship:

$$\ln A_{OH} = -25.95 \pm 0.32 + (0.191 \pm 0.025)E_{OH}; N = 24; R^2 = 0.724 \quad (12)$$

Then, the $k_{OH(T)}$ values for rest of OCs (PCDDs, PCDFs, and DDTs) were estimated using their $k_{OH(298)}$ values and parameters of the Eq. 12. Temperature trend of OH radical concentration (C_{OH} (molecule cm^{-3}) = 4,768.3 × T – 1.121 × 10^6, T in K) was taken from Halsall et al.[25] The

$\tau_{1/2(A)}$ and $\tau_{1/2(A,P)}$ quantities for the OCs at 25°C and −30°C are given in Table 2. From these estimations it follows that the gas–particle partitioning at low temperature greatly increases the gas phase half-lives for the OCs. This effect is more pronounced for heavy PCBs, PCDDs, and PCDFs congeners and for DDTs family. The loss of the OCs by degradation in cold Antarctic atmosphere is insignificant way of their removal and such sinks for the OCs on surface of Antarctic continent and in Southern Ocean as their burial in glaciers, ice caps and in deep oceanic water and shelf sediments would be more considerable.

3. Exchange of OCs between Abiotic Compartments of Antarctic Environment

3.1. SEA–AIR EXCHANGE

In order to understand the behavior and fate of OCs in the environment of Southern Ocean, a two-film resistance model of Whitman was employed to examine the flux by gas exchange.[26] Flux of contaminants (F) by gas exchange between air and water is expressed as

$$F = K_{\mathrm{ol}}\left(10^3 C_{\mathrm{W}} - \frac{C_{\mathrm{G}}}{K_{\mathrm{AW}}}\right) \tag{15}$$

where C_W and C_G are the dissolved solute concentration in water (pg L^{-1}) and the gas-phase concentration in air (pg m^{-3}), respectively, and K_{ol} is the overall mass transfer coefficient (m s^{-1}).

Positive value of F indicates the tendency of contaminant transfer from water to air. In this study, wind speed was approximated to 7 m s^{-1} for >60° S latitude. The water temperature was determined to be 0°C near Antarctica. The K_{ol} quantities for the OCs were calculated according to study.[26]

We estimated sea–air fluxes for α-HCH and γ-HCH in Atlantic Ocean in the line of Antarctica and the fluxes for these chemicals in cruise along Antarctic Peninsula using the C_W and C_G values from Lakaschus et al.[27] and Dickhut et al.[28] These fluxes are presented in Figure 4a and b. It is shown that net gas deposition is observed for both α-HCH and γ-HCH. This flux along Antarctic Peninsula is larger for γ-HCH than for α-HCH. For last isomer the air/water flux is near air–seawater equilibrium. The fluxes of HCHs near Neumayer Station/Ekstroem Ice Shelf are larger in comparison with the fluxes along Antarctic Peninsula. These data indicate that South Atlantic and Pacific act as global sinks for HCHs in present time.

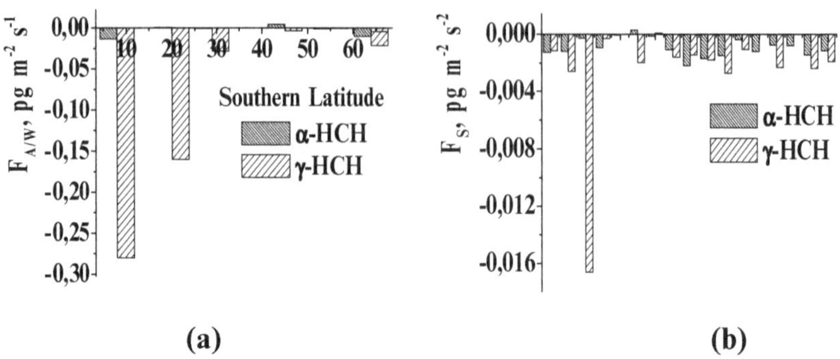

Figure 4. The air–seawater fluxes for α-HCH and γ-HCH in Atlantic Ocean in the line of Neumayer Station[27] (a) and along Antarctic Peninsula[28] (b)

The Weddell and Ross Seas in Southern Ocean represent main deepwater formation sites of the world ocean. The formation of the deep oceanic waters moves surface waters, including OCs, directly to deep Ocean. Using the temporal trends for water concentrations of OCs,[29] the rates of deepwater formation and areas of Weddell and Ross Seas,[30] we estimated specific total OCs fluxes in the seas due to deep water formation. The OCs fluxes are: 1.5 t year^{-1} for DDTs, 74 t year^{-1} for HCHs, 21 t year^{-1} for HCB, 3 t year^{-1} for CHLs and 7 t year^{-1} for PCBs.

3.2. SNOW–AIR EXCHANGE

The dry gas exchange flux at the surface was derived as the gradient between the air concentration and the concentration in the snow pore space multiplied by an exchange velocity, v[18]:

$$F_{\text{gas}} = v\left(\frac{C_{\text{S}}}{K_{\text{IA}} SA_{\text{s}}} - C_{\text{A}}\right) \tag{16}$$

where C_S is the chemical concentration in snow (pg L^{-1}) and C_A is the chemical concentration in air (pg m^{-3}).

The exchange velocities, v, for the OCs were calculated using the Whitman two-layer resistance model according to study.[18] In addition to C_A and C_S quantities, the snow specific surface area, snow density, water/air interface – air partition coefficients (K_{IA}) and air diffusion coefficients are required to estimate the snow/air fluxes. Most of snow samples were collected in coastal regions of Antarctica. As it follows from V. G. Aver'yanov[3] and from glaciological maps,[1] old snow in these regions has density from 0.35 to 0.40 g cm^{-3}. The snow specific surface area can be estimated from its dependence on the snow density. We obtained following empirical relation between these quantities for 9 snow samples[31]:

$$SA_s = 78.3 + 681 \times \exp(-\rho_{snow}/0.132); N = 9; r = 0.864 \quad (17)$$

Then, the SA_s quantity for coastal Antarctic snow may vary from 111 to 126 $cm^2\ g^{-1}$. In our estimations of the snow/air fluxes we used three testable values for the SA_s: 100, 1,000, and 10,000 $cm^2\ g^{-1}$. Mean wind speed, $U = 7$ $m\ s^{-1}$ and air temperature, $T = -10°C$ were used in these estimations. The K_{IA} values at this temperature for OCs were calculated using data from Table 1, whereas the air diffusion coefficients for OCs were calculated according to Schwarzenbach et al.[32] The air and snow concentrations of OCs[27,28,33,34] were used in the estimations of the snow/air fluxes. The fluxes for HCHs, DDTs and PCBs at SA_s = 1,000 $cm^2\ g^{-1}$ are presented in Figure 5a and b. All these fluxes demonstrate net deposition of the chemicals into snowpack. Strength of the fluxes for HCHs falls from 1981 to 2000. The volatilization of the chemicals from snowpack is predicted only for PCBs and HCHs in Lake Nurume (Figure 5b). Repetition of the calculations at SA_s = 100 and 10,000 $cm^2\ g^{-1}$ almost not changed these conclusions.

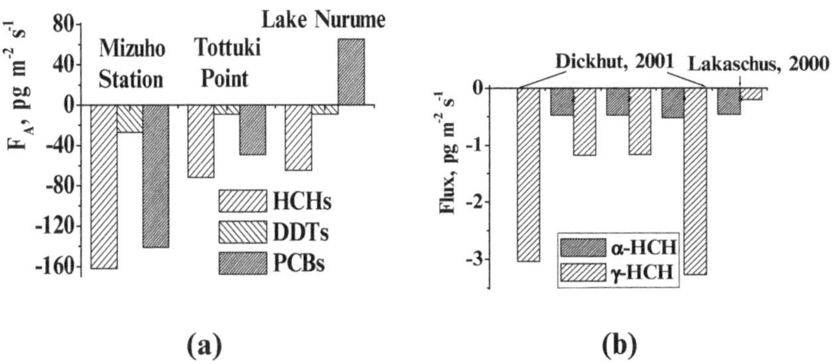

Figure 5. Air–snowpack fluxes for HCHs[28,34] (a) and for ΣHCH, ΣDDT, and ΣPCB[33] (b)

Therefore, Antarctic snowpack acts as global sink for the OCs. We estimated the amounts of the OCs deposited on whole surface of Antarctica during 1980–2006 using the temporal trends for OCs[29] and mean annually snow accumulation on the continent[3] (2.0×10^{15} kg $year^{-1}$). Following amounts were obtained from the OCs snow concentrations: 4.2 t for ΣDDT and 14.5 t for ΣHCH. Also, these amounts may be estimated using the temporal trends for OCs in air[29], their K_{IA} values and SA_s = 100 $cm^2\ g^{-1}$. Such estimations give 7.3 t for ΣDDT, 87 t for ΣHCH, 6 t for ΣPCB and 54 kg for HCB in Antarctic snowpack.

Annual melting of shelf glaciers in Antarctica is 370 $km^3\ year^{-1}$, loss of snow by wind is 20 $km^3\ year^{-1}$, liquid loss is 15 $km^3\ year^{-1}$, loss of icebergs is 2,200 km^3 $year^{-1}$.[3] As the annual snow accumulation in the coastal regions has reached from 30 to 60 cm $year^{-1}$,[3] from 15 to 30 m for upper part of the icebergs would be enriched by OCs during 50 years of their continuous emission in the environment. The melting of the icebergs would be appreciable source of secondary emission for the OCs into oceanic ecosystems.

3.3. SOIL–AIR EXCHANGE

It was proposed by Gouin et al.,[35] that a plot of log C_S versus log [OM] will have a slope of 1 at equilibrium between soil and air. The C_S is the concentration of OC in soil and [OM] is the content of organic matter in the soil. We used the concentrations of HCB, HCHs, DDTs, and 28 PCBs congeners in eleven soil samples collected on the Eastern coast of Antarctica[36] to verify the state of equilibrium between soil and air. The slopes for plots of log C_S versus log [OM] are presented in Figure 6. It is evident that the slopes are less than 1 for all OCs and the soils and air are far from the equilibrium. Although concentrations of ΣDDT and ΣHCH in the soils are not correlated with organic carbon content in the soils, such relationships were observed for HCB, ΣPCB, and individual PCBs congeners with correlation coefficients varied from 0.673 to 0.914. From other side, a relationship was found between the concentrations of ΣDDT and ΣHCH in eleven soils: $\Sigma HCH = 0.5 \pm 1.2 + (1.6 \pm 0.1) \times \Sigma DDT$; $r = 0.972$. This indicates on joint agricultural source of these chemicals and predominance of ΣHCH over ΣDDT in the soils.

Instead of air/sea and air/snow fluxes, we used the fugacity fractions in soil to estimate the direction of the air/soil fluxes. The fugacities (Pa) of the

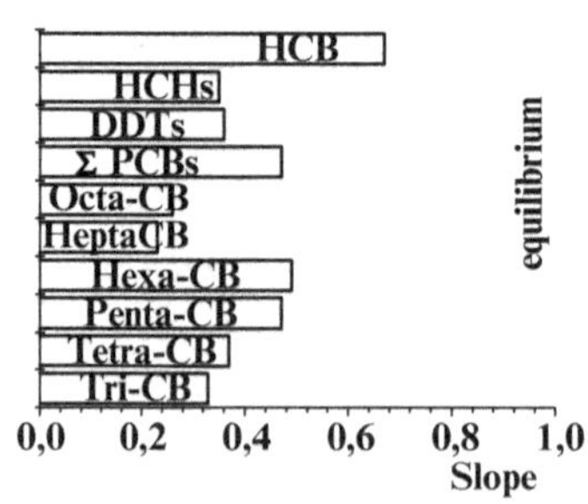

Figure 6. The slopes for plots of logarithm of OCs concentrations in soils from Eastern Antarctica[36] versus logarithm of organic matter content in the soils

OCs in soil (f_S) and air (f_A) were calculated using equations given by Harner et al.[37] and assuming that the fugacity capacity of soil is due to entirely to the organic matter fraction:

$$f_S = C_S RT / 0.411 \phi_{OM} K_{OA} \tag{18}$$

$$f_A = C_A RT \tag{19}$$

where C_S and C_A are the concentrations of OC in soil and air (mole m^{-3}), ϕ_{OM} is the fraction of organic matter in the soil (1.7 times the organic carbon fraction) and T is the average air temperature (K). Soil residue was converted from a mass to volume concentration basis assuming a soil bulk density of 1.5×10^6 g m^{-3}, intermediate between a loamy and sandy soil.

Results are expressed as the fraction of the total fugacity in soil (FF):

$$FF = f_S/(f_S + f_A) \tag{20}$$

Soil–air equilibrium is indicated by FF = 0.5. Net volatilization and deposition of gas-phase OCs are indicated by fugacity fractions > 0.5 and < 0.5, respectively.

The OCs concentrations in air (in 1998) were estimated using the temporal trends for OCs in air from,[29] whereas the K_{OA} coefficients at −10°C were calculated using data from Table 1. Possible contributions of OCs adsorption to mineral surfaces of soils were estimated using an approach from Goss et al.[38] The range from 1 to 50 m^2 g^{-1} for specific surface area of the mineral phase of soil was used in these estimations. It was found that this contribution gives not more 20% in overall sorption of OCs to soils with low content of organic matter and at highest specific surface area of the mineral phase of soil (50 m^2 g^{-1}). The fugacity fractions of the OCs for eleven soil samples collected on the Eastern coast of Antarctica[36] as the functions of organic matter content in the soils are presented in Figure 7. They demonstrate net volatilization flux for most volatile HCB from soils to air and fluxes for PCBs, DDTs, and HCHs in both directions (volatilization and deposition) for different soils.

As was suggested by Negoita et al.,[36] in addition to long-range transport of OCs by air from the continents of South Hemisphere, reasons for the soils contamination with OCs in Antarctica would be local contamination (preferentially by PCBs) due to human activities on scientific stations and local focusing of OCs, due to biotic activities (excrements, eggs, carcasses of birds). These activities are responsible for transport of the OCs to the Antarctic environment via migratory birds.

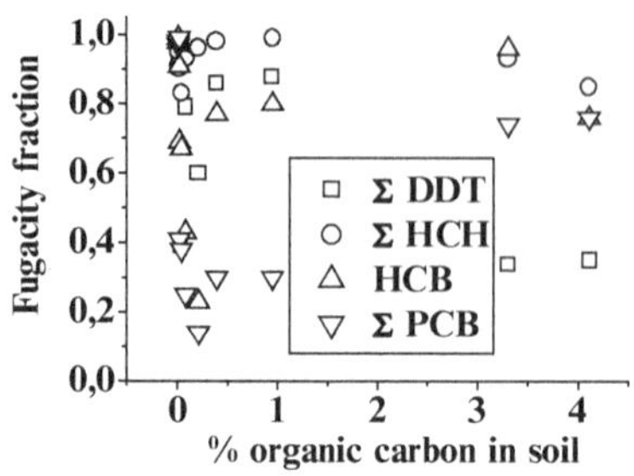

Figure 7. The fugacity fractions of the OCs for eleven soils on the Eastern coast of Antarctica[36] as the functions of organic matter content in the soils

Different directions of air-soil fluxes for the OCs confirm these suggestions. It is shown that soils in Antarctica are far from the equilibrium with the atmosphere. They have been "oversupplied" with the chemicals and have lost considerable quantities by evaporation. Also, the volatilization fluxes of OCs from most soils may be indication on consequence of their "global condensation" in the immobile Antarctic medium, however this case coincides with local "human" and "biotic" contaminations of the soils.

3.4. SEDIMENT–SEAWATER EXCHANGE

By analogy with soil–air equilibrium, seawater–sediment equilibrium is indicated by fugacity fraction = 0.5. Net desorption from sediments and deposition of dissolved OCs are indicated by fugacity fractions > 0.5 and <0.5. The fugacities (Pa) of the OCs in sediment (f_S) and seawater (f_W) were calculated as follows:

$$f_S = C_S RT / 0.411 \phi_{OM} K_{OW} \quad (21)$$

$$f_W = C_W RT . \quad (22)$$

C_S and C_W are concentrations of OCs in sediment and water (mole m^{-3}) and ϕ_{OM} is the fraction of organic matter in the sediment.

We used the C_S and C_W values for seven PCB congeners in Admiralty Bay, King George Island[39,40] and for HCB, γ-HCH, p,p′-DDT and p,p′-DDE near Galindez Island, Western Antarctic Peninsula[41] to estimate the directions of the sediment – sea fluxes. The log K_{OW} quantities at 0°C were calculated using the data from Table 1. The fugacity fractions for the OCs in the sediments are presented in Figure 8a and b.

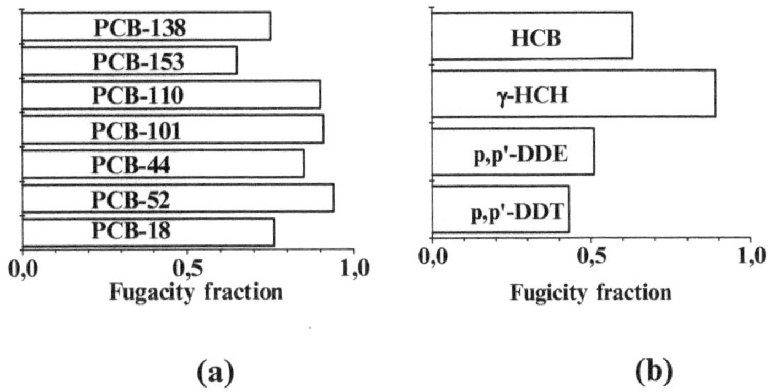

Figure 8. The fugacity fractions for the OCs in sediments of Southern Ocean: in Admiralty Bay, King George Island[39,40] (a) and near Galindez Island, Western Antarctic Peninsula[41] (b)

It is evident that net desorption of all PCB congeners, γ-HCH and HCB occurs from the sediments to seawater, whereas the equilibrium is maintained for DDTs. This would be evidence for contamination of the sediments from possible local anthropogenic sources of PCBs (diesel-powered electrical generators, garbage incinerators, waste disposal sites of the research stations) and local biotic sources of OCs (excrement and carcasses of sea fishes, birds and died algae, and phytoplankton). From other side, the burial of the OCs in shelf sediments is one of main final sinks for the OCs in Antarctica and the oversaturation of the sediments by PCBs congeners, γ-HCH and HCB would be result of their "cold condensation" in Antarctic environment. Here, as in case of soil – air exchange of OCs in Antarctica, accumulating the OCs in sediments after their "cold condensation" in the polar region as well as local anthropogenic and biotic pollutions of the sediments by OCs give similar results.

4. Relation of Climate Change in Western Antarctic Peninsula to the Release of OCs from Retreated Glaciers

Weather records in the Antarctic Peninsula indicate a 2.5°C warming trend in mean annual air temperature over the last 50 years. For example, on Faraday (Vernadsky) station (65° 15′ S, 64° 16′ W) this warming record is +5.7 ± 2.0°C 100 a^{-1}, mainly due to winter warming (+11 ± 9.0°C 100 a^{-1}).[42] Each degree of warming will result in a snow-accumulation rate of 12.5 mm $year^{-1}$.[3] One expected effect of global warming is therefore increased atmospheric cycling of OCs.

We suggest that meltwater from Antarctic ice sheet will release significant quantities of the deposited OCs into coastal ecosystems. To confirm this suggestion we used estimated and predicted total maximal glacier runoffs (GR) on Antarctic Peninsula (5.8 Gt a^{-1} in 1950 year, 20 Gt a^{-1} in 2000 year, and 56 Gt a^{-1} in 2050 year[43]) and obtained the relation between year and the GR value (GR = 0.5 × [year] – 972; $r = 0.968$). Then we estimated possible total amounts of the OCs released from glaciers into coastal waters of Antarctic Peninsula during 1980–2030 years. These amounts are ΣDDT = 760 kg (concentration in snow is taken from Risebrough et al. study[44]), and ΣDDT = 23 kg, ΣPCB = 304 kg and ΣHCH = 4,554 kg (concentrations in snow are taken from Tanabe et al., study[33]). Then, the melting of the glaciers in Antarctic Peninsula under current global warming would be essential source of the OCs secondary emission into the aquatic ecosystems.

5. Conclusions

It was found that heavy PCB congeners (starting from penta-CB), TCDDs, TCDFs and their more heavy congeners, and DDTs would be entirely particle-bounded at typical Antarctic temperature (–30°C). The snow scavenging at low Antarctic temperatures would be very effective way for removal of the OCs from atmosphere, with the exception of HCB and light PCB congeners. The gas-particle partitioning at low temperature greatly increases the gas phase half-lives of the OCs. This effect is more pronounced for heavy PCBs, PCDDs and PCDFs congeners, and for DDTs family.

From sea–air fluxes for HCHs in Southern Ocean it follows that net gas deposition is occurred for both α-HCH and γ-HCH. These data indicate that South Atlantic and Pacific act as global sinks for HCHs in present time. Deep water formation in Weddell and Ross Seas would be effective sink for removal of the OCs from the surface waters.

The snow–air fluxes of OCs demonstrate net deposition of the chemicals into snowpack. Strength of the fluxes falls for HCHs from 1981 to 2000. Antarctic snowpack acts as a global sink for the OCs. The melting of the icebergs calved from the continent would be appreciable source of secondary emission for the OCs into oceanic ecosystems.

The volatilization fluxes of OCs from most soils in Eastern coastal Antarctica would be indication on consequence of their "global condensation" or local "human" and "biotic" contamination in the immobile Antarctic medium.

As revealed by sediment fugacity fractions, net desorption for all PCBs congeners, γ-HCH and HCB occurs from the sediments to seawater, whereas the equilibrium is maintained for DDTs. This would be evidence for contamination of the sediments from possible local anthropogenic and biotic sources of OCs, or result of their "cold condensation" in Antarctic environment.

The glaciers melting in Western Antarctic Peninsula under current global warming would be essential source of the OCs secondary emission into the aquatic ecosystems in the present and in the future.

References

1. *Atlas of the Oceans, v. 6. Antarctic* (Ministry of Defense, Federation of Russian Navy, St. Petersburg, 2005).
2. F. Wania and D. Mackay, Global fractionation and cold condensation of low volatility organochlorine compounds in polar regions, *Ambio* 22, 10–18 (1993).
3. V. G. Aver'yanov, *Antarctic glacioclimatology* (Gidrometeoizdat, Leningrad, 1990).
4. A. Beyer, F. Wania, T. Gouin, D. Mackay, and M. Matthies, Selecting internally consistent physical-chemical properties of organic compounds, *Environ. Toxicol Chem.* 21, 941–953 (2002).
5. D. Mackay, W. Y. Shiu, and K. C. Ma, *Illustrated handbook of physical-chemical properties and environmental fate for organic chemicals*, Vols. 1–5 (Lewis Publishers, Boca Raton, FL, 1995).
6. N. Vulykh and V. Shatalov, *Investigation of dioxin/furan compositions in emissions and in environmental media. Selection of congeners for modelling* (Meteorological Synthesizing Centre – East, Moscow, 2001).
7. Th. E. M. ten Hulscher, L. E. van der Velde, and W. A. Bruggeman, Temperature dependence of Henry's Law constant for selected chlorobenzenes, polychlorinated biphenyls and polycyclic aromatic hydrocarbons, *Environ. Toxicol Chem.* 11, 1595–1603 (1992).
8. T. Harner and T. F. Bidleman, Measurement of octanol-air partition coefficients for polychlorinated biphenyls, *J. Chem. Eng. Data* 41, 895–899 (1996).
9. A. Malanichev, E. Mantseva, V. Shatalov, B. Strukov, and N. Vulykh, Numerical evaluation of the PCBs transport over the Northern Hemisphere, *Environ. Pollut.* 128, 279–289 (2004).
10. L. Liu and Q.-X.Guo, Isokinetic relationship, isoequilibrium relationship, and enthalpy – entropy compensation, *Chem. Rev.* 101, 673–695 (2001).
11. F. Wania, Potential of degradable organic chemicals for absolute and relative enrichment in the Arctic, *Environ. Sci. Technol.* 40, 569–577 (2006).
12. C. M. Roth, K.-U. Goss, and R. P. Schwarzenbach, Adsorption of a diverse set of organic vapors on the bulk water surface, *J. Colloid Interface Sci.* 252, 21–30 (2002).
13. Y. D. Lei and F. Wania, Is rain or snow amore efficient scavenger of organic chemicals? *Atmos. Environ.* 38, 3557–3571 (2004).
14. T. H. Nguyen, K.-U. Goss, and W. P. Ball, Polyparameter linear free energy relationships for estimating the equilibrium partition of organic compounds between water and the natural organic matter in soils and sediments, *Environ. Sci. Technol.* 39, 913–924 (2005).
15. J. A. Platts, D. Butina, M. H. Abraham, and A. Hersey, Estimation of molecular linear free energy relation descriptors using a group contribution approach, *J. Chem. Inf. Comput. Sci.* 39, 835–845 (1999).

16. F. Wania and G. L. Daly, Estimating the contribution of degradation in air and deposition to the deep sea to the global loss of PCBs, *Atmos. Environ.* 36, 5581–5593 (2002).
17. R. Lohmann, W. A. Ockenden, J. Shears, and K. C. Jones, Atmospheric distribution of polychlorinated dibenzo-*p*-dioxins, dibenzofurans (PCDD/Fs), and non-*ortho* biphenyls (PCBs) along a North – South Atlantic transect, *Environ. Sci. Technol.* 35(20), 4046–4053 (2001).
18. K. M. Hansen, C. J. Halsall, and J. H. Christensen, A dynamic model to study the exchange of gas-phase persistent organic pollutants between air and a seasonal snowpack, *Environ. Sci. Technol.* 40, 2644–2652 (2006).
19. E. C. S. Kwok, R. Atkinson, and J. Arey, Rate constants for the gas phase reactions of the OH radicals with dichlorobiphenyls, 1-chlorodibenzo-p-dioxin, 1,2-dimethoxybenzene and diphenyl ether: estimation of OH radical rate constants for PCBs, PCDDs, and PCDFs, *Environ. Sci. Technol.* 29, 1591–1598 (1995).
20. W. W. Brubaker Jr. and R. A. Hites, OH kinetics of polycyclic aromatic hydrocarbons and polychlorinated dibenzo-p-dioxins and dibenzofurans, *J. Phys. Chem.* 102, 915–921 (1998).
21. *Hazardous substances data bank* (HSDB), ChemKnowledge plus Ariel Global View, Vol. 48 (Micromedex Thompson SR, Englewood, 2001).
22. A. P. Altshuller, in: *Advances in Environmental Science and Technology*, edited by J. N. Pitts, L. Metcalf, and D. Grosjean (Wiley, New York, Vol. 10, 1980), pp. 181–219.
23. A. Beyer, F. Wania, T. Gouin, D. Mackay, and M. Matthies, Temperature dependence of the characteristic travel distance, *Environ. Sci. Technol.* 37(4), 767–771 (2003).
24. *Handbook of environmental degradation rates*, edited by P. H. Howard, R. S. Boethling, W. F. Jarvis, W. M. Meylan, and E. M. Mihalenko (Lewis Publishers, Chelsea, MI, 1991).
25. C. J. Halsall, A. J. Sweetman, L. A. Barrie, and K. C. Jones, Modelling the behaviour of PAHs during atmospheric transport from the UK to the Arctic, *Atmos. Environ.* 35, 255–267 (2001).
26. H. Iwata, S. Tanabe, N. Sakal, and R. Tatsukawa, Distribution of persistent organochlorines in the oceanic air and surface seawater and the role of ocean on their global transport and fate, *Environ. Sci. Technol.* 27, 1080–1098 (1993).
27. S. Lakaschus, K. Weber, F. Wania, R. Brun, and O. Schrems, The air-sea equilibrium and time trend of hexachlorocyclohexanes in the Atlantic Ocean between the Arctic and Antarctica, *Environ. Sci. Technol.* 36(2), 138–145 (2002).
28. R. M. Dickhut, A. Cincinelli, M. Cochran, and H. W. Ducklow, Atmospheric concentrations and air-water flux of organochlorine pesticides along the Western Antarctic Peninsula, *Environ. Sci. Technol.* 39, 465–470 (2005).
29. V. Bogillo and M. Bazylevska, Variations of organochlorine contaminants in Antarctica, this issue.
30. R. Lohmann, E. Jurado, M. E. Q. Pilson, and J. Dachs, Oceanic deep water formation as a sink of persistent organic pollutants, *Geophys. Res. Lett.* 33, doi: 10.1029/2006GL025953 (2006).
31. F. Domine, R. Salvatori, L. Legagneux, R. Salzano, M. Fily, and R. Casacchia, Correlation between the specific surface area and the short wave infrared (SWIR) reflectance of snow, *Cold Regions Sci. Technol.* 46, 60–68 (2006).
32. R. P. Schwarzenbach, P. M. Gshwend, and D. M. Imboden, *Environmental organic chemistry* (Wiley, New York, 2003).
33. S. Tanabe, H. Hidaka, and R. Tatsukawa, PCBs and chlorinated hydrocarbon pesticides in antarctic atmosphere and hydrosphere, *Chemosphere* 12(2), 277–288 (1983).
34. S. Lakaschus and O. Schrems, Air – snow/firn transfer of HCHs at Neumayer station/Antarctica, *Abstracts of SETAC-2002.*

35. T. Gouin, D. Mackay, K. C. Jones, T. Harner, and S. N. Meijer, Evidence for the "grasshopper" effect and fractionation during long-range atmospheric transport of organic contaminants, *Environ. Pollut.* 128, 139–148 (2004).
36. T. G. Negoita, A. Covaci, A. Gheorghe, and P. Schepens, Distribution of polychlorinated biphenyls (PCBs) and organochlorine pesticides in soils from the East Antarctic coast, *J. Environ. Monit.* 5, 281–286 (2003).
37. T. Harner, T. F. Bidleman, and D. Mackay, Soil-air exchange model of persistent pesticides in the United States Cotton Belt, *Environ. Toxicol. Chem.* 20, 1612–1621 (2001).
38. K. U. Goss, J. Buschmann, and R. P. Schwarzenbach, Adsorption of organic vapors to air-dry soils: Model predictions and experimental validation, *Environ. Sci. Technol.* 38, 3667–3673 (2004).
39. R. C. Montone, S. Taniguchi, and R. R. Weber, Polychlorinated biphenyls in marine sediments of Admiralty Bay, King George Island, Antarctica, *Marine Pollut. Bull.* 42(8), 611–614 (2001).
40. R.C. Montone, S. Tanihuchi, J. Sericano, R. R. Weber, and W. H. Lara, Determination of polychlorinated biphenyls on Antarctic Macroalgae *Desmarestia sp.*, *Sci. Total Environ.* 277, 181–186 (2001).
41. S. B. Bondar, L. G. Orlova, and V. P. Usenko, Ingredients of chemical pollution in the ecosystem shallow waters of the Argentine Archipelago (area of Ukrainian Antarctic station "Academician Vernadsky"), *Bull. Ukrainian Antarctic Centre* 3, 192–208 (2000).
42. V. E. Timofeev, Climatic indexes of Southern Hemisphere and their relation to tropospheric circulation, *Ukrainian Antarct. J.* 3, 85–92 (2005).
43. D. G. Vaughan, Recent trends in melting conditions on the Antarctic Peninsula and their implications for ice-sheet mass balance and sea level, *Arctic, Antarct. Alpine Res.* 38(1), 147–152 (2006).
44. R. W. Risebrough, W. I. Walker, T. T. Shmidt, B. W. Delappe, and C. W. Connors, Transfer of chlorinated biphenyls to Antarctica, *Nature* 364, 738–739 (1976).

Part IV

REMEDIATION METHODS

REMEDIATION OF CONTAMINATED SOILS AND WATER WITH ORGANIC CHEMICALS BY MEANS OF NATURAL, ANIONIC AND ORGANIC CLAYS

JUAN CORNEJO,* RAFAEL CELIS
Instituto de Recursos Naturales Y Agrobiología de Sevilla, CSIC, Avenida Reina Mercedes 10, P.O. Box 1052, 41080 Sevilla, Spain

Abstract: Clays and "anionic clays" are layered natural and synthetic compounds, respectively, with very special surface properties. Clay minerals are composed of negatively charged aluminosilicate layers whose surface charge is balance by hydrated inorganic cations located in the interlayer space. Smectites are a clay mineral group having swelling or expandable structure which makes the interlamellar space accessible for cationic or polar organic molecules. Anionic clays are positively charged layered double hydroxides (LDHs) where the surface charge is balanced by hydrated inorganic anions in the interlayer space. LDHs have a particularly interesting property, i.e., when heated at 500°C a mixed oxide is formed, which rehydrates in aqueous solution with reconstruction of the original LDH-layered structure. Clay minerals and LDHs are important adsorbents for ionic, ionizable and highly polar organic chemicals because of their charge characteristics and hydrophilic character of their surfaces. In this paper, examples of interactions of these materials with organic molecules are presented. Furthermore, replacement of the inorganic interlayer ions by organic ones has been shown to change the nature of the surface of clays and LDHs from hydrophilic to hydrophobic, increasing the affinity of these minerals for hydrophobic organic compounds. The selective modification of clay minerals with organic cations containing appropriate functional groups is a novel technique to maximize the affinity of the adsorbent for a given chemical. Clays and LDHs modified with organic ions can be used as adsorbents for diverse target molecules. These materials are particularly interesting for decontamination and contamination prevention purposes

* To whom correspondence should be addressed: E-mail: cornejo@irnase.csic.es

E. Mehmetli and B. Koumanova (eds.),
The Fate of Persistent Organic Pollutants in the Environment, 355–368.

in soil and water environments. One example of their application for preventing soil and water contamination is their use for the design of pesticide formulations. Factors influencing the release rate and extent of pesticides from clays and LDHs are still issues which need to be addressed. The characteristics of the mineral, the amount and nature of exchangeable ions, the mineral to pesticide ratio, and the procedure followed to prepare the formulation all affect the interaction of the pesticide with the adsorbent and in turn its release from the formulation.

Keywords: adsorption, clay minerals, contamination, layered double hydroxides, organic pollutants, pesticide formulations, pesticides, soil, water

1. Introduction

Increasing contamination of soils and water by organic pollutants has raised concerns for the protection of the environment and has stimulated research to look for suitable materials to be used as adsorbents, containment barriers, and pollutants stabilizers (Brixie and Boyd, 1994; Zheng et al., 2005). Adsorbent materials are also being proposed for the design of controlled release formulations of organic pesticides to minimize the impact of these compounds once incorporated into the environment (Gerstl et al., 1998; Hermosín et al., 2001; Cardoso et al., 2006).

Natural clays, particularly 2:1-type layered silicates, have very good adsorbent properties due to their low particle size and reactivity of their surfaces. Because of their hydrophilic, negative character, clays are efficient adsorbents for cationic and highly polar organic compounds, whereas their adsorption capacity for highly hydrophobic and anionic organic compounds is usually low (Mortland, 1970; Jaynes and Vance, 1996). Layered double hydroxides (LDHs), also called anionic clays or hydrotalcite-like compounds, can be considered antitypes of 2:1 clay minerals. Naturally occurring hydrotalcite consists of $Mg(OH)_2$ brucite layers with Al-for-Mg substitution $[Mg_{1-x}Al_x(OH)_2)^{x+}]$, which imparts a positive layer charge that is compensated with interlayer carbonate anions. A number of synthetic hydrotalcite-like compounds or LDHs can be easily prepared in the laboratory by varying the nature of the divalent (M^{2+}) and trivalent (M^{3+}) cations, the M^{2+}/M^{3+} molar ratio, and the nature of the interlayer inorganic anion (Inacio et al., 2001). Because the interlayer anions can be exchanged, LDHs are potential adsorbents for anionic organic compounds (Miyata, 1983; Meyn et al., 1990; Ulibarri et al., 1995). Consequently, as a result of their structural characteristics, clays and LDHs have complementary adsorptive properties with regard to the charge characteristics of organic pollutants (Ulibarri et al., 1995; Celis et al., 1999).

The mechanism most widely used to enhance the affinity of clays for hydrophobic, nonionic organic compounds consists of replacing the clay natural inorganic exchange cations with organic cations through ion exchange reactions (Lagaly, 2001). This simple modification changes the nature of the surface from hydrophilic to hydrophobic increasing its affinity for nonionic organic molecules (Mortland et al., 1986; Boyd et al., 1988). Because of their similarity to clays, the inorganic exchange anions of LDHs can also be replaced with organic anions to increase the affinity of these materials for nonionic organic molecules (Celis et al., 2000; Bruna et al., 2006).

In this paper, interaction mechanisms of unaltered and modified clays and LDHs with organic compounds are discussed. Several examples of interactions of these materials with organic molecules are presented to illustrate the potential usefulness of clays and LDHs as adsorbents to prevent and remediate environmental contamination by organic pollutants.

2. Natural and Organic Clays as Adsorbents of Organic Pollutants

2.1. NATURAL CLAYS

Due to isomorphous substitution in their structural framework, most clay minerals, in particular 2:1-type clay minerals, carry a permanent negative charge which, in their natural state, is compensated by exchangeable hydrated inorganic cations. Accordingly, cationic organic compounds are readily adsorbed to clay minerals by ion exchange processes. Thus, different cationic organic pollutants such as chlordimeform, diquat, paraquat and monobuthyltin have been shown to be extensively removed from aqueous solution by clay minerals (Weber and Weed, 1968; Hermosín and Rodríguez, 1981; Hermosín et al., 1993).

Anionic organic compounds can interact with positively charged clay edges or with certain multivalent metal ions at exchange sites, but usually the number of such adsorption sites are very limited. Thus, repulsions between organic anions and the negatively charged aluminosilicate layers predominate and organic anions are often weakly retained by clay minerals. Negligible adsorption by clay minerals has been observed in adsorption experiments with organic anions such as 2,4-D, imazamox and imazethapyr anions (Weber et al., 1965; Celis et al., 1999; Cruz-Guzmán et al., 2005).

TABLE 1. Freundlich coefficients (K_f) for the adsorption of the herbicide thiazafluron on Wyoming montmorillonite saturated with different inorganic cations

Saturating cation	Ionic potential (nm^{-1})	K_f
K^+	7.3	150
Na^+	9.8	122
Ca^{2+}	20.2	80

Nonionic but highly polar organic compounds can extensively interact with interlayer cations either directly or through water bridges. In both cases, the nature of the interlayer inorganic cation greatly influences the adsorption process. For instance, the pyridinium nitrogen atom is directly coordinated to the Cu^{2+} ion but bound by a water molecule to the Mg^{2+} cation in montmorillonite (Farmer and Mortland, 1966; Lagaly, 2001). Similarly, the adsorption of the herbicide thiazafluron in the interlayers of montmorillonite, when occurring by substitution of water molecules hydrating the exchangeable cation, is favored when the exchangeable cation has a small ionic potential (Table 1). In all cases where water bridges are formed between the exchangeable cations and the pesticide, hydrogen bonds play an important role. Hydrogen bonds can also involve OH groups at the clay mineral edges and siloxane oxygen atoms of the silicate layers (Lagaly, 2001). Adsorption of polar molecules on clay minerals may also catalyze their hydrolysis rendering different degradation products. For instance, Fe^{3+} has been shown to increase the adsorption capacity of montmorillonite for *s*-triazine herbicides, such as atrazine and simazine, because the high polarizing power of Fe^{3+} promotes the ionization of water molecules at the clay mineral interlayer and in turn the protonation of triazine molecules, which are readily adsorbed by montmorillonite through a cationic exchange mechanism (Figure 1). In addition, protonated atrazine and simazine are more susceptible of being chemically degraded to hydroxyatrazine and hydroxysimazine than the unprotonated triazine molecules, whereby the presence of Fe^{3+} in the interlayers of montmorillonite has a catalytic effect on the degradation of these herbicides (Laird, 1996; Celis et al., 1997).

The adsorption capacity of clays for nonionic, highly hydrophobic organic compounds is often considered to be very limited due to the highly hydrophilic character of the clay surfaces (Mortland, 1970; Jaynes and Vance, 1996). Nevertheless, significant adsorption of nonionic, very hydrophobic organic compounds, such as phenanthrene, by certain smectitic clays has been reported and related to the presence of "hydrophobic" surface on these clay minerals (Laird, 1996; Hundal et al., 2001). This hydrophobic surface is due to basal oxygens distal from the charge sites in the clay mineral, for which organic compounds can effectively compete with water molecules (Laird, 1996). The presence of hydrophobic regions in smectites varies with the distribution and character of the surface charge sites. Hydrophobic microsites are more abundant in low charge density smectites, because the surface charge sites are widely separated (Laird, 1996).

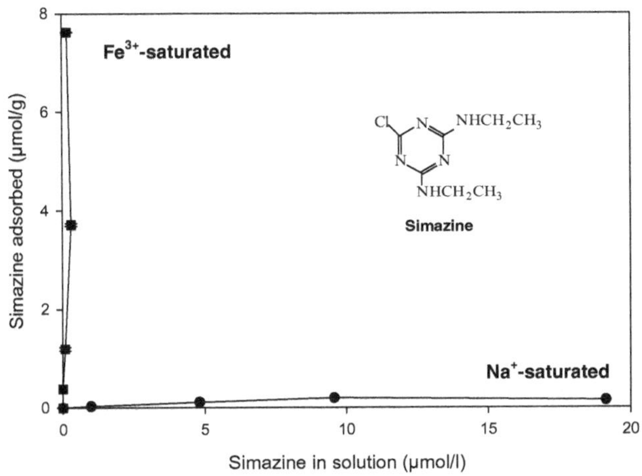

Figure 1. Adsorption isotherms of the herbicide simazine on Na^+- and Fe^{3+}-saturated Wyoming montmorillonite (SWy-2)

2.2. ORGANIC CLAYS

The adsorptive properties of natural clays for organic pollutants can be increased after different physical or chemical treatments, but the mechanism most widely used to enhance the affinity of clays for hydrophobic, nonionic organic compounds (NOCs) consists of replacing the clay natural inorganic exchange cations with organic cations through ion exchange reactions. As mentioned above, because of the hydrophilic, negative character of their surfaces, natural clays, in particular 2:1-type phyllosilicates, are good adsorbents for cationic and highly polar organic compounds, but their adsorption power for nonionic, highly hydrophobic organic molecules is usually low (Mortland, 1970; Jaynes and Vance, 1996). The strong hydration of natural inorganic exchange cations produces a hydrophilic environment at the clay mineral surface that reduces considerably the adsorptive capacity of clays for such compounds. Replacement of natural-exchange cations with organic cations through ion exchange reactions has been shown to change the nature of the surface from hydrophilic to hydrophobic. This simple modification has been proposed for the improvement of the adsorptive properties of clay minerals for hydrophobic organic chemicals (Mortland et al., 1986; Boyd et al., 1988). In the last two decades, there has been an increasing interest in the use of alkylammonium-exchanged clays to remediate environmental contamination by organic pollutants (Boyd et al., 1991; Hermosín and Cornejo, 1992; Brixie and Boyd, 1994; Xu et al., 1997; Carmody et al., 2007).

Adsorption of organic pollutants by alkylammonium-exchanged clays has been mainly attributed to hydrophobic interactions between the organic compound and the alkyl chains of the alkylammonium cation (Jaynes and Boyd, 1991; Lemke et al., 1998). Adsorption of the uncharged fungicide triadimefon by Arizona montmorillonite increases from 0 to >90% after modification with hexadecyltrimethylammonium cations. A major factor determining the adsorptive power of these organoclays is the organic cation arrangement in the clay mineral interlayers, since it determines the amount of space available to host the organic pollutant (Aguer et al., 2000; Celis et al., 2002). This organic cation arrangement is a function of the size of the hydrocarbon chains of the alkylammonium cation, the surface charge density of the clay mineral, and the amount of organic cation in the interlayer (Jaynes and Boyd, 1991; Jaynes and Vance, 1996; Cruz-Guzmán et al., 2005).

Large quaternary alkylammonium cations form an interlayer organic phase that functions as a partition medium for nonionic organic compounds and effectively removes such compounds from water (Boyd et al., 1988; Jaynes and Boyd, 1991; Brixie and Boyd, 1994; Jaynes and Vance, 1996). They can form monolayers, bilayers, pseudotrimolecular layers, or paraffin complexes depending on the surface charge density of the clay mineral and the amount of organic cation incorporated into the clay mineral (Lagaly, 1982; Jaynes and Boyd, 1991; Jaynes and Vance, 1996). In contrast, small quaternary alkylammonium cations, such as tetramethylammonium, do not form an organic partition phase (Jaynes and Boyd, 1991; Brixie and Boyd, 1994; Roberts et al., 2006). They act as nonhydrated pillars that prop open the clay layers exposing the abundant siloxane surface area for adsorption of organic pollutants (Lee et al., 1990; Roberts et al., 2006). Although small organic cations exist as discrete species in both low- and high-charge density smectites, the presence of "free" space available for adsorption of organics decreases with the charge density of the clay and the amount of organic cation present, as a result of the proximity of adjacent organic cations in the clay mineral interlayers. As an example, high-charge Arizona bentonite modified with trimethylphenylammonium (TMPA) cations adsorbed less herbicide hexazinone than TMPA-modified low-charge Wyoming bentonite (Celis et al., 2002). Figure 2 shows different examples of possible arrangements of organic cations in clay mineral interlayers.

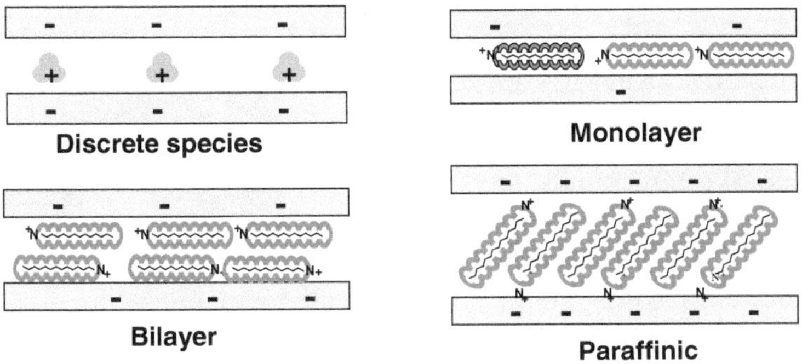

Figure 2. Examples of possible arrangements of organic cations in clay mineral interlayers

Clay minerals can also be selectively modified with organic cations containing appropriate polar functional groups to maximize their affinity for a given organic pollutant (Cruz-Guzmán et al., 2004, 2005). For example, the chemical nature of the interlayer organic cation greatly influences the adsorption of the herbicide simazine by modified montmorillonites (Figure 3). Notably high is the adsorption of simazine by montmorillonite modified with L-carnitine, due to interaction of the weakly basic simazine molecule with the –COOH functionality of montmorillonite-associated carnitine (Cruz-Guzmán et al., 2004).

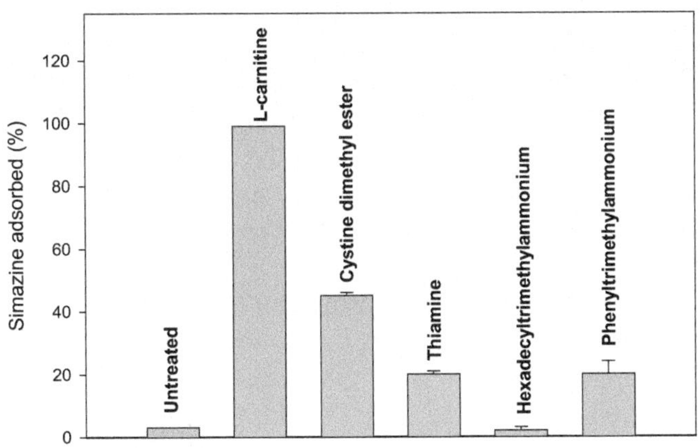

Figure 3. Percentage of simazine adsorbed on Wyoming montmorillonite (SWy-2) modified with different organic cations. The initial simazine concentration was 1 mg/L and the adsorbent to solution ratio was 20 mg:8 ml

3. Adsorptive Properties of Anionic Clays

3.1. INORGANIC ANIONIC CLAYS

Anionic clays, also referred to as layered double hydroxides (LDHs) or hydrotalcite-like compounds, are synthetic materials that can be considered as antitypes of 2:1 pyllosilicates. They have a structural formula $[M^{II}_{1-x} M^{III}_{x}(OH)_2]A^{m-}_{x/n}\cdot nH_2O$, and consist of positively charged brucite-like layers neutralized with interlayer anions, and water molecules occupying the interlayer space. Because these anions and water molecules can be replaced with organic anions and/or polar molecules, LDHs are potential adsorbents for different types of anionic and polar organic pollutants (Hermosín et al., 1996). In addition, the calcined product (500°C) of LDHs is a mixed oxide which has the peculiarity that it can rehydrate from water containing anions, reconstructing the original layered structure of LDHs.

On the basis of their structural characteristics, there are two possible mechanisms to incorporate anionic organic pollutants in the interlayers of LDHs: (i) exchange of the original inorganic anions in LDHs by the anionic organic pollutant; and (ii) reconstruction of the layered structure of the LDH from its calcined product in aqueous solutions containing the anionic organic pollutant (Hermosín et al., 1996; Celis et al., 1999). One interesting example of the anion exchange mechanism is the adsorption of the herbicide MCPA by Mg/Al LDHs (Inacio et al., 2001). At low concentration of MCPA, only the more external exchange sites are involved, whereas internal exchange sites are also occupied at high MCPA/LDH ratios. The adsorption isotherms are S-type with a tendency to L-type at high MCPA equilibrium concentrations. This change from S to L isotherms can be ascribed to changes from external to interlayer adsorption. The extent of adsorption is affected by the pH of the solution, the charge density of the LDH layers, the nature of the original interlayer anions, and the morphology of the adsorbent.

A representative example of the adsorption of anionic organic pollutants by LDHs and their calcined product is the removal of phenolic compounds, such as trichlorophenol (TCP) and trinitrophenol (TNP), from aqueous solution. Adsorption of TCP and TNP by Mg/Al LDH takes place through the exchange of the original carbonate anions by the phenolate anions, whereas adsorption by the corresponding calcined LDH takes place through the reconstruction of the layered structure with incorporation of TCP and TNP anions in the interlayers. The adsorption of TCP and TNP by reconstruction is much greater than that by anion exchange (Table 2). This is due to the fact that carbonate anions are very difficult to replace because they greatly stabilize the LDH-layered structure (Ulibarri et al., 1995).

Adsorption of TCP and TNP by calcined LDH is greater at lower pH levels because the reconstruction process is a proton-consuming reaction (Table 2). Similar high adsorptions by calcined LDHs have been observed for other anionic organic pollutants, such as the herbicides 2,4-D, clopyralid, and picloram (Pavlovic et al., 2005).

TABLE 2. Amount of trinitrophenol (TNP) adsorbed by an Mg/Al LDH and its calcined product at different pH levels. The initial TNP concentration was 1 mM and the adsorbent to solution ratio was 0.1 g:5 ml

Adsorbent	Initial pH	Final pH	TNP adsorbed
	2.0	8.2	60
LDH	6.0	9.4	22
	12.0	12.7	15
	2.0	8.5	2,250
Calcined-LDH	7.0	11.7	1,875
	12.0	12.0	1,350

3.2. ORGANIC ANIONIC CLAYS

The inorganic hydrated anions of LDHs can also be replaced through exchange and reconstruction mechanisms with large organic anions to enhance the affinity of LDHs for hydrophobic organic pollutants (Celis et al., 2000; Bruna et al., 2006). Incorporation of dodecylsulphate and dodecylbenzenesulphonate anions into an Mg/Al LDH renders organo-LDHs with enhanced affinity for the herbicide triadimefon (Celis et al., 2000). Dodecylsulphate-exchanged Mg/Al LDH is also an excellent adsorbent for the pesticides carbetamide and metamitron (Bruna et al., 2006). Although the existing information on the adsorption of organic compounds by organo-LDHs is much more limited than that on adsorption by organoclays, these preliminary results indicate that both organoclays and organohydrotalcites may find application in the removal of hydrophobic organic pollutants from contaminated soil and water.

4. Pesticide Formulations Based on Clays and LDHs

The use of clays and LDHs for the development of pesticide formulations to reduce the environmental impact of the pesticide once incorporated into the environment is one example of the application of these materials for preventing soil and water contamination. Among the various materials proposed as carriers in pesticide formulations, recently there has been a renewed interest in the use of natural soil constituents, like clays, iron oxides or humic acids (Gerstl et al., 1998; Johnson and Pepperman, 1998;

Fernández-Pérez et al., 2000). Clays have unique properties, such as their high specific surface areas associated with their small particle size, low cost, and ubiquitous occurrence in most soil and sediment environments. In addition the possibility exists to modify the clay mineral surface through the incorporation of organic cations in the interlayers to increase the affinity of the clay mineral for a given pesticide and to control the desorption rate once added to the environment. LDHs have also been proved to be efficient supports for the slow release of anionic pesticides, such as picloram, MCPA and 2,4-D (Cardoso et al., 2006).

Different types of pesticide formulations based on organoclays, particularly alkylammonium-exchanged clays, have been proposed (Hermosín et al., 2001; Celis et al., 2002; Undabeytia et al., 2000). However, while much attention has been given to describe the diversity of organoclay–pesticide interactions, information on the factors controlling the release rate and extent of pesticides from organoclay formulations is less abundant. Particularly limited are studies that validate the behavior of organoclay formulations of pesticides under real, field conditions.

One major obstacle of many clay–pesticide formulations assayed has been observed to be the high amount of pesticide not released from the formulation, which would result in soil contamination and would require higher application rates for the same amount of active ingredient to be released. Thus, formulations of hexazinone supported on Fe(III)-montmorillonite released from 75% to only 20% of the herbicide associated with the clay mineral (Celis et al., 2002). Gerstl et al. (1998) also observed that a large portion (35–70%) of the active ingredient remained adsorbed in controlled release formulations of alachlor at the end of release experiments. Similarly, Johnson and Pepperman (1998) found that from 5% to 27% of alachlor and atrazine formulated with bentonite, montmorillonite, kaolinite, and iron (III) oxide was not released, with the greatest retention by bentonite formulations.

To circumvent the above-mentioned limitation, factors influencing the release rate and extent of pesticides from clays and organoclays need to be considered. The characteristics of the clay mineral, the amount and nature of exchangeable cations, the clay to pesticide ratio, and the procedure followed to prepare the formulation all affect the interaction of the pesticide with the adsorbent and in turn the release rate and extent from the resulting formulation. Thus, El-Nahhal et al. (1998) found that organoclay complexes with 0.5 mmol benzyltrimethylammonium/g of montmorillonite gave larger adsorbed amounts and better formulations of alachlor as compared to benzyltrimethylammonium preadsorbed up to the cation exchange capacity of the clay (0.8 mmol/g). The release of fenuron from organoclays is inversely proportional to the adsorbent power of the organoclay and the fenuron–organoclay mixing time (Hermosín et al., 2001). Similarly, loosely bound preparations of hexazinone with hexadecyltrimethylammonium-montmorillonite (i.e., simple mechanical mixtures) lead to greater amounts of herbicide released than formulations where an intimate association was promoted by the use of organic solvent (Figure 4). Hexazinone formulations based on hexadecyltrimethylammonium-montmorillonite display slow release properties in water, retard herbicide leaching through soil columns and maintain a herbicidal efficacy similar to that of the free form of the herbicide, whereas formulations based on phenyltrimethylammonium-montmorillonite release the herbicide instantaneously and do not display slow release properties. High organoclay to herbicide ratios make the interaction of the herbicide with the organoclay more intimate and reduce the release rate of hexazinone as well as its leaching through soil columns. Therefore, the possibility exists to select all these variables to optimize the performance of the formulation for practical applications.

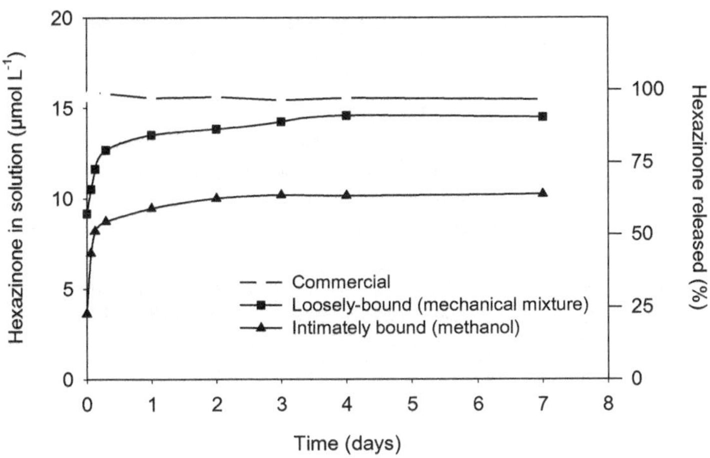

Figure 4. Release kinetics of the herbicide hexazinone into water from an immediately available commercial formulation, and loosely and intimately bound complexes with hexadecyltrimethylammonium–Arizona montmorillonite

Acknowledgments

This work has been supported by the Spanish Ministry of Education and Science with FEDER funds through the project AGL2005-05063-C02-01 and by Junta de Andalucía through the project 2006/RNM-523 and Research Group RNM-124.

References

Aguer, J.P., Hermosín, M.C., Calderón, M.J., and Cornejo, J., 2000, Fenuron sorption by homoionic natural and modified smectites, *J. Environ. Sci Health* **B35**:279–296.

Boyd, S.A., Shaobai, S., Lee, J.F., and Mortland, M.M., 1988, Pentachlorophenol sorption by organo-clays, *Clays Clay Miner.* **36**:125–130.

Boyd, S.A., Jaynes, W.F., and Ross, B.S., 1991, Immobilization of organic contaminants by organo-clays: application to soil restoration and hazardous waste containment, in: *Organic Substances and Sediments in Water*, Vol. 1, R.A. Baker, ed., CRC Press, Boca Raton, FL, pp. 181–200.

Brixie, J.M. and Boyd, S.A., 1994, Treatment of contaminated soils with organoclays to reduce leachable pentachlorophenol, *J. Environ. Qual.* **23**:1283–1290.

Bruna, F., Pavlovic, I., Barriga, C., Cornejo, J., and Ulibarri, M.A., 2006, Adsorption of pesticides carbetamide and metamitron on organohydrotalcite, *Appl. Clay Sci.* **33:**116–124.

Cardoso, L.P., Celis, R., Cornejo, J., and Valim, J.B., 2006, Layered double hydroxides as supports for the slow release of acid herbicides, *J. Agric. Food Chem.* **54**:5968–5975.

Carmody, O., Frost, R., Xi, Y.F., and Kokot, S., 2007, Adsorption of hydrocarbons on organo-clays: implications for oil spill remediation, *J. Colloid Interface Sci.* **305**:17–24.

Celis, R., Cornejo, J., Hermosín, M.C., and Koskinen, W.C., 1997, Sorption-desorption of atrazine and simazine by model soil colloidal components, *Soil Sci. Soc. Am. J.* **61**: 436–443.

Celis, R., Koskinen, W.C., Cecchi, A., Bresnahan, G., Carrisoza, M.J., Ulibarri, M.A., Pavlovic, I., and Hermosín, M.C., 1999, Sorption of the ionizable pesticide imazamox by organoclays and organohydrotalcites, *J. Environ. Sci. Health* **B34**:929–941.

Celis, R., Koskinen, W.C., Hermosín, M.C., Ulibarri, M.A., and Cornejo, J., 2000, Triadimefon interactions with organoclays and organohydrotalcites, *Soil Sci. Soc. Am. J.* **64**:36–43.

Celis, R., Hermosín, M.C., Carrizosa, M.J., and Cornejo, J., 2002, Inorganic and organic clays as carriers for controlled release of the herbicide hexazinone, *J. Agric. Food Chem.* **50**:2324–2330.

Cruz-Guzmán, M., Celis, R., Hermosín, M.C., and Cornejo, J., 2004, Adsorption of the herbicide simazine by montmorillonite modified with natural organic cations, *Environ. Sci. Technol.* **38**:180–186.

Cruz-Guzmán, M., Celis, R., Hermosín, M.C., Koskinen, W.C., and Cornejo, J., 2005, Adsorption of pesticides from water by functionalized organobentonites, *J. Agric. Food Chem.* **53**:7502–7511.

El-Nahhal, Y., Nir, S., Polubesova, T., Margulies, L., and Rubin, B., 1998, Leaching, phytotoxicity, and weed control of new formulations of alachlor, *J. Agric. Food Chem.* **46**:3305–3313.

Farmer, V.C. and Mortland, M.M., 1966, An infrared study of the co-ordination of pyridine and water to exchangeable cations in montmorillonite and saponite, *J. Chem. Soc. A* **3**:344–351.

Fenández-Pérez M., González-Pradas, E., Villafranca-Sánchez M., Flores-Céspedes F., and Ureña-Amate, M.D., 2000, Bentonite and humic acid as modifying agents in controlled release formulations of diuron and atrazine, *J. Environ. Qual.* **29**:304–310.

Gerstl, Z., Nasser, A., and Mingelgrin, U., 1998, Controlled release of pesticides into soils from clay-polymer formulations, *J. Agric. Food Chem.* **46**:3797–3802.

Hermosín, M.C. and Cornejo, J., 1992, Removing 2,4-D from water by organo-clays, *Chemosphere* **24**:1493–1503.

Hermosín, M.C. and Rodriguez, J.L.P., 1981, Interaction of chlordimeform with clay minerals, *Clays Clay Miner.* **29**:143–152.

Hermosín, M.C., Martín, P., and Cornejo, J., 1993, Adsorption mechanisms of monobutyltin in clay minerals. *Environ. Sci. Technol.* **27**:2606–2611.

Hermosín, M.C., Pavlovic, I., Ulibarri, A.A., and Cornejo, J., 1996, Hydrotalcite as sorbent for trinitrophenol: sorption capacity and mechanism, *Water Res.* **30**:171–177.

Hermosín, M.C., Calderón, M.J., Aguer, J.P., and Cornejo, J., 2001, Organoclays for controlled release of the herbicide fenuron, *Pest Manag. Sci.* **57**:803–809.

Hundal, L.S., Thompson, M.L., Laird, D.A., and Carmo, A.M., 2001, Sorption of phenanthrene by reference smectites, *Environ. Sci. Technol.* **35**:3456–3461.

Inacio, J., Taviot-Guého, C., Forano, C., and Besse, J.P., 2001, Adsorption of MCPA pesticide by MgAl-layered double hydroxides, *Appl. Clay Sci.* **18**:255–264.

Jaynes, W.F. and Boyd, S.A., 1991, Clay mineral type and organic compound sorption by hexadecyltrimethylammonium-exchanged clays, *Soil Sci. Soc. Am. J.* **55**:43–48.

Jaynes, W.F. and Vance, G.F., 1996, BTEX sorption by organo-clays: cosorptive enhancement and equivalence of interlayer complexes, *Soil Sci. Soc. Am. J.* **60**:1742–1749.

Johnson, R.M. and Pepperman, A.B., 1998, Release of atrazine and alachlor from clay-oxamide controlled-release formulations, *Pestic. Sci.* **53**:233–240.

Lagaly, G., 1982, Layer charge heterogeneity in vermiculites, *Clays Clay Miner.* **30**:215–222.

Lagaly, G., 2001, Pesticide-clay interactions and formulations, *Appl. Clay Sci.* **18**:205–209.

Laird, D.A., 1996, Interactions between atrazine and smectite surfaces, in: *Herbicide Metabolites in Surface Water and Ground Water*, M.T. Meyer and E.M. Thurman, eds, ACS Symposium Series, Washington, DC, pp. 86–100.

Lee, J.F., Mortland, M.M., Chiou, C.T., Kile, D.E., and Boyd, S.A., 1990, Adsorption of benzene, toluene, and xylene by two tetramethylammonium-smectites having different layer charge densities, *Clays Clay Miner.* **38**:113–120.

Lemke, S.L., Grant, P.G., and Phillips, T.D., 1998, Adsorption of searalenone by organophilic montmorillonite clay, *J. Agric. Food Chem.* **46**:3789–3796.

Meyn, M., Beneke, K., and Lagaly, G., 1990, Anion-exchange reactions of layered double hydroxides, *Inorg. Chem.* **29**:5201–5207.

Miyata, S., 1983, Anion-exchange properties of hydrotalcite-like compounds, *Clays Clay Miner.* **31**:311–319.

Mortland, M.M., 1970, Clay-organic interactions, *Adv. Agron.* **23**:75–117.

Mortland, M.M., Shaobai, S., and Boyd, S.A., 1986, Clay-organic complexes as adsorbents for phenol and chlorophenols, *Clays Clay Miner.* **34**:581–585.

Pavlovic, I., Barriga, C., Hermosín, M.C., Cornejo, J., and Ulibarri, M.A., 2005, Adsorption of acidic pesticides 2,4-D, clopyralid and picloram on calcined hydrotalcite, *Appl. Clay Sci.* **30**:125–133.

Roberts, M.G., Li, H., Teppen, J., and Boyd, S.A., 2006, Sorption of nitroaromatics by ammonium- and organic ammnoium-exchanged smectite: shifts from adsorption/complexation to a partition-dominated process, *Clays Clay Miner.* **54**:426–434.

Ulibarri, M.A., Pavlovic, I., Hermosín, M.C., and Cornejo, J., 1995, Hydrotalcite-like compounds as potential sorbents of phenols from water, *Appl. Clay Sci.* **10**:131–145.

Undabeytia, T., Nir, S., and Rubin, B., 2000, Organo-clay formulations of the hydrophobic herbicide norflurazon yield reduced leaching, *J. Agric. Food Chem.* **48**:4767–4773.

Weber, J.B. and Weed, S.B., 1968, Adsorption and desorption of diquat, paraquat and prometone by montmorillonitic and kaolinitic clay minerals, *Soil Sci. Soc. Am. Proc.* **32**:485–

Weber, J.B., Perry, P.W., and Upchurch, R.P., 1965, The influence of temperature and time on the adsorption of paraquat, diquat, 2,4-D and prometone by clays, charcoal, and an anion-exchange resin, *Soil Sci. Soc. Am. Proc.* **29**:678–688.

Xu, S., Sheng, G., and Boyd, S.A., 1997, Use of organoclays in pollution abatement, *Adv. Agron.* **59**:25–62.

Zheng, W., Papiernik, S.K., Guo, M.X., Dungan, R.S., and Yates, S.R., 2005, Construction of a reactive surface barrier to reduce fumigant 1,3-dichloropropene emissions, *Environ. Toxicol. Chem.* **24**:1867–1874.

THE ROLE OF NATURAL ORGANIC MATTER (HUMIC SUBSTANCES) ON ADSORPTION OF PESTICIDES POSSESSING ENDOCRINE DISRUPTOR ACTIVITY

E. LOFFREDO, N. SENESI*

Dipartimento di Biologia e Chimica Agroforestale e Ambientale, Università di Bari, Via Amendola, 165/A, 70126 Bari, Italy

Abstract: In this paper an overview is provided of adsorption processes of representative persistent organic pollutants (POPs), in particular atrazine, alachlor and *cis*- and *trans*-chlordane pesticides with endocrine disruptor activity, by humic acids (HAs), the most abundant and chemically and biochemically active fractions of humified natural organic matter in soil. Endocrine disruptor compounds (EDCs) are hormone-like substances able to alter, i.e., disrupt, the normal endocrine functions in animals and humans. The estrogenic risk of EDCs is generally related to their distribution and speciation in the various soil phases, in which adsorption processes play a very important role. Adsorption kinetics of atrazine and adsorption isotherms of atrazine, alachlor and *cis*- and *trans*-chlordane onto HA samples isolated from various soil sets were determined using a batch equilibrium method and the HPLC technique with various detectors. Adsorption of atrazine onto HAs occurs rapidly within the first 4 h of contact. Experimental adsorption data were best fitted in a linear isotherm or a nonlinear, L-shaped Freundlich isotherm for atrazine, a linear isotherm for alachlor, and a nonlinear, S-shaped isotherm for *cis*- and *trans*-chlordane. Thus, no limiting adsorption is observed for any pesticide adsorbate onto any HA sorbent examined over the concentration range tested. In general, the adsorption capacity of HAs for chlordane is much larger than that for atrazine and alachlor, which are adsorbed at almost the same extent by HAs. The carboxylic and phenolic hydroxyl group content, the organic free radical concentration, and the aromaticity and humification degrees appear to be the most important chemical properties of HAs affecting atrazine adsorption, whereas it is not clear which compositional, structural and functional features of HAs may affect adsorption of alachlor and chlordane.

* To whom correspondence should be addressed: E-mail: senesi@agr.uniba.it

E. Mehmetli and B. Koumanova (eds.),
The Fate of Persistent Organic Pollutants in the Environment, 369–383.

Keywords: persistent organic pollutants, endocrine disruptor compounds, pesticides, atrazine, alachlor, chlordane, soils, humic substances, adsorption kinetics, adsorption isotherms

1. Introduction

1.1. PERSISTENT ORGANIC POLLUTANTS POSSESSING ENDOCRINE DISRUPTOR ACTIVITY

An “organic pollutant” may be defined as “any organic chemical that is foreign to the natural ecosystem and may adversely affect, either directly or indirectly, the natural physical, chemical, and biological equilibria and processes in the global environment or a portion of it,” e.g., the soil. A persistent organic pollutant (POP) is an organic chemical particularly resistant to chemical and photochemical decomposition and biodegradation, which can persist unaltered in the environment (and soil) for long time spans.

Several POPs, including various classes of pesticides, are known or suspected to act as “endocrine disruptor compounds” (EDCs), i.e., to interfere either directly or indirectly with the normal functioning of the endocrine system of animals and humans by acting as hormone-like substances (Colborn et al., 1993; Kavlock et al., 1996; Lintelmann et al., 2003). Exposure to potential EDCs may interfere with human reproductive function and success (Nolan, 1998), and can lead, especially during certain stages of life, such as development, pregnancy, and lactation, to serious and long-lasting disturbances of endocrine functions (Colborn and Clement, 1992). Despite the high inputs and potential toxicity of EDCs, and the increasing social concern for a reliable risk assessment of the actual exposure of animals and humans to EDCs, until now the attention on this issue has mainly focused on water contamination, whereas relatively few data are available on soil contamination by EDCs.

Potential POPs can reach the soil: (i) by purpose, e.g., various types of pesticides, including herbicides, fungicides, insecticides, and nematicides, currently used in large amounts in common agricultural practices, and belonging to various classes of organic compounds of extremely variable physical and chemical properties; or (ii) incidentally, e.g., chemicals of widely differing classes and properties, including polynuclear aromatic hydrocarbons, phtalic acid diesters, and several organic solvents, detergents and surfactants (Senesi, 1993a, b).

Once on soil surface, the parent POPs and their degradation products may be subjected to various fates, e.g., they can be transported by soil runoff and/or erosion to surface waters, or undergo volatilization and/or photodecomposition, can enter the soil and be leached, eventually reaching the groundwater table, or be subjected to partial or total chemical decomposition and/or biological degradation, uptake by plant roots, and physical and chemical interactions with soil inorganic and organic solid and colloidal components (Senesi, 1993a, b; Senesi and Miano, 1995).

The behavior and performance of POPs in soil is controlled by several factors including the physical and chemical properties of the POP, the thickness, hydrological status and nature of soil, and the type and extent of interactions of the POP with the various inorganic, organic and biological soil components. Most studies have suggested that POPs show affinity for soil organic colloids greater than that for mineral colloids. Thus, the content and nature of soil organic matter, and especially its humified fractions, play a major role in determining the fate of POPs in soil (e.g., Stevenson, 1994; Senesi and Miano, 1995).

The risk assessment of potential environmental hazards of POPs in soil and the definition of possible remediation measures require an accurate evaluation and quantification of the soil response to these compounds. In particular, adsorption of POPs onto soil organic and inorganic components represents a key factor in their fate in soil.

1.2. SOIL HUMIC SUBSTANCES

Approximately 60–70% of the total soil organic carbon occurs as humic substances (HS), which represent the most ubiquitous and widespread natural nonliving organic materials in soil, as well as in all terrestrial and aquatic environments. HS consist of a physically and chemically heterogeneous mixture of relatively high-molecular mass, yellow-to-black-colored organic compounds of mixed aliphatic and aromatic nature, formed by secondary synthesis reactions, i.e., humification, of products of microbial and chemical decay and transformation with recalcitrant residues of biomolecules originated from organisms during life and after death (e.g., Stevenson, 1994). HS are exclusive of undecayed plant and animal tissues, their partial decomposition products, and the soil biomass.

On the basis of their solubility in water solutions at various pH, HS are divided into two main fractions, that are: (i) humic acids (HAs), which are soluble in dilute alkaline solution and are precipitated upon acidification to pH 1–2; and (ii) fulvic acids (FAs), which are soluble at any pH values, even below 1–2 (e.g., Stevenson, 1994). HAs and FAs cannot be regarded as single chemical individuals described by unique, chemically defined molecular formulas, but they can be operationally described by model structures constructed on the basis of available chemical and physicochemical data (e.g., Stevenson, 1994; Senesi and Loffredo, 1999).

A "typical" model macromolecule of soil HA basically consists of aromatic, phenolic, quinonic and heterocyclic "building blocks" that are randomly condensed or linked by aliphatic, oxygen, nitrogen, or sulphur bridges, and bear aliphatic, glucidic, aminoacidic, and lipidic surface chains and chemically reactive functional groups of various nature (mainly acidic, e.g., carboxylic and phenolic, but also alcoholic hydroxyls, carbonyls, etc.) (e.g., Senesi and Loffredo, 1999). FAs generally feature structure and composition less complex than those of HAs, have a lower molecular mass and aromaticity and higher solubility, aliphatic character and content of O-containing functional groups. HA and FA are rich in hydrophilic and hydrophobic sites, exhibit a polydispersed and polyelectrolitic character, possess surface activity, present a relatively open, flexible, sponge-like structure rich of holes (Schnitzer, 1978; Hayes and Swift, 1978), and contain a variable amount of highly reactive organic free radical moieties of prevalent semiquinonic nature (Senesi and Loffredo, 1999). All the above described properties qualify HS, and especially HAs, as privileged natural organic compounds in the interaction with POPs.

1.3. INTERACTIONS BETWEEN HUMIC SUBSTANCES AND PERSISTENT ORGANIC POLLUTANTS

HS may interact with POPs in several ways and thus influence their behavior, performance and fate in soil with important implications in their persistence, immobilization and accumulation, mobility and transport, bioavailability and biotoxicity, degradability, volatilization and leaching, and residue monitoring in soil. In particular, HS are shown to be able to modify water solubility of POPs, exert catalytic activity on some POP transformations, act as photosensitizers in promoting POP photodegradation, and, especially, adsorb and partition POPs (e.g., reviews by Senesi and Chen, 1989; Senesi, 1992, 1993a, b; Senesi and Miano, 1995).

Adsorption represents probably the most important mode of interaction of POPs with soil components, and especially with HS. POPs can be adsorbed to various extents by HS through specific physical and chemical binding mechanisms and forces with varying degrees and strengths. These include ionic, hydrogen and covalent bonding, charge-transfer or electron donor–acceptor mechanisms, dipole–dipole and Van der Waals forces, ligand exchange, and cation and water bridging (e.g., reviews by Senesi and Chen, 1989; Senesi, 1992, 1993a, b; Senesi and Miano, 1995). However, adsorption of nonpolar (hydrophobic) POPs can be better described in terms of nonspecific, hydrophobic, or partitioning processes between soil water and the HS organic phase.

The mechanism and extent of adsorption will depend on the amount and the physical and chemical nature and properties of both the POP and the HS. These include: the molecular structure, the number and type of functional groups, the size, shape and configuration, the polarity, polarizability and charge distribution, and solubility of both HS and POP, and the acidic or basic or neutral, ionic or nonionic, polar or nonpolar nature of POP. The conditions of the medium, such as pH, ionic strength, redox potential, amount of water, will also greatly influence adsorption of POPs onto soil HS. Further, for any given POP, several possible adsorption mechanisms may operate simultaneously and/or in sequence.

The main objective of this paper is to discuss adsorption kinetics and adsorption isotherms of some representative POPs behaving as EDCs onto some selected soil HAs. However, a comprehensive study of adsorption processes of POPs onto HS should include, beside the evaluation of quantitative aspects, the qualitative and molecular aspects and the determination of adsorption mechanisms.

2. Materials and Methods

2.1. CHEMICALS

The POPs with EDC behavior (Keith, 1998) used in this study are the following: (i) atrazine (2-chloro-4-ethylamino-6-isopropylamino-1,2,5-triazine), a nonionizable, slightly basic, polar, low water soluble, widely and abundantly used preemergence and postemergence herbicide for the control of broadleaf and grassy weeds in major crops, especially corn; (ii) alachlor (2-chloro-2′-6′-diethyl-*N*-(methoxymethyl) acetanilide), a nonionizable, moderately water-soluble, extensively used preplant or preemergence herbicide for weed control in several important agricultural crops including corn, soybean and peanut; and (iii) *cis*- and *trans*-chlordane (1,2,4,5,6,7,8,8-octachloro-2,3,3a,4,7,7a-hexahydro-4,7-methano-indene), a nonpolar, low water soluble, high lipophilic, broad-spectrum, nonsystemic insecticide used in land applications against ants, coleopters, cutworms, and termites.

2.2. SOILS AND HUMIC ACIDS

The HAs used in this study were isolated from the surface horizons (0–30 cm deep) of the three following sets of soils: (i) five reference Eurosoils of the first generation, a Vertic Cambisol (E1) from Sicily, Italy, an Orthic Rendzina (E2) from Peloponnesus, Greece, a Dystric Cambisol (E3) from Wales, UK, an Orthic Luvisol (E4) from Normandy, France, and an Orthic

Podzol (E5) from Schleswig-Holstein, Germany, and an alluvial soil (AS) from Northern Germany (Senesi et al., 2003); (ii) two loamy carbonatic Terra Rossa soils (SO1 and SO2) sampled in Puglia, Southern Italy (Senesi et al., 1993); and (c) two sandy soils provided by the Rothamsted Exp. Station, Harpenden, UK, and the Universidade Nova de Lisboa, Lisbon, Portugal (UK1 and PO1, respectively) (Loffredo et al., 1999).

A conventional procedure (Schnitzer, 1982) was used to isolate the HAs. Briefly, each soil sample was extracted three successive times by a 0.5 M NaOH + $Na_4P_2O_7$ solution. After centrifugation, the supernatants were unified and acidified with 6 M HCl to pH 1–2. The precipitated HA was separated by centrifugation, then purified by three alternate alkaline dissolution and acid precipitation treatments. The HA was then washed and recovered with distilled water, dialyzed until free of chloride ions, and finally freeze dried to obtain a fine-powdered HA sample. More details on the isolation procedure of HAs can be found in Senesi et al. (2003).

The principal chemical and physicochemical properties of the HA samples used, which include moisture and ash contents, elemental composition, total acidity, carboxyl and phenolic hydroxyl group contents, the E_4/E_6 ratio, the organic free radical (OFR) content, and Fourier transform infrared and fluorescence spectra have been described in detail previously (Senesi et al., 1993, 2003; Loffredo et al., 1999).

2.3. ADSORPTION KINETICS

Adsorption kinetics were measured only for atrazine on HAs isolated from the first set of five Eurosoils (EU1–EU5) and sample AS. Aliquots of 10 mg of each HA were suspended in 5 mL of a 10 mg L^{-1} aqueous solution of atrazine. The mixtures were magnetically stirred for 0.5, 1, 2, 3, 4, 8, and 16 h, and centrifuged. The supernatant solutions were then analyzed by high-performance liquid chromatography (HPLC) to determine the concentrations of atrazine using the same procedure as for determining atrazine adsorption isotherms as described below. All experiments were conducted in triplicate at a temperature of 20 ± 2°C.

2.4. ADSORPTION ISOTHERMS

Adsorption isotherms of each POP-EDC onto each HA were obtained using a batch equilibrium method in glass centrifuge tubes. Volumes of 5 mL of aqueous solutions at concentrations ranging from 0.5 to 30 mg L^{-1} for atrazine and from 25 to 200 mg L^{-1} for alachlor were added to 10 mg of each HA, whereas aliquots of 30 mL of 7.5% ethanol–water solutions of an equimolar mixture of *trans*- and *cis*-chlordane at concentrations ranging from 0.01 to 1 mg L^{-1} for each isomer were added to 5 mg of each HA. Equilibration was achieved by magnetic stirring of mixtures for 16 h at 20 ± 2°C in the dark. All experiments were conducted in triplicate. The supernatant solutions were then removed by centrifugation at 14,600 or 17,600 *g* for 25 or 20 min, and the equilibrium concentrations, C_e, of free POP-EDC were measured by HPLC using: either (i) a Perkin Elmer Model Series 2 Liquid Chromatograph equipped with an ultraviolet (UV) detector at 254 nm and a Supelcosil LC-8-DB column for atrazine; or (ii) a Thermo Separation Products Liquid Chromatograph equipped with an ultraviolet (UV) detector at 265 nm and Supelco C-18-reverse phase column for alachlor. The mobile phase used was an isocratic solution of: (i) 60% (v/v) acetonitrile and 40% water for atrazine; and (ii) 70% (v/v) acetonitrile and 30% of a 5% (v/v) water solution of acetic acid for alachlor. No measurable effect of wall adsorption and/or degradation of POP-EDCs was observed in control blank experiments run in parallel at each concentration used for each molecule.

In the case of chlordane, the molecule was extracted from adequate aliquots of the supernatant solutions using solid phase extraction (SPE) with Supelclean Envi 18, 3 mL (0.5 g) cartridges, and the Merck Lichrolut Extraction Unit, according to a previously described procedure (Loffredo et al., 1997). Chlordane analyses were then performed with a Fisons HRGC Mega 2 Series Gas Chromatograph equipped with a GC Mega SE-52 column and an electron capture detector (ECD) containing a Ni^{63} beta-emitting radioactive source at operating conditions described previously (Loffredo et al., 1997).

The amount of each POP-EDC adsorbed by each HA was calculated as the difference between the initial and the equilibrium amounts of POP-EDC in solution. Adsorption isotherms were constructed by fitting tentatively experimental adsorption data to a linear model:

$$x/m = K\,C_e \tag{1}$$

a nonlinear Freundlich equation:

$$x/m = K\,C_e^{1/n} \tag{2}$$

and a Langmuir equation:

$$x/m = (KbC_e)/(1 + KC_e) \tag{3}$$

where x/m is the amount of POP-EDC adsorbed in μg g^{-1} or μmol kg^{-1}; C_e is the equilibrium concentration of POP-EDC in solution in μg mL^{-1} or μmol L^{-1}; the constant K is a measure of the magnitude of adsorption, i.e., the adsorption capacity of the HA substrate; the constant 1/n indicates the degree of nonlinearity between solution concentration and amount adsorbed; and b represents the Langmuir adsorption maximum (Giles et al., 1960, 1974; Hamaker and Thompson, 1972).

The magnitude of adsorption was also evaluated by using the distribution coefficient, K_d, which is defined as the mean value of the amount of POP-EDC sorbed, x/m, at each equilibrium concentration, C_e, and can be calculated by:

$$K_d\ (\mathrm{L/kg^{-1}}) = \frac{x/m\ (\mu\mathrm{mol\ kg^{-1}})}{C_e\ (\mu\mathrm{mol\ L^{-1}})} \tag{4}$$

3. Results and Discussion

3.1. ADSORPTION KINETICS

Figure 1 shows a representative example of adsorption kinetics curves of atrazine onto AS-HA. In any case, about 100% adsorption occurs in the first 4 h of contact, which is followed by attainment of an apparent steady-state equilibrium. The initial rapid adsorption phase (about the first hour) would occur on the most reactive and/or accessible sites of HAs, whereas the subsequent adsorption would reflect the involvement of less reactive and/or more sterically hindered sites.

3.2. ADSORPTION ISOTHERMS

On the basis of the calculated correlation coefficients (r) values over the whole concentration range tested, experimental adsorption data of atrazine fit best in a linear, C-type isotherm for samples E1-HA, E2-HA, and E3-HA, and in a Freundlich nonlinear L-shaped ($1/n < 1$) isotherm for samples E4-HA, E5-HA, and AS-HA (Figures 2 and 3, respectively). According to Giles et al. (1960, 1974) a linear, C-type isotherm indicates that a constant partition of the herbicide occurs between the solution and the HA substrate, that is, adsorption is directly proportional to the herbicide concentration in solution and no saturation occurs over the whole concentration range examined. A nonlinear, L-shaped isotherm indicates that the sorbent HA has a moderately high affinity for atrazine in the initial stages of adsorption, which occurs with increasing difficulty as adsorption sites are filled, but never reaches saturation up to the liquid-phase solubility (Giles et al., 1960, 1974).

The correlation coefficients, r, the constants 1/n, and the adsorption and distribution coefficients, K and K_d, of atrazine onto the six HAs examined are listed in Table 1. The greatest extent of adsorption for atrazine is apparently shown by AS-HA whose K and K_d values are from about 2–15

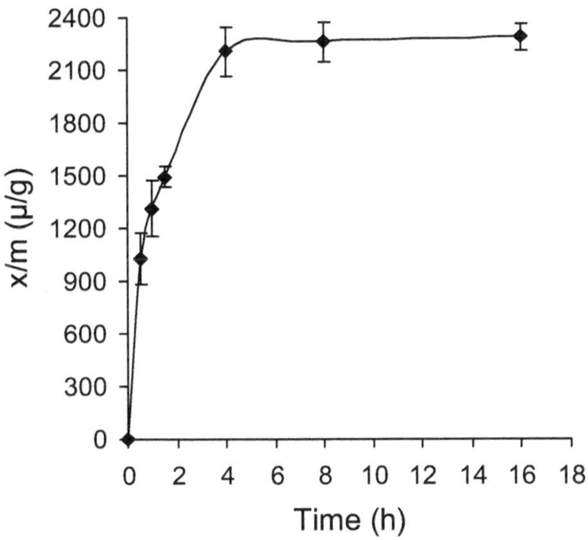

Figure 1. Adsorption kinetics curve of atrazine at 10 mg L^{-1} onto AS-HA. The vertical line on each marker point indicates the standard error ($n = 3$)

times higher than that of any other HA examined. According to the K values, the adsorption capacity of Eurosoil HAs would follow the order: E5-HA > E4-HA >> E1-HA > E3-HA > E2-HA, with the K value of E5-HA about one order of magnitude higher than that of E2-HA. However, the K_d values and their trend are different from those of K values, thus suggesting that the use of K values or K_d values, alone, as an index of adsorption capacity of an adsorbate to a substrate, may be misleading. In general, the adsorption capacity of HAs for atrazine appears to be related positively to carboxylic and phenolic hydroxyl group contents, organic free radical concentration, aromaticity, and humification degree of sorbent HA.

Adsorption data of alachlor onto the two HAs examined, SO1-HA and SO2-HA, fit best in a linear (C-type) model (Figure 4). Thus, also for alachlor no limiting adsorption (saturation) is observed over the concentration range measured. The correlation coefficients, r, and the adsorption and distribution coefficients, K and K_d, of alachlor onto the two HAs examined are shown in Table 2. Both K and K_d values show that the extent of alachlor adsorption is similar onto SO1-HA and SO2-HA, as it can be expected by their similar composition and structural and functional properties. Further, both K and K_d values fall in the range of those measured for atrazine adsorption onto the other set of HAs examined.

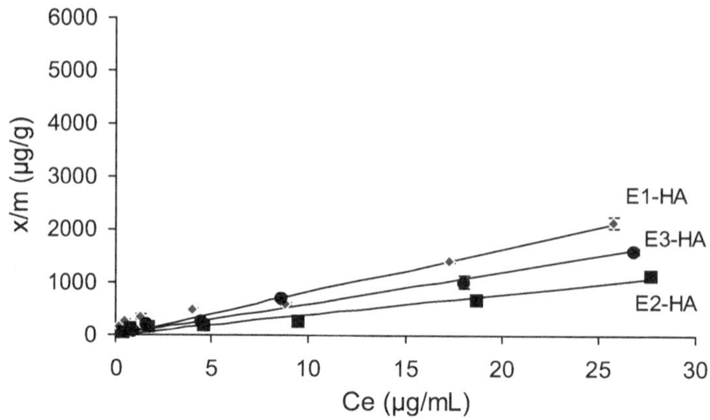

Figure 2. Linear (C-type) adsorption isotherms of atrazine onto the three HA indicated. The vertical line on each marker point indicates the standard error ($n = 3$)

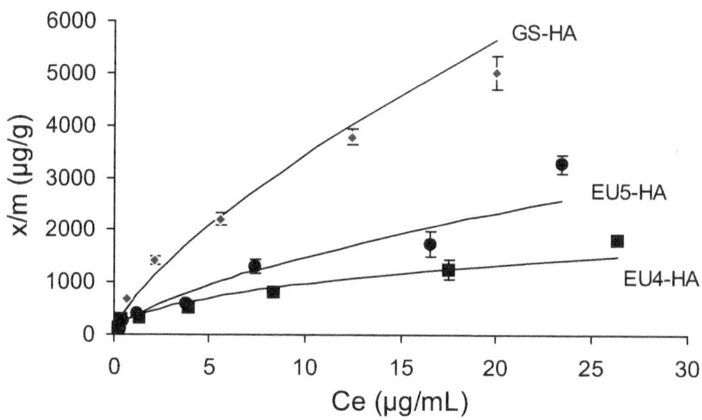

Figure 3. Freundlich (L-type) adsorption isotherms of atrazine onto the three HA indicated. The vertical line on each marker point indicates the standard error ($n = 3$)

TABLE 1. Correlation coefficients, r, adsorption coefficients, K and 1/n, and distribution coefficients, K_d for atrazine onto the six humic acids examined

Origin of HA	r	K ($L\ kg^{-1}$)	1/n	K_d ($L\ kg^{-1}$)
E1	0.970	82.4		321.8
E2	0.982	39.2		80.4
E3	0.985	60.4		78.3
E4	0.989	257.8	0.56	122.9
E5	0.980	335.4	0.65	279.5
AS	0.988	646.1	0.72	592.9

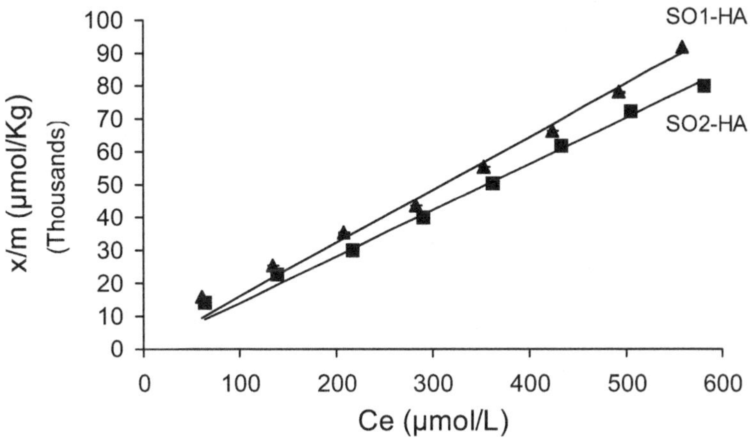

Figure 4. Linear (C-type) adsorption isotherms of alachlor onto the HAs examined

TABLE 2. Adsorption and distribution coefficients, K and K_d, for alachlor onto the humic acids examined

Origin of HA	r	K ($L\ kg^{-1}$)	K_d ($L\ kg^{-1}$)
SO1	0.992	161.2	177.4
SO2	0.995	140.8	152.4

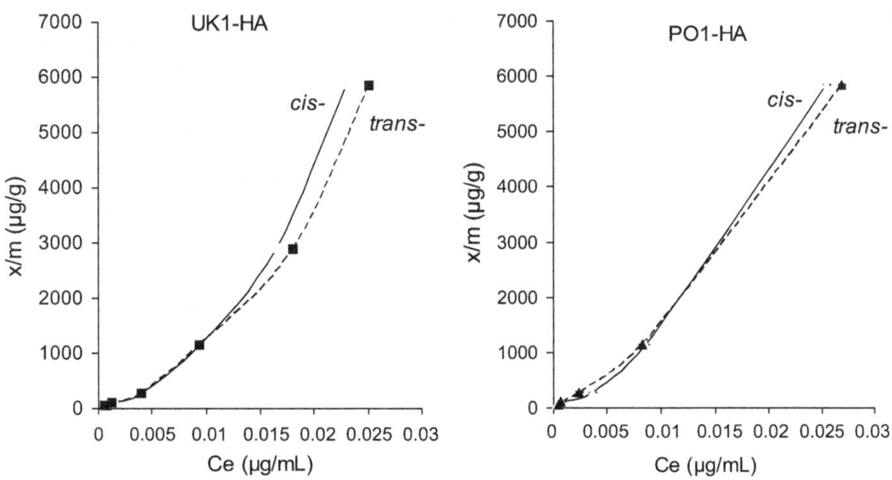

Figure 5. Freundlich (S-type) adsorption isotherms of *cis*- and *trans*-chlordane onto the HA examined

In all cases, adsorption data of *cis*- and *trans*-chlordane onto UK1-HA and PO1-HA best fit a nonlinear, S-shaped ($1/n > 1$) isotherm (Figure 5). This type of isotherms suggests that the adsorbent has a higher affinity for the solvent than for the adsorbate at low concentration, while the opposite effect occurs at high adsorbate concentration when multimolecular layers may be forming on the sorbent surface (Giles et al., 1960, 1974).

The correlation coefficients, r, the constants 1/n, and the adsorption and distribution coefficients, K and K_d, of *cis*- and *trans*-chlordane onto the two HAs examined are shown in Table 3. According to K values, *cis*- and *trans*-chlordane are both much more adsorbed by UK1-HA than by PO1-HA, whereas K_d values suggest the opposite trend, although the two HAs feature similar chemical and physicochemical properties. Further, K values indicate that UK1-HA adsorbs more *cis*-chlordane than *trans*-chlordane, and PO1-HA adsorbs a little more *trans*-chlordane than *cis*-chlordane, whereas very similar K_d values are measured for UK1-HA or PO1-HA for both chlordane isomers. Thus, similar to atrazine, also in the case of chlordane the trend of K values differ from that of K_d values, and the use of K values or K_d values, alone, may lead to uncorrect evaluation of the adsorption capacity of the substrate for the adsorbate. As a whole, adsorption and distribution coefficients of both *cis*- and *trans*-chlordane onto the two HAs examined are much greater (about three orders of magnitude) than those of atrazine and alachlor onto the correspondent HAs. Thus, even if K and K_d values obtained for different HA substrates cannot be quantitatively compared rigorously, the adsorption capacity of HAs for chlordane is generally very much higher than that for atrazine and alachlor.

TABLE 3. Correlation coefficients, r, Freundlich adsorption coefficients, K and 1/n, and distribution coefficients for *cis*- and *trans*-chlordane onto the two humic acids examined

Origin of HA	r	K (L kg^{-1})	1/n	K_d (L kg^{-1})
		cis-chlordane		
UK1	0.978	562,031	1.32	119,356
PO1	0.975	155,919	1.01	154,608
		trans-chlordane		
UK1	0.976	298,638	1.19	119,932
PO1	0.989	174,736	1.10	154,381

4. Conclusions

Adsorption data onto HAs fit best either in a linear or a nonlinear Freundlich isotherm for atrazine, in a linear isotherm for alachlor, and a nonlinear Freundlich isotherm for chlordane. In general, the adsorption capacity of HAs for chlordane is much larger than that for atrazine and alachlor, which are adsorbed at almost the same extent by HAs.

The soil HAs examined are slightly different one from another in their compositional, structural and functional properties, on dependence of soil origin and horizon. These differences are shown to affect variously the adsorption processes of POPs onto HAs. In particular, the carboxylic and phenolic hydroxyl group content, the organic free radical concentration, and the aromaticity and humification degrees appear to be the most important chemical properties of HAs affecting atrazine adsorption, whereas it is not clear which compositional, structural and functional features of HAs may affect adsorption of alachlor and chlordane.

The K_d values and their trend are often different from those of K values, which suggests that the use of K values or K_d values, alone, as an index of adsorption capacity of an adsorbate to a substrate, may be misleading. Thus, in general it is advisable to use both parameters for a better evaluation of adsorption capacity of POPs onto HAs.

The type and extent of interaction may change with time and may ultimately result in the immobilization or reversible retention of the POP which may result in a time-delayed risk of contamination. A better knowledge of the chemical nature and reactivity of HS and a better understanding of the mechanisms of their interactions with POPs are expected to help in the progress of the description of POP fate in soil.

References

Colborn, T. and Clement, C., 1992, *Chemically-Induced Alterations in Sexual and Functional Development: the Wildlife/Human Connection*, Princeton Scientific Publishing, Princeton, NJ.

Colborn, T., vom Saal, F. S., and Soto, A. M., 1993, Developmental effects of endocrine-disrupting chemicals in wildlife and humans, *Environ. Health Perspect.* **101**:378–384.

Giles, C. H., MacEwan, T. H., Nakhwa, S. N., and Smith, D., 1960, Studies in adsorption. Part XI. A system of classification of solution adsorption isotherms and its use in diagnosis of adsorption mechanisms and in measurement of specific surface areas of solids, *J. Chem. Soc.* **3**:3973–3993.

Giles, C. H., Smith, D., and Huitson, A., 1974, A general treatment and classification of the solute adsorption isotherm. I. Theoretical, *J. Colloid Interface Sci.* **47**:755–765.

Hamaker, J. W. and Thompson, J. M., 1972, Adsorption, in: *Organic Chemicals in the Soil Environment*, Goring, C. A. and Hamaker, J. W., eds., Marcel Dekker, New York, pp. 49–143.

Hayes, M. H. B. and Swift, R. S., 1978, The chemistry of soil organic colloids, in: *The Chemistry of Soil Constituents*, Greenland, D. J. and Hayes, M.H.B., eds., Wiley, New York, pp. 179–320.

Kavlock, R. J., Daston, G. P., Derosa, C., Fenner-Crisp, P., Earl Gray, L., Kaattari, S., Lucier, G., Lustre, M., Mac, J. M., Maczka, C., Miller, R., Moore, J., Rolland, R., Scott, G., Sheehan, M., Sinks, T., and Tilson, H. A., 1996, Research needs for the risk assessment of health and environmental effects of endocrine disruptors: a report of the U.S.-EPA-sponsored workshop, *Environ. Health Perspect.* **104** (Suppl.):715–740.

Keith, L. H., 1998, Environmental endocrine disruptors, *Pure and Appl. Chem.* **70**:2319–2326.

Lintelmann, J., Katayama, A., Kurihara, N., Shore, L., and Wenzel, A., 2003, Endocrine disruptors in the environment, *Pure and Appl. Chem.* **75**:631–681.

Loffredo, E., Senesi, N., and D'Orazio, V., 1997, Chlordane adsorption onto soils and pig slurry, *Int. J. Environ. Anal. Chem.* **66**:163–174.

Loffredo, E., D'Orazio, V., Brunetti, G., and Senesi, N., 1999, Adsorption of chlordane onto humic acids from soils and pig slurry, *Org. Geochem.* **30**:443–451.

Nolan, C., 1998, *Ecosystem Research Reports Series No. 29: Endocrine-Disrupters Research in the EU, Report EUR 18345,* Environment and Climate Research Programme, Office for Official Publication of the EC, Brussels, Luxembourg.

Schnitzer, M., 1978, Humic substances chemistry and reactions, in: *Soil Organic Matter*, Schnitzer, M. and Khan, S. U., eds., Elsevier, Amsterdam, pp. 1–64.

Schnitzer, M., 1982, Organic matter characterization, in: *Methods of Soil Analysis, Part 2, Chemical and Microbiological Properties*, Page, B. L., Miller, R. H. and Keeney, R. D., eds., 2nd edn., Agronomy Monograph No. 9, Soil Science Society of America, Madison, WI, pp. 581–594.

Senesi, N. and Chen, Y., 1989, Interactions of toxic organic chemicals with humic substances, in: *Toxic Organic Chemicals in Porous Media*, Gerstl, Z., Chen, Y., Mingelgrin, V. and Yaron, B., eds., *Ecological Studies*, Vol. 73, Springer-Verlag, Berlin, pp. 37–90.

Senesi, N., 1992, Binding mechanisms of pesticides to soil humic substances, *Sci. Total Environ.* **123/124**:63–76.

Senesi, N., 1993a, Organic pollutant migration in soils as affected by soil organic matter. Molecular and mechanistic aspects, in: *Migration and Fate of Pollutants in Soils and Subsoils*, Petruzzelli, D. and Elfferich, F. G., eds., *NATO-ASI Series*, Vol. G32, Springer-Verlag, Berlin, pp. 47–74.

Senesi, N., 1993b, Nature of interactions between organic chemicals and dissolved humic substances and the influence of environmental factors, in: *Organic Substances in Soil and Water: Natural Constituents and their Influence on Contaminant Behavior*, Beck, A. J., et al., eds., Royal Society of Chemistry, London, Ch. 4, pp. 73–101.

Senesi N., La Cava P., Miano T. M., and Brunetti G., 1993, Adsorption isotherms and binding mechanisms of alachlor to soils and soil humic acids, in: *Proceedings of IX Symposium Pesticide Chemistry, Degradation and Mobility of Xenobiotics*, Del Re, A. M., et al., eds., Biagini Editor, Lucca, pp. 309–319.

Senesi, N. and Miano, T. M., 1995, The role of abiotic interactions with humic substances on the environmental impact of organic pollutants, in: *Environmental Impact of Soil Component Interactions. Natural and Anthropogenic Organics*, Vol. I, Huang, P.M.,

Berthelin, J., Bollag, J. M., McGill, W. B., and Page, A. L., eds., CRC Press/Lewis, Boca Raton, FL, pp. 311–335.

Senesi, N. and Loffredo, E., 1999, The chemistry of soil organic matter, in: *Soil Physical Chemistry*, 2nd edn., Sparks, D. L., ed., CRC Press, Boca Raton, FL, pp. 239–370.

Senesi, N., D'Orazio, V., and Ricca, G., 2003, Humic acids in the first generation of EUROSOILS, *Geoderma* **116**:325–334.

Stevenson, F.J., 1994, *Humus Chemistry: Genesis, Composition, Reactions*, Wiley, New York, pp. 1–443.

Part V

HEALTH EFFECTS

PCB-CONTAMINATED AREAS IN KAZAKHSTAN AND ANALYSIS OF PCB IMPACT ON HUMAN HEALTH EXPERIENCE

MARAT ISHANKULOV*
Natural and Technical Department, Eurasian National University named after L.N. Gumilev, 7, Munaitpasov Street, 010010 Astana, Republic of Kazakhstan

Abstract: There are eight "hot spots" polluted with PCB identified in the territory of Kazakhstan. The volume of the PCB-containing equipment is approximately estimated at 980 t, and the total volume of wastes is 250,000 t. A populated area Ablaketka suburb, a part of Ust-Kamenogorsk city polluted with PCB due to the former activities of the Ust-Kamenogorsk Capacitor Plant, was chosen as exemplary in order to assess the threat of the PCB pollution to the health of the population. The research does not exclude a possibility of PCB impact on the exposed population.

Keywords: PCB, pollution, exposure, hot spots, human health, impact, Kazakhstan

1. Introduction

At present Kazakhstan has revealed the polychlorinated biphenyls (PCB)-containing equipment as following: 116 transformers and nearly 50,000 capacitors. The volume of PCB contained in the equipment is approximately estimated at 980 t. The total volume of the PCB-containing wastes reaches 250,000 t. The stocks of the PCB wastes have placed the Republic of Kazakhstan (RK) on the second place among the Eastern and Central European countries, after the Russian Federation.

* To whom correspondence should be addressed. E-mail: marat.ishankulov@mail.ru

E. Mehmetli and B. Koumanova (eds.),
The Fate of Persistent Organic Pollutants in the Environment, 387–403.

2. PCB Hot Spots

As a result of past activity in the sphere of such chemical substances "management" eight PCB-polluted "hot spots" have appeared in Kazakhstan:

- The territory of the Ust-Kamenogorsk Capacitor Plant (UKCP). From 1968 to 1990, TCB had been used as a capacitor fuel in the UKCP. In 1989, the republican commission of the Ministry of Health of the Kazakh SSR while working in the plant banned the TCB and developed an action plan for rehabilitation of the plant territory. TCB leftovers (about 6–9 t) and the soil layer from the plant territory were taken out and buried in the gathering pond. The technology itself was redirected to the saturating agent – DOF produced in Japan.
- The UKCP gathering pond. The plant's gathering pond is located on the mountain side in the opposite part of the city and it constantly accepts new current wastes from the treatment facilities of the plant and effluent pits. The pond is filled with snowmelt waters (the height of water varies from 2 to 6 m, according to various data). The pond is not fenced, there is no decent security, and since it is located above the city, there is a danger of PCB getting into Irtysh River with ground waters. There are wells around the pond but nobody monitors them. The TCB leftovers (about 6–9 t) and the soil layer taken from the plant territory during the rehabilitation works were dumped into the pond. The analysis of the soil layer from the beach and pond water revealed that the PCB concentration on the beach is 12,438 mg/kg, and in the water – 0.19 mg/kg. Thus, the pond presents a threat of air pollution with PCB due to their evaporation during the warm season and of PCB entry into the Irtysh River with the ground waters.
- The territory of the electric power substation of Ekibastuz city. The substation construction began back during the Soviet regime with the purpose of the electric power transmission from the Ekibastuz thermal power plants to the European regions of the Soviet Union and the countries of the Council for the Mutual Economic Assistance. The substation was supposed to rectify the alternating current into direct current with the help of the bank of capacitors. By the time of the Soviet Union collapse, the substation already had about 15,000 capacitors assembled on the open sites of the rectifying substation. During the economic crisis, the population broke and unsealed many capacitors with the purpose of extracting nonferrous metal – copper bars. In 2001, an extraordinary commission was created to eliminate the ecological danger due to the evaporation of the TCB (there is a dacha suburb and the Irtysh-Karaganda canal nearby). In 2002 during

the liquidation works, a new owner of the substation estate dissembled and sealed capacitors with sealing foam. The part of soil contaminated with TCB during the dissembling was taken out and packed into the sacks. Later, the capacitors and sacks with soil were taken and buried in the Semipalatinsk nuclear testing ground. But the actions had not been taken in full. The soil under the docks on which the capacitors were installed was not taken out. The PCB concentration under the docks reaches 26,200 mg/kg; therefore it is necessary to dissemble the docks, take out the soil layer and temporarily bury it either in the closed facility or in the Semipalatinsk nuclear testing ground until the issue of its utilization is solved.

- Workshop for soft cable and footwear compound production of the Pavlodar Chemical Plant
- Derzhavinsk polygon for liquidation of military equipment
- Zhangiztobinski polygon for liquidation of military equipment
- Territories of former military bases in the Northern Balkhash
- Territory of the electrical power substation in Kostanai city

There are no relevant regulations, methodological guidelines, or measures to improve PCB situation in Kazakhstan due to the novelty of the subject.

3. Analysis of PCB Impact on Human Health Experience

A populated territory – Ablaketka suburb, a part of Ust-Kamenogorsk city, polluted with PCB was chosen as exemplary to estimate the significance of the PCB pollution threat to the population health. The PCB pollution of the environment in this suburb is directly connected with the activity of the Ust-Kamenogorsk Capacitor Plant (UKCP) which is situated on its territory. The control territory is a suburb of the Combine of Silk Fabrics (CSF) situated outside the PCB polluted zone in Ust-Kamenogorsk on the opposite bank of Irtysh.

3.1. THE SITUATION WITH POLLUTION IN THE 1980s

The Ust-Kamenogorsk Capacitor Plan has been functioning since 1968. According to the archival data, the UKCP annually discharged 188–227 t of TCB with industrial wastes alone. 12–14 t/year (6–7%) was discharged through the vent systems, the rest amount, more than 85% (mass), was liquid and semiliquid wastes.

The analyses of the samples of precipitation event (snow, water, bottom silt, aquatic vegetation, ground vegetation, and fish from the fish-rearing ponds) were taken through the gas–liquid chromatography method by the experts of the Institute of the Applied Geophysics named after E. K. Fyodorov (Moscow), who carried a complex research of the life support environments (excluding soil). Also, the samples of breast milk were taken from 9 women who did not contact PCB and from 3 women who had the PCB contact.

PCBs were identified in all analyzed samples (Table 1–5). Their content in the wastewater of the capacitor plant varied from 12 to 46 mkg/L. The water in the bayou in the distance of 100 m below the discharge had a very high PCB concentration – 1,770 mkg/L. In the distance of 400 m there was a sudden drop of PCB content. Water in the fishing ponds also was contaminated with PCB. The bottom silt and aquatic vegetation turned out to be powerful accumulators of such hydrophobic compounds as PCB, therefore the bottom silt and the aquatic vegetation had such high levels of their content. The accumulation rates of their content were 10^3–10^4· Being accumulators of persistent organic compounds, the bottom silt became a secondary source of the long-term water pollution.

TABLE 1. TCB and PCB content in the water (wastewater, river water) and bottom silt in the zone of the Ust-Kamenogorsk Capacitor Plant

Site of the sample collection	Date of the sample collection	Unit of measure	TCB	PCB
Wastewater outlet; wastewater	February 1985	mkg/L	280.000	46.000
Wastewater outlet; wastewater	August 1985	mkg/L	190.000	12.000
100 m below the wastewater outlet; water from the branch (stagnant zone)	August 1985	mkg/L	1,770.000	12.000
300 m below wastewater outlet; water from the branch	August 1985	mkg/L	40.000	0.650
Pond entry; water	August 1985	mkg/L	2.100	0.072
Fishing pond No. 3; water	August 1985	mkg/L	0.350	0.015
Irtysh River, 200–250 m below the wastewater outlet; bottom silt	December 1984	mg/kg	1,920.000	185.000
Irtysh River, 300–400 m below the wastewater outlet; bottom silt	February 1985	mg/kg	67.500	154.000
Irtysh River, 300–400 m below the wastewater outlet; bottom water	February 1985	mkg/L	600.000	44.000
Branch of the Irtysh River, northwest 800 m, stagnant water; aquatic vegetation	July 1985	mg/kg	3,250.000	60.500
Branch of the Irtysh River, northwest 800 m, stagnant water; aquatic vegetation	July 1985	mg/kg	564.000	8.300
Branch (below the dam); water	July 1985	mkg/L	1.000	0.350

TABLE 2. TCB and PCB concentration in the snow in the zone of the Ust-Kamenogorsk Capacitor Plant

Number	Site of the sample collection		Concentration (mkg/L)		Precipitation per season (g/km^2)	
	Direction	Distance from the plant, m	TCB	PCB	TCB	PCB
1	Northwest	250	60.40	1.68	5,080.00	140.00
2	Northwest	400	16.97	1.91	790.00	88.00
3	Northwest	410	52.80	1.61	5,717.00	173.00
4	Northwest	450	3.29	0.28	160.50	19,70
5	North–northwest	480	19.90	0.45	1,200.00	19,20
6	North–northwest	500	20.85	0.97	1,590.00	74.10
7	North–northwest	530	40.30	0.76	2,225.00	42.10
8	North–northwest	580	80.64	3.24	3,680.00	147.00
9	West–southwest	500	0.13	0.92	16.50	11,70
10	West–southwes	600	1.43	1.39	91.20	88.60
11	Southeast	300	3.33	0.95	280.00	79.60
12	Southeast	1,000	47.22		6,920.00	
13	East–southeast	250	0.51	4.56	68.30	610.00

TABLE 3. TCB and PCB content in the vegetation selected in the zone of the Ust-Kamenogorsk Capacitor Plant, mg/kg of the dry weight (August 1985)

Site of the sample collection		Plant	TCB	PCB
Direction	Distance from the factory (m)			
North	220	Sage-brush – leaves	31.80	3.76
North	220	Sage-brush – stalk	8.00	1.70
North	220	Sage-brush – root	24.30	8,90
North	220	Poplar – leaves	9.20	1.90
Northeast	330	Poplar – leaves	1.90	0.67
Northeast	1,400	Motley grass	1.60	0.12
Northwest	400	Motley grass	8.20	0.67
Northwest	1,000	Motley grass	1.00	0.18
Northwest	1,000	Willow-herb	8.90	0.86
West	270	Motley grass	4.70	0.67
West	270	Poppy – leaves	0.33	0.22
West	1,000	Maple – leaves	0.57	0.66
Southwest	330	Motley grass	1.20	0.14
Southwest	700	Motley grass	1.30	0,26
Southwest	700	Poplar – leaves	1.18	0.18
Southwest	1,600	Poplar – leaves	0.77	0.47
Southwest	1,600	Sage-brush – all plant	4.20	0.90
Southwest	1,500	Willow-herb	3.40	0.40
West–southwest	800	Raspberry – leaves	1.80	0.19
West–southwest	1,300	Poplar – leaves	0.70	0.22
South	400	Thyme – leaves	0.54	0.12
South	400	Thyme – stalk	0.29	0.08
South	400	Thyme – root	0.91	0.08
Southeast	380	Thyme – leaves	2.10	0.24
Southeast	380	Birch – leaves	3.00	0.71
Southeast	1,400	Raspberry – leaves	0.22	0.11

TABLE 4. TCB and PCB content in the organs and tissues of carp, caught in the ponds and branch of the Irtysh River in Ust-Kamenogorsk (mg/kg of raw weight)

Site of the sample collection	Age	Organ	TCB	PCB
	Pond culture 8–10 km from the source			
1 Fish-rearing pond		Gills	0.260	0.048
	2 years	Muscles	0.160	0.017
	2 years	Liver	0.320	0.032
	2 years	Kidney	0.220	0.016
	2 years	Gonad	0.430	0.016
	2 years	Brains	0.770	0.030
	2 years	Gills	0.480	0.063
	2 years	Skin	0.500	0.017
	2 years	Muscles	0.200	0.034
	2 years	Muscles	0.077	0.010
	2 years	Muscles	0.084	0.012
	2 years	Liver	0.260	0.011
	2 years	Liver	0.290	0.012
	2 years	Gills	0.078	0.032
	2 years	Skin	0.430	0.011
	Underyearling	All fish	0.150	0.020
	Underyearling	All fish	0.220	0.016
	Underyearling	Liver	0.740	0.075
	Underyearling	Brains	0.820	0.093
Irtysh branch		All fish	9.000	22.000

TABLE 5. TCB and PCB content in the breast milk (pg/g of fat)

Number	Date of the sample collection	Age, years	PCB Chlorophen A-56	TCB Chlorophen A-30	ΣPCB, TCB
1	18.02.85	35	20.0	–	20.0
2	18.02.85	24	51.0	–	51.0
3	18.02.85	23	45.0	–	45.0
4	18.02.85	28	39.0	–	39.0
5	18.02.85	27	28.0	–	28,0
6	20.03.85	27	68.0	–	68.0
7	20.03.85	27	50.0	–	50.0
8	20.03.85	27	54.0	–	54,0
9	20.03.85	27	20.0	–	20.0
10*	13.08.85	24	146.0	700.0	846.0
11*	13.08.85	22	145.0	709.0	855.0
12*	13.08.85	23	84.0	348.0	432.0

*Women, who work at the UKCP

3.2. THE CURRENT SITUATION OF THE ENVIRONMENTAL PCB POLLUTION OF THE PLANT'S TRANSFORMERS AND PRODUCTION WORKSHOPS OF THE ENTERPRISE

In 2005 during addressed chemical and analytical works at the UKCP[1] the current functioning Plant's transformers and workshops turned out to be extremely polluted with PCB (Tables 6 and 7).

TABLE 6. PCB content in the leakage from the transformers and plaster

Number	Sample characteristics	Summary concentration (mg/g)
1	Plaster from the walls of the assembly and impregnating workshop of the OS UKCP	9.96
2	Mixture of dust and organic matter from a transformer	0.17
3	Mixture of dust and organic matter from a transformer	0.02

TABLE 7. Isomeric composition of PCB in the samples, mg/g

Chemical	Number of the sample (according to the enumeration in Table 6)		
	1	2	3
Monochloro – PCB	0.0080	0.0000	0.0000
Dichloro – PCB	0.684	0.0000	0.0000
Trichloro – PCB	4.7280	0.0730	0.0360
Tetrachloro – PCB	2.6050	0.0750	0.0106
Pentachloro – PCB	0.9470	0.0230	0.0025
Heptachloro – PCB	0.7730	0.0027	0.0007
Octachloro – PCB	0.0120	0.0009	0.0000
PCB summary	9.9640	0.1750	0.0170

Even during the substandard sample selection (leakage scrape from the crack of the transformer mixed with dust), the submitted samples NN 2 and 3 possess 170 and 20 ppm of PCB which confirms the fact that these transformers are filled with "pure" PCB.

In compliance with the UNEP guidelines,[2] in accordance with the Stockholm Convention on POPs PCB-containing transformers after their detection should be reclassified and ultimately, liquidated. The guidelines also specify that many transformers that have been sold and marked as PCB-free transformers may actually contain them due to the crossed pollution and a transformer marked as a regular-oil transformer may contain PCB in the amount far exceeding 50 ppm (the limit allowed by the Annex A of the Stockholm Convention). In such cases retrofilling is a practical method of PCB reduction down to less than 50 ppm.

A special sample is a sample of plaster wall from the Ust-Kamenogorsk Capacitor Plant taken in the assembly and impregnating workshop on the impregnating site. As a consequence of the long-term pollution, the plaster contains about 10 mg/g of PCB (10 mg/kg or 10,000 ppm) which is 200 times higher than the limit (50 ppm). Thus, the material of the researched production building is subject to the classification as an extremely dangerous one and after liquidation of the building, all necessary measures on liquidation and storage of this material should be taken as wastes of danger class 1.

3.3. THE MODERN SITUATION OF THE ENVIRONMENTAL POLLUTION

In 2003, 114 soil samples in 35 areas adjacent to the Plant were collected to estimate the current soil pollution.[3] 10 samples of the bottom silt were collected and analyzed for the integral estimation of the waterway pollution. In spite of the rehabilitation works carried out in 1990–1991, according to the analysis of the soil from the Plant territory and soil adjacent to it – Ablaketka suburb – the PCB content in the bottom silt (Figure 1) and soil (Figure 2–5) remains very high even almost 20 years later.

Figure 1. Pollution of bottom silt in the vicinity of Ablaketka suburb. Range of variation of the isoconcentrate from 250 to 5,200 mkg/kg

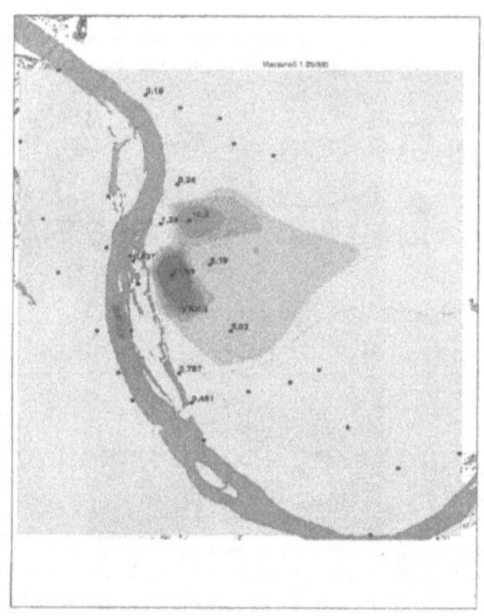

Figure 2. Soil pollution with PCB in the vicinity of UKCP in the layer 0–10 cm. Range of variation of the isoconcentrate from 0.3 to 21.6 mg/kg

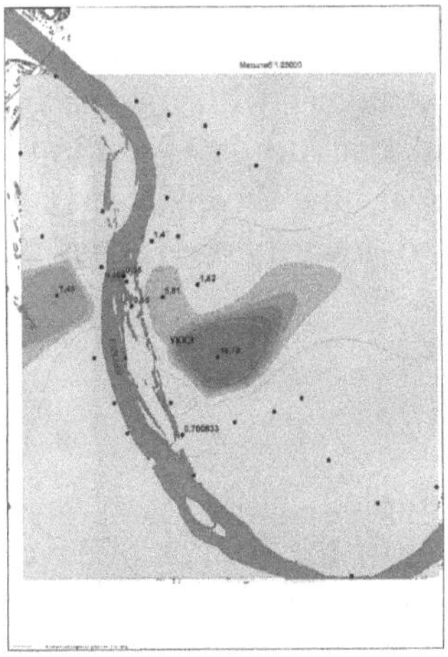

Figure 3. Soil pollution with PCB in the vicinity of UKCP in the layer 10–30 cm. Range of variation of the isoconcentrate from 0.3 to 16.8 mg/kg

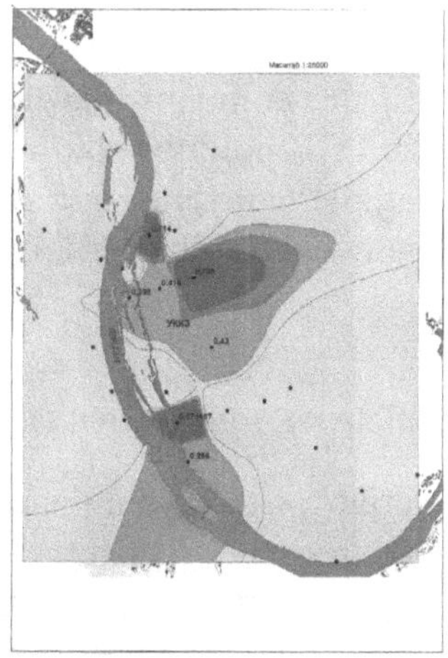

Figure 4. Soil pollution with PCB in the vicinity of UKCP in the layer 30–50 cm. Range of variation of the isoconcentrate from 0.06 to 0.73 mg/kg

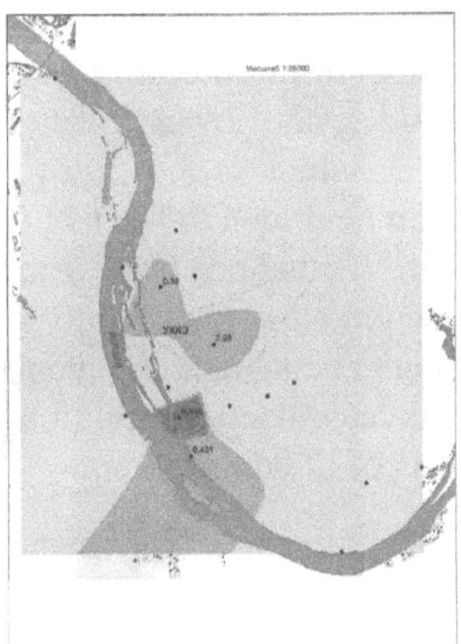

Figure 5. Soil pollution with PCB in the vicinity of UKCP in the layer 50–100 cm. Range of variation of the isoconcentrate from 0.06 to 0.88 mg/kg

PCB content in the soil beyond the Plant territory is 35.7 mkg/kg, direct current of the basement of the assembly and impregnating workshop – 98.3 mkg/kg; the pollutant content in the treatment facilities, in the sludge basins and in the soil where the PCB containers are unloaded makes 1,296 mkg/kg and 1,720 mkg/kg, respectively. Layer-by-layer analysis of the PCB content in the soil allowed to reveal a regularity of the total soil pollution of the Plant territory and its vicinity, with a clear tendency of the gradual penetration of the pollutant into deeper soil layers with the distance from the Plant. The PCB polluted soil in Ablaketka suburb is 15–20 km^2. Unlike the usual sample collection of surface soil, the layer-by-layer method (covering the whole soil profile) allows to identify the actual scope of pollution.

3.4. PRELIMINARY EPIDEMIOLOGICAL ANALYSIS

A basic material for the preliminary epidemiological analysis was the data of the republican cancer register and register of the congenial development pathology with addressed survey as well as the data on the reproductive health and demographic indicators collected in the family outpatient sites of suburbs of Ablaketka and Combine of Silk Fabrics covering 5 years – from 1999 to 2003.

3.4.1. Birth Rate

During a 5-year period the birth rate was permanently lower than the republican indicators. Besides, the lowest birth rate was in Ablaketka: in 1999 this indicator was 4.3, which is 3.4 times lower than the republican rate. Also, the attention can be drawn to the fact that in Ablaketka there is a more frequent birth rate of infants born both with a high-birth weight (over 4,000 g) and low-birth weight (less than 2,500 g) (Figures 6 and 7). Such differentiation of the biological constant (frequent birth of infants with low- and high-birth weight) indicates deep morphofunctional health situation of the generation in conditions of ill environment.

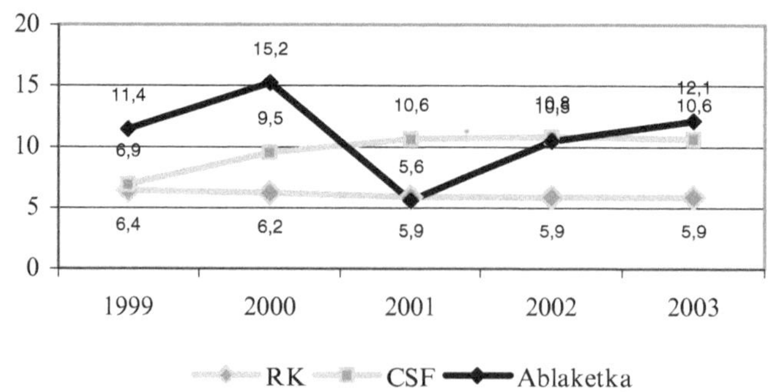

Figure 6. Birth rate of infants born with low-birth weight (%)

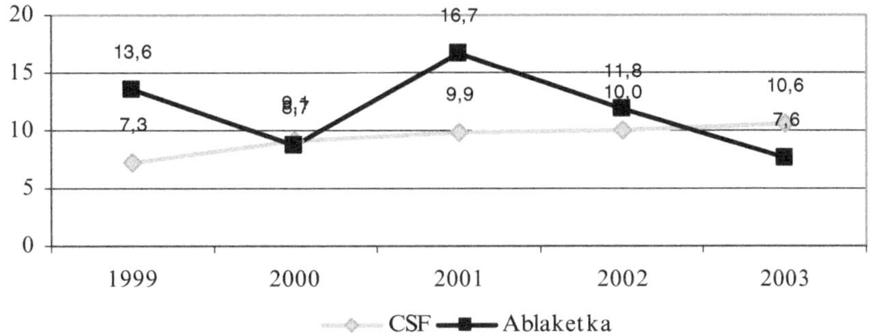

Figure 7. Birth rate of infants born with high-birth weight (%)

3.4.2. Sex Ratio

In recent years, an infant sex ratio has been used as an indicator of the ecological ill-being (boys/girls ratio). Ecological ill-being is accompanied by the increase of girls' rate. In Ablaketka, an average rate for the 5 research years was 0.93 against 1.03 in the area of the Combine of Silk Fabrics. In 2001, the rate decreased down to 0.85. In Kazakhstan during the 5 research years, the sex ratio was 1.06, i.e., 100 girls per 106 boys.

3.4.3. Congenial Malformation

In Ablaketka suburb during 1999–2003 the infants were born with congenial malformation (Figures 8 and 9). This is 3.1 times more than in the Republic of Kazakhstan and 1.5 times more than in the control area of the Combine of Silk Fabrics of Ust-Kamenogorsk.

Among the infants of Ablaketka suburb, the frequency of the congenial malformation during the 5-year period was 8.8%, whereas in the area of the Combine of Silk Fabrics this rate did not exceed 3.3%.

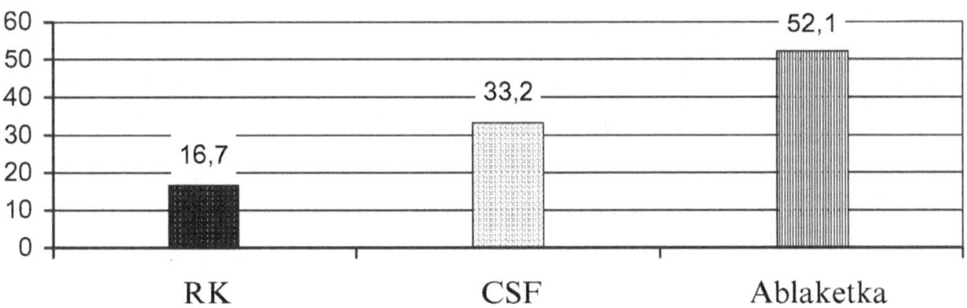

Figure 8. Rate of the congenial malformation (per 1,000 infants)

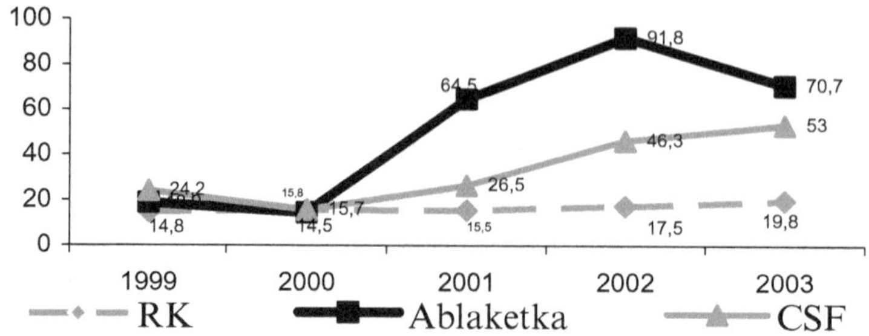

Figure 9. Frequency of the congenial malformation of the infants (per 1,000 infants)

In the PCB polluted Ablaketka suburb a high rate of the sexual development pathology as well as minor development deformities such as cryptorchism and hypospadias was discovered. It is detected that androgen-dependent congenial malformation, to which cryptorchism and Scrotal hydrocele belong, is twice higher in Ablaketka suburb than in the area of the Combine of Silk Fabrics.

In Ablaketka suburb (14.8%) the rate of hormone-dependent congenial malformation is twice higher of the republican and control rate (Figure 10).

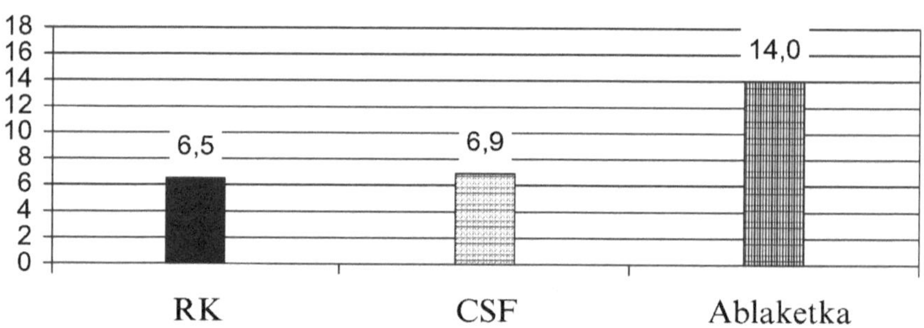

Figure 10. Rate of malignant neoplasms (per 100,000 people) – intensive indices

Since the major part of the congenial malformation diagnosed during the birth is mutation de novo, this reflects the ecological ill-being of Ablaketka suburb.

3.4.4. Malignant Neoplasms

The total rate of the malignant neoplasms (cancer) in Ablaketka suburb is 1.2 times higher of the analogous rate in the control area of the Combine of Silk Fabrics. Among the cancer pathology this rate is significantly lower than in general in Kazakhstan (Figure 11).

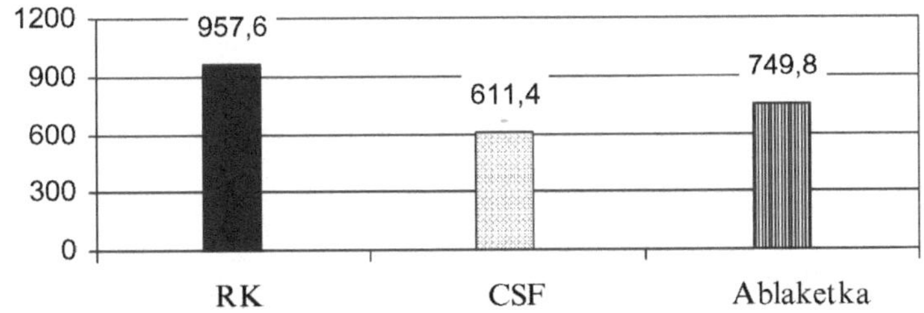

Figure 11. Rate of malignant neoplasms (per 100,000 people) – intensive indices

With the elimination of sex and age influence by the standardization method, the cancer rate among the population of Ablaketka suburb turned out to be 1.96 times higher than in the area of the Combine of Silk Fabrics (Annex 1).

3.4.5. Hormone-dependent Malignant Heoplasms

However, the analysis of the hormone-dependent malignant neoplasms rate revealed that such cancer localization as malignant tumor of female genital sphere, breast cancer, prostate cancer, bladder cancer, and thyroid carcinoma is definitely higher in Ablaketka suburb than in the Republic of Kazakhstan (Figures 12–14, Table 8).

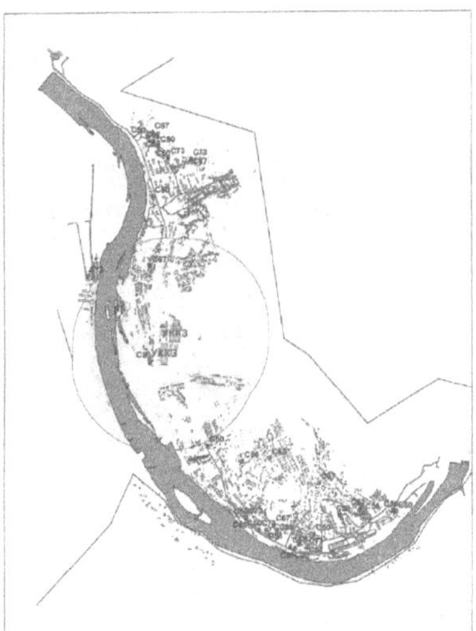

Figure 12. Spread of hormone-dependent types of cancer in Ablaketka suburb
• Morbid event

Figure 13. Spread of malignant tumor of female genital sphere in Ablaketka suburb
• Morbid event

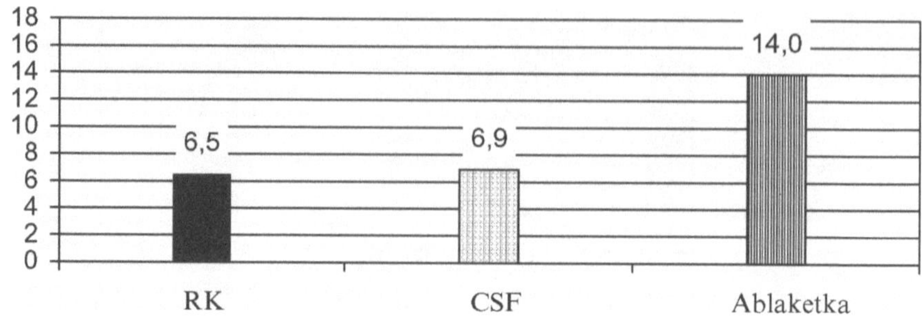

Figure 14. Frequency of the hormone-dependent congenial malformation among infants, %

TABLE 8. Hormone-dependent types of cancer (per 100,000 female and male population

Matter	Malignant tumors of female genital sphere	Breast cancer	Prostate cancer	Bladder cancer	Thyroid cancer
Republic of Kazakhstan	344.9	170.1	35.3	19.1	14.0
Combine of silk fabrics, Ust-Kamenogorsk	235.3	130.7	29.3	17.3	17.3
Ablaketka suburb	355.0	185.9	48.2	57.7	24.7
Balkhash city	431.5	205.7	31.8	33.6	30.9

In Ablaketka suburb the intensive rate of malignant neoplasms is the following:

- Female genial sphere is 1.5 times higher than the average rate in Kazakhstan and 2.2 times higher than the rate in the control area of the Combine of Silk Fabrics.
- Breast cancer is 1.5 times higher than the republican rate and 2.6 higher than the control area.
- Prostate cancer is 1.4 times higher of the republican rate and 1.6 times higher of the control area.
- Thyroid carcinoma is 3.5 times higher of the republican rate and 2.9 higher of the rate of the control area.

Among the hormone-dependent types of cancer, the bladder cancer rate among the population of Ablaketka reaches the highest numbers: it is 3.9 times exceeds the republican rate and 4.3 times higher than the rate of the control area.

Frequency and structure of the malignant neoplasms of breast and female genital sphere were separately analyzed. The malignant neoplasms of cervical cancer is numerically dominating (Figure 15).

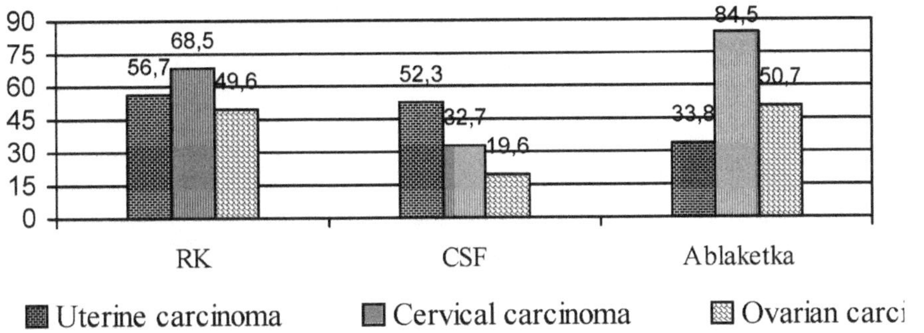

Figure 15. Frequency and structure of the malignant neoplasms of the breast and female genital sphere

Thus, the carried out research does not exclude a possibility of PCB impact on the exposed population health and its reproductive functions. Such data is not enough and every specific case represents not only a national interest: its conclusions can also be transferred to the local conditions. At the same time it is necessary that a more through study of the problem should be performed. It would be a good thing to do cohort analysis, questionnaire design, risk assessment, biomonitoring and so on.

The world experience of dealing with the PCB impact on the environment and population health is very little. One should take into account the fact that it is extremely difficult to find a regularity of the PCB impact on the population health and the environment in the conditions of practical lack of reliable statistical data on the subject, addressed collection of such data, lack of mass basic data on the environmental pollution with persistent organic pollutants. Nevertheless, any attempts to reveal them should be welcome.

References

1. Report on Research and Development "Assessment of the pollution of the industrial enterprises of the Republic of Kazakhstan with polychlorinated dibenzo-para-dioxins and dibenzo-furans as well as pollution of the soil and transformer oil with polychlorinated biphenyls," Team-leader Z.K. Amirova. Ufa, 2006 (unpublished).
2. PCB transformers and capacitors: from service and regulation to reclassification and liquidation/Guidelines UNEP, 2002.
3. Report on analysis of the ecological samples within the framework of the UNDP/GEF Project No. 00013224: Initial Assistance to the Republic of Kazakhstan on obligations taken under the Stockholm Convention on Persistent Organic Pollutants. Team-leader V.A. Solomin, Almaty, 2004 (unpublished).

Annex 1. Incidence rate of the malignant neoplasms (standard indicators on sex and age)

Age, sex	Ablaketka		CSF		Incidence rate per 100,000		Total number	Expectant number of sick	
	Number	Incidence	Number	Incidence	Abla-ketka	CSF		Abla-ketka	CSF
0–4 Male	196	0	491	0	0,0	0.0	687	0	0
Female	165	0	497	0	0,0	0.0	662	0	0
5–9 Male	326	0	734	0	0,0	0.0	1,060	0	0
Female	275	0	674	0	0,0	0.0	949	0	0
10–14 Male	555	1	1,072	0	180.2	0.0	1,627	3	0
Female	512	0	1,111	3	0.0	270.0	1,623	0	4
15–19 Male	575	3	1,657	2	521.7	120.7	2,232	12	3
Female	542	1	1,629	2	184.5	122.8	2,171	4	3
20–24 Male	436	0	1,179	1	0.0	84.8	1,615	0	1
Female	409	1	1,506	1	244.5	66.4	1,915	5	1
25–29 Male	363	0	1,045	2	0.0	191.4	1,408	0	3
Female	347	4	1,351	2	1,152.7	148.0	1,698	20	3
30–34 M ale	339	3	871	3	88.,0	344.4	1,210	11	4
Female	385	6	1,117	0	1,558.4	0.0	1,502	23	0
35–39 Male	321	1	813	0	311.5	0.0	1,134	4	0
Female	360	11	1,127	10	3,055.6	887.3	1,487	45	13
40–44 Male	406	4	1,222	4	985.2	327.3	1,628	16	5
Female	486	12	1,620	9	2,469.1	555.6	2,106	52	12
45–49 Male	441	2	1,234	10	453.5	810.4	1,675	8	14
Female	500	14	1,597	16	2,800.0	1,001.9	2,097	59	21
50–54 Male	358	4	976	2	1,117.3	204.9	1,334	15	3
Female	469	8	1,184	15	1,705.8	1,266.9	1,653	28	21
55–59 Male	229	3	497	5	1,310.0	1,006.0	726	10	7
Female	254	14	615	10	5,511.8	1,626.0	869	48	14
60–64 Male	208	14	377	5	6,730.8	1,326.3	585	39	8
Female	269	24	514	13	8,921.9	2,529.2	783	70	20
65–69 Male	226	5	335	6	2,212.4	1,791.0	561	12	10

Female	372	5	442	8	1,344.1	1,810.0	814	11	15
70–74 Male	135	9	189	5	6,666.7	2,645.5	324	22	9
Female	281	10	376	14	3,558.7	3,723.4	657	23	24
75–79 Male	118	1	55	11	847.5	20,000.0	173	1	35
Female	269	3	386	10	1,115.2	2,590.7	655	7	17
80–84 Male	28	0	27	1	0.0	3,703.7	55	0	2
Female	111	0	194	4	0.0	2,061.9	305	0	6
85–89 Male	10	0	11	2	0,0	18,181.8	21	0	4
Female	49	2	72	1	4,081.6	1,388.9	121	5	2
90–94 Male	1	0	2	0	0.0	0.0	3	0	0
Female	26	0	35	0	0.0	0.0	61	0	0
95–99 Male	0	0	0	0	0.0	0.0	0	0	0
Female	1	0	8	0	0.0	0.0	9	0	0
Total	11,353	165	28,842	177	1,453.4	613.7	40,195	552	282
							1,000	1,373.5	702.7

Indicators	Ablaketka	CSF	Comparison results			
Intensive	1,453.4	613.7	Ablaketka > CSF			
Standard	1,373.5	702.7	Ablaketka > CSF			

IMPACT OF POPs ON THE REPUBLIC OF MOLDOVA ENVIRONMENT AND PUBLIC HEALTH

NICOLAE OPOPOL*
State Medical and Pharmaceutical University "Nicolae Testemitsanu," 165 Stefan-cel-Mare Bul., MD-2004 Chisinau, Republic of Moldova

Abstract: Republic of Moldova POPs problem is determined by chlorineted pesticide and polychlorinated biphenyls (PCBs) and linked to their use in the past. Scientific studies brought out direct negative impacts of excessive pesticide application, as well as of uncontrolled use of PCBs, on both environment health and on human health. It is concluded that there is an urgent necessity to create a comprehensive chemical safety management system in Republic of Moldova and completely to put into practice the National Implementation Plan for the Stockholm Convention on Persistent Organic Pollutants.

Keywords: persistent organic pollutants, dioxins, polychlorinated biphenyls, chlorinated pesticides, polycyclic aromatic hydrocarbons, impact on environment, impact on human health

1. Introduction

Republic of Moldova has some severe public health and environmental problems linked to the POPs. Regarding the national importance and impact, POPs problematic could be classified in three different groups: chlorinated pesticides, polychlorinated biphenyls, and unintended POPs by-products.

Persistent Organic Pollutants (POPs) persist in the environment, bioaccumulation through the food web, and pose a risk of causing adverse effects to human and the environment. They can be transported over long

*To whom correspondence should be addressed. E-mail: nopopol@sanepid.md

E. Mehmetli and B. Koumanova (eds.),
The Fate of Persistent Organic Pollutants in the Environment, 405–424.

distances in the atmosphere, by watercourses and living organisms, resulting in widespread distribution across the earth.

The Stockholm Convention is the United Nations treaty negotiated to eliminate persistent organic pollutants worldwide. Under the Convention, countries commit to reduce and/or eliminate the production, use, and/ or release of the 12 POPs of greatest concern to the global community including nine pesticides, polychlorinated biphenyls (PCBs), and unintended by products such as dioxins. The POPs of relevance to Republic of Moldova in terms of the Stockholm Convention have been found to be obsolete pesticides and PCBs. In terms of obsolete pesticides, one residual effect of the past extensive use of pesticides in Moldova is an extremely high amount of stockpiled obsolete pesticides (are banned, deteriorated, expired, or are not wanted by the current owner). Regarding PCBs, in Moldova they were primary used in the energy sector as dielectric fluids in power installations (transformers and capacitors).

In spite of fact that Republic of Moldova never produced POPs and prohibited POPs pesticides in the early 1970s, over 50 years of overuse of pesticides and other toxic chemicals left a lot of chlorinated pesticides (dichloro-diphenyl-trichloroethane, hexachlorocyclohexane, polychlorinated biphenyls, etc.) and polychlorinated biphenyls. The pesticide load on Republic of Moldova population was of high proportions: professional contact, incorporation with food, water, atmospheric air, etc. Republic of Moldova population was affected by the pesticides, first of all chlorinated ones. Investigations undertaken in the period of intensive pesticide application demonstrated that their load on human organism was significant. The mentioned fact was demonstrated through the pesticide influence on the most vulnerable group of people.

2. The Persistent Organic Substances Negotiated for Worldwide Elimination

POPs are organic compounds that are resistant to environmental degradation through chemical, biological, and photolytic processes. They have been observed to persist in the environment, to be capable of long range transport, bioaccumulation in human and animal tissue, biomagnification in food chains, and to have potential impacts on human health and the environment.

2.1. DICHLORO-DIPHENYL-TRICHLOROETHANE (DDT)

DDT is a chlorinated compound, insecticide and a persistent organic pollutant. DDT and its metabolites are ubiquitous contaminants of the ecosystems. These compounds are readily adsorbed to sediments and soil that can act as sink and as long term sources of exposure. DDT is sequestered in the soil so as to reduce its bioavailability to microorganisms, extractability with solvents, and toxicity to some organisms (Alexander, 1997; Robertson and Alexander, 1998). The high lipid solubility and the low water solubility lead to the retention of DDT and its stable metabolites in fatty tissue. Organisms can accumulate these chemicals from the surrounding medium and from food, and these is clear evidence that the residues of DDT and its metabolites persists in the organism for long period after exposure has ceased. In aquatic organisms uptake from the water is generally more important, whereas in terrestrial fauna food provides the major source (WHO, 1989). Once inside the body, DDT residues accumulate in fatty tissues and are slowly released. The rates of accumulation into organisms vary with the species, the duration, and concentration of exposure, and the environmental conditions. The high retention of DDT metabolites means that toxic effects can occur in organism remote in time and geographical area from the moment of DDT use (WHO, 1989).

DDT and its degradation products are detected in food samples in every country in the world, and absorption through digestive tract is likely the greatest source of exposure for the general population.

DDT transfers freely across the placenta from mother to fetus, and is also excreted in breast milk. Lactation increase the release of DDT to the blood, and therefore the presence of DDT in breast milk raises serious concerns regarding potential health effects in developing infants. DDT and other organochlorine pesticides were measured in samples of breast milk taken from donors of the general population (Волнянская А.; Hooper et al. 1997). Levels of DDT were between 240 and 10,540 ng/g fat, with mean of 1,960 ng/g fat. Levels of DDE were between 75 and 1,030 ng/g fat, with mean of 300 ng/g fat. However, in a few countries average values for total DDT have been reported to range from 1 to 5 mg/L, and the highest value observed was 12.21 mg/L (WHO, 1979). The proportion of the mother's DDT intake that could be recovered from milk varied from 12.6% to 30.2%, and averaged 24.6% in rats receiving the compound in their diet at an average dose of 32.4 mg/kg bw per day. Under these conditions, the dosage of the young was somewhat less than half that of their mothers (Smith, 1991).

Many health effects in humans due to DDT impact were discussed. DDT is an endocrine-disrupting chemical. The carcinogenic risk from DDT was evaluated by WHO in 1991. The final evaluation states that there is inadequate evidence in humans and sufficient evidence in experimental animal of the carcinogenicity of DDT, and that therefore DDT is possibly carcinogenic to humans (group 2B) (IARC, 1991).

2.2. HEXACHLOROCYCLOHEXANE (HCCH)

HCCH is a synthetic chemical that exists in eight chemical forms called isomers; one of these forms, commonly called lindane (γ-HCCH), has been widely used as an insecticide. HCCH does nor occur as a natural substance.

HCCH present in soil can leach to groundwater, adsorb to soil particulates or volatilize to atmosphere. Owing to the persistence of HCCH, rapid bioconcentration takes place in invertebrates, fish, birds, and humans. γ-HCCH can be detected in air, water, soil, sediment, aquatic, and terrestrial organisms and food throughout the world, although the concentrations in these different compartments are generally low and are gradually decreeasing. Humans are exposed daily through food, and γ-HCCH can has been found in blood, adipose tissue, and breast milk (Health Risks of Persistent Organic Pollutants from Long-Range Transboundary Air Pollution, 2003).

More than 90% of human of exposure to all HCCH isomers originates from food sources, particularly those that are animal-based (WHO, 1991). Historically, the largest source of HCCH released to the air was agricultural use of the pesticide lindane (γ-HCCH). In the early 1980s, mean concentrations of γ-HCCH in human adipose tissue were very high due to high exposure form different sources. Anyway, more than 90% of human exposure to all HCCH isomers originates from food sources, particular those that are animal-based (WHO, 1991). The intake of lindane in food decreased significantly in the 1990s.

HCCH is absorbed following oral exposure in humans and animals. Various studies have demonstrated the rapid absorption of γ-HCCH from the gastrointestinal tract. The distribution of HCCH isomers in humans and animals is primarily in the adipose tissue but also in the brain, kidney, muscle, lungs, heart, spleen, liver and blood (Siddiqui et al., 1981). Placental transfer of HCCH in humans has been well documented (Saxena et al., 1981). Numerous studies have demonstrated hepatic (Oldiges et al., 1983), neurotoxic (Muller et al., 1981), immunotoxic (Banergee et al., 1996), and reproductive (Dalsenter, 1997) effects of lindane.

2.3. HEXACHLOROBENZENE (HCB)

HCB is a chlorinated hydrocarbon industrial chemical. Although it is not currently manufactured as a commercial end-product, i.e., if formed as a waste product in the production of several chlorinated hydrocarbons such as tetrachloroethylene, trichloroethylene, and carbon tetrachloride. HCB is contained as a contaminant in some pesticides. The presence of HCB in the environment is also due to its previous use as a fungicide (Beyer, 1996). Another minor source of HCB releases to the air comes from the use of pyrotechnic mixtures that produce obscurant screening smokes used by the military to obscure vision and hide targets.

HCB is banned as a chemical for production and use. However, HCB also occurs as combustion by-product, and its emission and reemission from soil and water are still relatively high in some regions. HCB is one of the most persistent environmental pollutants. If released in atmosphere, HCB exists primarily in the vapor phase and degradation is extremely low. Physical removal of HCB from the air may occur via washout or snowfall, or through dry deposition. If released to water, HCB is partitioning from the water column into sediment and suspended particulate matter. The half-life of HCB is very long: from 2.7 to 5.7 years in surface water and from 5.3 to 11.4 years in groundwater.

The bioaccumulation of HCB in organisms is high. The highest levels of HCB were found in adipose tissue, adrenals, bone marrow and liver (Foster, 1993; Smith et al., 1987).

The critical effects induced by HCB in experimental animals comprise both nonneoplastic and neoplastic effects. The WHO Task Group on Environment Health Criteria for HCB (WHO, 1997) concluded that the available data were sufficient to develop guidance values for nonneoplastic and neoplastic effects. Most data on the effects of HCB on humans originate from accidental poisoning that took place in Turkey in 1955–1959, in which more than 600 cases of porphyria cutanea tarda were identified (Health risks of POPs from long-range transboundary air pollution, 2003). Also, there have been case reports of workers developing porphyria cutanea tarda as a result of direct contact with HCB (Currier et al., 1980).

2.4. HEPTACHLOR

Heptachlor is a nonsystemic contact insecticide, used primarily against insects. Heptachlor is present as an impurity and is also breakdown product of the pesticide chlordane. The use of heptachlor has been banned or severely restricted in many countries around the world and current environment concentrations are therefore principally the result of environment recycling of historical use of the compound.

Heptachlor is quite persistent in soil, where it is mainly transformed to its epoxide. The latter is very resistant to further degradation. Heptachlor and heptachlor epoxide bind to soil particles and migrate very slowly (WHO, 1993b). The persistence of heptachlor, combined with a high partition coefficient, provides the necessary conditions for it to bioconcentrate in terrestrial and aquatic foodchains.

Heptachlor is readily absorbed via all routes of exposure, and readily metabolized to heptachlor epoxide by mammals. Heptachlor epoxide is metabolized slowly and is the most persistent metabolite; it is mainly stored in adipose tissue, but also in liver, kidney, and muscle.

Animal studies have reported effects on the liver, kidney and immune and nervous systems from oral exposure of heptachlor. There is evidence that heptachlor and heptachlor epoxide are associated with infertility and improper developing of offspring (ATSDR, 1993; WHO, 1996).

2.5. POLYCHLORINATED BIPHENYLS (PCBs)

PCBs are mixtures of more than 200 individual chlorinated compounds used as coolants and lubricants in transformers, capacitors, and other electrical equipment because they do not burn and are good insulators. There are no known natural sources of PCBs.

PCBs do not readily break down in the environment and can travel long distances in the air and be deposited in areas far away from were they were released. In water, a small amount of PCBs may remain dissolved, but most stick to organic particles and bottom sediments. PCBs also bind strongly to soil. PCBs can be taken up by small organisms and fish in water and can accumulate in fish and marine mammals, reaching levels that may be thousands of times higher than in water. The international Agency for Research on Cancer (IARC) has determined that PCBs are probably carcinogenic to humans.

PCBs entered the air, water, and soil during their manufacture, use, and disposal, as well as from accidental spills and leaks during their transport, and from leaks of fires in products containing PCBs. PCBs can still be released to the environment from hazardous waste sites; illegal or improper disposal of industrial wastes and consumer products; leaks from old electrical transformers containing PCBs; and burning of some wastes in incinerators.

At present, the major source of PCBs exposure for the general population appears to be as a consequence of the redistribution of PCBs previously introduced into the environment. Their redistribution involves volatilization from soil and water into atmosphere, with subsequent transport in air and removal from atmosphere through wet or dry deposition, and then revolatilization. The concentration of PCBs in precipitation range from 0.001 to 0.25 μg/L. Since the volatilization and degradation rates of PCBs vary among the different congeners, this redistribution leads to an alteration in the composition of PCBs mixtures in the environment.

2.6. DIOXINS AND DIOXIN-LIKE POLYCHLORINATED BIPHENYLS

Polychlorinated dibenzo-*p*-dioxins and polychlorinated dibenxfuranes (PCDD/PCDFs). PCDD/PCDFs or dioxins consist of two groups of tricyclic aromatic compounds with similar chemical and physical properties. The number of chlorine atoms in each molecule can vary from one to eight. The number of chlorine atoms and their positions are of utmost importance for the toxicological potency of each congener. PCDD/PCDFs have never been produced intentionally, except for pure substances used as references in analytical and toxicological research. They have also been shown to be formed in the environment by volcanoes, forest fires, etc.

Primary sources of essential environmental contamination with PCDD/PCDFs was determined by the production and use of organic chemicals containing chlorine (herbicides, polychlorinated biphenyls, etc.), metal industry, paper industry, waste incineration, and in a less extent by home heating. Secondary sources of PCDD/PCDFs, their reservoirs, are those matrices where they are already present, either environment or as product. Produce reservoirs include pentachlorphenol-treated wood, transformers containing polychlorinated biphenyls, sewage sludge. Reservoirs in the environment are, for example, landfills and waste dumps containing soils from former chemical production and contaminated sediments.

PCDD/PCDFs are very persistent and they very intensively are absorbed on to particles in air, soil and sediment and accumulate in fat-containing tissues. Owing to such properties, PCDD/PCDFs are able to remain in the environment for a long time (Health risks of POPs from long-range transboundary air pollution, 2003). The high lipid solubility and the low water solubility lead to the retention of PCDD/PCDFs, PCBs and their metabolites in fatty tissues. Protein binding may also contribute to their retention. The rates of accumulation into organisms vary with the species, the duration and concentration of exposure, and the environment conditions.

The high retention of PCDD/PCDFs and PCBs, including their metabolites, implies that toxic effects can occur in organisms spatially and temporally remote from the original release. PCDD/PCDFs are today found in almost all compartments of the global ecosystem in at least trace amounts. They are ubiquitous in soil, sediments and air. Most human background exposure to dioxins and polychlorinated biphenyls occurs through the diet, with food of animal origin being the major source, as they are persistent in the environment and accumulate in animal fat. Once absorbed, the compounds are readily distributed to all body compartments, where the storage rate is proportional to the fat content of the organ. The main rout of excretion is via faeces, urine, and breast milk. Excretion through breast milk results in transfer to breast-fed infants, who therefore are highly exposed. There is also transfer across the placenta, thus causing fetal exposure. Perinatal exposure is a major concern with regard to human health effects, even at present background exposure levels.

As PCDD/PCDFs and coplanar PCBs are believed to act through a common toxicological mechanism, a toxic equivalency factor concept has been established. The human exposure to PCDD/PCDFs and PCBs results primarily from transfer along the pathway: atmospheric emissions → air → deposition → terrestrial/aquatic food chains → human diet. Information from food surveys in industrialized countries indicates a daily intake of PCDD/PCDFs in the order of 50–200 pg I-TEQ/person per day for a 60 kg adult, or 1–3 pg I-TEQ/kg bw per day. If dioxin-like PCBs are also included, the daily total TEQ intake can be higher by a factor of 2–3.

A plethora of effects have been reported from multiple studies following exposure to PCDD, PCDFs and polychlorinated biphenyls. Among the most sensitive endpoints are: endometriosis, developmental neurobehavioural (cognitive) effects, hearing loss, developmental reproductive effects (sperm count, female urinogenital malformations) and immunotoxic effects, both adult and developmental. The 2,3,7,8-TCDD isomer has been shown to be carcinogenic in several long-term studies at multiple sites in several species and in both sexes. Several PCDDs, PCDFs, non-ortho and mono-ortho PCBs have also been show to be tumor promoters (Health risks of POPs from long-range transboundary air pollution, 2003).

Epidemiological studies on people exposed in connection with the accident in Seveso generated valuable information. Excess risks were observed for ovarian and thyroid cancers and for some neoplasia of the haematopoietic tissue. The relative risk for all cancers combined in the most highly exposed and longer-latency subcohorts is 1.4 (Bertazzi et al., 1998). Studies on noncancer effects in children have indicated neurodevelopmental delays and neurobehavioural effects, including neonatal hypotonia.

2.7. PENTACHLOROPHENOL (PCP)

One major use of PCP is a wood preservative, primarily applied to protect timber from fungal wood-boring insects, but it has been also used as a preharvest defoliant in cotton, as a general preemergence herbicide, and as a biocide in industrial water systems. Volatilization (between 30% and 80% of the applied) from treated wood poles and other outdoor wood occurs. At present, the most extensive use of PCP is the production of the ester, pentachlorophenyl laurate, which is less toxic by an order of magnitude. Production of PCP in European countries is ceased now. PCP can be formed during the incineration of chlorine-containing waste material.

Technical grade PCP has historically contained dioxins and hexachlorobezene as manufacturing by-products (Health risks, 2003).

Increasing concentration tends to lengthen the process of biodegradation, probably owing to the possible toxicity of PCP for bacteria. PCP released into the atmosphere from treated wood can be transported to surface waters and soils via wet and dry deposition. Atmospheric PCP is transformed via photolysis; the compound may slowly undergo free radical oxidation with an estimated half-life of approximately 2 months (ATSDR, 2001).

The bioaccumulation of PCP in aquatic organisms is high. Bioconcentration factor of 100–1,000 have been reported by a number of authors (Bude et al., 1985; Devillers et al., 1996, etc.). The bioconcentration depends on the pH and increases with falling pH.

PCP is rapidly and completely absorbed by the digestive tract. After entering the blood, the higher concentrations of PCP are in the liver, kidneys, and brain, but the bioaccumulation remains very low. Regarding human health effects, the experimental data related to PCP are well documented for the oral low-dose chronic exposures and indicate: impact on the liver characterized by biochemical, functional and histopathological changes; impact on immune system; and significant alteration of thyroid hormone levels at exposures of 1–2 mg/kg bw per day.

2.8. POLYCYCLIC AROMATIC HYDROCARBONS (PAHs)

PAHs are a group of compounds composed of two or more fused aromatic rings. Most direct releases of PAHs to the environment are to atmosphere, both from natural and from anthropogenic sources, with emissions from human activities predominating. The primary natural sources of airborne PAHs are forest fires and volcanoes. The residential burning of wood is the largest source of atmospheric PAHs, the others being industrial power generation, incineration, the production of asphalt, coal tar and coke, petroleum catalytic cracking, etc. PAHs are present in the atmosphere in the gaseous phase or absorbed to particulates. Photolysis is the most important factor in the decay of particle-sorbed PAHs in the atmosphere (Berg et al., 2001).

Aquatic organisms that metabolize PAHs to little or no extent, such as algae, mollusks and the more primitive invertebrates accumulate high concentrations of PAHs, whereas organisms that metabolize PAHs to a great extent, such as fish and higher invertebrates, accumulate little or no PAHs. The concentration of PAHs in vegetation is generally considerably lower than that in soil. Organisms at higher trophic levels in food chains show the highest potential for biotransformation (WHO, 1998b). PAHs are highly lipid soluble. In humans, the major routs of uptake of PAHs are through the lungs and the respiratory tract after inhalation of PAH-containing aerosols or particulates, the gastrointestinal tract after ingestion of contaminated food or water, and the skin as a result of contact with PAH-containing materials.

The acute toxicity of PAHs appears to be moderate or low. Benzo[α]anthracene, dibenzo[α,h]anthracene, and naphthalene were embriotoxic. WHO (1987) considers that the most appropriate indicator for the carcinogenic PAHs in air seems to be benzo[α]pyrene. PAHs have generally been reported to have immunosuppressive effects. Most of the 32 PAHs covered by the International Programme on Chemical Safety (IPCS) in 1998 (WHO, 1998b) are genotoxic, 26 compounds are or are suspected to be carcinogenic. As binding of electrophilic PAH metabolites to DNA is thought to be a key step in the initiation of cancer, measurement of DNA adducts could be an indicator of exposure to PAHs and also of cancer risk.

2.9. POLYCHLORINATED TERPHENYLS (PCTs)

PCTs are chlorinated aromatic compounds that are structurally and chemically similar to PCBs. The chemical and thermal stability and electrical properties of PCTs made them favored for several industrial uses: as caulking compounds, vapor suppressants, sealants, carbonless copying paper, waxes, fire retardants, plasticizers, hydraulic fluids, and lubricants.

PCTs are resistant to both biodegradation and photodegradation. Owing to their lipophylicity and stability, PCTs can accumulate in organisms by means of the food chain (de Boer, 2000). PCTs have been identified in river sediment, in tissues of some animals, fishes, birds etc. High doses of PCTs have been reported to stimulate the hepatic microsomal enzyme system in test animals and in vitro systems, to increase levels of cytochrome P-450, aniline hydroxilase, aminopyridine-*N*-demethylase, esterases, and to reduce levels of glucose-6-phosphatase and aromatic hydrolase (de Boer, 2000). Some earlier data show evidence of fatty degeneration in the liver (WHO, 1993b). Some PCTs were many times more estrogenically active in rats than some PCBs (de Boer, 2000). Data on health effects of PCTs in humans were not found in the available literature.

2.10. POLYBROMINATED DIPHENYL ETHERS (PBDEs)

PBDEs are described as novel environmental contaminants. Although more than 200 congeners can theoretically be produced, the commercial PBDEs usually include a limited number of them. PBDEs have been detected in a variety of environmental matrices, including air, sediment and biota.

Despite the fact that many PBDEs are detectable in terrestrial samples (soil, sludge), it appears as though the aquatic environment represents the greatest potential for human exposure, especially waterways receiving direct input from industrial sources. In women reporting the highest frequency of fish consumption (5–6 meals per week), breast milk concentrations of PBDEs were significantly higher (1.7 ng/g lipid) compared to those reporting consuming only 1–2 fish meals per week (0.77 ng/g lipid). The developing fetus and breast-fed infant are considered to be the main "at risk" groups from potential adverse effects due to exposure to PBDE congeners found in commercial penta-brominated diphenyl ether mixtures.

2.11. POLYBROMINATED DIBENZO-P-DIOXINS AND POLYBROMINATED DIBENZOFURANS (PBDD/PBDFs)

PBDD/PBDFs consist of two groups of tricyclic aromatic compounds with similar chemical and physical properties. The number of bromine atoms and their positions are important for the toxicological potency of each congener. PBDD/PBDFs exist as intentional by-products in chemical processes, but can also be formed during various combustion processes and photolytic degradation of PBDFs and bromphenols (WHO, 1998a). All 2,3,7,8-substituted PBDDs and PBDFs show the same type of biological and toxic response as PBDD/PBDFs. Generally, their physicochemical properties suggest similarities to PCCD/PCDFs. Therefore, they would most probably accumulate in carbon- or fat-rich compartments.

Based on the mentioned similarities with PCCD/PCDFs, it is possible that PBDD/PBDFs could resist degradation, bioaccumulation and be transported through the atmospheric air and watercourses.

2.12. SHORT-CHAIN CHLORINATED PARAFFINS (SCCPs)

SCCFs are chlorinated derivates of n-alkanes, having carbon chain length ranging from 10 to 38 and a chlorine content ranging from about 30–70% by weight. There are over 2,000 of commercial products. The very low solubility in water and low vapor pressure of SCCFs would predict low mobility, but monitoring data in Sweden and the UK indicate widespread low level of contamination in water, sediments, aquatic and terrestrial organisms and even commercial foods (Government of Canada, 1993). Chlorinated paraffins are generally considered persistent. In the aqueous phase, rates of hydrolysis, photolysis with visible or near UV radiation, oxidation and volatilization are significant under ambient temperatures (Government of Canada, 1993). Nevertheless, some studies have shown that degradation by microorganisms is possible.

Bioaccumulation factors for SCCFs homolog groups in western Lake Ontario trout range from 21,250 to 114,444 (Muir et al., 2000). The European Union assessment also describes several accumulation studies in fish and mollusks (Fisk et al., 1996; Madeley and Thompson, 1983).

There is no information available on the effect of repeated exposure to SCCFs in humans. In animals the liver and thyroid were identified as target in the oral studies. Limited information in bacteria indicates that SCCFs are not mutagenic. In rodent carcinogenicity studies, the chlorinated paraffins produced toxicologically significant, dose-related increase in the incidence of several tumor types.

In such a way, there is a limited database regarding environmental fate, levels in food and toxicology of SCCFs. The European Union risk assessment report (European Commission, 2000) concludes with that there is no significant risk to humans exposed via the environment. Compared to other chlorinated organic pollutants, SCCFs seem to have lower acute and chronic toxicities.

3. Overview of Persistent Organic Pollutants in Republic of Moldova Environment

Some severe Republic of Moldova' public and environmental problems are determined by the POPs. Regarding the national importance and impact, POPs problematic is determined, first of all, by pesticides and polychlorinated biphenyls and linked to the intensive their use in the past. The pesticide load on Republic of Moldova inhabitants was of high proportions – through professional contact and through incorporation with food, water and atmospheric air. Moldova inhabitants were affected by the pesticide chemical compounds, inclusive of chlorinated compounds.

3.1. PESTICIDE USE

In the last decades of the 20th century in Republic of Moldova the average pesticide pressure on soil has been very high (the highest was 19.7 kg/ha of active substance), and it exceeded many times the average amount used in the former USSR (Moldova national chemical management case study, 2006). The intensive pesticide use has begun long ego. The necessity of intensive use was determined by the agriculture specialization in growing the monocultures. The pesticide application was a continuous process during the entire vegetation period. As long as more affected were perennial plants, they where treated 12–16 times per season. The intervals among treatments were insufficient for the pesticide degradation and that fact determined increasing environment pollution. At the same time pesticides were largely used in the private sector. The process of pesticides use on the private plots for a long time was accompanied by multiple violations of toxic substances management regulations, such as:

- Incorrect storing of pesticides (in incidental places)
- Incorrect pesticides application, frequent treating the cultivated plants from neighborhood which have had not to be treated
- Performing the pesticide treatments by occasional people, including adolescents, expectant mothers, and persons suffering from chronic diseases
- Not respecting the waiting period from the last treatment up to the harvesting
- Delivering the pesticides to private persons untrained in pesticide treatment methods and correct their stocking
- Employment by the peasants of different pesticide packing means (sacks, sheet iron, or plastic containers, etc.) for water, feed, and food storing
- Wide utilization of DDD and HCCH for domestic animals and fowls parasite control, etc.

Pesticide utilization (stocking, storing, transport, treatment, etc.) on the whole application process was characterized by the facilitate peasant access to all pesticide groups despite of preparative form, specific action, toxicity, etc.

Epidemiological investigations conducted in 1980–1985 denoted a negative impact of the exceeding and long time pesticide use on the public health. The results were communicated to decision factors and public. A decision to decrease the level of pesticide use was adopted in 1987. Further the doses of pesticide use went down.

Together with the USSR collapse, the collective farmers were dissolved. In the majority of villages the pesticide storehouses were abadoned and they remained without owner and control in spite of the fact in storehouses were certain quantities of pesticides.

In spite of the fact that Republic of Moldova never produced POPs and prohibited POPs pesticides in the early 1970s, over 50 years of overuse of pesticides and other toxic chemicals left, as was estimated in a damaging legacy: more than 5,650 t of obsolete and prohibited pesticides, with more than 1,700 t remain scattered all over the country and stored in 347 ramshackle storage facilities and over 4,000 t buried in a 1974 built landfill, including 560 t of DDT (Environmental impact assessment and environmental management plan, 2005).

The stockpiles of obsolete pesticides are a continuous threat to the health of thousands of people.

3.1.1. Human Exposure to the Pesticides

DDT and its metabolites are identified in foods in all countries of the world. The recent investigations show the frequency and pollution intencity are mentained (Laden et al., 1999). It was established that DDT incorporation is more considerable in the developing countries. În 1996, WHO evaluated the DDT daily exposure for a large number of countries and estimated the dietary risk, comparing the Estimated Daily Intake with Acceptable Daily Intake.

The general frequency of pesticide residue finding out in food samples in 1984–1989 years in Republic of Moldova was from 5.6% (1984) up to 19.8% (1990). The share of samples with the exceeding maximal admitted concentrations was 0.7 (1984) – 1.9% (1988), in 1990 being 1.26%. The more contaminated were fruits (1.7–4.0%), tins (0.4–8.6%), dairy products (0.1–1.6%), meat (0.03–0.8%). For example, in apples in 1987 were found simultaneously 11 pesticide molecule, five from which exceeded maximal admissible concentrations.

The residual content of HCCH and DDT in foods drawn from Moldova collective farms in the same period of time was high. DDT contamination varied from 1.9% to 6.7% and was, in comparison with the other pesticides, was more frequent.

It is known that DDT easy penetrates the placenta barrier from mother to foetus. Being incorporated in organism, the DDT residues are accumulating in adipose tissues. The lactation is reducing pollutant transition in blood, as well as its metabolization which is conducting to the transition of DDT in milk and child body.

In conditions when in mother's body were incorporated definite amounts of pesticide residues, the last pass with the milk in child's body. This process starts simultaneously with the outset of suckling. It should be mentioned that DDT and DDE are found in concentrations much higher in maternal milk in comparison with cow milk (Health Risks, 2003). One of such investigations was undertaken in Republic of Moldova as well (Volneanschi and Romanciuc, 1992).

The obtained results are of high interest even now. There were investigated 84 specimens of maternal milk drawn from two rural villages in which the total media pesticide use was 6.3 kg/ha (in the first) and 18.8 kg/ha (in second), and respectively 0.12 and 0.26 kg/ha of organochlorine pesticides. Maternal milk turned out to be contaminated with HCH and DDT. The frequency of both pesticides finding out was quite high (96.2% in the first and 87.5% in second locality). In such a way practically all children were constantly incorporating significant amounts of organochlorine pesticides with the maternal milk. The average concentrations of the HCH and DDT in milk were higher in the second locality, the element which confirm that the pesticide loading on human body was increasing at one time with the level of pesticide use in agriculture.

3.1.2. The Pesticide Pollution Impact on Human Health

Epidemiological investigations carried out in previous years in Republic of Moldova demonstrated the baneful influence on people health, children first of all, of the excessive pesticide application. It was proved the noxious impact expression on children depends, first of all, on the total pesticide application level. Immune system of a number children and adolescent was oppressed. A direct relationship of the immune deficiencies with the pesticide use was established. Different grade of retard was brought to light in a significant number of children both in biological and physical development. For example, in areas with 6.0 kg/ha pesticide use the share of children with biological delay made 9.7% when in areas with 20.0 kg/ha it rosed up to 31.2%. The integrated health estimation revealed an evident change for the worse in health indices of chilrden in settlments were was a multiannual use of high level of pesticide doses in agriculture. Simultanneously with the increasing of pesticide intencity it was noticed a reducing of the cildren share in the genuine and satisfatory health groups and the increasing of the share those with the health defficiences.

In the whole, due to the longtime, excessive and unjustified pesticide use in agriculture the health of at least two people generations in rural areas of Republic of Moldova was affected. Actually those generations are of fertile age and bring to the life a new generation with the undermined health or already ouverfulfilled the fertile age being charaterized by high indeces both of chronic morbidity and mortality in ableboded age.

3.2. POLYCHLORINATED BIPHENYLS

No data are available as to when PCB-containing equipment was first imported into Moldova from Russia but it was probably sometime in the early 1950s. The management of PCBs and PCB containing equipment in Moldova appears to be a joint but obscure, responsibility of both Government and the private sector. Given the criticality of transformers and capacitors to the energy sector, the fear of PCB leakage in the food sector and the relationship of electrical transmission equipment to the future economic performance of the country, control programs for PCBs became a more higher priority then other environmental issues (Buxton, 2001). The general public, especially in villages and local communities was reportedly not aware of the risks associated with PCBs. This was a serious regulatory gap with respect to the management of PCB releases from industry (unintentional industrial emissions) whether they be spills, accidental releases, operational losses, effluent releases, or air emissions.

Neither the companies (particularly electroproduction and distribution sector) nor government (missing of the legal background and control) are aware about the problem. It was estimated that in Republic of Moldova there are over 20,000 PCB-containing power capacitors are stockpiled in 20 electric substations, with a substation housing about 12,000 – these are estimated to contain 380 t of PCBs (the content of PCB in over 22,000 transformers if still unknown). A pilot investigation showed that soil PCB contamination at the main site stockpile exceeded national maximal allowable concentrations for over 1,000 times (Environmental impact assessment and environmental management plan, 2005).

Most of the contaminated equipment with polychlorinated biphenyls as well as waste oils are remnants from the Soviet Union Times. Republic of Moldova has no plants for producing polychlorinated biphenyl-containing materials or equipment. All present compounds or equipment were imported from abroad. Their use in some areas was discontinued or prohibited in the last decades of the 20th century. However, due to the lack of knowledge about the consequences, polychlorinated biphenyls continue to be used as dielectrics in power installations (transformers and capacitors), as thermal agents and other types of equipment. PCBs emissions from generation, transportation and distribution transformers in the power system amount 14.417 g ET/year (Moldova national chemical management case study, 2006). Losses through oil leakage amount 9.611 t/year. They are only a share of the total emissions, since there is no data on the PCB content in all capacitors in Moldova (including about 2.000 transformers owned by private consumers).

4. Unintended POPs By-Products

PCDD/PCDF releases to the atmosphere, water, land, and residues in products amounted 777.3 g TEQ in 2001 (Moldova national chemical management case study, 2006). Over the period 1990–2001 PCDD/PCDF releases have decreased by 86.6% due to economic decline and restructuring of the economy. The main source of PCDD/PCDF releases was dumping of water (96.6% in 1990 and 97.4% in 2001).

HCB releases amounted to 0.025 kg/year in 1990. The main emission source was cement production (99%). In 2001 HCB releases were estimated at 1 kg/year. The main emission source was secondary aluminum production (99%).

PAH emissions in 1990 were estimated at 2,708.8 kg/year. The main emission were coal – 33%, firewood – 35%, ethylated gasoline – 17%, diesel oil – 12%. PAHS release registered in 2001 amounted 2,215.4 kg. The main source of PAH emissions were firewood (92%). In the last years the emissions of specific polluting substances in atmospheric air by stationary sources have aggravated (Table 1).

TABLE 1. Emissions of some specific polluting substances in atmospheric air by stationary sources are presented, tons. (From Statistical yearbook of the Republic of Moldova, 2006. National Bureau of Statistics of the Republic of Moldova. Chsinau, 2006.)

	2000	2001	2002	2003	2004	2005
Aromatized multiring hydrocarbon of which	1.8	9.0	16.0	11.4	16.4	9.2
Benzo(α)pyrene	0.4	0.4	7.8	4.1	7.5	4.0
Persistent organic pollutants of which	–	–	0.3	4.2	0.5	2.1
Dioxins and furans	–	–	0.1	0.1	0.1	0.1
Hexachlorobenzene	–	–	0.0	0.0	0.2	0.2

The statistical records from the Ministry of Health (2006) indicate that the principal atmospheric polluters in Republic of Moldova are transport, energy producers and some localized industry. Benzo(α)pyrene is among the principal air pollutants. In addition, monitoring results from the State Geological Laboratory has indicates the benzo(α)pyrene as being among the substances which are measured above their limits (Moldova national chemical management case study, 2006).

5. Conclusions

Cleanup activities should be undertaken in Republic of Moldova. Improving environment conditions by mitigating POPs-related problem can help to stimulate economic growth, reduce poverty, and contribute to sustainable development, especially in rural areas.

The main vector-borne diseases remain an exclusively important factor for the public health; therefore, the chemical vector control is still needed. It is necessary to develop monitoring schemes in order to ensure systematic data collecting of the residues of POPs in various environmental media, living organisms, and humans.

On the basis of Stockholm Convention, the primary task of Moldova is to enforce National Implementation Plan on POPs.

References

Alexander, M. (1997) Sequestration and bioavailability of organic compounds in soil. In: Linz, D.G., Nakles, D.U., eds. *Environmentally acceptable endpoints in soil.* Annapolis, MD: American Academy of Environmental Engineers, 43–136.

ATSDR (1993) *Toxicological profile for heptachlor and heptachlor epoxide.* Atlanta, GA: Agency for Toxic Substances and Disease Registry (http://www.atsdr.cdc.gov/toxprofiles/tp12.html).

ATSDR (2001) *Toxicological profiles: PCP.* Atlanta, GA: Agency for Toxic Substances and Disease Registry (http://www.atsdr.cdc.gov/toxpro2.html).

Banergee, B.D. et al. (1996) Influence of subchronic exposure to lindane on humoral immunity in mice. *Indian Journal of Experimental Biology*, **34**: 1109–1113.

Berg, T. et al. (2001) *Heavy metals and POPs within the EMER region, 1999.* Kjeller: Norvagian Institute for Air Research (EMER/CCC Report 9/2001).

Bertazzi, P.A. et al. (1998) The Seveso studies on early and long-term effects of dioxin exposure: a review. *Environmental Health Perspectives*, **106** (Suppl. 2): 625–633.

Beyer, W.N. (1996) Accumulation of chlorinated benzenes in earthworms. *Bulletin of Environment Contamination and Toxicologt*, **57**: 729–736.

Bude, J. et al. (1985) Influences of selected parameter bioconcentration of 14C-2,5,4-trichlorophenyl and pentachlorophenyl. *Chimica Acta Turcica*, **13**: 235–252.

Boul, H.L. et al. (1994) Influence of agricultural practicies on the level of DDT and its residues in soil. *Environmental Science and Technology*, **28**: 61–66.

Buxton, V. G. (2001) *Towards the Republic of Moldova's Cost-effective and Timely Compliance with the Stockholm Convention on Persistent Organic Pollutants (POPs).* A preliminary assessment of Moldova's state of readiness to meet the Stockholm POPs Convention requirements. World Bank Mission Report, 10–14 September 2001.

Chemicals, Industrial Ptrocesses and Industries Associated with Cancer in Humans (1982) IARC Monographs on the Evaluation of the Carcinogenic Risk of Chemicals to Man. Suppl. 4. IARC, Lyon, France.

Currier, M.F. et al. (1980) Hexachlorobenzene blood levels and the health status of men employed in the manefature of chorinated solvents. *Journal of Toxicology and Environmental Health*, **6**: 367–377.

Dalsenter, P.R. (1997) Reproductive toxicity and toxicokinetics of lindane in the male offspring of rats exposed during lactation. *Human and Experimental Toxicology*, **16**: 146–153.

De Boer, J. (2000) Polychlorinated terphenyls. In: Paasivirta, J., ed. *New types of persistent halogenated compounds.* Berlin, Heidelberg: Springer-Verlag (Handbook of Environmental Chemistry, Vol. 3, Part K).

Devillers, J. et al. (1996) Comparison of BCF models based on log P. *Chemosphere*, **33**: 1047–1065.

Environmental impact assessment and environmental management plan (2005) GEF PAD Grant for preparation of Sustainable Persistent Organic Pollutants Stockpiles Management Project, E1198, V. 1, 30 June 2005.

Technical appendix to the main report (2005) Final version. Environmental impact assessment and environmental management plan GEF PAD Grant for preparation of Sustainable Persistent Organic Pollutants Stockpiles Management Project, 27 June 2005.

European Commission (2000) *European Union risk assessment report. Alkanes, C_{10-13} chloro.* Brussels, European Commission.

Fisk, A.T. et al. (1996) Dietary accumulation of C-12- and C-16-chlorinated alkanes by juvenile rainbow trout *(Oncorhynchus mykiss). Environmental Toxicology and Chemistry* **15**: 1775–1782.

Foster, W.G. et al. (1993) Body distribution and endocrine toxicity of hexachlorobenzene in the female rat. *Journal of Applied Toxicology*, **13**: 79–83.

Government of Canada (1993) *Priority substances list supporting document. Chlorinated paraphyns.* Ottawa, Ontario, Environment Canada and Health and Welfare Canada.

Health Risks of Persistent Organic Pollutants from Long-Range Transboundary Air Pollution (2003). Joint WHO/Convention Task Force on the Health Aspects of Air Pollution.

Hooper, K. et al. (1997) Analysis of breast milk to assess exposure to chlorinated contaminants in Kazahstan: PCBs and organochlorine pesticides in southern Kazahstan. *Environment health perspectives*, **105**: 199–295.

IARC (1991) *Occupational exposures in insecticide application and some pesticides.* Lyon, International Agency for Research on Cancer (IARC monographs on the evaluation of carcinogenic risks to humans). Vol. **53**, pp. 179–249.

Laden, F. et al. (1999) Predictors of plasma concentrations of DDT and PCBs in a group of US women. *Environmental Health Perspectives*, **107**: 75–81.

Madeley, J.R., Thompson, R.S. (1983) *The bioconcentration of a chlorinated paraffin in the tissue and organs of rainbow trout (Salmo gairdneri).* Brixham, Imperial Chemical Industries (ICI Report BL/B/2310).

Moldova National Chemical Management Case Study (2006) Preliminary Draft Consultatn Report. Milieu – Environmental Law and Policy." Contract for intellectual services 2006/JA00040791.

Muir D.G. et al. (2000) Short chain chlorinated paraffins: are they persistent and bioaccumulative?. In: Lipnick R. et al., ed. *Persistent bioaccumulative and toxic substances,* Washington, DC, ACS Books, Vol. **2**, pp. 184–202.

Muller, D. et al. (1981) Electroneurophiziological studies on neurotoxic effects of hexachlorocyclohexane isomers and gamma-pentachlorocyclohexane. *Bulletin of Environment Contamination and Toxicology*, **27**: 704–706.

Oldiges, H. et al. (1983) *90-day inhalation study with lindane.* Schmallenberg, Fraunhofer-Institute for Toxicology and Aerosol Research (Celamerck document No.111AC-435-005).

Peterson, J.R. et al. (1971) Soil properties influencing DDT bioactivity. *Soil Science Society of America proceedings*, **35:** 73–79.

Robertson, B.K., Alexander M. (1998) Sequestration of DDT and dieldrin in soil: disappearance of acute toxicity but not the compounds. *Environmental Toxicolology and Chemistry*, **17**: 1034–1038.

Robinson, A.K., Sirbasku D.A., Stancel G.M. (1985) DDT promotes the gowth of an estrogen-resposive tumor. *Toxicology Letters*, **27**: 109–114.

Samuel, T., Pillai M.K.K. (1990) Effect of temperature and solar radiations on volatilization, mineralization and degradation of [14C] DDT in soil. *Environment Pollution*, **57**: 63–77.

Saxena, M.C. et al. (1981) Organochlorine pesticides in specimens from women undergoing spontaneous abortion, premature of full-term delivery. *Journal of Analitical Toxicology*, **5**: 6–9.

Siddiqui, M. et al. (1981) Storage of DDT and BHC in adipose tissue of Indian mmales. *International Journal of Environmental Analitical Chemistry*, **10**: 197–204.

Smith, A.G. (1991) Chlorinated hydrocarbon insecticides. In: Hayes, W.J., Laws, E.R., eds. *Handbooke of pesticide toxicology*. San Diego, CA: Academic Press, pp. 731–791.

Smith, A.G. et al. (1987) Goitre and wasting induced in hamsters by hexachlorobenzene. *Archives of Toxicology*, **60**: 343–349.

Social Assessment Study for the CEF PAD Grant for Preparation of Sustainable Persistent Organic Pollutants (POPs) Stockpiles Management Project, 2005.

Some Organochlorine Insecticides (1974) *IARC Monographs on the Evaluation of Carcinogenic Risk to Man*. V.5. International Agency for Research on Cancer, Lyon, France.

Stanley, C. et al. (1971) Mesurement of atmospheric levels of pesticides. *Environment Science and Technology*, **5**: 430–435.

The WHO Recommended Classification of Pesticides by Hazard and Guidelines to Classification 2000–2002 (2002) UNEP, ILO, WHO.

Volneanschi, A., Romanciuc, P. (1992) Evaluarea igienică a conţinutului de pesticide organoclorurate în laptele mamar. *Probleme actuale ale igienei*. Tezele referatelor congresului III al igieniştilor, microbiologilor, epidemiologilor şi parazitologilor din Republica Moldova, 23–24 octombrie 1992, Chişinău, 68–70.

Wenning, R., Kirsch, N. (1988) In: Sieler, H.G., Sigel, H., eds. *Handbook on toxicity of inorganic compounds*. New York: Marcel Dekker, pp. 731–738.

WHO (1979) *DDT and its derivatives*. Geneva, World Health Organization (Environment Health Criteria, No. 9).

WHO (1987) Polynuclear aromatic hydrocarbons (PAH). In: *Air quality guidelines for Europe*. Copenhagen, WHO Regional Office for Europe (WHO Regional Publications, European Series, No. 23), pp. 105–117.

WHO (1989) GEMS/Food International Dietary Survey. *Infant exposure to certain organochlorine contaminants from breast milk. A risk assessment.* Geneva, WHO (document WHO/FSF/FOS/98.4).

WHO (1991) *Lindane*. Geneva, World Health Organization (Environmental Health Criteria, No. 124).

WHO (1993a) *Guidelines for drinking-water quality*, 2nd edn. *Vol. 1 Recommendations*. Geneva, World Health Organization.

WHO (1993b) *Polychlorinated biphenyls and terphenils*, 2nd edn. Geneva, World Health Organization (Environmental Health Criteria No. 140).

WHO (1996) *Guidelines for drinking-water quality*, 2nd edn. *Vol. 2 Health criteria and other supporting information*. Geneva, World Health Organization.

WHO (1998a) *Polybrominated dibenzo-*p*-dioxins and dibenzofurans*. Geneva, World Health Organization (Environmental Health Criteria, No. 205).

WHO (1998b) *Selected non-heterocyclic polycyclic aromatic hydrocarbons*. Geneva, World Health Organization (Environmental Health Criteria, No. 202).

Woodwell, G. et al. (1971) DDT in the atmosphere: were does it go? *Science*, **174**: 1101–1107.

Wolff, M.S. (1993) Blood levels of organochlorine residues and risk of breast cancer. *Journal of National Cancer Institute*, **85**: 648–652.

PERSISTENT ORGANIC POLLUTANTS (POPs) AT NANOTECHNOLOGY AND THEIR IMPACT ON PEOPLE HEALTH

OLEKSIY KHARLAMOV*,
Frantsevich Institute for Problems of Materials Science, National Academy of Science of Ukraine, 3, Krjijanovskogo Street, 03680 Kiev, Ukraine

GANNA KHARLAMOVA, NATALIA KIRILLOVA
Kiev National Taras Shevchenko University, 64 Volodimirska Street, 03001 Kiev, Ukraine

VENEAMIN FOMENKO
National University of Food Technology, 68 Volodimirska Street, 03001 Kiev, Ukraine

Abstract: We demonstrate that at reactionary conditions of the growth of nanostructures of different substances can be by-products of most incredible on composition, structure, and morphology objects. Two main innovations are presented in this article. (i) Novel molecular crystals of transparent carbon in the shape of isotropic and anisotropic (thread-like) micro particles (by diameter more 5 mkm) are presented first. Many transparent carbon threads are painted in various colors. (ii) We believe nanochemistry is the science mostly about nanochemical objects which the properties depend on their size, morphology and structure. Different nanoobjects as nanophases can be the subject also of nanochemistry but as it high border. A new mechanism common for carbon nanotubes, onions and fullerenes formation by hydrocarbons molecules conversations is proposed.

*To whom correspondence should be addressed. E-mail: dep73@ipms.kiev.ua; akharlamov@ukr.net

E. Mehmetli and B. Koumanova (eds.),
The Fate of Persistent Organic Pollutants in the Environment, 425–441.

Keywords: nanochemistry, nanochemical objects, carbon transparent threads, silicon carbide nanothreads, nanotube, onions, benzene polycondensation mechanism, self-accelarating reaction, sublimation

1. Introduction

The most fundamental and lasting objective of synthesis is not production of new compounds, but production of properties

George S. Hammond
Norris Award Lecture, 1968

Has not passed and 30 years as these words have found a perfect embodiment – it is revealed, that already well known substances can show new, unusual properties because of change of the size and morphology of their particles. First of all it is connected to the discovery of the new unique form of a matter, such as spatial (spherical and tubular) molecule, in particular, carbon (fullerenes and nanotubes), and also one-dimensional uncarbon nanostructures. Certainly, the similar anomaly in change of properties does not concern directly to liquid and gaseous substances as well to molecular crystals, which in the majority are organic pollutants of an environment. However appearance (revealing and the synthesis) in a nature of carriers of in principle new properties, naturally, results in necessity of deep and all-round study of these objects, including, and from the point of view of their ecology. In the ecological relation these new objects are extremely unpredictable also because of the contents, in particular, on a surface of carbon nanostructures of adsorbed persistent organic pollutants. So, agrees [1], on a surface of nanosized particles of carbon formed during burning hydrocarbons, contain various carcinogenic polycyclic aromatic hydrocarbons (PAN) such as circobiphenyl, perylene, and benzopyrene.

Really, on the singularity of chemical behavior of spatial molecules of carbon show the mutually contradictory data about toxicity, in particular, fullerenes and nanotubes on people health. So the researchers from universities Rice and Georgia (USA) have shown [2–4], that in water fullerene solutions are formed colloidal "nano-C_{60}" particles, which even at extremely their low concentration (approximately two molecules of fullerenes on 108 molecules of water) negatively influence on the cells of a skin and liver. It is marked, that the toxicity of water dispersion of "nano-C_{60}" are comparable with toxicity of dioxins. On the other hand, the researchers from Ukraine represent the opposite data [5], demonstrating, that fullerenes C_{60} is harmless and, on the contrary, has unique antioxidant activity. In [6] was reveal also that C_{60} prepared by a variety of methods (concentration up to 200 μg/ml) is not toxic to a number of cell types. The similar contradiction between data of different authors is caused, probably, different composition and dispersion investigated "nano-C_{60}" particles.

It is necessary also to take into account that new objects and the new methods of their preparation, undoubtedly, will result in formation and new by-products, some of which can be more harm for people health. Therefore alongside with intensive study of various methods of synthesis nanostructures of carbon it is necessary the special attention to pay on research of a chemical nature and fate parallel formed organic pollutants.

2. Nanochemistry and Nanochemical Objects

However before to present original experimental results on this question it is expedient to determine (especially among the experts of different directions) a general and clear terminology in this new field of knowledge. New nanodimensional objects already today are extremely various concerning their chemical nature, composition, structure, and morphology and consequently it is necessary to classify them (Figure 1).

At the end of the 20th century the remarkable experimental results of epoch-making importance were obtained and the epoch of "nano" began: nanochemistry, nanotechnology, nanomedicine, nanobiology, and many others "nano." In general, it is possible to tell, the epoch nanophilosophy began, as beginnings to change itself paradigm all scientific and industrial activity of the man. New paradigm – "from below – upwards", i.e., from separate atoms to a production gradually comes on change of paradigm of traditional technology, where, as is known, the production obtain by means of processing of huge massive half-finished product, i.e. "from above – downwards." The prefix nano pointed out earlier only on the size of object, namely on one milliard part (10^{-9} m), today has got practically magic sense. It is considered, that most advanced, nanorevolutionary technological achievements can be supplied by means of nanoactions only.

There is a question – why the dimensional parameter of object changes so deeply and fundamentally our representations not only about object, but also about all scientific and industrial activity of the man? Why these changes occur only today, though about a "magic" role of dimensional effect, for example, particles of metals (silver and gold) in glass (cup of Likurga), it was known very much for a long time? First of all, it is connected to two very important discoveries at the end of the last century, when the opportunity has appeared not only to detect and to measure individual molecules and particles by the size considerably less 100 nm (nanoparticles), but also, that is very essential, to study their physical and chemical behavior. It was established that the properties of individual molecules differ from behavior of these molecules in neutral molecular nanoensembles (clusters), in which molecule and the atoms are connected by means of only weak Van der Waals bonds (table 1). As against neutral, charged (ionic or ionic-molecular) clusters are formed owing to donor–acceptor interaction of cations with molecules.

But most surprising has appeared the discovery of a new type of molecules such as monocyclic unflat (spheroidal) molecules of carbon (fullerenes and nanotubes). In subsequent were obtained fullerene-similar and tubular nanostructures of many other simple substances and compounds. Thus, in many cases the electronic and physicochemical properties of these individual molecule-similar structures in principle differ from properties of corresponding 3D materials (micro phases).

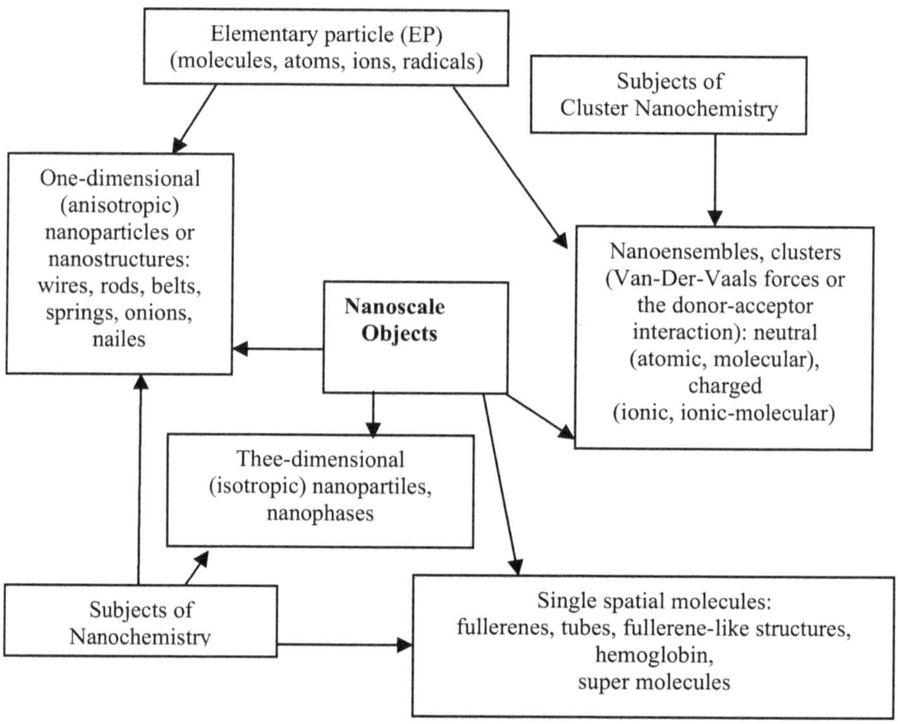

Figure 1. Classification of objects of nanochemistry

On the other hand, unusual nanoparticles (nanophases) were found out which properties essentially differ from properties micro phases. Moreover, in some cases nanoparticles (nanophases) have other, distinct from micro phase, crystal structure. So nanoparticles of gold by the size 3–5 nm have icosahedral structure, whereas micro phase of gold, as is known, has a cubic lattice.

It is important to specify, that a nanoparticle is a part whole (3D material), the part of a microphase. The top border of three-dimensional (isotropic) nanoparticle is the quantum point. By the top border of anisotropic nanoparticles such as rods, tape, the nails and film are a quantum wire or hole. Anisotropic nanoparticles (as against isotropic) because of internal structurized, for example, in one direction can refer to as also nanostructures.

Just detection nanoobjects (nanophases, spatial molecules, and anisotropic nanostructures), which properties are abnormal concerning properties of micro phases and change depending on their size, structure and morphology, has become determining in appearance of new area of chemistry is a nanochemistry. Just these nanoobjects are the basic subject of nanochemistry and consequently it is necessary to name them "nanochemical". Hence, nanochemistry is science about synthesis and properties first of all "nanochemical" of objects, which scaled parameter is the important factor of their electronic structure and properties. (In the world of microsystems of property of substance, as is known, depend only on a nature of its atoms and structure of their construction). By subject of nanochemistry, as bottom border it, can be also nanodimensional objects (nanoparticles), which properties are not abnormal micro phase and, hence, do not depend on their size and morphology, but such objects are not "nanochemical." So, such concepts as nanoparticle display only a scaled parameter of particle. The term "nanochemical" (particle, structure, or the phase) specifies the dependence of properties of object on its size, structure and morphology. It is important to note, that in some cases the term "nanochemical" can be and object of the size more 100 нм, but it, as a rule, is an exception. Clusters of metal it is necessary also to name nanoparticles or nanophases as between atoms of metal (even if them only five) are kept strong metal bond. Nanoparticles of the majority of metals are, as a rule, nanochemical (or nanophases) and in it their special role and not only in catalysis. So melting temperature (T_m) of nanoparticles (5–10 nm) of gold on hundreds degrees is lower T_m of microphase of gold.

Clusters are also nanodimensional as neutral (atomic or molecular), and charged (ionic or ionic-molecular) complexes or ensembles from molecules, atoms or ions and consequently they can be a subject of clusters nano-chemistry. The top border of nanochemistry can be nanochemistry of individual (single) molecules, as these molecules now not only are detected, but their behavior is studied which, as already was marked, is different from properties of clusters from the given kind of molecules.

The objects of nanochemistry, submitted in the table, being, on the one hand, a nanodimensional part of microobjects, have, on the other hand, essentially other physicochemical and electronic properties. The study of influence of these new untraditional for microsystems of properties on an environment and, first of all, on the man is represented extremely important.

3. Novel Unusual Products of Pyrolysis of Hydrocarbons

Among nanostructures already of synthesized substances of various chemical nature (sulfides, chlorides, carbides, oxides, and carbon) especially unsafe are nanostructures of carbon and carbon-containing compounds (carbides), at which synthesis with participating of hydrocarbons the formation most improbable on composition of carcinogenic compounds can be carried out (Figure 2). It is caused by a unique capability of carbon to form at temperatures above 1,000°C various aromatic multinuclear carbon molecules and molecule-like structures.

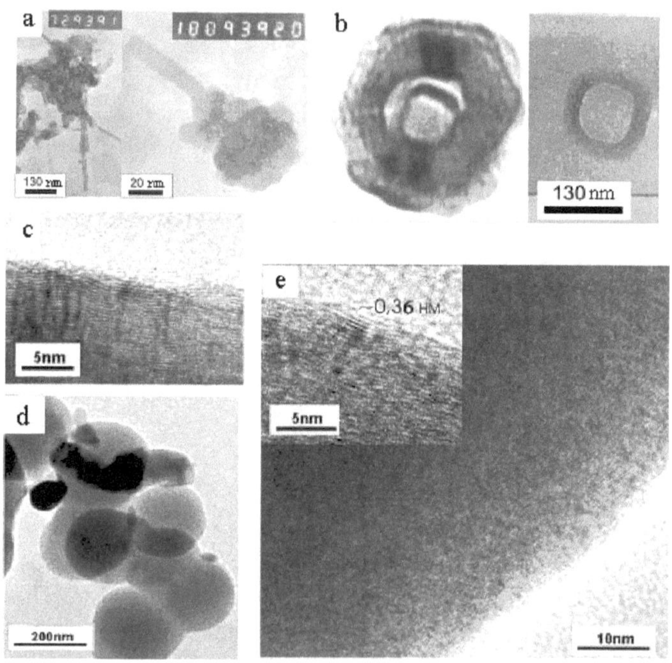

Figure 2. The images of transmission electronic microscopy of carbon nanotubes (a), closed structures in the form of rings and polyhedrons (б), onions (г) and electronic microscopy of high resolution of a fragment of a wall of polyhedron (b) and spherical particle of onion (d)

So authors [7–9] still in the first investigations shown, that at process of catalytic growth of carbon nanotubes polynuclear aromatic hydrocarbons on their surface are formed. In [1] was noted also that nanoparticles from combustion contain thousands of different chemicals, which chemical characterization is impossible because of their complexity and very tiny amount.

Thus products of pyrolysis process of hydrocarbons contain both nanodimensional particles and adsorbed PAH on their surface. Therefore nanoobjects can exert double influence on health of the people and, hence, ecological insecurity of nanoobjects is determined both morphology and structure of nanoparticle, and chemical nature of adsorbed chemicals. In [1] are presented very troubled data, which show the mortality from cardiopulmonary disease increases by 1% when loading of the air with respiratory particles formed at combustion of organic materials increases by 10 $\mu g/m^3$.

Here we represent original experimental results, which convincingly demonstrate, that in reactionary conditions of growth of nanostructures, in particular, carbon and silicon carbide most improbable on structure, morphology and composition of substance can parallel be formed. So, us for the first time it was revealed [10–12], that at thermal conversion of hydrocarbons alongside with carbon nanotubes are formed also spherical multiwalled (fullerene-like) structures in the form of toroids and onions (Figure 1).The size of carbon toroids is less 50 nm basically, whereas a diameter of onions reaches 250 nm. Onions and toroids, as it is visible, have a high-degree structural ordering. The high degree of graphitization nanoonions (diameter less 40 nm) is confirmed by the data of Raman spectra [13].

By us is established, that these three kinds of carbon structures (Figure 2) are formed in different temperature zones of reactor. Besides by means of a variation of a regime of hydrocarbons pyrolysis rather large multiwalled tubular (Figure 3a) and fullerene-like (Figure 3b) structures were obtained. It is difficult to present, that nanoparticle of metal accepted any participation in growth of given multiwalled carbon micro tubes. Besides, all our experiences were carried out without using of metallic catalysts. As it is visible (Figure 3), partially destroyed multiwalled microstructures can be excellent containers for a storage simultaneously formed in process of pyrolysis of PAN and, probably, new carbon molecules. Really, the processsing of synthesized by us of products samples of pyrolysis of hydrocarbons by the various organic solvents has shown, that on their surface extremely unusual molecules are adsorbed.

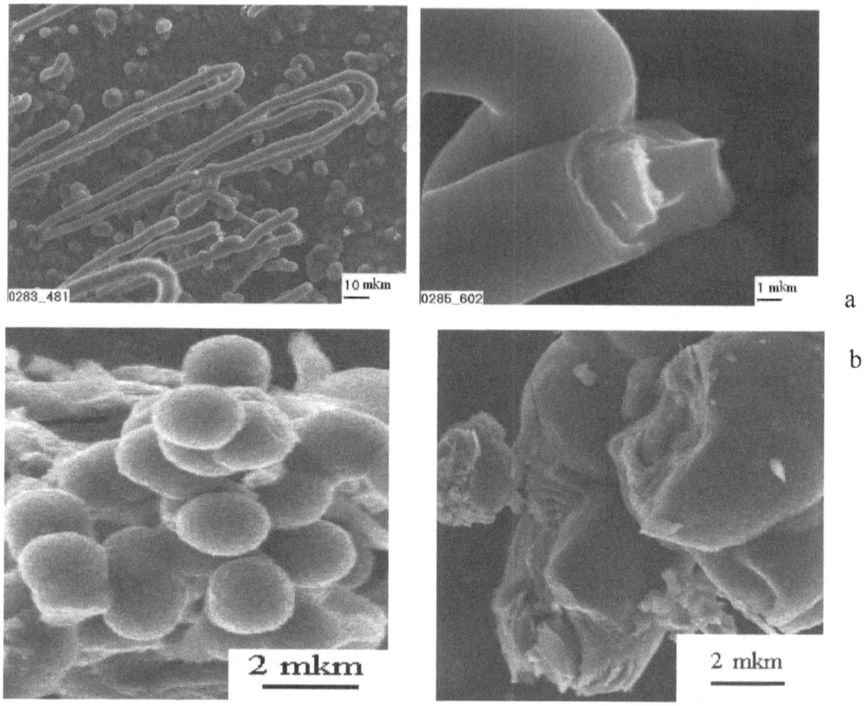

Figure 3. Images of scanning electron microscopy of multiwalled tubular (a) and fullerene-like (b) structures of carbon

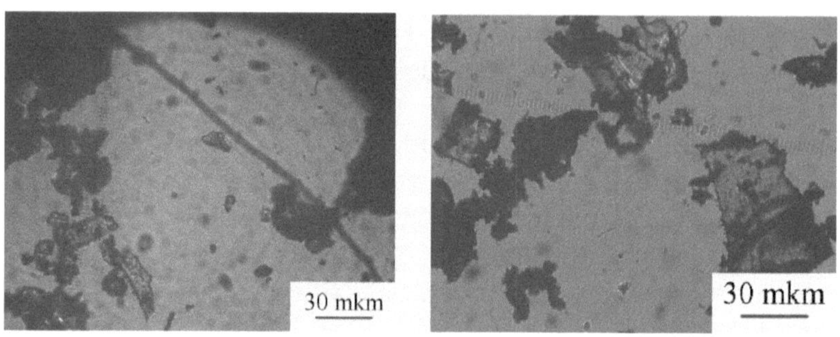

Figure 4. Images of optical microscopy in polarized light of transparent isotropic crystals of sample 1. (From Table 1.)

Some of these molecules are precipitated from organic dissolvent as isotropic (Figure 4) and anisotropic (thread-like) particles (Figure 5).

It is remarkable, that many threads have the diversified colors. The diameter of these threads sometimes reaches 40 mk, and their length more 50 mm. Some of these threads have extremely original one of the ends.

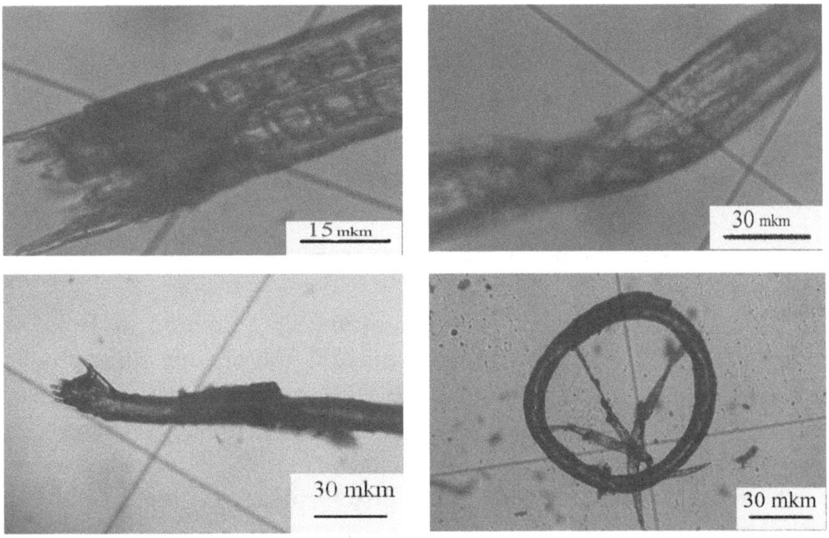

Figure 5. Images of optical microscopy in polarized light of color transparent threads of carbon

In Table 1 are submitted chemical and hypothetical formula compositions of substances, which parallel with carbon nanostructures in conditions of high-temperature conversion of hydrocarbons, are formed. Let's note that these of substances were obtained owing to using of an original technique of process of pyrolysis of hydrocarbons.

TABLE 1. Chemical composition of precipitated from organic dissolvent of products of highly temperature conversion of hydrocarbons

Sample	Chemical composition, % (mass)				Formula
	O	N	H	C	
1	14	–	–	86	$C_{8,2}O_{1,0}$
2	4,19	–	3,47	92,34	$C_{29,6}O_{1,0}H_{13,34}$
3	1,92	1,2	–	96,9	$C_{67,3}O_{1,0}N_{10}$
4	7,30	0,3	–	92,4	$C_{16,9}O_{1,0}N_{0,04}$

According to the chemical analysis (on burning of substance) new solid products of conversion of hydrocarbons though are heteroatomic, however the contents of carbon in them is prevalent. The qualitative chemical analysis as small isotropic crystals (sample 1), and threads (the sample 3) by means of x-ray spectral analyzer (Camebax) has shown, that they consist of carbon (Figure 6).

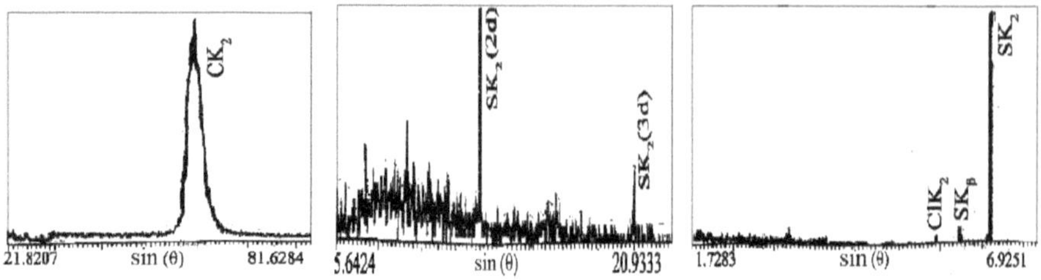

Figure 6. X-ray spectral analysis of red thread of sample 3

It is important to note, that the substances, submitted in Table 1, are transparent crystals and poorly are dissolved in the organic solvents, such as benzene and toluene, and also in alcohol. Research of transparent crystals of sample 1 by means of x-ray photoelectronic spectroscopy (Figure 7) has shown that they contain as well nitrogen (Table 2). According to crystal optical analysis these crystals are anisotropic with high (more 1, 6) coefficient of two-refraction.

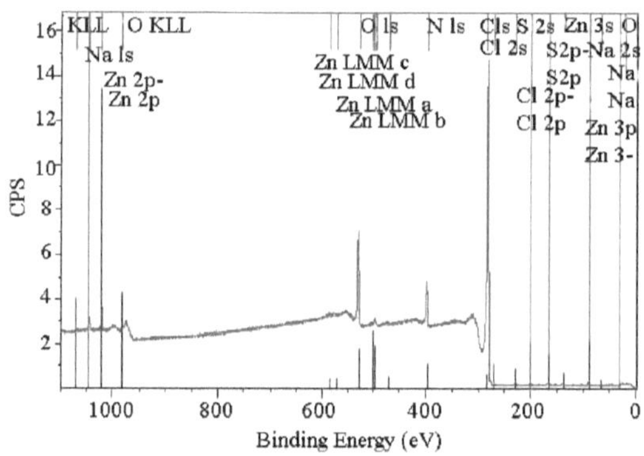

Figure 7. X-ray photoelectron spectrum of transparent crystals of sample 1

Pulverized threads of carbon are investigated by means of x-ray diffractional analysis (Figure 8) and infrared (IR) spectroscopy (Figure 9). Threads have original hexagonal crystalline structure with the following parameters: a = 0.498 nm and c = 0.826 nm. In IR spectrum the group of lines from 1,157 up to 1,724 sm^{-1} is visible which are usually responsible for skeletal fluctuations of C–C – bonds in aromatic hydrocarbons. (In IR spectrum fullerene C_{60} [15] in this range there are three lines: 1,182; 1,428; and 1,539 sm^{-1}). The lines in the range less than 900 sm^{-1} can be referred to deformative fluctuations of aromatic C–H – bonds.

TABLE 2. Chemical composition of sample 1 (Table 1) according to XPS

Chemical elements	Orbitals	BE/eV	% (atomic mass)
Na	1s	1,070.95	0.2
Zn	2p	1,020.85	2.31
O	1s	530.95	10.55
N	1s	398.05	6.91
Ca	2p	345.55	0.33
C	1s	283.45	78.42
Cl	2p	198.55	0.27
S	2p	162.85	0.62
Si	2p	101.95	0.39

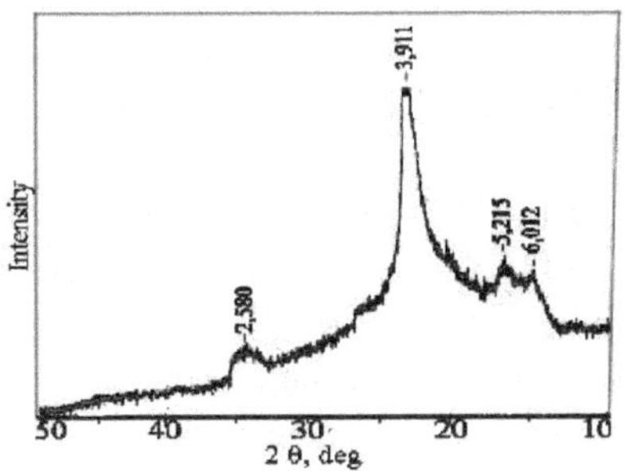

No	2θ	d/n	
1	14.75	6,012	001
2	16.65	5,215	100
3	22.74	3,911	101
4	34.77	2,580	200, 102

Figure 8. X-ray diffraction patterns of color transparent threads of carbon

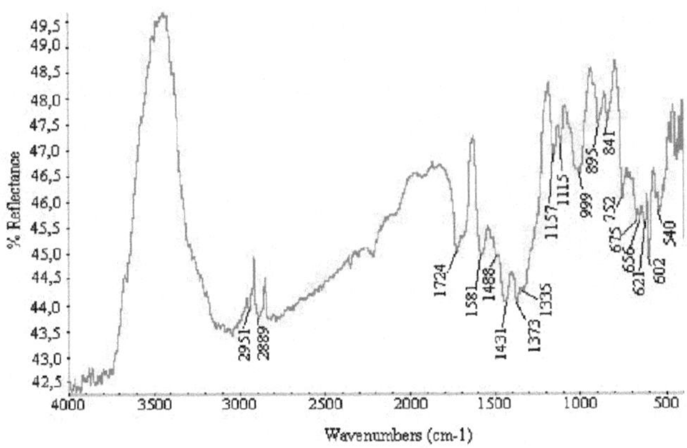

Figure 9. IR spectrum of pulverized transparent threads of carbon

4. Thread-Like Transparent Crystals of Carbon as By-Product at the Growth of Nanostructures of Silicon Carbide from Powdery Carbon and Silicon

Recently [15–19] we developed a new type of chemical reactions, namely self-accelerating reactions of growth of anisotropic nanoparticles from powdery initial reagents, such as carbon, silicon, and boron. The new mechanism of reactions between solids was offered also, according to which the product, in particular silicon and boron carbides nanothreads, is formed not on diffusion mechanism (as it was accepted to consider it), and on gasphase one with participation of atoms of evaporating reagents. Accelerating growth of nanothread-like carbides is carried out at the expense of essential firing and, as a consequence, reactionary evaporation of reagents in microvicinities of local nanocenter for the account exothermal vaporphase reaction. It is possible to assume, that reactionary sublimation of silicon and carbon as the indispensable stage of this reaction can promote also the formation of anisotropic particles of silicon and carbon.

Here we for the first time represent results of research by means of optical microscopy of powdery samples of silicon carbide obtained at realization of this new reaction. The experiences on obtaining of nanostructures of silicon carbide from powders of silicon and carbon were carried out at temperatures below 1,200°C according to a technique [15]. The study of sample of silicon carbide by means of TEM has shown that it practically completely consists from nanothreads. At the same time, in a sample are visible as well separate tubular nanostructures (Figure 10).

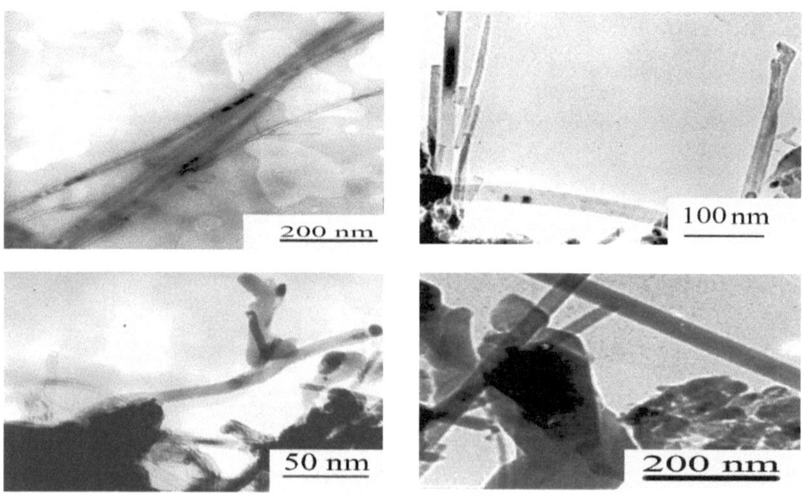

Figure 10. SEM images thread-like and tubular nanostructures of silicon carbide

The same samples of SiC by means of optical microscopy were examined also. It was revealed, that at the majority of such samples are present of various color of transparent anisotropic particles (thread, tape) long up to 30 mm and diameter up to 50 mkm (Figure 11). The investigation of threads in polarized light has shown that most of them are transparent anisotropic crystals with mainly directed crystallization. Some of threads become dim simultaneously and, consequently, are single crystalline. The direction of crystallization on length of a thread sometimes cardinally varies, that has an effect for change of morphology of a particle. Factor of two-refraction (n) of threads is about equal 1,575–1,580, that, basically, is closest meaning n of quartz (1,554 n_g and 1,543 n_p). However qualitative analysis of threads by means of microanalyzer (Camebax) has shown (Figure 12), that they contain mainly only carbon. In micro-Raman spectra of threads (Figure 13) the most intensive peak (approximately 125 sm^{-1}), characteristic for quartz, also is absent. Unique in morphology of the found out threads of carbon the structure of one of their ends is, that, undoubtedly, specifies the special mechanism of their formation.

Figure 11. Optical microscopy in transparent light images of color threads formed at synthesis of silicon carbide

5. Discussion of Mechanism of Growth of Carbon Nanostructures

The essential (inevitable) elementary stage for a carbon molecule formation, as used to consider, is the generation of atomic carbon by graphite evaporation (arc-discharge and laser ablation methods) or hydrocarbon molecules decomposition (CVD method) on a metal-containing particle. Carbon atoms (clusters) then considered as the only building elements for as macromolecules (fullerenes) as nanotubes. However, within the framework of the above mentioned mechanism it is not possible to describe one important experimental phenomenon of formation of polynuclear aromatic hydrocarbons. It is undoubtedly indicates about existence of another principally different mechanism of hydrocarbons conversations. On the support of this point of view can be presented observed by us phenomenon [10–12] of simultaneous carbon nanotubes and onions formation by pyrolysis of benzene (Figure 2). If carbon nanotubes, as used to consider, are nucleated and then fixed on a metal nanoparticle than onions with fullerene as the growth center undoubtedly should be nucleated within the space of the reactor. Further, continuously growing onion-like particles are transported by gas flow and deposited in a low-temperature zone of a reactor. (Condensation of individual fullerene molecules not suffered from cracking on the reactor surface is quite possible). It is necessary to notice that under worked out by us conditions of benzene pyrolysis we managed to obtain not only singlewalled and multiwalled carbon nanotubes but two onion-like forms of carbon differing in structure (Figure 9). Toroidal structures are more thermostable because are formed in a reaction zone with higher temperature. It is not possible to describe these experimental results within VLS mechanism which is usually proposed for the growth of carbon nanotubes, for instance. There is still open question about a role of sulfur as a catalytic addition for nanotubes growth rate (elongation) increase in spite of the fact that sulfur as well as palladium is efficient dehydration catalysts of aromatic hydrocarbons, in particular. Building up of spherical and cylindrical surfaces of carbon is likely to precede uniformly but from different, in terms of composition, carbon elements of structure. At the same time such stages as hydrocarbon molecules destruction on a particle of metal with formation of metal-carbon eutectic and further generation on its surface of carbon atoms as building elements for a grapheme network could not be considered as an intermediate one during the growth process of nanostructures.

It was proposed by us earlier [10, 12] that the main element taking place in the formation of the graphene network could be a molecule of benzene in which molecules often used as initial hydrocarbons such as saturated (hexane, methane) and unsaturated (ethylene, acetylene) are dehydrocycled within an experimental temperature range. So, in contrast to the destruction process of the strongest C–H bonds (96 kkal) and C–C bonds (200 kkal) in molecules, for instance, of methane and acetylene (usually proposed for a generating of atomic carbons of which nanostructures than formed) the development of C–C bonds, in particular, in the process of acetylene trimerization into benzene is thermodynamically more favorable. Moreover, at temperatures higher 600°C, benzene readily dehydrates with following diphenyl formation which can be the first stage (1) of a graphene network formation. Further dehydropolymerization (condensation) of benzene and diphenyl leads to the increase of the hexagon numbers of which as it is well known the graphene network consists. Further condensation of benzene and formed diphenyl leads to the increase of hexagons numbers (stage 2) in a developing grapheme sheet of which as it is well known the graphene network consists. Sulfur as a dehydration catalyst facilitates to the condensation process and, consequently, to the elongation of nanotubes which is observed in most of experiments. Spheroid surface of fullerene and onion-like structures in contrast to nanotubes contain the both hexagons and pentagons with their minimal ratio equal to 2, for instance, in the simplest fullerene C_{60}. Formation of such network is also possible by means of the condensation reaction among different molecules but not only aromatic hydrocarbons. The network formed of planar molecules of hydrocarbons taking place in the condensation can bend slightly because of absence of the coplanarity of all atoms in a molecule of at least one of the hydrocarbons. It is likely that by high temperature pyrolysis of hydrocarbons as well as coal coking not only benzene forms but its closest homolog – toluene and cyclopentadien molecules of which contain non-coplanar atoms, in particular, of hydrogen. Interaction of benzene with both cyclopentadien and toluene leads to the formation (as shown in schemes 1 and 2) of a specific fragments of the fullerene network.

Scheme 1

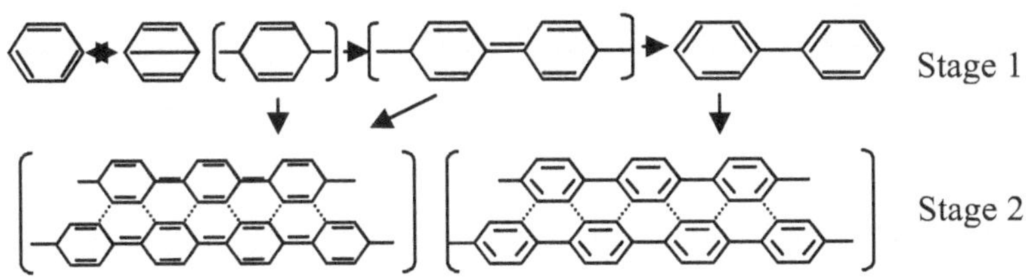

Scheme 2

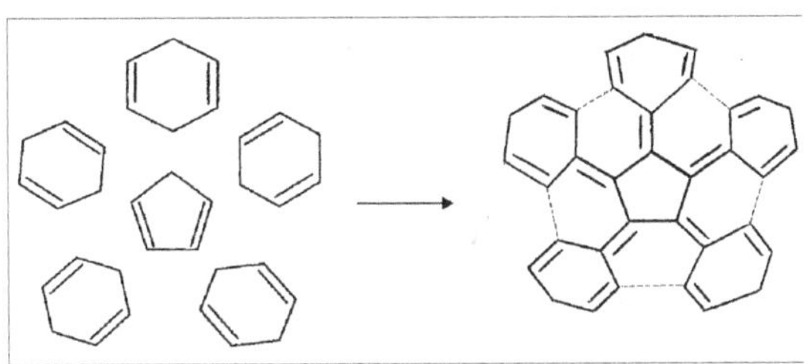

References

1. Siegmaan, K. and Siegmann, H. C. (1997) The formation of carbon in combustion and how to quantify the impact on human health. *Europhysics News* **28**, 50–57.
2. Sayes, C. M., Fortner, J. L., Guo, W., Lyon, D., Boyd, A. M, Ausman, K. D, Tao, Y. J, Sitharaman, B., Wilson, L. J, Hughes, J. B, West, J. L., and Colvin, V. L. (2004) The differential cytotoxicity of water-soluble fullrenes. *Nano Letter* **4**, 1881–1887.
3. Obedorster, E. (2004) Manufactured nanomaterials (fullerenes, C_{60}) induce oxidaticve stress in the brain of juvenile largemouth bass. *Environmental Health Perspectives* **112**, 1058–1062.
4. Sayes, C. M., Gobin, A M., Ausman, K. D., Mendez, J., West, J. L., and Colvin, V. L. (2005) Nano-C_{60} cytotoxicity is due to lipid peroxidation. *Biomaterials* **26**, 7587–7595.
5. Andrievsky, G. V., Klochkov, V. K., Bordyuh, A. B., and Dovbeshko, G. I. (2002) Comparative analysis of two aqueous-colloidal solutions of C- 60 fullerene with help of FTIR reflectance and UV-Vis spectroscopy. *Chemical Physics Letters* **364**, 8–17.
6. Levi N., Hantgan R. R., Lively M. O., David, L., Carroll, D. L., and Prasad, G.L. (2006) C_{60}-Fullerenes: detection of intracellular photoluminescence and lack of cytotoxic effects. *Journal of Nanobiotechnology* **4**(14).
7. Biro, P., Bernardo, C. A., Tibbetts, G. G., and Lambin, Ph. (2000) Carbon Filaments and Nanotubes: Common Origins Differing Applications? Ed. NATO Science Series, Series E: Applied Sciences, vol 372.
8. Van Huttum, F. W. J., Bernardo, C. A., and Tibbetts, G. G. (2000) Mechanical properties of vapor grown carbon fibers and VGCF-thermoplastic composites In: *Carbon Filaments and Nanotubes: Common Origins Differing Applications?* Ed. NATO Science Series, Series E: Applied Sciences, vol 372, pp. 372–366.
9. Chung, D. D. L. (2000) Applications of submicron diameter carbon filaments In: *Carbon Filaments and Nanotubes: Common Origins Differing Applications?* NATO Science Series, Series E: Applied Sciences, vol. 372, pp. 275–288.
10. Kharlamov, A. I., Kirillova, N. V., and Ushkalov, L. N. (2006) Simultaneous growth of spheroidal and tubular carbon structures during the pyrolysis of benzene. *Theoretical and Experimental Chemistry* 42(2), 90–95.
11. Kharlamov A. I., Ushkalov, L. N., Kirillova, N. V., Fomenko, V. V., Gubareny, N. I., and Skripnichenko, A. V. (2006) Synthesis of onion nanostructures of carbon at pyrolysis of aromatic hydrocarbons. Report of Academia of Science of Ukraine (3), pp. 97–103 (Russian).
12. Kharlamov, A. I., Loythenko, S. V., Kirillova, N. V., Kaverina, S. V., and Fomenko, V. V. (2004) Toroidal nanostructures of carbon. Single-walled 4-, 5– and 6 hedrons and nanorings. Report of Academia of Science of Ukraine (1), pp. 95–100 (Russian).
13. Wanga, X., Xua, B., Liua, X., Jiaa, H., and Hidekib, I. (2005) The Raman spectrum of nano-structured onion-like fullerenes. *Physica B: Condensed Matter* **357** (3–4), pp. 277–281.
14. Pul, Ch. and Ouens, F. (2005) Nanotechnology. Moscow. Texnosfera (Russian).
15. Kharlamov, A. I., Kirillova, N. V., Karachevtseva, L. A., and Kharlamova, A. A. (2003) Low-temperature reactions between vaporizing silicon and carbon. *Theoretical and Experimental Chemistry* **39**(6) 374–379.
16. Kharlamov, A .I. and Kirillova, N. V. (2002) Gas-phase reactions of formation of silicon carbide nanofilaments from silicon and carbon powders. *Theoretical and Experimental Chemistry* **38**(1) 59–63.

17. Kharlamov, A. I., Kirillova, N. V., and Kaverina, S. V. (2003) Hollow and threadlike nanostructures of boron carbide. *Theoretical and Experimental Chemistry*. **39**(3) 141–146.
18. Kholmanov, I., Kharlamov, A. I., and Milani, P. et. al. (2002) A simple Method for the Synthesis of Silicon Carbide. Nanorods *Journal of Nanoscience and Nanotechnology* **2**(5), 453–456.
19. Kharlamov, A. I., Kirillova, N. V., and Kaverina, S. V. (2002) Hollow nanostructures of silicon carbide. *Theoretical and Experimental Chemistry* **38**(4), 232–236.

PRELIMINARY ASSESSMENT OF SITUATION RELATED TO DIOXINS AND FURANS RELEASES IN KAZAKHSTAN

MARAT ISHANKULOV*
Eurasian National University named after L.N. Gumilev, 7, Munaitpasov Street,010010 Astana, Republic of Kazakhstan

Abstract: In 2005 for the first time in Kazakhstan a limited analysis of the samples for dioxins and furans content in the environment was carried out. A high level of air pollution and dust in the workshops of selected nonferrous enterprises was revealed. There was no active soil pollution through the atmospheric transfer detected. However, there were circumstances that raised doubts to the lack of soil pollution. They are based on the results of the landscaping and geochemical as well as ecological and epidemiological studies of Balkhash city where a Mining and metallurgical integrated works is situated. Strategic objectives of the NIP in the sphere of unintentionally produced products are being considered.

Keywords: dioxins, furans, releases, pollution, human health, impact, Kazakhstan

1. Introduction

The development of the Stockholm Convention on Persistent Organic Pollutants (POPs) triggered an interest and attention to the dioxins and furans problem in Kazakhstan. Earlier there were no any R&D or applied works on unintentional POPs releases. Launching a UNDP/GEF project "Initial assistance to the Republic of Kazakhstan to meet its obligations under the Stockholm Convention on Persistent Organic Pollutants" in Kazakhstan right after signing the Stockholm Convention in 2001 provided an opportunity to preliminary assess the scope of pollution and implement selected analyses of dioxins and furans samples.

*To whom correspondence should be addressed. E-mail: marat.ishankulov@ mail.ru

E. Mehmetli and B. Koumanova (eds.),
The Fate of Persistent Organic Pollutants in the Environment, 443–456.

2. Potential Sources of the Environmental Pollution with Industrial Releases of Dioxins and Furans

In Kazakhstan, enterprises of almost all industrial sectors are potential industrial sources of dioxins and furans releases. The leading sectors are energy, ferrous and nonferrous metallurgy, chemical, oil, pulp and paper, and cement industries (Figure 1).

Majority of the Kazakhstani enterprises has been built during the Soviet times and were supposed to meet the needs of the whole Soviet Union. Their productive capacities are enormous. Chiefly, they are high energy- and material intensive productions. Many enterprises have their own energy producing stations and electric substations with a lot of oil-filled equipment. In addition, the secondary electric equipment was installed in the workshops and production facilities that feeds the specific technological equipment such as electrically heated furnaces in the metallurgical enterprises, rolling trains in the machine-building enterprises, e.g., in OJSC "Mittal Steel Temirtau," Aksuiski Ferroalloy Plant, MNC "KazKhrom," etc.

Small and medium businesses that provide services usually do not have their own electric equipment and therefore, they have to use services of the organizations that supply them with electric power (enterprises or residential communal unions). All this presented only aggravates the situation with the unintentional POPs releases.

Figure 1. Enterprises of Kazakhstan – potential sources of unintentional POPs releases

Probability of unintentional POPs in the releases of the enterprises:

Black dots mean a high probability; grey dots – average probability; white dots – weak probability.

Enterprises: 1. "Plant Polygraphist," Ltd, OJSC "AIK," OJSC "Ural-Pliplast" – Uralsk city. 2. OJSC "ANPZ" – Atyrau city. 3. JSC "AKPO" – Aktau city. 4. OJSC "AZKHS" – Aktubinsk city. 5. OJSC "Kazakhmys" – Zheskazgan city. 6. JSC "Achpolimetal" – Kentau city. 7. OJSC "Intercomshina," JSC "Uzhpolymetal," OJSC "ShNOS" – Shymkent city. 8. "Kazphosphat," Ltd. – Taraz ciyt. 9. CJSC "Kazsabton," "Plant of household cleaning products," Ltd, "Gaudi Point," Ltd, "Rainbow Point," Ltd, OJSC "KazKhrom," JSC "Altyn-Almaz," "ANT-ALCOM," Ltd, JSC "Heat and Energy Equipment," CJSC "EuroTeck Industries Limited," "Anna," Ltd, "ShYZ" Ltd, OJSC "KazRemEnergo" – Almaty city. 10. OJSC "Taldykurgan Accumulator Plant," "Kuat," Ltd – Taldykurgan city. 11. "SaranRezinoTechnika," Ltd – Saran city. 12. "KaragandaChimProm," Ltd, OJSC "Mittal Steel Temirtau," Chemical and metallurgical plant – Temirtau city. 13. Mining-and-metallurgical Integrated Works – Balkhash city. 14. JSC "Zhairemski GOK" – Astana city. 15. Stepnogorsk mining-and-chemical Integrated works, OJSC "Herbicides" – Stepnogorsk city. 16. "Plant of household cleaning products," Ltd – Kokshetau city. 17. JSC "MaikainZoloto" – Pavlodarskaya oblast. 18. OJSC "PKhZ," OJSC "Aluminium of Kazakhstan," OJSC "PKhNZ," OJSC "KazEnergoCable" – Pavlodar city. 19. "KazakhCable," Ltd – Semipalatinsk city. 20. OJSC "KazakhZinc," JSC "VKKhMK," JSC "IrtyshChMZ," JSC "UKTMK" – Ust-Kamenogorsk city. 21. "KazStroiPolimer," Machine-building Plant – Karaganda city. 22. "TransEnergo," Ltd – Shakhtinsk city. 23. GRES-2 (Hydroelectric station) – Ekibastuz city.

3. Assessment of the Dioxins and Furans Releases

3.1. ESTIMATED ASSESSMENT OF THE DIOXINS AND FURANS RELEASES

Based on the payroll of the enterprises that are potential sources of unintentional POPs releases in 2004 the level of their releases was assessed within the framework of the UNDP/GEF project. The level showed 340 g-TEQ/year (Table 1).

The final estimated figure that equals 340 g-TEQ/year is definitely understated. The estimation did not include the data on incineration of medical wastes, fires on the trash dumps, unsanctioned burning of wastes on the enterprises and households (i.e., those categories that usually have the biggest dioxins and furans releases).

TABLE 1. Dioxins and furans releases in the selected sectors of the Kazakhstani economy (calculation is based on the UNEP Chemicals methodology)

Sector of the economy	Air	Water	Soil	Slag
Production of electric power and heating energy	315.981	0.000	0.000	0.0
Production of ferrous and nonferrous metals	3.324	0.000	0.000	9.1
Production of goods made of mineral raw materials	17.819	0.000	0.000	2.1
Uncontrolled combustion processes	2.829	0.000	0.051	2.7
Production and usage of chemical agents and consumer goods	0.000	0.000	0.000	0.0
Other	0.002	0.000	0.000	0.0
Total	340.000	0.000	0.100	13.9

3.2. SELECTED DATA ON LEVEL OF DIOXINS AND FURANS IN THE COMPONENTS OF THE ENVIRONMENT

Within the framework of the project "Initial assistance to the Republic of Kazakhstan to meet its obligations under the Stockholm Convention on Persistent Organic Pollutants" (2003–2006), in 2005 for the first time in Kazakhstan a sample analysis on dioxins and furans content was carried out by the Bashkir Republican R&D Ecological Center.[2] Their toxic isomers were defined in six samples of the air mainly selected on the worksites of the metallurgical enterprises and seven samples of urban soil and scrapes taken from the workshop walls (Table 2).

TABLE 2. PCDD/PCDF in the samples of the air and dust (scrapes taken from the walls)

Site of selection	Air (pg/m^3)		Dust (pg/g)	
	Concentration	TEQ-WHO	Concentration	TEQ-WHO
Balkhash city, Mining-and-metallurgical integrated works, copper workshop	51.33	4.08	5,377.06	263.78
Karaganda city, machine-building plant	17.80	1.84	46.12	4.06
Temirtau city, "Mittal Steel Temirtau" plant, Sinter machine № 5	42.64	3.77	5,419.70	607.70
Ust-Kamenogorsk city, "KazZinc," Hydrometallurgical production	6.33	0.47	289.35	19.83

3.2.1. Dioxins in the Air of the Working Sites

PCDD/PCDF content in the samples of the air taken in the working sites is high in the copper workshop of the Mining-and-metallurgical integrated works in Balkhash city (over 4 pg/m^3), near the sinter machine in the plant "Mittal Steel Temirtau" (about 4 pg/m^3) and in the machine-building plant in Karaganda city (about 2 pg/m^3), because the results are eight and four times respectively higher than the maximum acceptable concentration in the air which is 0.5 pg/m^3. Also, the working conditions are barely tolerable in the hydrometallurgical enterprise JSC "KazZinc."

Low concentration in the by-product coking production "Mittal Steel Temirtau" is explained by the technological specifics due to the high concentration of the duff dust in the air that adsorbs PCDD/PCDF. However, this does not exclude a possibility of a long-term pollution of the territory near the coke-oven batteries, the level of which could have been identified through the soil and dust analysis.

The air of the sanitary protection zone of JSC "Ust-Kamenogorsk titanium and magnesium integrated plant (UKTMIP) meets the norms of other countries (USA – 0.02, the Netherlands – 0.024, Italy – 0.04 pg/m^3) concerning the dioxin pollution.

3.2.2. Dioxins in the Dust of the Industrial Enterprises

The results of the polluted air in the working zone of the metallurgical enterprises were confirmed by the analyses of the scrape dust taken from the walls of these workshops (Table 2). The highest figures of the current inflow (air) are correlating with the data on a long-term pollution (dust). Additive effects of both aerosols and dust particles on the workers, PCDD/PCDF emission especially in the summer time can cause the accumulation in the biotissues. Utilization of this dust during the cleaning process in the workshops requires the same treatment as the wastes of the 1st class of danger and observance of special safety measures. It is obvious that the wastes of these enterprises can also be a source of pollution of the environment and population.

The proof of the high level of exposure to dioxins in these workshops can be an analysis of their content in the blood of the workers. This data can be a basis for improvement measures of the working conditions.

3.2.3. Dioxins in the Urban Soil

The findings for the soil selected in 1–3 km away from the industrial zone of Balkhash city show less 1 pg/g which means that there is no active pollution through the atmospheric transfer from the chimneys of the Heat and Power station and the plant. At the same time, the soil pollution indicator within the city, in the central park of Balkhash city is more than 6 pg/g and it can imply the existence of local sources of pollution. For comparison, the approximate safe level of PCDD/PCDF content in the soil for the Russian Federation is 0.33 pg/g; norms of dioxin content in the agricultural soil of the European countries are significantly higher – 5 pg/g (Italy), 10 pg/g (the Netherlands, Germany).

4. Preliminary Epidemiology Analysis

However, there are circumstances prompting doubt that the situation of the environmental components of Balkhash city (namely, soil) concerning dioxins and furans releases is not that good, as it can be judged by the individual random samples.

To proof the competence of such an idea we would like to present the materials of the landscape and geochemical map of the city[3] and actual data collected under our supervision within the UNDP/GEF project "Initial assistance to the Republic of Kazakhstan to meet its obligations under the Stockholm Convention on Persistent Organic Pollutants" on the epidemiology of the environment in 2004 (Figures 2–7).

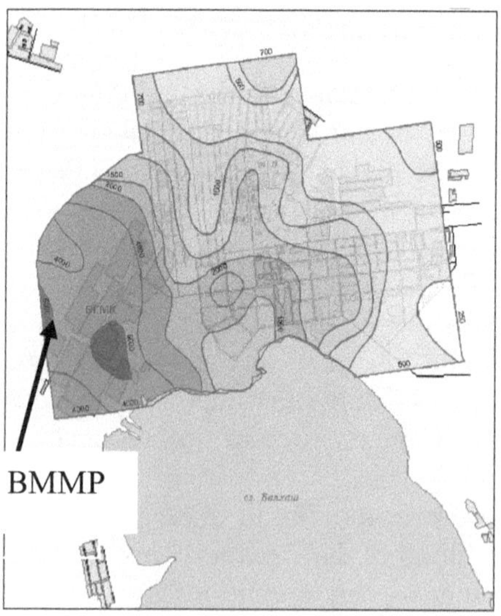

Figure 2. Pollution of the soil surface layer with copper in Balkhash city in 1992. Reference: Pollution in the units of background $C_f = 60$ mg/kg. Range of variation from 100 to more than 4,000.)

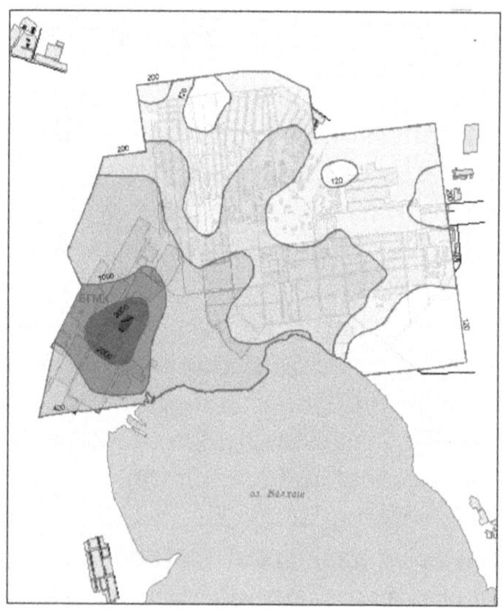

Figure 3. Pollution of the soil surface layer. total zinc in Balkhash city in 1992. Reference: Pollution in the units of background $C_f = 25$ mg/kg. Range of variation from 120 to more than 200

Priority pollutants of the Balkhash Mining-and-metallurgical plant (BMMP) as an enterprise of the nonferrous metallurgy are copper, lead, zinc, and cadmium (Figures 2 and 3).

The soil of the city is a source of weak radioactive radiation (Figures 4 and 5).

Figure 4. Specific activity of uranium (Ra^{226}) of the soil of Balkhash city in 1992 Reference: Level of the radiation hazard, becquerel/g. Range of variation of the isoconcentrate from 30 to 63

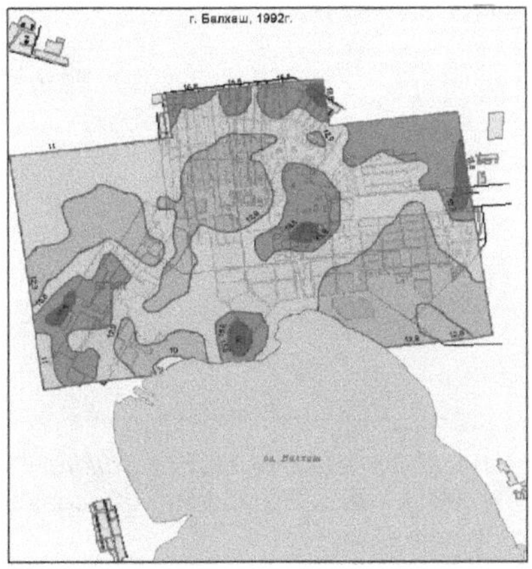

Figure 5. Intensity of the exposition doze of gamma rays of the soil of Balkhash Reference: Gamma rays, A/kg. Range of variation of the isoconcentrate from 10 to 20

A basic material for the assessment of the city population incidence rate was the data of the republican cancer register and register of addressed congenial development pathologies; the data on reproduction health and demographic indicators collected in the family outpatient medical sites for 5 years – from 1999 to 2003. In the aspect interesting for us in accordance with the literature data[3–5], in the basis of the analysis of the chemical pollution impact on human health lie the ideas of impact of the unintentional POPs releases, first of all, on the endocrine system of the exposed population. The addressed data were transferred onto the map (Figures 6 and 7).

It turned out that in Balkhash city neither copper, nor lead which are the major pollutants of the urban environmental components (Figures 2 and 3), nor data on natural radioactivity (Figures 4 and 5) reveal a pattern of the pathology areas that is characteristic to the emissions of the ferrous enterprises (Figures 6 and 7) following after areas of pollutants transfer ways.

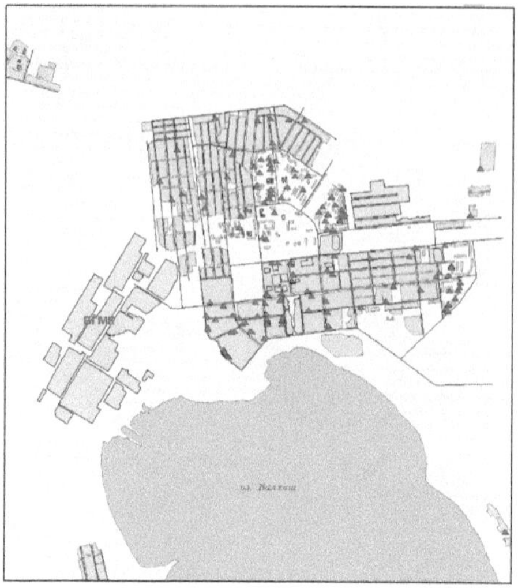

Figure 6. Rates of *congenial in Balkhash city* (1999–2003). ▲ – incident of congenial development pathology

Figure 7. Rates of cancer pathologies in Balkhash city (1999–2003). • – incident of cancer

We found this pattern in 1987 while studying the impact of the Shymkent Lead Plant's (SLP) activity on the rates of infant mortality and incidence rate of diabetes. The areas of rates of infant mortality (weight of infants deceased before 12 months old and gestational age) and individuals suffering from diabetes (Figures 8–12) coincide with the vectors of atmospheric transfer of the priority pollutants.

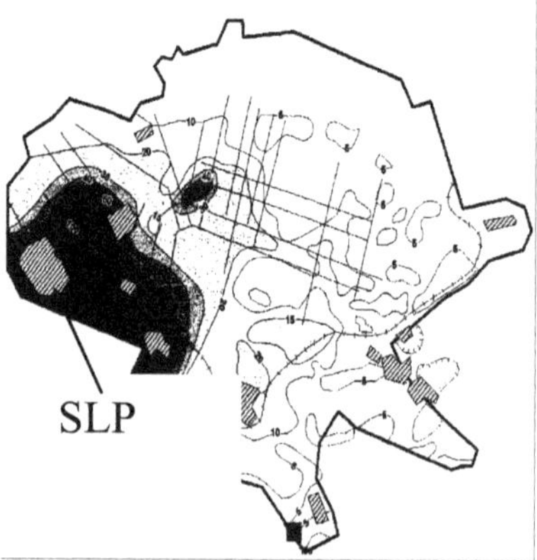

Figure 8. Pollution of soil surface layer with mobile forms of lead in Shymkent – 20 – isoconcentration fold MPC (Maximum permissible concentration is 6 mg/kg)

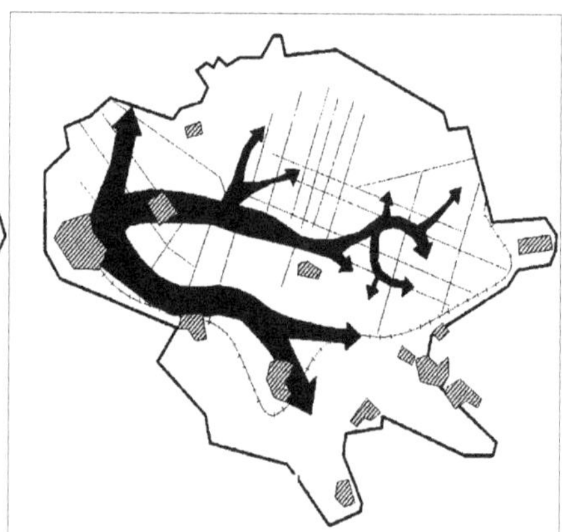

Figure 9. Vector of atmospheric transfer of lead identified on the lead pollution through the analysis of the geographical field

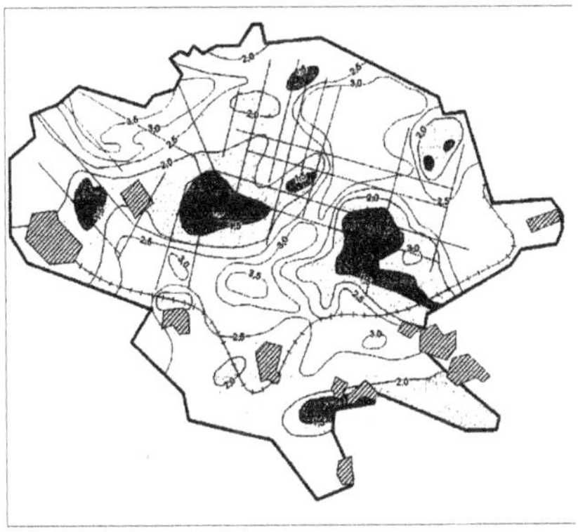

Figure 10. Birth weight of deceased before 12-month-old newborns

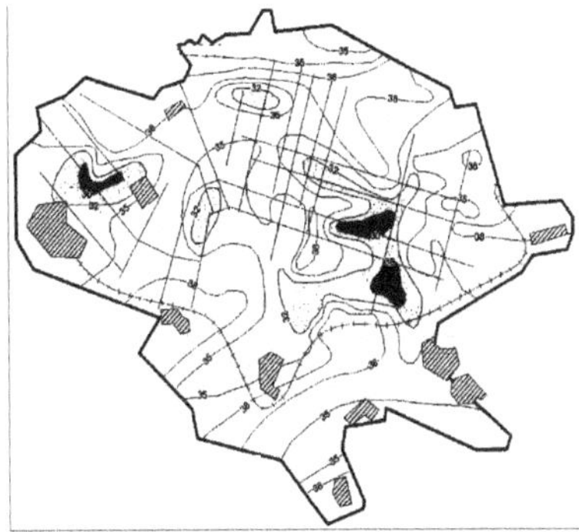

Figure 11. Gestational period of newborns in weeks (children deceased before 12 months old)

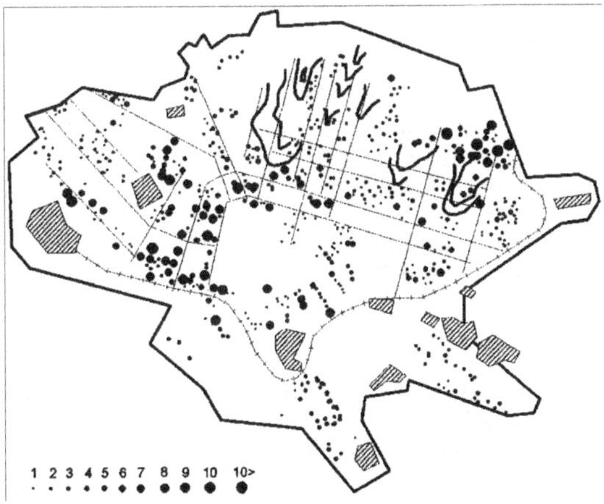

Figure 12. Cases of diabetes in Shymkent city. ● Quantity of people suffering from diabetes in one house, registered in 1985

Also, it is crucially important to know that the major pollutants of Balkhash city are not chemical substances that can cause congenial pathology or malignant neoplasms (Figure 13).

Balkhash city in Kazakhstan stands out with its high indicators of congenial development pathologies. For the 5-year analyzed period the rate of newborns with congenial development pathologies was 2.7 times higher than in the Republic of Kazakhstan.

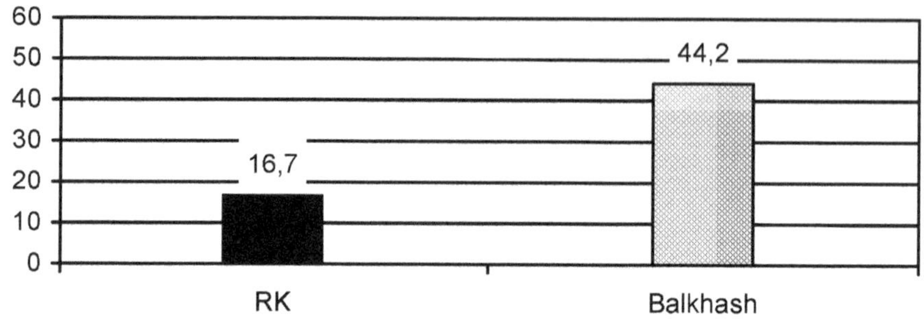

Figure 13. Level of congenial pathology (per 1,000 newborns)

There was also noticed a high cancer incidence rate in the city (Figure 14).

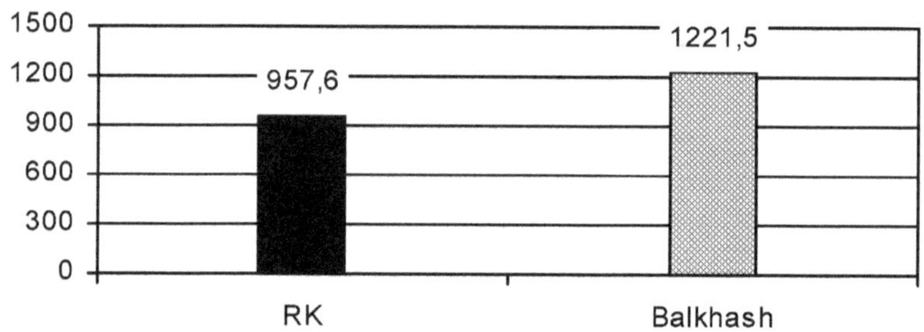

Figure 14. Level of the malignant neoplasms (per 100,000 of population) – intensive index

Incidence rate of selected hormone-dependent malignant neoplasms is higher than in the Republic of Kazakhstan (Table 3).

TABLE 3. Hormone-dependent types of cancer (per 100,000 female and male population) – intensive index for 1999–2003

Object	Neoplasms of female genital sphere	Female breast cancer	Bladder cancer	Thyroid carcinoma
Republic of Kazakhstan	344.9	170.1	19.1	14.0
Balkhash city	431.5	205.7	33.6	30.9

Frequency and structure of the malignant neoplasms of breast and female genital sphere (uterine carcinoma, cervical cancer, and ovarian cancer). Uterine carcinoma and cervical cancer dominate in the structure of the malignant neoplasms of the female genital sphere (Figure 15).

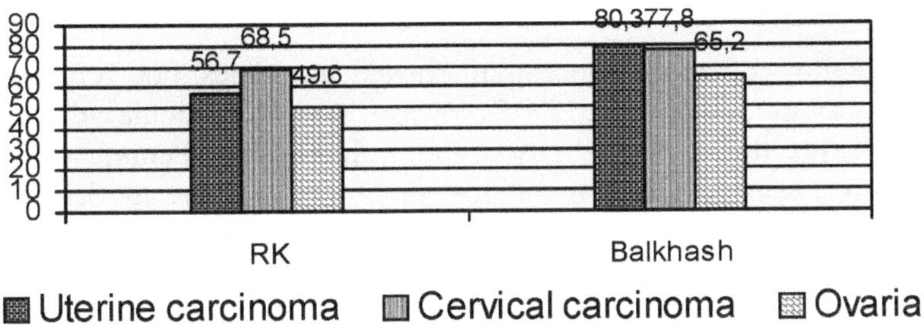

Figure 15. Frequency and structure of the malignant neoplasms of the female genital sphere

Thus, the actual material on the environmental epidemiology collected in Balkhash during 1999–2003 does not explain the adverse situation with hormone-dependent congenial and cancer pathologies in the city only by the impact of pollutants that are of priority for ferrous enterprises or natural radioactive factors of the environment. It is possible that the explanations lie in the impact of the unintentional POPs releases and the collected epidemiological materials speak for themselves. Further more elaborate research is necessary.

5. Strategic Purposes of the National Implementation Plan in The Sphere of Unintentionally Formed Products

The strategic purpose of the NIP in the sphere of the unintentionally formed POPs during various thermal processes is the reduction of PCDD/PCDF, PCB, and HCB (hexachlorobenzene) releases and, where possible, prevention of their formation. To achieve the purpose, the establishment of a management system for unintentional POPs and their monitoring will be urgent in the nearest future. Therefore, the following actions are necessary:

- Establishment of the normative and legislative base
- Detailed inventory of the sources of dioxins and furans releases
- Establishment of a dioxin laboratory
- Monitoring and compilation of the annual register of dioxins and furans releases
- Actions on reduction of dioxins and furans releases

5.1. ESTABLISHMENT OF THE NORMATIVE AND LEGISLATIVE BASE

The legislation of the Republic of Kazakhstan does not reflect issues connected to the unintentional POPs releases and, in particular, dioxins and furans releases, due to the novelty of the problem for the country. It can be referred to a special Law on dioxins and furans on the model of Japan thanks to which the threat of these emissions impact on the population health was eliminated.

5.2. DETAILED INVENTORY OF THE SOURCES OF DIOXINS AND FURANS RELEASES

During a detailed inventory it is planned to cover the following categories of the releases of unintentional POPs, on which there is little information:

- Small enterprises of the energy complex (all small boiler and gas-turbine plants)
- Enterprises of the machine-building complex (with smelting furnaces)
- Enterprises of the wood-processing industry (waste incineration)
- Enterprises of cable thermal processing
- Enterprises of the food industry (with smokehouses)
- Medical facilities – clinics, hospitals, medical sites, veterinary stations (that carry out incineration of the medical wastes and carcasses)
- Enterprises of the autotransport service (utilization of the exhausted machine and capacitor oils)
- Urban and rural communal trash dumps, industrial waste polygons
- Stoves in the private households
- Cases of intentional burning of the agricultural properties and agricultural wastes

5.3. ESTABLISHMENT OF A DIOXIN LABORATORY

Kazakhstan does not have any laboratory on dioxins and furans identification certified on the international level. At the same time, during the implementation of the obligations of the RK under the Stockholm Convention there will be a necessity even in regular analysis of the unintentional releases of the enterprises to monitor the furans and dioxins release. In addition, having established an enterprise on POPs liquidation (pesticides with POPs properties, PCB-containing equipment) on the territory of RK, a control of the releases on dioxin and furan presence will be necessary. Thereupon, the establishment of the dioxin laboratory that in the perspective can provide service to the whole Central Asian region seems to be most advisable.

5.4. MONITORING AND COMPILATION OF THE ANNUAL REGISTER OF THE DIOXINS AND FURANS RELEASES

Monitoring and compilation of the annual register of the dioxins and furans releases shall become a prerogative of the dioxin laboratory. The monitoring shall be financed by the republican budget.

5.5. ACTIONS ON REDUCTION OF THE DIOXINS AND FURANS RELEASES

Actions on reduction of the PCDD/PCDF releases, chiefly, consist of replacement of the basic used materials and raw materials, modification of the technological process (including monitoring of service and usage of the equipment and reequipment of the current production). The following belong to the possible and available actions that can be taken separately and in combination:

- Primary precautionary measures during the material selection
- Primary precautionary measures during the process selection
- Precautions due to the gas release
- Technology for gas emission purification
- Procedures for the leftovers processing

An analysis of technologies for waste gas purification applied in various enterprises is necessary in order to introduce the best available technologies (BAT) and best ecological practice (BEP) as well as to reconstruct the current productions that are the sources of the dioxins and furans releases. This kind of analysis, recommendations development and study of the advanced foreign experience and innovation technologies, search and transfer from the foreign investors or donors to the domestic enterprises should be carried out by the National Center on the Stockholm Convention on POPs, the establishment of which was provisioned in the National Action Plan on obligations under the Convention.

The issue of unintentional POPs was included into a list of such governmental documents as "Conception of the ecological safety of the Republic of Kazakhstan for 2004–2015," "Environmental protection of the Republic of Kazakhstan for 2005–2007," "Environmental protection of the Republic of Kazakhstan for 2008–2012," and "Ecological Code of the Republic of Kazakhstan." This action indicates the determination of the state to solve the problem.

References

1. Methodological guidelines on detection and quantitative assessment of the PCDD/PCDF releases, 2001.
2. Report on R&D: assessment of pollution of the industrial enterprises of the Republic of Kazakhstan with polychlorinated dibenzo- para-dioxins and dibenzofurans as well as pollution of the soil and capacitor oil with polychlorinated biphenyls. Teamleader Z. K. Amirova. Ufa, 2006 (unpublished).
3. Ecological and Geochemical Atlas of the Cities and Industrial Centers of Kazakhstan. Almaty. Vol. 1, 2001.
4. V. F. Levshin, The environment and cancer. Persistent organic pollutants and human health, 2001, pp. 18–20.
5. Khudolei V. V., Mizgirev I. V., Ecologically dangerous factors, St. Petersburg, 1996.
6. Revich B. A. et al., On impact of dioxins on malignant neoplasms and reproduction health disorder, *Hygiene and sanitary*, 2002, 1, pp. 8–12.

Part VI

ADVANCES IN THE MEASUREMENT OF POPs

DETERMINATION OF POLYBROMINATED DIPHENYL ETHERS (PBDEs) AS PERSISTENT ORGANIC POLLUTANTS (POPs) IN POLISH FOOD USING SEMIPERMEABLE MEMBRANES (SPMs)

DOMINIKA WOJTALEWICZ, ADAM. GROCHOWALSKI,* MALGORZATA WĘGIEL
Faculty of Analytical Chemistry, Department of Chemical Engineering and Technology, Krakow University of Technology, Warszawska 24, 31-155 Krakow, Poland

Abstract: The aim of this study was to determine the PBDE contents in selected food products available on the Polish market: butter, eggs, chocolate, pork fat, beef fat, and three kinds of fish: carp, salmon and cod. Therefore, a two-stage (for fish) and three-stage (for all the other food products) methods of sample cleanup was developed with the use of the GC-MS technique, using semipermeable membranes (SPM) in one of the stages. In all of the samples six dominating PBDE congeners: BDE-28, BDE-47, BDE-99, BDE-100, BDE-153, and BDE-154 were determinated. PBDE contents in the Polish butter ranged from 55 to 174 pg/g of fat. The results for the fish samples (salmon and cod) were shown to be highly variable. The analysed salmon samples contained for example 377–5,340 pg/g of the wet muscle tissue (1,850–26,700 pg/g of fat, respectively).

Keywords: POPs, food, PBDE, HCB, dioxin, SPME, gas chromatography, pollution

1. Introduction

Polychlorinated dibenzodioxins (PCDDs), dibenzofurans (PCDFs), biphenyls (PCBs), hexachlorobenzene (HCB), and polybrominated diphenyl ethers (PBDEs) are ubiquitous in the environment and are usually

* To whom correspondence should be addressed: E-mail: agrochow@chemia.pk.edu.pl

E. Mehmetli and B. Koumanova (eds.),
The Fate of Persistent Organic Pollutants in the Environment, 459–469.

determined at a ppt level in various matrices using GC-MS and GC-ECD techniques. Therefore, when a chromatographic trace analysis has to be done, removal of matrix compounds that may interfere is critical. In this work SPM techniques have been introduced for analyte isolation from matrix compounds as an effective cleanup method in the determination of PBDE in food.

PBDE are flame retardants that are added to polymers, which are used in plastics, textiles, electronic circuitry and other materials to slow down the burning process and prevent fires. Some of technical flame retardant products contain brominated organic compounds. Beside PBDE, the most used brominated flame retardants (BFRs) are hexabromocyclododecane (HBCD), tetrabromobisphenol-A (TBBP-A) and polybrominated biphenyls (PBBs).[1]

The commercial PBDEs products consist predominantly of penta-, octa-, and decabromodiphenyl ether products. At present, decabromodiphenyl ether is the most used PBDEs product. Polybrominated diphenyl ethers (PBDEs) are added to wide range of materials at concentrations up to 30% by weight.[2] They are commonly used additive of high impact polystyrene, flexible polyurethane foam, textile coatings, cable insulation as well as electrical and electronic equipment.[3]

Polybrominated diphenyl ethers are nonreactive flame retardants, so they are not incorporated into the polymeric materials by bonding between the polymer and the flame retardant. Therefore, they may separate from the surface of their product applications into the environment.[4]

PBDE were first discovered in pike, eel and sea trout from Sweden in 1981. Since then, researchers from Europe, Canada, and the USA have confirmed the presence of PBDEs in various environmental samples, detecting the primary lower brominated isomers such as tetra- and penta-BDEs.[5]

Polybrominated diphenyl ethers are persistent, lipophilic and bioaccumulating chemicals. They are toxic to human and animals.[6] Sample preparation plays an important role in the analysis of PBDEs in food samples. Food samples require highly efficient methods of purification before the final analysis. Dialysis with semipermeable membranes in an organic solvent leads to separation of organohalogen contaminants such as PBDEs from lipids.

The aim of this study was application of SPM in PBDEs determination. In the next step SPM with silica gel and alumina chromatography were used to PBDEs determination in fatty matrices. In this contribution new sample preparation technique for PBDE determination are reported.

SPM is a nondestructive method of separation, used for the preconcentration of organic chemicals in animal or plant fat samples. Membranes are made of low-density polyethylene foil, which have pore sizes of about 1 nm. This makes the restriction for dialysis of molecules bigger than this size. The membrane is formed in a tube, sealed in one of its end. The polyethylene foil thickness is about 80 μm.[7,8]

2. Experimental

2.1. MATERIALS, REAGENTS, AND APPARATUS

Certified PBDE standard solutions BDE-MXA lot MBDEMXA1199 and BDE-MXB lot MBDEMXB0300 of 5 μg/ml concentrations were used (Wellington Laboratories, Ontario, Canada). The standard solution of PBDE contained 6 native congeners as follows: Tri-BDE 28, Tetra-BDE 47, Penta-BDE 99, Hexa-BDEs 153, 154, and Hepta-BDE 183 was prepared. All compounds were in nonane at concentrations of 0,5 μg/ml. Certified PBDE standard solutions MBDE-MXA lot MBDEMXA1199 and MBDE-MXB lot MBDEMXB0200 were used in preparation labeled standard solution. The standard solution contains the same six PBDE congeners as native standard solution. All compounds were in nonane at concentrations of 50 ng/ml.

For GC/MS/MS an experiment was carried out using a calibration solution set BDE-CVS-A Brominated Diphenyl Ether (Wellington Laboratories, Ontario, Canada) comprising five individual solutions each containing 20 individual native brominated diphenyl ethers (Mono-BDE 3, Di-BDEs 7 and 15, Tri-BDEs 17 and 28, Tetra-BDEs 47, 49, 66, 71 and 77, Penta-BDEs 85, 99, 100, 119 and 126, Hexa-BDEs 148, 153 and 154, Hepta-BDE 183 and Deka-BDE 209), and 10 selected 13C12 labeled PBDEs (Mono-BDE 3L, Di-BDE 15L, Tri-BDE 28L, Tetra-BDE 47L, Penta-BDE 99L, Hexa-BDEs 139L, 153L, 154L, Hepta-BDE 183L, deka-BDE 209L. All native compounds (Mono-BDE to Hepta-BDEs) were in nonane at concentrations of 1.0–400 ng/ml, Deka-BDE 209 concentration was 10 times higher. All labeled congeners were at concentrations of 100 ng/ml. Tribromobenzene (TBB) at concentration 1 μg/ml as syringe standard was used.

Certified PCDD standard solution DFISS of 200 ng/ml concentration as syringe standard for GC/MS/MS was used. The standard solution contained two native PCDD congeners as follows: 1,2,3,4-TCDD and 1,2,3,7,8,9-HxCDD.

SPM membranes were purchased from Exposmeter, Sweden. All solvents were purchased from POCH, Gliwice, Poland. Sulfuric acid, sodium hydroxide and anhydrous sodium sulfate were purchased from J.T. Baker, the Netherlands.

Silica gel and aluminum oxide were purchased from Merck. Aluminum oxide and silica gel were activated overnight in 130°C and 200°C, respectively.

Acidic silica gel is prepared by mixing 44% of the total mass of H_2SO_4 with appropriate mass of silica gel to obtain a uniform powder. Basic silica gel is prepared by mixing 33% of the total mass of 1 g/mol NaOH solvent with appropriate mass of silica gel.

Varian CP3800 gas chromatograph equipped with electron capture detection (GC/ECD) was used. A CP-Sil5 CB capillary column (30 m* 0.32 mm i.d., 0.25 μm film thickness) was employed for the chromatographic separations. Splitless injection was used and the split was opened 1 min after injection. The column oven temperature was programmed as follows: 90°C (1 min), 10°C/min up to 160°C, 20°C/min up to 250°C, 10°C/min up to 300°C (5 min).

TRACE GC 2000 SERIES Thermo Quest coupled to GCQ Thermo Quest mass spectrometry was used. A DB5 MS capillary column (30m* 0.25 mm i.d., 0.25 μm film thickness) was employed for the chromatographic separations. The column oven temperature was programmed as follows: 100°C (3 min), 80°C/min up to 220°C and 3°C/min up to 300°C (10 min).

2.2. SAMPLES

Salmon and cod originated from the Baltic Sea and carp from the south of Poland. Chicken eggs, pork, beef, chocolate, and butter were bought in Cracow's market. Eggs were cooked and the yolks were separated. Fish tissue samples and egg yolks were extracted with dichloromethane in a Soxhlet apparatus for 8 h then freeze-dried. Butter and adipose tissues of pork and beef were heated for 5 h in glass beaker in 150°C. After the fat had melted it was transferred to a glass bottle, which was weighed, sealed, and stored under refrigeration. The chocolate was dissolved in *n*-hexane in a glass beaker. After deposition of the solid particles (~24 h) the upper, transparent yellowish layer was transferred to a rotary evaporation flask and the solvent was removed. The residual fat was weighed.

2.3. PROCEDURE

The membranes were prewashed before use by placing in a bottle containing *n*-hexane for 72 h. Although the highest recovery of PBDE was obtained after washing the membranes for 72 h, acceptable recovery was obtained after washing for 48 h.

Fat samples (5 g) were dissolved in 15 mL dichloromethane–*n*-hexane, 10:90 (v/v), and transferred quantitatively into the SPM membrane by use of a Pasteur pipette (Figure 1) The membrane was then sealed at its open end and placed in a bottle containing 80 mL *n*-hexane.

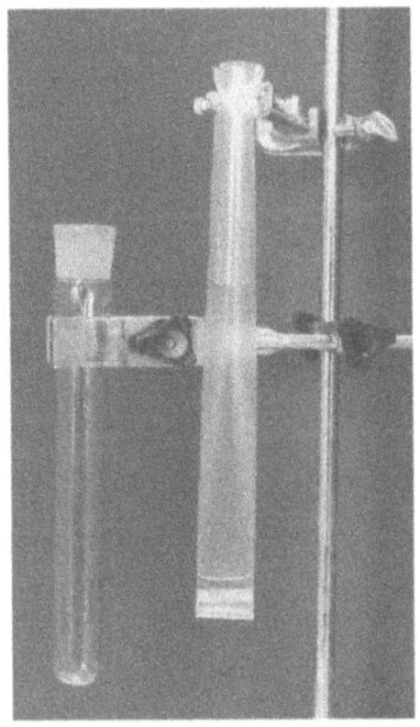

Figure 1. Photograph of sample preparation by SPM

Dialysis was conducted for 24 h, after which the dialysate was concentrated to ca 2 mL by rotary evaporation. A new portion of 80 mL *n*-hexane was then added to the bottle containing the membrane and dialysis was conducted for another 24 h. The procedure was repeated a third time to give three samples of dialysate from a total dialysis time of 72 h.

Further cleanup was achieved by column chromatography in a glass column containing 0.5 g anhydrous Na_2SO_4, 2 g basic silica gel, 0.5 g neutral silica gel, 8 g acidic silica gel, and 0.5 g anhydrous Na_2SO_4, from bottom to top. Concentrated dialysate (2 mL) was applied to the silica column. The analytes were eluted from this column with 60 mL *n*-hexane and the sample was evaporated to ca 1 mL by rotary evaporation.

Final cleanup was achieved on neutral alumina in a column containing 0.5 g anhydrous Na_2SO_4, 5 g aluminium oxide, and 0.5 g anhydrous Na2SO4 (Figure 2). After transfer of the sample to the column it was eluted with 15 mL *n*-hexane and 30 mL dichloromethane–*n*-hexane, 2:98 (v/v). These fractions were discarded. The analytes were eluted with 50 mL 1:1 (v/v) dichloromethane–*n*-hexane, to furnish a third fraction. Finally, 50 μL *n*-decane, as a keeper, and 2 μL DFISS standard were added to the third fraction and excess solvent was evaporated with a stream of inert gas.

The [M − 2Br]+ ion was monitored for quantitative and qualitative determination of PBDE by GC–MS–MS. The internal standard method was used for calculation of results from the appropriate PBDE congener peak areas in the sample analysed and the peak area for the corresponding PBDE congener in the BDE-CVS-A standard solution.

To monitor analyte recovery, the chocolate and butter fat solutions before SPM dialysis were fortified by addition of 2.5 ng of each of the selected six native PBDE congeners – BDE28, BDE47, BDE99, BDE153, BDE154, and BDE183. Amounts of these compounds in the samples investigated were below the detection limit. In this work we selected PBDE congeners from tri to heptabrominated diphenyl ethers because these groups of congeners are the most abundant in animal adipose tissue.

Lipid carryover (%LC) through the SPM membrane was investigated. Butter, chocolate, chicken egg, salmon, and pork lipid samples were prepared as described above and dialysed for 24 h. After this time the dialysate was transferred to a round-bottomed flask and concentrated to 1 mL by rotary evaporation. The residual solvent was finally evaporated by a stream of inert gas in a small conical flask and the fat obtained was weighed.

3. Results and Discussion

Recovery of PBDE after dialysis for 24, 48, and 72 h is depicted in Figure 2.

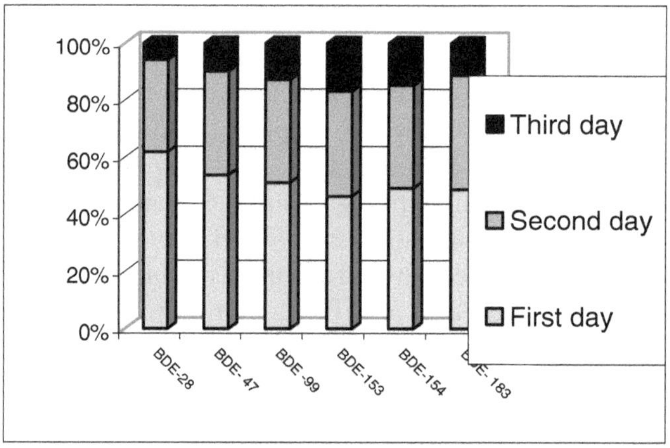

Figure 2. Dependence of PBDE recovery on dialysis time

Approximately 50% recovery of PBDE standards was obtained in the first day of dialysis. In the third day recovery for all six PBDE congeners was below 10%. These studies indicated that acceptable recovery of PBDE standards could be obtained by dialysis of food samples for 48 h. To investigate recovery chocolate and butter fat were used. For both matrices analyte recovery after SPM dialysis for 48 h was in the range 50–70% (Table 1).

TABLE 1. Recovery of PBDE standards from butter and chocolate fat (dialysis for 48 h)

Congener	Recovery of PBDES standards	
	Butter fat (%)	Chocolate fat (%)
2,4,4′-TriBDE (28)	64	70
2,2′,4,4′-TeBDE (47)	66	70
2,2′,4,4′,5-PeBDE (99)	67	66
2,2′,4,4′,5,5′-HxBDE (153)	64	54
2,2′,4,4′,5,6′-HxBDE (154)	69	62
2,2′,3,4,4′,5′,6-HpBDE (183)	51	57

Use of prewashed and unwashed membranes was investigated. Dialysis was conducted with a standard solution containing six PBDE congeners and the dialysate was analysed after dialysis for 24 h. Comparison of the PBDE chromatograms (Figure 3) reveals the signal-to-noise ratio was better for dialysate obtained by use of prewashed membranes. All subsequent dialysis was therefore conducted with membranes prewashed by placing in bottles filled with *n*-hexane for 3 days or more.

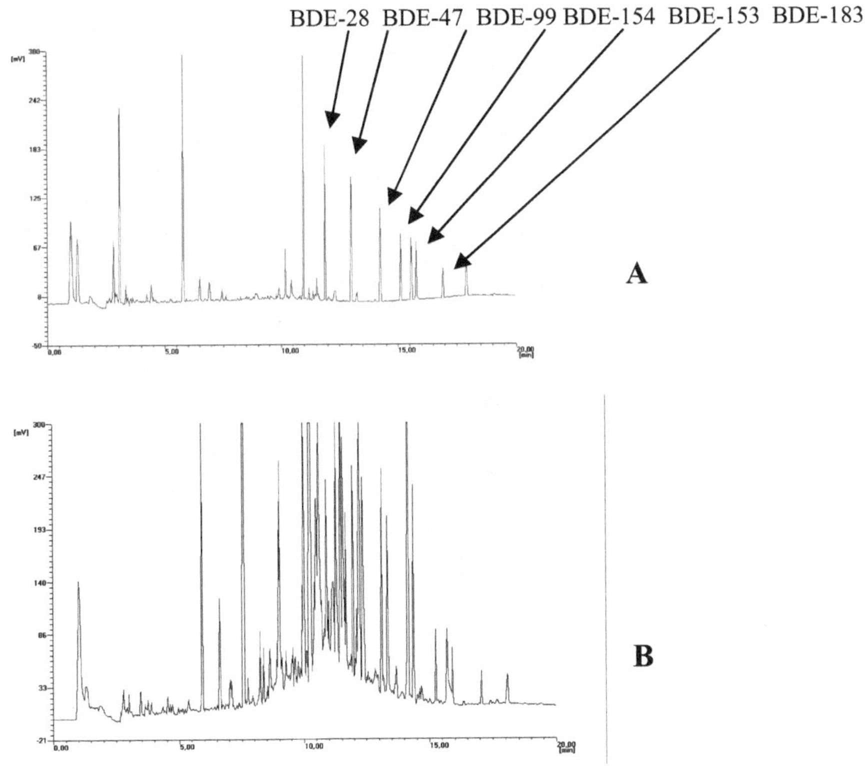

Figure 3. ECD chromatogram obtained after dialysis with (A) a membrane prewashed in *n*-hexane for 7 days and (B) an unwashed membrane

All fat samples were dissolved in 10:90 dichloromethane–hexane and dialysed with *n*-hexane. The results obtained showed that the best recovery of PBDE standards was obtained by use of this solvent mixture for dialysis. If pure dichloromethane was used inside and outside the membrane, recovery of the standards was between 34% and 50% after dialysis for 24 h. With *n*-hexane inside and outside the membrane recovery was between 76% and 97% but addition of 10% dichloromethane to fat samples improved solvation of the lipids.

Lipid carryover values (% LC) obtained for animal and plant fats are listed in Table 2.

TABLE 2. Percentage lipid carryover (% LC) for 5 g animal fat using the SPM method

Food sample	Fat mass breakthrough (g)	% LC	Average % LC
Butter	0.0162	1.31	
Chocolate	0.0079	0.88	
Chicken egg	0.0120	1.15	**1.10 ± 0.3**
Salmon	0.0102	0.83	
Pork	0.0170	1.35	

Lipid masses obtained after dialysis were negligible and caused no problems with subsequent cleanup steps. Use of SPM as an efficient method for lipid removal thus makes the whole procedure for cleanup of fat samples more effective. Results obtained from real animal fat and plant fat samples indicated that all the samples investigated contained PBDE. Analytical results from determination of PBDE in fat samples are presented in Figure 4.

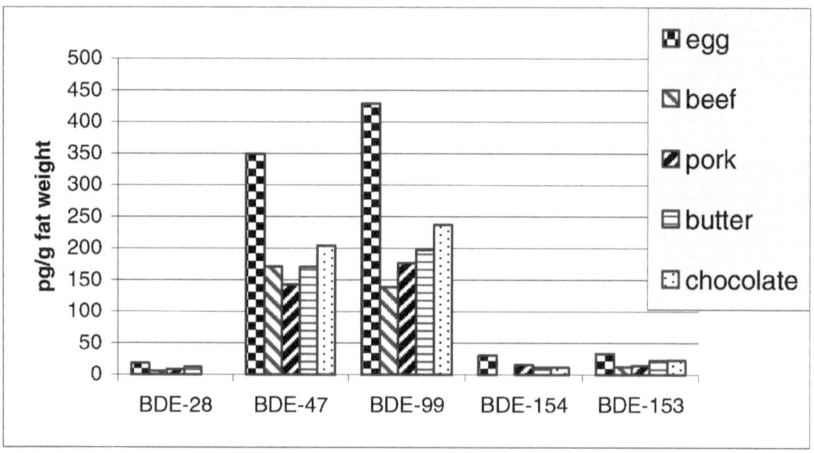

Figure 4. PBDE concentrations in five types of food

The concentrations of PBDE in beef, pork, butter, and chocolate fat samples, were similar for each congener. For egg samples PBDE concentrations were approximately twice as high. Among the PBDE congeners determined in food samples, BDE-47 and BDE-99 were approximately in 90% of the total mass of PBDE. The concentrations of PBDE in edible fillet tissue from three species of fish are presented in Figure 5.

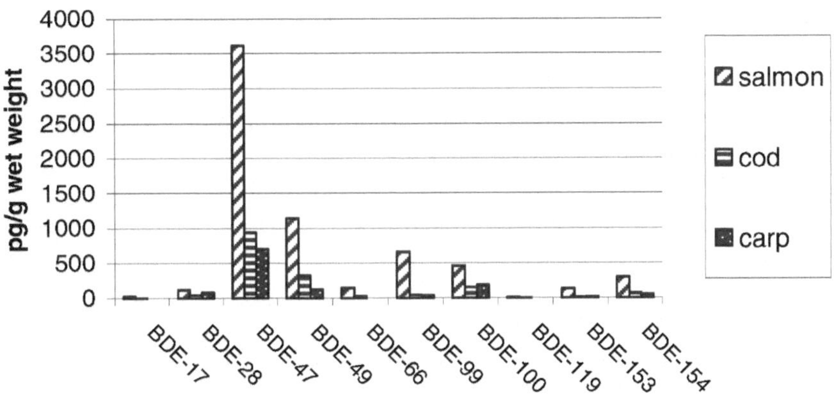

Figure 5. PBDE concentration in three types of fish

The highest concentrations were measured in salmon tissue. The figure shows there are large differences between PBDE concentrations in the two types of marine fish. The greater fat content of salmon tissue compared with cod tissue is the reason for this difference. PBDE are very lipophilic and poorly water soluble and therefore tend to accumulate in marine, fatty fish. The dominance of BDE-47 and BDE-49 (tetra) over the other PBDE congeners may indicate that tetrabrominated diphenyl ethers are bioaccumulated to a greater extent than the more brominated congeners.[9–13] Although the deca-BDE mixture constitutes 82% of global PBDE production, BDE-209 has rarely been reported in wildlife.[14] The PBDE congeners most often detected in food and fish samples are the tri to hexa-BDE congeners.[5] In our research only tri to hexa-BDE congeners were determined in the samples collected; BDE-209 was not detected in these samples.

4. Conclusions

PBDE were found in all the samples investigated. Tetra to hexa-BDE were detected in all the samples; among these BDE-47 was most abundant.

Qualitative determination of PBDE in fish and food products indicates that tetra and penta-BDE congeners are the most abundant in fatty matrices.[1] PBDE concentrations were highest in marine fish sam-ples. PBDE tend to accumulate in marine, fatty fish.[5,9,11]

Dialysis with organic solvent using polyethylene membranes is a good method for cleaning animal fat samples. The SPM technique is easy to handle and requires no advanced or costly equipment. Sample preparation with the SPM technique and silica gel and alumina chromatography takes approximately 3 days. This seems a long time in contrast with other methods, but in serial analysis use of a SPM results in good time saving. The SPM technique can be recommended for routine determination of PBDE in fatty matrices. Recovery for this method is between 50% and 70%. These recovery values are not acceptable for residue analysis, for which isotope dilution should be used.

References

1. de Wit C.A., An overview of brominated flame retardants in the environment, Chemosphere 46 (2002) 583–624
2. M. Alaee, R.J. Wenning, The significance of brominated flame retardants in the environment: current understanding, issues and challenges, Chemosphere 46 (2002) 579–582
3. R.C. Hale, M.J. La Guardia, E. Harvey, T.M. Mainor, Potential role of fire retardant-treated polyurethane foam as a source of brominated diphenyl ethers to the US environment, Chemosphere 46 (2002) 729–735
4. K. Jakobsson, K. Thuresson, L. Rylander, A. Sjödin, L. Hagmar, A. Bergman, Exposure to polybrominated diphenyl ethers and tetrabromobisphenol A among computer technicians, Chemosphere 46 (2002) 709–716
5. S. Ohta, D. Ishizuka, H. Nishimura, T. Nakao, O. Aozasa, Y. Shimidzu, F. Ochiai, T. Kida, M. Nishi, H. Miyata, Comparison of polybrominated diphenyl ethers in fish, Vegetables, and meats and levels in human milk of nursing women in Japan, Chemosphere 46 (2002) 689–696
6. P.O. Darnerud, Toxic effects of brominated flame retardants In man and In wildlife, Environment International 29 (2003) 841–853
7. B. Strandberg, P-A. Bergqvist, C. Rappe, Dialysis with semipermeable membranes as an efficient lipid removal method in the analysis of bioaccumulative chemicals, Analytical Chemistry 70 (1998) 526–533
8. P.-A., Bergqvist, B. Strandberg, C. Rappe, Lipid removal using semipermeable membranes (SPMs) in PCDD and PCDF analysis of fat-rich environmental samples, Chemosphere 38/5 (1999) 933–943
9. S. Burreau, Y. Zebühr, D. Broman, R. Ishaq, Biomagnification of PBDEs and PCBs in food webs from the Baltic Sea and the northern Atlantic Ocean, The Science of the Total Environment 366/2-3 (2006) 659–672

10. H. Wolkers, M.O. Hammill, B. van Bavel, Tissue-specific accumulation and lactation transfer of polychlorinated biphenyls, chlorinated pesticides, and brominated flame retardants in hooded seals (Cistophora cristata – from Gulf of St. Lawrence, Application for monitoring, Environmental Pollution 142 (2006) 476–486
11. H. Knutsen, C. Bergsten, C. Thomsen, A. Sletta, G. Becher, J. Alexander, H. Metzer, Preliminary assessment of PBDE exposure from food in Norway, Organohalogen Compounds 67 (2005) 1624–1627
12. Sjődin, D.G. Patterson, Å. Bergman, A review on human exposure to brominated flame retardants-particularly polybrominated diphenyl ethers, Environment International 29 (2003) 829–839
13. J. She, M. Petreas, J. Winkler, P. Visita, M. McKinney, D. Kopec, PBDE in the San Francisco Bay Area; measurements in harbor seal blubber and human breast adipose tissue, Chemosphere 46 (2002) 697–707
14. R.C. Hale, M. Alaee, J.B. Manchester-Neesvig, H.M. Stapleton, M.G. Ikonomou, Polybrominated diphenyl ether flame retardants in the North American environment, Environment International 29 (2003) 771–779

INDEX